The Untold Stories of African Agriculture

To Melko, Emuye and Tita

The Untold Stories of African Agriculture: Lessons from Ethiopia

Tsedeke Abate

Homegrown Vision (HGV), Addis Ababa, Ethiopia

CABI

CABI is a trading name of CAB International

CABI
Nosworthy Way
Wallingford
Oxfordshire OX10 8DE
UK

Tel: +44 (0)1491 832111
E-mail: info@cabi.org
Website: www.cabi.org

CABI
200 Portland Street
Boston
MA 02114
USA

Tel: +1 (617)682-9015
E-mail: cabi-nao@cabi.org

A catalogue record for this book is available from the British Library, London, UK.

ISBN-13: 9781800626362 (Hardback)
9781800626379 (ePDF)
9781800626386 (ePub)

DOI: 10.1079/9781800626386.0000

Commissioning Editor: Ward Cooper
Editorial Assistant: Helen Elliott
Production Editor: Shankari Wilford

Typeset by Straive, Pondicherry, India
Printed and bound in the USA by Integrated Books International, Dulles, Virginia

Contents

Contributors

Tsedeke Abate, Homegrown Vision (HGV), Addis Ababa, Ethiopia (tsedeke@homegrownvision.org)
Melaku Admasu, Ethiopian Seed Association, Addis Ababa, Ethiopia (kmadmasu@gmail.com)
Getinet Alemaw, Ethiopian Institute of Agricultural Research (EIAR), Addis Ababa, Ethiopia (getalebir@gmail.com)
Dessie Ambaw, University of South Australia, Adelaide, Australia (dessie.ambaw@unisa.edu.au)
Gashawbeza Ayalew, EIAR, Melkassa, Ethiopia (gashawbezachekol@gmail.com)
Workneh Ayalew, ICIPE-Ethiopia, Addis Ababa, Ethiopia (workneh.ayalew@gmail.com)
Amsalu Ayana, Independent Consultant, Addis Ababa, Ethiopia (aga.amsaluayana@gmail.com)
Ashenafi Ayano, EIAR, Jimma, Ethiopia (ashenafiayano@gmail.com)
Ferdu Azerefegne, Hawassa University, Hawassa, Ethiopia (azeref@gmail.com)
Endashaw Bekele, Addis Ababa University, Addis Ababa, Ethiopia (endashawbw@gmail.com)
Getachew Belay, formerly COMESA, Lusaka, Zambia (getab62@gmail.com)
Terefe Belehu, formerly Agricultural Transformation Agency, Addis Ababa, Ethiopia (tbelehu@gmail.com)
Mintewab Bezabih, Policy Research Institute, Addis Ababa, Ethiopia (mintewabayele@gmail.com)
Olubunmi Bossey, IITA, Ibadan, Nigeria (b.bossey@cgiar.org)
Asmare Dagnew, EIAR, Melkassa, Ethiopia (asmaredm@gmail.com)
Maru Degefa, Ethiopian Agricultural Business Corporation, Addis Ababa, Ethiopia (maru.degefa@yahoo.com)
Mulat Demeke, formerly FAO, Nairobi, Kenya (recently deceased)
Minyahil Fikadu, MS Consult, Addis Ababa, Ethiopia (minyahilfikadu@gmail.com)
Daneal Fekersillassie, Addis Ababa Institute of Technology, Addis Ababa, Ethiopia (dfeker321@yahoo.com)
Regassa Feyissa, Ethiopian Organic Seed Association, Addis Ababa, Ethiopia (regfey88@gmail.com)
Asnake Fikre, Ethiopian Institute of Agricultural Research, Debre-Zeit Research Center, Bishoftu, Ethiopia (fikreasnake@yahoo.com; tataw71@gmail.com)
Messele Fisseha, HGV, Addis Ababa, Ethiopia (messele.fisseha@gmail.com)
Amare Fufa, EIAR, Addis Ababa, Ethiopia (amarefufa@gmail.com)
Muluken Goftishu, Haramaya University, Haramaya, Ethiopia (mulukengoftishu@yahoo.com)
Endeshaw Habte, EIAR, Addis Ababa, Ethiopia (endhabte@gmail.com)
Konjit Hailu, Addis Ababa University, Addis Ababa, Ethiopia (konjithg@gmail.com)
Karta Kaske Kalsa, EIAR, Addis Ababa, Ethiopia (kartakaske@gmail.com)
Girma Tesfahun Kassie, ICARDA-Morocco, Rabat, Morocco (g.t.kassie@cgiar.org)

Friew Kelemu, formerly EIAR, Melkassa, Ethiopia (friewkelemu@gmail.com)
Seid Ahmed Kemal, ICARDA-Morocco, Rabat, Morocco (s.a.kemal@cgiar.org)
Tolera Keno, University of Queensland, Queensland, Australia (tolekeno@gmail.com)
Rahwa Kidane, Wageningen University and Research, Wageningen, The Netherlands (rahwa.kidane@wur.nl)
Berga Lemaga, formerly CIP-Ethiopia, Addis Ababa, Ethiopia (berga7lemaga@gmail.com)
Abebe Menkir, IITA, Ibadan, Nigeria (a.menkir@cgiar.org)
Wondimagegnehu Mersie, Virginia State University, Petersburg, VA, USA (wmersie@vsu.edu)
Takele Negewo, EIAR, Addis Ababa, Ethiopia (takelenegewo27@gmail.com)
Hawani Negussie, Early Childhood Education Ethiopia, Addis Ababa, Ethiopia (earlyed20@gmail.com)
Mesfin Shenkut, MS Consult, Addis Ababa, Ethiopia (mesfin_shenkut@yahoo.com)
Amare Tegbaru, formerly IITA, Ibadan, Nigeria; current address: Nybhovsbacken 31, 11763, Stockholm, Sweden (a.tegbaru@gmail.com)
Kindie Tesfaye, CIMMYT-Ethiopia, Addis Ababa, Ethiopia (K.TesfayeFantaye@cgiar.org)
Taye Tessema, EIAR, Addis Ababa, Ethiopia (tayetessema67@gmail.com)
Melaku Wale, Bahir Dar University, Bahir Dar, Ethiopia (melakuwale68@gmail.com)
Zerihun Yemataw, South Agricultural Research Institute, Hawassa, Ethiopia (zerihunyemataw2@gmail.com)
Firehun Yirefu, Kwale International Sugar Company, Nairobi, Kenya (firehunyirefu@yahoo.com)
Mikael Fesseha Yohannes, Homegrown Vision, Addis Ababa, Ethiopia (mikaelyohannes@gmail.com)
Engida Zemedagegnehu, MS Consult, Addis Ababa, Ethiopia (engidazemed2006@yahoo.com)

Foreword

The Untold Stories of African Agriculture: Lessons from Ethiopia is a remarkable and timely contribution to the global discourse on agriculture, development and sustainability. Authored by Dr Tsedeke Abate, a seasoned leader and researcher with over four decades of experience, this book brings together the collective wisdom of more than 40 contributors who have lived and breathed the challenges and triumphs of Ethiopian agriculture. It is a testament to the power of local knowledge and expertise, and it underscores the critical importance of African voices in shaping the future of the continent's agriculture.

This book is not just a collection of stories; it is a comprehensive and evidence-based assessment of Ethiopian agriculture, rooted in decades of research and experience. It offers invaluable insights that extend far beyond Ethiopia's borders, providing lessons that resonate with policymakers, researchers and practitioners across the African continent and beyond.

One of the standout features of this book is its African perspective. While numerous initiatives and foreign interventions have attempted to address the challenges of African agriculture, *The Untold Stories of African Agriculture* stands out as a unique and authoritative work told by Africans, for Africans. It emphasizes the importance of a homegrown approach, where qualified Africans lead agricultural research and development, and take accountability for it. It is a call to action, reminding us that African development cannot be outsourced, and sustainable change comes through hard work and local leadership.

This book is not about polemics or finger-pointing; it is about presenting facts and building compelling cases. It offers a synthesis of knowledge-based arguments, free from maligning institutions or governments. Its chapters are organized into eight parts, covering a wide range of topics from crop productivity gains and modern science to crop protection and agricultural inputs. Each chapter offers concrete recommendations that are rooted in real-world experiences, making this book a valuable resource for shaping policy decisions, learning and priorities in African agriculture.

From its insights into the transformation of cereals productivity to the potential of root and tuber crops, grain legumes and oilseeds, this book paints a comprehensive picture of Ethiopia's agricultural landscape. It also delves into the challenges and opportunities in areas like fruit crops, spices and coffee production. It addresses critical issues such as climate change adaptation, irrigation water management and the role of indigenous farming knowledge in education.

The Untold Stories of African Agriculture is a call to action, a testament to resilience, and a roadmap for agricultural transformation. It is a reminder that Africa's agricultural potential is vast, but unlocking it requires a deep understanding of local contexts and a commitment to sustainable practices.

As we navigate the complex and interconnected challenges of the 21st century, this book serves as a beacon of hope and a source of inspiration. It is a must-read for anyone interested in the future of African agriculture, and it has the potential to shape the trajectory of agricultural development on the continent and beyond. We are proud to introduce this ground-breaking work from Dr Tsedeke Abate and his esteemed colleagues, and we believe it will have a lasting impact on the global conversation about agriculture, food security and sustainable development.

Professor Johnson Ehima Onyibe
Former Director of Research and Planning
NAERLS, Ahmadu Bello University, Zaria, Nigeria

Foreword

When I began my career as an agricultural economist working in the US Peace Corps in Africa in the 1970s, a serious concern of my friends and I was what we would do after cadres of local agriculturalists were trained and Africa's food problems were solved, which would certainly occur before we reached middle age. But fifty years later, food insecurity and malnutrition are still widespread across Africa and in Ethiopia, with over 20 million people, one-fifth of the nation's population, requiring food aid. Civil strife and climate change are partly to blame, but as Dr Tsedeke Abate points out in this book, the strategies and policies of donor agencies, international organizations and African governments have often been ineffective or detrimental.

But this volume is by no means only about what development practitioners got wrong over the years. Dr Tsedeke also highlights successes in agricultural development, many of which are little known except to those rural households that benefited from them. As he points out, key elements of the success stories were combinations of functioning national and local institutions, sustained investments and adequate and timely supply of high-quality inputs. A notable finding is that African policy makers cannot outsource development to the international community.

Dr Tsedeke is in a unique position to tell the story of the successes and shortcomings, having worked in Africa his entire career. More importantly, he worked in and led national institutions as well as large research and development programs for international organizations. I first met Dr Tsedeke when he was a young entomologist/integrated pest management specialist in the Ethiopia Institute of Agricultural Research and he rose to serve as research centre director and then as director general of the Institute.

His international positions included coordinating desert locust research in northeast Africa and the Arabian Peninsula, leading the tropical legumes project for the International Crops Research Institute for the Semi-Arid Tropics and leading programmes on seed systems and drought tolerant maize for the International Maize and Wheat Improvement Center across Africa. Carefully assessing progress on farmers' fields was always as important to Dr Tsedeke as the leadership and coordination duties of his positions.

Most books on African agriculture are written by expatriates and those by Africans tend to be by academics and social scientists. This volume is thus a welcome and unique addition and deserves a wide readership.

Steven Franzel
Principal Agricultural Economist (retired)
World Agroforestry Center
September 2023
Orlando, Florida, USA

Preface

I consider myself to be truly lucky in life for many reasons. My upbringing was a rural setting in a pristine valley of two perennial rivers – Wogayi to the south and Gashi to the north – that my father and grandfather jointly developed as a semi-commercial farm in south-western Ethiopia soon after the end of the Italo-Ethiopian war. They named the valley Dalba (land of abundance). Its borders were a public road to the west and the confluence of the rivers in the east. We did not share borders with anyone.

Destiny had it that my father died from a preventable infectious disease just a few years later, when I was only 3 years old. My ever-demanding grandfather had me do the actual farming and many other chores from my childhood until I ran away to pursue modern education at the only modern government school at Waka, then capital of Dawro Zone, when I was 11 years old. I became more and more fascinated with farming and continued to be engaged in it during summer vacations until I went to college where my grandfather chose for me agriculture over medicine.

My grandfather, *Balambaras* (lowest title bestowed by the Emperor) *Erasha* (second highest title bestowed by the Dawro kings) Dilbo Alilo, was the reference point in crop husbandry to the farmers of the surrounding areas. They came to him for advice on which varieties of *teff*, maize, or sorghum and when to start planting. The community returned the favour by volunteering their oxen and labour so that his fields would be planted and harvested before anyone else's.

I was also blessed to have the opportunities to serve both as a researcher and leader of agricultural research and development at the national, regional and international levels. My learning experience about agricultural research and development spans many, many years. It started from a young Ethiopian University Service student at Holetta Research Station of EIAR after 3 years of fully government sponsored college education. This was followed by service as a researcher and centre director at Melkassa, to Director General of EIAR, to tropical legumes research leader and senior scientist at ICRISAT, and to principal scientist and leader of the drought tolerant maize for Africa project at CIMMYT.

After taking my retirement in 2017, I had the opportunity to think about finding the best way to serve African agriculture. Subsequently, a group of like-minded colleagues and myself established HGV (Homegrown Vision), an independent non-profit (family financed) think tank (www.homegrownvision.org) where I am serving as Leader at present, registered in Ethiopia. We wanted to anchor HGV in a solid platform and the stories told in this book are part of that effort.

As alluded to above, this book took nearly 7 years to complete. My original plan was to prepare one broad journal article on Ethiopian agriculture including crops, livestock and natural resources. When the first version, that included data through 2016, was completed in mid-2018, I realized that it was not possible to tell the stories in one article. The second version (containing data through 2018) was completed in late 2020, but I finally settled for the current version that included data through 2020. The hope is that there will be sequels to this volume.

Tsedeke Abate, Lead Author
September 2023, Addis Ababa

Acknowledgements

At the outset, I would like to pay particular tributes to our dear friend and colleague Dr Mulat Demeke, one of the founding members of HGV and coauthor of one of the chapters, who is no longer with us. We miss him so preciously.

My first appreciations go to my eternal friends Drs. Bekele Shiferaw 'Beke' and Tesema Chekol 'Tes' who have worked with me from the very first day to the conclusion of this project. Life would have been impossible for me without the love and relentless support from my children. They also played a major role in aligning my phraseology with the style of the younger generation. Emuye's prowess as a writer has helped in finessing some of the expressions and illustrations; I am so grateful for her overall editorial support and advice. All of these people invested their precious time during numerous iterations of each version of each chapter. Melko, Emuye and Tita also provided generous material support for running the HGV website and the overall project.

I am indebted to Dr Gregory Edmeades for reviewing all chapters and providing priceless advice. My appreciations also go to H.E. Ato Mersie Ejigu, Prof. Johnson Ehima Onyibe, Drs Seife Ayele, Tony Fischer, Steven Franzel, Jeff Herrick, Walter Knausenberger, Assefa Gebre-Amlak, Bedada Girma and Tesema Megenasa for peer reviewing one or more chapters. Drs Tseganesh Wubale Tamrat, Moti Jaleta and Yohannes Ayele provided useful feedback on Chapters 24 and/or 25. My appreciations to Drs Kassa Semagn, Melaku Gedil, Woldeyesus Sinebo, Yoseph Beyene and Zerihun Tadele for their feedback and advice during the early stages of developing the biotechnology chapter; and Dr Girma Haile-Michael on spices. I have also received feedback on the earliest version of the book from Dr Berhanu Amsalu, Dr Brhanne Gebrekidan, Prof. Sebsebe Demissew, Drs Seme Debela (may his soul rest in eternal peace), Yilma Kebede, Mulugettta Mekuria and Alastair Orr.

Dr Abebe Menkir and Ms Olubunmi Bossey helped with the statistical analyses for all crops; Dr Dagne Wegary and Ato Zerihun Tadesse provided advice on statistical analysis procedures. Ato Alemu Asres Merso, Dr Chilot Yirga, Ato Damene Darota, Drs Fentahun Mengistu, Kebebew Assefa, Mohammed Dawd, Endalkachew Wolde-Meskel, Mohammed Yusuf, Nigussie Alemayehu, Ato Laeke Berhanu, Wrt. Mahlet Alemayehu and Ato Tamru Habte provided data and information on different chapters. Ms Aberrash Tsehai gave information on small machinery. Ato Guta Bukero, Ato Regassa Dibaba and Wrt. Zeytuna Abe organized agricultural sample surveys data from CSA and made them available for analysis.

My appreciations go to the HGV Founding Members (the late Dr Mulat Demeke, Drs Amare Tegbaru and Mintewab Bezabih, Engineer Mesfin Shenkut, Drs Workneh Ayalew and Konjit Hailu, Wro. Hosana Negash, Ato Techalew Negash) and the advisory group (Drs Aberra Debelo, Belay Demissie, Habtu Assefa, Wro. Martha Hirpa, Ato Messele Fisseha, Prof. Mitiku Haile and Prof. Wubshet Mamo) for advice and oversight.

A large number of people have invested their emotions in this project, as if it was their own business, encouraged me to get it published as quickly as possible and provided moral support. Lieutenant Lemma Asfaw, Ato Abraham Zeleke, Prof. Temesgen Haile-Mariam, Ms Venus Easwaran Narayan, Ato Amha Mersae Hazen, H.E. Prof. Eyasu Elias, Dr Tesfaye Kumsa, Mrs Tigist Tesfaye Haverkort, Prof. Afework Kassu, Ato Brook Worku, Wro. Hiwot Gebre-Medhin, Ato Tenagashaw Mengistu, Wro. Yeshihareg Assefa, Dr Getachew Gebru, Wro. Alemtsehay Wolde-Selassie, Ato Aklilu Mulat, Ato Elias Tesheberu, Wro. Elisabeth Baslyos, Wro. Genet Melaku, Ato Getinet Anteneh, Ato Kidanu Chekol, Wro. Konjit Kidane, Dr Said Silim, Ambassador Teruneh Zenna and Ato Tsegaye Tesfaye top this list. Following is an alphabetical list of the rest: Ato Abebe Areru, Ato Abrham Kassa, Wro. Almaz Kassaw, Wro. Ades Beyene, Ato Amare Molla, Dr Amsal Tarekegne, Wro. Bertukan Gebre-Wold, Ato Assefa Adefris, Dr Badege Bishaw, Wro. Bosena Bizuneh, Dr Chalachew Aninteneh, Dr Fisseha Asfaw, Wrt. Helinna Ayalew, Ato Kidus Asfaw, Dr Million Belay, Mrs Njeri Okono, Ms Jenny Peters, Dr Said Silim, Mrs Million Skoda, Dr Solomon Haile-Mariam, Ato Solomon Kebede and Dr Teshome Hunduma.

I am so grateful to Wro. Meaza Birru of FM 102.1 Sheger Radio, Ato Tariku Tessema of Deutsche Welle Radio (Amharic Service), Prof. Minga Negash of FESP and Dr Erik Alexandersson of Swedish University of Agricultural Sciences for giving exposure to this project to the general public and the outside world. I am so grateful to Ato Bahakal Abate and Ato Tameru Wondimagegnehu for legal advice.

I would like to pay my tributes to the individuals who impacted my professional life. First, there was Ato Tilahun Wondim-Tsega, my elementary and junior high school director, who oversaw my promotions from grade one to grade five within one academic year and my double promotion from grade six to grade eight. Then there was Ato Degu Nigussie, my sixth-grade science teacher who was the first person to note that I had a bright future in science after grading my assignment paper. Ato Abebe Hailiye, my high school director, gave me a double promotion from grade ten to grade twelve. My dream to get a college education could not have been realized without strong encouragement from my uncle, Grazmach Aweke Aemro, and a dear friend, Ato Agize Sahle-Mariam. Mr T.J. Crowe (FAO entomologist at EIAR) taught me to have a deep understanding of the science of entomology. Dr Zemedu Worku (then general manager at EIAR) and Mr Gaylord Walker (of USAID in Ethiopia) played a vital role in my transfer as a senior year student from Alemaya College of Agriculture to the University of Florida. I am so indebted to Dr Haile-Michael Kidane-Mariam ('Gash Haile') for his lifetime advice, wisdom during the formative years of my career and being one of my greatest fans. My appreciations also go to Prof. Mesfin Abebe and Prof. Manfred Mackauer for mentorship and to Dr Thomas Lumpkin for his kind support during my tenure at CIMMYT.

Finally, I don't have enough words to express my appreciation to my dearest sister, Geremush Abate, for the love and support since our early childhood to this day.

1 Critical Issues in African Agriculture: A Synthesis Report

Tsedeke Abate*
Homegrown Vision, Addis Ababa, Ethiopia

Abstract

This book presents the results of a comprehensive in-depth assessment of Ethiopian agriculture and draws lessons from it to generate unambiguous recommendations that can inform policy decisions and priority setting for agricultural transformation in Africa. The current chapter provides a synthesis of the entire book so that the reader is equipped with adequate background information for delving into specific chapters in the rest of it. The book consists of eight parts, comprising 25 chapters. The parts include the introductory chapters, measuring productivity gains, biotechnology, crop protection, input use, environment, extension and long-term socioeconomic benefits of investing in agriculture. It provides summaries of partnerships, missed opportunities and unfinished agenda for African agriculture. The key lesson from the book is that development cannot be outsourced.

Introduction

Rationale and conceptual premises

Africa is blessed with ample agricultural land (Annex Fig. 1.1) and has good potential for irrigated agriculture. And yet, as of the early1980s, its agricultural commodities import costs have outpaced its export earnings from agricultural products (Fig. 1.1). In 2021, the continent imported agricultural products valued at US$111 billion, compared to the export earnings of less than US$66 billion. The international donor community and development partners have tried to help improve Africa's agricultural productivity since the early 1970s, but so far, any changes in production have come mainly from increases in the cultivated area rather than from increases in productivity. For example, between 2000 and 2021, the cultivated area grew at an annual rate of 2.44% vs 0.88% for yield (FAO, 2023c). This means that about 74% of the increase in production was attributed to area expansion whereas the contribution of increase in yield was 26%. The unchecked land area expansion contributes to a decline in biodiversity and environmental degradation and leads to an obvious decline in soil fertility.

The African agricultural research and development (AR&D) system, as perceived by the international development partners, has undergone several paradigm shifts – from alley cropping to food security, to poverty alleviation, to conservation agriculture, to agriculture

* Corresponding author: tsedeke@homegrownvision.org

 Untold Stories of African Agriculture: Lessons from Ethiopia (T. Abate ed.)
DOI: 10.1079/9781800626386.0001

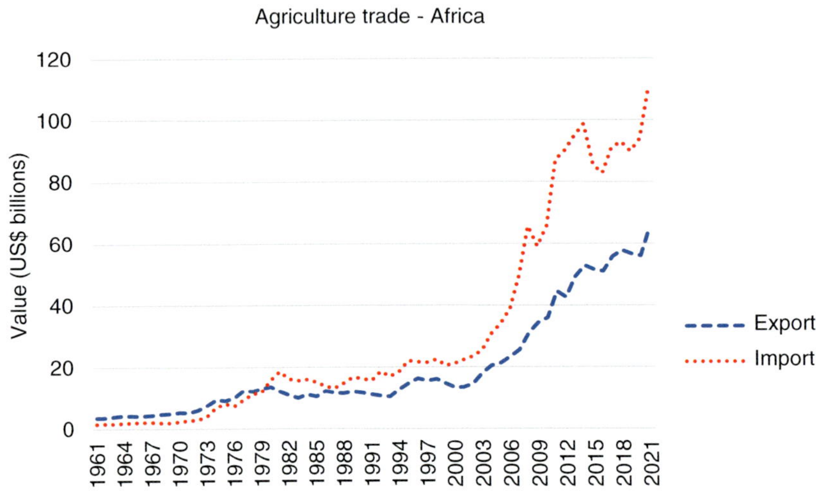

Fig. 1.1. Export and import values of agricultural products in Africa. Source: calculated and constructed by the author from FAO (2023a).

intensification, and now regenerative agriculture – over the past five decades. However, there has been no benefit from comprehensive and coherent in-depth studies to isolate the root causes of the problem, and as a result no powerful, clear and unambiguous recommendations are available. Nor has it strongly embraced an African voice. We therefore intend to offer an African perspective in this study and forward suggestions on the changes that need to be implemented to jumpstart African agriculture, using the Ethiopian example to illustrate what went well and why, and what did not go well and why.

Ethiopia has a long history of agriculture, starting with the domestication of *enset* (*Ensete ventricosum*) some 8000 years ago and cereals about 1000 years later. The use of the *maresha* (single-tine plough), drawn by draft animal power, for land preparation goes back to more than 1500 years. However, this country's agriculture has remained traditional for a long time.

Concerted efforts to 'modernize' Ethiopian agriculture started with the establishment of the Imperial Ethiopian College of Agriculture and Mechanical Arts (now Haramaya University) in 1952, followed by the Institute of Agricultural Research (now Ethiopian Institute of Agricultural Research [EIAR]) in 1966, which was mandated to develop and demonstrate improved technologies suited to the diverse agroecologies of the country. The Ministry of Agriculture (MOA), established in 1907 under the reign of Emperor Menelik II, was charged with the responsibility for delivering the improved technologies.

Successive governments gave AR&D high priority and made significant investments. Initially, the imperial government (through to 1974) focused on building the infrastructure and a critical mass of researchers. The military-socialist regime (1974–1991) expanded the agroecological coverage. The EPRDF (Ethiopian Peoples' Revolutionary Front) that ruled Ethiopia between 1991 and 2018, envisioned agricultural transformation, capitalizing on the technological progress made by its predecessors and concentrated on delivery. The current Prosperity Party government is embarking on transformation of Ethiopian agriculture.

Like many of its peers in Africa, and elsewhere in the developing world, Ethiopia has experimented with various models of AR&D over the past nearly six decades and has registered varying levels of progress. These cumulative investments are now paying dividends. A large number of crop varieties (Fig. 1.2) and production technologies have been developed for various crops over the years, and these have been vital to the advances being made in agricultural productivity in the country over the past two decades. Ethiopia has achieved significant improvements in cereals (particularly maize) and grain legumes (mainly

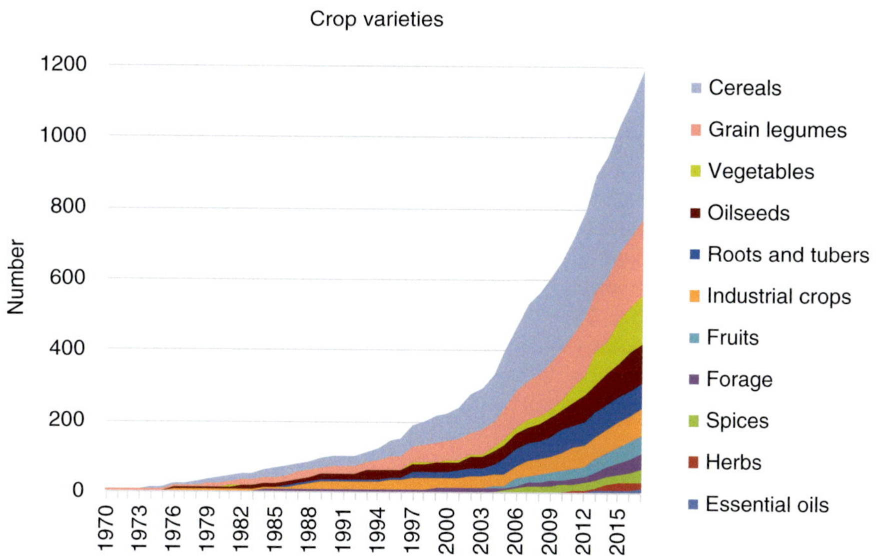

Fig. 1.2. Running totals of major crop varieties released in Ethiopia between 1970 and 2017. Source: constructed by the author from MOALR (2017).

common bean and chickpea) productivity. The cereal and grain legume agriculture in many of the major producing areas in Ethiopia can now be characterized as 'smallholder commercial production' in the sense that it is primarily market rather than subsistence led. Yields of both groups of crops have more than doubled between 2000 and 2020. In the second most populous country in Africa, these led to substantial improvements in the calorie intake from cereals and revenues from foreign export of grain legumes. These changes (despite some shortcomings) represent some of the untold positive stories from Ethiopia that can provide valuable insights for agricultural transformation in Africa.

Through its Growth and Transformation Plan (GTP I and GTP II), the EPRDF government has made efforts to introduce agricultural transformation during the two decades before 2018. However, the GTP has not benefited from the results of comprehensive and coherent scholarly studies that can help to tackle the complex problems of Ethiopian agriculture. Also, it followed a top-down approach, without the participation of key stakeholders. The recent literature dealing with agricultural transformation in this country dwelt more on the transformative aspects of markets, value chains, processing and shifts in dietary preferences rather than the bigger issue of sustainability at the farm level.

Even though the country has made substantial productivity gains with its main cereals and grain legumes, there have been missteps and missed opportunities that could have been avoided. Productivity of oilseeds and the majority of other crops cultivated in Ethiopia has remained low and increases in the production of these came largely from increases in area expansion rather than improvements in yield. Their production has remained at subsistence levels. By focusing heavily on the cereals for food security and on selected grain legumes for export earnings, Ethiopia has missed the opportunities to exploit the full potential of its own sustainable indigenous/traditional crops such as *enset*, oilseeds like Ethiopian mustard and safflower, and even coffee. Even for major cereals like wheat, Ethiopia remained a net importer despite the strong potential for import substitution. Despite the continued focus on food security through selected staples, the current government is attempting to change this with a national effort to improve productivity of wheat.

Ethiopia (and Africa as a whole) faces the dual challenges of ensuring food and nutrition security and economic wellbeing for its rapidly growing population, while maintaining the integrity of its natural resources and enhancing resilience against climate change. Diversifying and improving agricultural productivity through demand-led and climate-responsive approaches is the only way for Ethiopia to achieve these goals, thereby avoiding crises and maintaining social and political stability. Much remains to be done to keep pace with the population growth rate (estimated at 2.8% per annum) and the rapid land degradation and decline of biodiversity that are threatening to create crises for food and nutrition security and economic wellbeing of the country. Implementation of the past agricultural transformation agenda as it stands today has not delivered the desired results – especially in terms of increasing national capacity for sustainable development – and is an opportunity that has been missed.

Cognizant of the challenges Ethiopian agriculture is facing at present and motivated by the strong desire and commitment to bring about sustainable positive change to the development of Ethiopia (and other African countries), I conceived this book in mid-2017 and solicited support from a group of like-minded professionals in AR&D as contributing authors, under the umbrella of Homegrown Vision (HGV), a not-for-profit independent think tank (www.homegrownvision.org). As the name suggests, HGV strongly believes that Ethiopia (and Africa as a whole) should seek homegrown solutions for their sustainable development, and for this to happen, governments, policy makers and research organizations should take the leadership of their development agenda and be accountable. One of the main aims of HGV is to support Ethiopian (and other African) government agencies and other institutions to develop and implement agricultural transformation agendas that are commensurate with the demands of the 21st century.

Objective

The overall objective of this book is, therefore, to develop a comprehensive in-depth assessment to generate unambiguous recommendations that would inform policy decision and priority setting for agricultural transformation in Africa.

The perception about African agriculture by the general public both within and outside Africa has not been so positive. Authoritative stories describing African agriculture and told by Africans are virtually non-existent. On the other hand, researchers from outside Africa (e.g. Bindraban *et al.*, 2009; Haggblade and Hazell, 2010; van Ittersum *et al.*, 2016; Christiansen and Demery, 2018; Schurman, 2018; Giller, 2020; Giller *et al.*, 2021; Dixon *et al.*, 2021), not to mention various foreign initiatives like Millennium Villages and the One Acre Fund, have attempted to promote debate on the future of African agriculture. In this book, an attempt is made to tell credible and evidence-based stories and lessons learnt from AR&D initiatives over the past 50 years. Our stories and lessons are not something we have learnt from the literature, but something we have lived through and worked with.

It is my hope to change the negative narrative about African agriculture by using stories from Ethiopia and other countries on the continent. I believe that researchers and policy makers in Africa can draw lessons from the Ethiopian experience when they prioritize their AR&D agenda. They should be asking several questions, including what went so well and what did not, and why. How could they adapt and adopt some of the good practices from the Ethiopia experience and avoid the pitfalls?

Organization of the book

This book consists of eight parts and comprises a total of 25 chapters. For brevity, this current volume mainly focuses on crop-based agriculture and does not include in-depth analysis of livestock agriculture or natural resources.

- Part I contains the introductory chapters: the introduction and synthesis (Chapter 1) and then the resource base for Ethiopian agriculture (Chapter 2).
- Part II contains assessments of productivity gains of major crops cultivated in the country: cereals (Chapter 3), roots and tubers (Chapter 4), grain legumes (Chapter 5), oilseeds (Chapter 6), vegetables (Chapter 7), fruit crops (Chapter 8), spices (Chapter 9) and coffee and other plantation/industrial crops (Chapter 10). A chapter on apiculture (Chapter 11), considering the importance of honeybees for crop pollination, is also included in this part.
- Part III is a single chapter on biotechnology (Chapter 12), which assesses what modern

science can offer to enhance agricultural productivity.

- Part IV considers crop protection. It includes entomology (Chapter 13), plant pathology (Chapter 14) and weed science (Chapter 15).
- Part V discusses agricultural inputs, comprising seed systems (Chapter 16), fertilizer use trends (Chapter 17), pesticide use patterns (Chapter 18) and agricultural mechanization (Chapter 19).
- Part VI deals with the environment. It contains a chapter each on climate change (Chapter 20) and irrigation water use (Chapter 21).
- Part VII consists of two chapters, one on extension (Chapter 22) and the other on early childhood and agricultural education (Chapter 23).
- Part VIII discusses the long-term socioeconomic benefits of investing in agriculture. It contains two chapters; one is on the nutritional wellbeing of the country (Chapter 24) and the other on agricultural commodity trade (Chapter 25).

Summaries

The resource base

As in many countries in Africa, Ethiopia is a country blessed with a huge agricultural resource base. This includes (among others) nearly 38 million ha of agricultural land (FAO, 2022b), diverse agroecologies, climates and farming systems, diverse crop species for many of which it is the primary or secondary centre of origin (Westphal, 1975), adequate surface water and groundwater for irrigation, a large youthful population, and a well-established AR&D system. About 16 million ha of the available agricultural land is currently arable, suggesting that there is about 22 million ha of land that can be harnessed in the future with well-planned public/private investment.

Successive governments of Ethiopia over the last seven decades have attempted to improve agricultural productivity. The initial efforts by the imperial government (1950–1974) concentrated on institution building. The military-socialist government (1974–1991) expanded agroecological coverage of research centres. The most recent past government (1991–2018) embarked on extending and delivering available technologies developed thus far. The current government (2018–present) is trying to transform Ethiopian agriculture, as an engine for economic development of the country.

As of 2018, there were more than 80 agricultural research centres covering various agroecologies and more than 3000 researchers working in the country. The Ethiopian Agricultural Research and Development System has succeeded in establishing the culture of AR&D. The government increased the number of FTCs (farmer training centres) and DAs (development agents). The research system introduced new crop species and released nearly 1200 crop varieties between 1970 and 2017, along with production technologies suited to Ethiopian agriculture. It has achieved many milestones, starting in the 1970s with successful identification and introduction to smallholder farmers of coffee lines resistant to the coffee berry disease. A large number of technologies have been developed and a large amount of scientific knowledge has been accumulated.

Past efforts on agricultural transformation in this country have not benefited from well thought out comprehensive and coherent studies. There are no reviews that measure the successes or inadequacies and lessons learnt from existing programmes that can be used to feed policy and priority setting for the future. Cognizant of these challenges, HGV launched this study in 2017 to address these critical issues in Ethiopian agriculture.

Measuring yield gains

Ethiopia has registered remarkable productivity growth for its cereals (Fig. 1.3). The aggregate yield more than doubled between 2000 and 2020. The overall yield gain was 92 kg/ha/yr, the third largest gain (after Brazil and the USA) and ahead of Canada, China, Indonesia, India, Russian Federation, Nigeria, Australia and Kazakhstan, among the top ten cereal-producing countries of the world. The estimated aggregate yield of about 2.9 MT/ha in 2020 was second only to South Africa among those African countries where the crops are produced mainly under

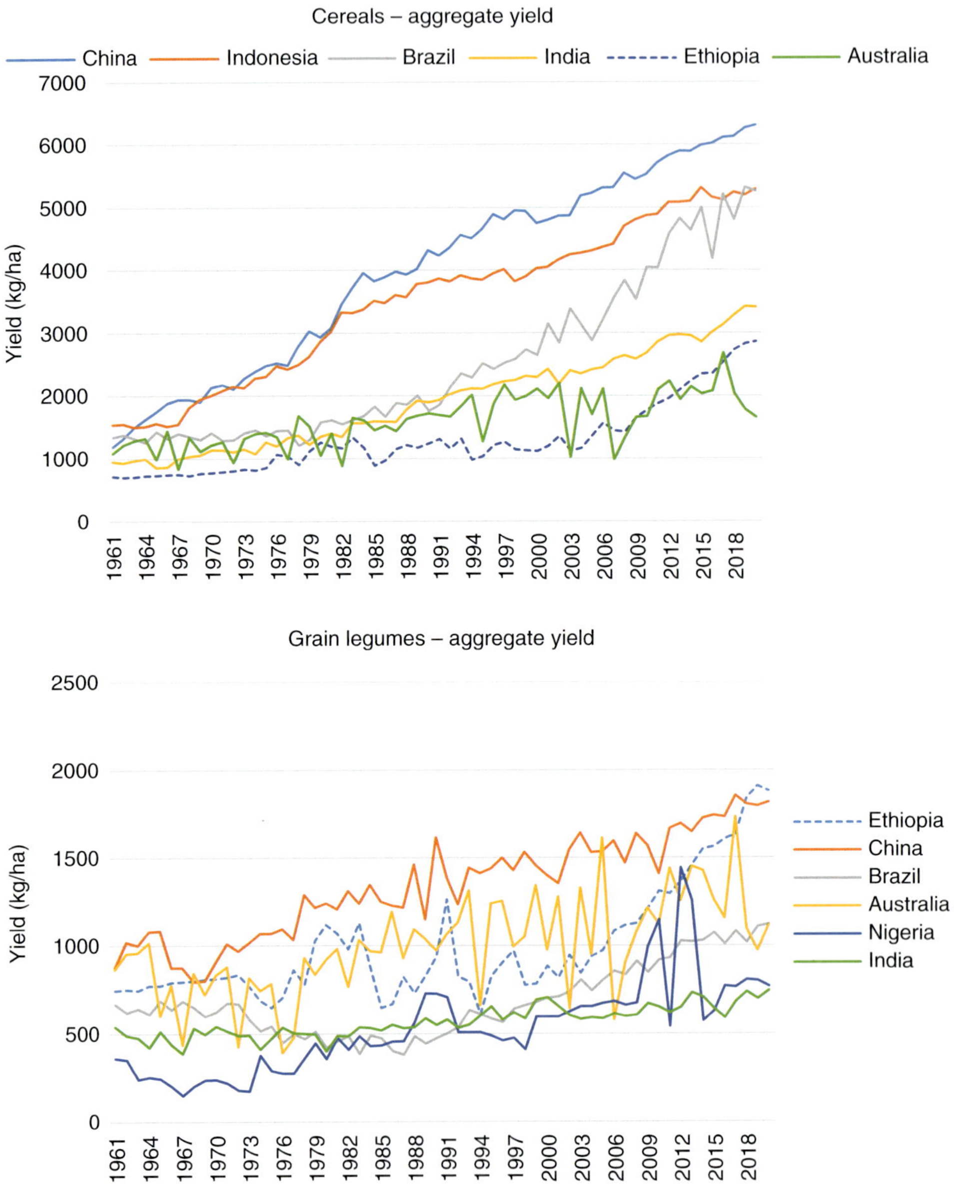

Fig. 1.3. Aggregate grain yields of cereals and grain legumes in selected major producing countries of the world. Source: constructed by the author from FAO (2022b).

rainfed conditions. The 2020 yield for maize was >4.2 MT/ha, the second highest in Africa after South Africa. Its yield gain was 127 kg/ha/yr, fourth highest (after Ukraine, Indonesia and Brazil) and followed by the US, China, Argentina, Mexico, India, Nigeria and Tanzania, among the world's top ten cereal-producing countries (where Ethiopia is ranked 12th). Investment in extension systems and improved access to modern varieties suitable for the different agro-ecologies, as well as yield-enhancing inputs, contributed to improving yields of smallholder agriculture.

As presented in Fig. 1.3, the country has also done well with grain legumes (or pulses), common bean and chickpea, in particular. The aggregate grain legumes yield in Ethiopia rose from ~880 kg/ha in 2000 to ~1880 kg/ha in 2020. The yield gain of 56 kg/ha/yr was the highest among the top ten producing countries in the world (superior to Canada, Brazil, China, Niger, Australia, Nigeria, Mexico, India and Russian Federation). Ethiopia has overtaken China's first place in its grain legumes yield (Fig. 1.3).

Even though there are at least nine species of cultivated root crops in Ethiopia, vital statistics (area, yield and production) are available only for taro, potato, sweet potato and yam. There are no reliable data for *enset*, the most important root crop in the country. It is grown in densely populated areas and remains relevant to the local economies of about 39 million people. This is an indigenous sustainable system with great potential for the future of Ethiopia and beyond. It has the highest fresh weight yield per unit area of cultivated land and has many other benefits, including its service as soil cover, preventing erosion and land degradation. In addition, *enset* has the highest content of Ca, Fe, K, Mg, and Zn among all staples in the country. Some progress has been made with the productivity of potato and sweet potato, but it is not nearly as robust as the cereals or grain legumes mentioned above. The area occupied by yam is declining fast. Overall, policy makers in Ethiopia have not given roots and tubers the attention commensurate with their socioeconomic contribution to the country.

All oilseeds, except sesame, showed highly significant ($P<0.01$) yield gains between 2000 and 2020. Leading the group was soybean, with 110 kg/ha/yr. Yields of all oilseeds except sesame and *gomenzer* (Ethiopian mustard) have doubled during this period, even though the overall yield was still relatively low when compared to the top ten largest producing countries of the world. Soybean was the fastest growing oilseed in area and yield. The area occupied by the traditional oilseeds of linseed, safflower, *gomenzer* and *noug* (*Guizotia abyssinica*) declined during the above period.

One major achievement with vegetables in Ethiopia is the increase in the number of species cultivated, from only a few through the 1990s to more than 20 species in recent years. Increased involvement of the private sector, starting in the early 1990s, appeared to have helped the country to make good progress with increasing the diversity of vegetables for food and nutrition security as well as for export diversification. Results with productivity gains of vegetables in Ethiopia varied from crop to crop. There were positive yield gains for green pepper, miscellaneous vegetables, green bean and cabbages, whereas lettuce, dry onion, tomato, garlic and carrot had declining yield gains. Overall, the estimated national average yield of about 6 MT/ha fresh yield is much lower than the world average of ~17 MT/ha. Inefficient irrigation systems and lack of access to high-yielding and disease and pest resistant cultivars remain major limiting factors on the supply side. Underdeveloped marketing systems and logistics that lead to significant crop losses along the supply chain are key challenges on the demand side.

Ethiopia grows more than a dozen species of cultivated fruits. Just as in vegetables, productivity gains in fruit crops varied according to crop. With a yield gain figure of 146 kg/ha/yr, banana was the only fruit crop that showed highly significant ($P<0.01$, $R^2=0.69$) yield gain increases with time. Although the yield gain for papaya of 346 kg/ha/yr was non-significant, it was the only fruit crop for which increases in production was the result of increase in yield (65%) rather than increase in area (33%). Miscellaneous fresh fruits, grape, avocado, mandarin and mango all suffered non-significant yield losses. Ethiopia's average yield of just over 6 MT/ha is much lower than the world average of 11 MT/ha. This country has been a net importer of fruits and fruit products since 2003. Export earnings from fruit crops in Ethiopia were estimated at US$7 million, compared to import costs of nearly US$24 million annually in the 2016–2020 period. Biotic and abiotic factors are responsible for the relatively poor performance of this sector. Some of the limiting constraints for fruit production in Ethiopia include: diseases such as leaf anthracnose on mango and pseudocercospora leaf/fruit spot on orange and other citrus species; insects such as the white mango scale on mango and pineapple mealybug on pineapple; generally poor crop management practices; and lack of market access due to lack of transportation connecting remote producers with markets, cold storage and underdeveloped processing capacity in remote rural areas.

Ethiopia embarked on diversification of its agricultural exports away from heavy reliance on coffee in the early 1970s and found that modernizing its rich spice diversity could contribute significantly to this goal. This country grows more than 50 species of spices and herbs, but data are available only for six of these and a miscellaneous category. These are 'spices, nes' (i.e. various spices not elsewhere specified, or miscellaneous spices), dry chilli, fenugreek, peppers (*Piper* spp.), coriander, ginger and cardamoms. Production of all spices has been increasing over the last two decades (2000–2020), but with the exception of fenugreek, it was due more to area expansion rather than to increases in productivity. Yield gains were highly significant ($P<0.01$) for 'spices, nes', fenugreek, coriander, peppers and cardamoms, but non-significant for dry chilli. Ginger suffered a highly significant yield loss of about 26 kg/ha/yr ($P<0.01$, $R^2=0.65$). In general, average yields for all spices and herbs in Ethiopia were low when compared to their potential, the world average and averages for many of the top ten producing countries in the world. Depending on the crop, the major limiting factors for spice production in Ethiopia are declines in forest cover (as most of the spice crops are adapted to forest conditions) and biodiversity. This was linked also with declining soil fertility, mainly due to removal of organic matter, poor agronomic practices and lack of widespread use of improved varieties. In the case of dry chilli, farmers in the central rift valley (a major producing region), are switching to *teff* (*Eragrostis tef*) and maize, because the latter are more profitable. The chilli crop has an additional burden of maintaining the purity of varieties because of its high level of outcrossing.

Ethiopia produces exclusively *arabica* coffee – the preferred type worldwide because of its superior cup quality – whereas its major competitors, including Viet Nam, Uganda, Malaysia, Cote d'Ivoire and Thailand produce only the *robusta* type with its higher yields per unit land area. Unlike many of those countries, the yield of coffee, the largest export earner in Ethiopia, declined at a non-significant 3 kg/ha/yr. By contrast, the area grew at an annual rate of 7.45% per annum. Ethiopia's national average yield of coffee is estimated at 650 kg/ha, which is much lower than its potential and less than the world average of 760 kg/ha. Traditionally, coffee in this country was grown 'wild' in the forest where there is plenty of organic matter, adequate moisture and abundant biodiversity that kept insect pests and diseases in check. At present, forest coffee accounts for only 10% of production, whereas individual garden, semi-managed forest and plantation coffee covered ~50%, ~30% and ~10% of coffee-growing areas, respectively. There has been heavy attrition of senior researchers from coffee over the past two decades; at present, there are few seasoned researchers to provide the technical leadership, future direction and mentoring of young scientists.

Tea is one of the success stories in Ethiopian agriculture. Though first introduced in 1927, initiatives to expand cultivation started in 1981, under the military-socialist government. The country produced more than 10,000 MT of tea annually during the 2016–2020 period. The area, yield and production grew at an annual rate of 5.45%, 0.84% and 6.34%, respectively, indicating that the bulk of increases in production were the result of increases in area rather than increases in productivity. The yield gain was estimated at 8 kg/ha/yr, lower than the top ten producers but still statistically significant ($P<0.01$, $R^2=0.81$). The average yield is estimated at 1.04 MT/ha, which is below the world average of 1.88 MT/ha but similar to that of Indonesia, Taiwan Province of China, Nepal, Mainland China and Mozambique. Much higher yields are reported for Argentina, Iran and Thailand.

Sugarcane cultivation area has increased, whereas its yield and production has declined steadily over the years. In 1976–1980, Ethiopia had the highest yield in the world of 177 MT/ha compared to 45 MT/ha in 2016–2020. The ROG (annual rate of growth) for area grew at 1.70% per annum while the yield and production declined at 5.08% and 3.47%, respectively, between 2000 and 2020. This means a yield loss of fresh cane of about 3.6 MT/ha/yr, which was highly significant ($P<0.01$, $R^2 = 0.54$). These losses are attributed to the loss of skilled technical and managerial level senior staff after the nationalization of the private sector by the military-socialist government in the mid-1970s. There has been continued bad governance and mismanagement ever since, e.g. senior management assignments have not been based on merit. The government of Ethiopia spent ~US$255 million on sugar imports annually during the 2016–2020 period. The sugarcane case is a good example of negative policy effects.

Cotton has also faced an almost similar situation to that of sugarcane, although most production indicators increased in the 2000–2020 period. They varied considerably with time: area, yield and production all saw a steady fall through the early 2000s, following nationalization. The yield and production rose from early 2000s but fell sharply from the early 2010s onwards. Ethiopia cultivated ~70,000 ha of cotton and produced ~148,000 MT of seed cotton, with an average yield of 2.1 MT/ha in the 2016–2020 period. This yield is much lower than the world average of nearly 3.0 MT/ha. Overall, the area, yield and production increased at an annual rate of 1.91%, 4.17% and 6.16%, respectively, between 2000 and 2020. Ethiopia is a net exporter of cotton. A simple regression analysis showed that this country registered a yield gain of about 61 kg/ha/yr during the above period. The yield gain compares favourably with several of the top ten cotton-producing countries of the world. With 169 kg/ha/yr, China, followed by Myanmar, Türkiye, Brazil and Australia, made by far the highest gains. Ethiopia's yield gain in cotton was similar to those of the USA and India. Yield gains were negative in Turkmenistan, Greece, Benin, Cote d'Ivoire, Chad and Mali, but all were statistically non-significant.

Ethiopia has a long tradition of apiculture. With more than 6 million hives and 52,000 MT of annual production, it ranks fifth and tenth in the world in the number of colonies and honey production, respectively. About 95% of beehives in smallholder apiaries are traditional, with the modern and transitional systems accounting for a little over 3% and 1%, respectively. The average honey yield was 9 kg/colony, compared to the world average of 23 kg/colony. Honey and wax yields declined at an annual rate of 0.56% and 0.67%, respectively, between 2002 and 2019 while the number of hives grew by 3.04%, and production showed a positive annual growth rate of 1.91% for honey and 2.47% for beeswax.

The external parasitic mite (*Varroa destructor*) has sharply reduced honey production in the European honeybee (*Apis mellifera* subspecies *scutellata*) in North America and Europe. Its presence has been recorded in parts of Ethiopia, but it is very unlikely to be the major factor for the poor performance of honeybees in this country where a different honeybee subspecies, *A. m. simensis*, that is resistant or tolerant to the mite is utilized. Long-term pesticide exposure is also unlikely to play a major role in the loss of production in Ethiopia due to low levels of application (<0.3 kg/ha active ingredients (AI) of arable land, compared to 2.6 kg/ha for the USA). The most likely major factors could be loss of abundance and diversity of flowers, nutrition, inadequate management practices, low stock diversity of bees and environmental change. However, well-designed scientific studies are needed to generate information that would help to isolate major factors affecting honeybee populations and its performance and provide appropriate recommendations.

Biotechnology in Ethiopian agriculture

Ethiopia has made substantial progress in building its agricultural biotechnology institutions, the centrepiece of which is the National Biotechnology Research Center at Holetta. The centre is staffed by close to 300 biotechnology-related researchers, of which nearly 46%, 40% and 15%, respectively, are PhD, MSc and BSc degree holders. Applications of tissue culture, disease cleaning, molecular markers and bio-fertilizer development have been successfully used in 32 plant species.

EIAR is now working on an additional four crops, including genetically modified (GM) *enset*, cotton, maize and finger millet. However, proponents of the use of GM technology in Ethiopia have not presented a compelling case. There is no objectively proven evidence that either the Bt (*Bacillus thuringiensis*) cotton nor Bt maize have any advantage over the existing locally developed varieties or hybrids under insect attack. The work on GM *enset* has not considered the priority needs for the crop, and there is no indication of the outcome considering the lack of previous work on this or any related crops outside the country, or the current level of expertise within the country. The emerging biotechnology tools (such as gene editing using CRISPR-Cas9 or TALENs) could have great perspectives to help break the low-yield deadlock of important traditional oilseeds such as *noug*, safflower, *gomenzer*, linseed and many other crops, without the attendant market issues raised by the use of transgenics.

Crop protection

Past efforts in crop protection research and development focused on integrated pest management (IPM), dealing mostly with smallholder agriculture that is the dominant crop production system at the present time in Ethiopia. However, there are indications that commercial farming on a larger scale is going to be expanding in the coming years and decades, following the current government policy to modernize agriculture as the major engine of economic development. This, plus a changing climate, would mean changes in the pest complex – including resurgence of existing pest species or appearance of new ones. The research system needs to gear itself towards tackling those problems using existing and emerging scientific advances and initiate a scouting system to give advance warning of changes in pest numbers and pest species.

Pest dynamics are bound to change more quickly with the government's intent to expand and intensify agriculture in the coming years. The research community needs to be ready to provide practical solutions as quickly as possible. The research system gave particular emphasis to diseases associated with cereals and grain legumes while many of the diseases on other crop groups did not receive the attention they deserve. This needs to change. It is very likely that priorities will change as the government enters into a new phase of expanded and intensified agricultural development, and researchers should make themselves ready to keep ahead of the evolving dynamics of both pests and farmers.

The need for prioritizing pest problems based on the country's development agenda and available resources is stressed by listing pest species that need to be given special emphasis by research in the immediate future. These include the pea aphid (*Acyrthosiphon pisum*) on pea; white mango scale (*Aulacaspis tubercularis*) on mango; fruit flies (*Bactrocera invadens*) on avocado and mango; Russian wheat aphid (*Diuraphis noxius*) and the green bug (*Schizaphis graminum*) on wheat; sesame seed bug (*Elasmolomus sordidus*) on sesame; African bollworm (*Helicoverpa armigera*) on cotton and tomato; pink bollworm (*Pectinophora gossypiella*) on cotton; fall armyworm (*Spodotera frugiperda*) on maize and sorghum; and tomato leaf miner (*Tuta absoluta*) on tomato. The criteria used in arriving at this list of priority arthropod pests included: (i) the government priority for the crop; (ii) the current and potential importance of the crop (in terms of food security, export earning, import substitution and raw material supply for the local industry); (iii) the importance of the pest; and (iv) the availability or the lack thereof of control measures currently in place.

The available literature shows that the research system in Ethiopia has accumulated adequate knowledge and skills to tackle plant disease problems, especially in major diseases of high-priority crops. Management practices are based largely on host plant resistance and cultural practices. Although information on the control of many diseases by chemical sprays is available, there are limitations in the availability, affordability, knowhow, logistics, human hazard risk and environmental concerns that constrain chemical sprays in smallholder conditions.

The top 20 priority diseases identified are: root rot/decline (*Phytophthora cinnamomi*) on avocado; purple blotch (*Alternaria porri*) and white rot/bulb rot (*Sclerotium cepivorum*) on onion/shallots; garlic rust (*Puccinia alli*) on garlic; wilt complex (*Fusarium* sp. and *Phytophthora* sp.) on hot pepper; leaf and fruit spot (*Pseudocercospora angolensis*) on citrus; anthracnose (*Colletotrichum gloeosporioides*) on mango; halo blight (*Pseudomonas phaseolicola*) and brown leaf spot (*Colletotrichum boltshauseri*) on mungbean; ascochyta blight (*Ascochyta phaseolorum*) on cowpea; chocolate spot (*Botrytis fabae*) and faba bean gall (*Physoderma viciae*) on faba bean; phyllody (mycoplasma-like organism) and bacterial blight (*Xanthomonas campestris* pv. *sesami*) on sesame; Dubti syndrome (unidentified) on cotton; coffee berry disease (*Colletotrichum kahawae*) and coffee rust (*Hemileia vastatrix*) on coffee; and stem/black rust (*Puccinia graminis* f. sp. *tritici*) and yellow/stripe rust (*Puccinia striiformis*) on wheat.

Weed science research in Ethiopia has contributed significantly to the areas of identifying major species, developing the national database, determining yield losses and delivering management practices both for smallholders and commercial farms in the country. Nearly 1100 species of weeds representing 552 genera and 126 families have been registered so far. Losses range from 15% to total crop failure in the worst-case scenario.

There have also been successes in the biological control of the whitetop weed (*Parthenium hysterophorus*); integrated management of the witchweed (especially *Striga hermonthica*) on sorghum, using host plant resistance and improved cultural practices; and using crop rotation in the management of herbicide-resistant annual and perennial grasses in commercial wheat farms in the Arsi and Bale highlands. Future work will focus on the further development of the national database, understanding weed biology, risk analysis, monitoring herbicide resistance in weeds, herbicide residues and environmental safety, among others.

Agricultural inputs use

The Ethiopian Seed Enterprise, now under the Ethiopian Agricultural Business Corporation (EABC) was the first and sole formal seed source of the country since its establishment in 1979 through to the 1990s. The number of private seed companies has grown rapidly since the 1990s following the rolling out of direct seed sales (i.e. private companies selling seed directly to the smallholders rather than through the government apparatus) in the first half of the 2010s. As of December 2021, there were a total of 54 seed companies licensed to produce seed in Ethiopia. These include 4 public, 2 transnational, 1 farmers' cooperative union and 47 small private seed companies that are owned by Ethiopian nationals. Research institutes are solely responsible for the production and maintenance of breeder seed, whereas the public entities produced 98% of the total EGS (early generation seed). Cereals, grain legumes and oilseeds constituted 89%, 9% and <3%, respectively, of the total EGS production (2016–2018 average). Oromia Seed Enterprise and EABC each accounted for about 40% of the total certified seed production, with Corteva contributing about 14%; the small national companies were responsible for the remaining 6% (2019–2020 average).

Out of the overall estimated average annual demand of about 729,000 MT of certified seed between 2016 and 2018, the formal seed system was able to supply about 60%, of which 57% was distributed. The supply (and distribution) for cereals, grain legumes and oilseeds, respectively, was 64% (60%), 24% (21%) and 18% (17%). Maize and wheat among cereals, mungbean and common bean among grain legumes, and soybean and sesame among oilseeds received the highest percentage distribution of certified seed. In general, the proportion of distribution was higher for crops of high priority – whether for food security or the export market. The lack of recognition by policy makers of organized farmers' (community) seed production contributed to the low supply of improved seed in Ethiopia.

Cereals, particularly hybrid maize, dominate certified seed production in Ethiopia. For example, 17 out of 22 (77%) seed companies for which there are data were engaged in certified maize seed production, whereas wheat and *teff* seeds were produced by 8 certified companies. Three companies produced chickpea and onion seeds and two produced common bean, soybean, *noug*, forages, pepper and mango seeds; malt barley, black cumin, tomato, vegetables, fruits, papaya, sesame, potato, pigeonpea, alfalfa and sesbania were each all available from one certified producer. Corteva, a US-based seed company, supplied 25% of the total maize seed (2015–2019 average).

Even though the EARS (Ethiopian Agricultural Research System) has released nearly 1200 cultivars of cultivated crops between 1970 and 2017, it has not put in place strict rules over the rate of variety turnover. Consequently, many cultivars that are essentially obsolete are still under production. For example, the average age of all crops cultivars in Ethiopia is 23 years (range: 9–42). The average age of *teff* is 22 years (range: 14–37); maize is 20 years (range: 9–33); and wheat 18 years (range: 9–26). The malt barley cultivar 'Beka' was released in 1976 and another malt barley cultivar, 'Holker', in 1979. The onion cultivar 'Adama Red' was released in 1980. All of these and many others were reported to be under production as recently as 2017. Continued cultivation of obsolete cultivars reduces the economic impact of AR&D investments. Their replacement would require awareness creation for new and better cultivars, focusing on their superior economic benefits and increased production and delivery of certified seed to meet the demand.

The total area under certified seed varied widely among major cereals over the years. For example, our calculation of seed replacement rate shows that 82% of maize area was covered under certified seed in the 2020 *meher* season, compared to 15% in 2006. By contrast, the sorghum area under certified seed stayed at less than 1% and changed little over the above period. Overall, the area under certified seed for wheat, *teff* and barley was about 17%, 7% and 5%, respectively, in 2020, compared to about 5% for wheat and <1% each for *teff* and barley in 2006. This indicates the remaining gaps in delivering modern inputs and the underutilized potential to increase production of *teff*, barley and dryland cereals such as sorghum and millets. While more rigorous analysis is needed, there seems to be a strong economic rationale to expand production to meet the growing unmet demand and potential for exports (especially for *teff*).

It is concluded that small private seed companies, farmers' cooperatives and community seed producers have yet to build their capacity to produce an adequate amount of quality improved seed to satisfy Ethiopia's growing demand. Public sector and established transnationals will continue to be the major sources of certified seed for the foreseeable future in Ethiopia.

The use of mineral fertilizer to enhance crop productivity in Ethiopia began in the early 1960s following the imperial government's efforts to modernize agriculture. It has since become an essential component of farm management practices for both smallholder and commercial farming. All of the national demand is met through imports, which makes it highly vulnerable to price volatility similar to the situation after the COVID-19 pandemic and the war in Ukraine. The import in 2020 of mineral fertilizers (by product) was 1.6 million MT, valued at US$500 million, making it the third largest importer in Africa after Morocco and South Africa. Mineral fertilizer use (by nutrient) showed progressive growth, especially after the mid-2000s. The total consumption was 157,000 MT in 2000 compared to 586,000 MT in 2020, a 4.38-fold increase (or annual growth rate of 9.83%). Mineral fertilizer nutrients in Ethiopia consisted of roughly 70% N and 30% P, with negligible amounts of K.

The application rate (kg of NPK nutrients per ha of arable land) followed similar trends to national consumption; it was 12 kg/ha in the 2000–2004 period vs 35 kg/ha in 2016–2020. This makes Ethiopia the sixth highest country (after Zambia, South Africa, Morocco, Kenya and Cote d'Ivoire, followed by Zimbabwe, Malawi, Mali and Senegal) among the top ten countries with the largest amount of arable land in Africa. Of those, only the first four achieved the African Union (AU) recommendation of 50 kg/ha of nutrients per ha of arable land. Even though Ethiopia is one of the countries in Africa where the overall application rate is below the AU recommendation, high-priority crops (including maize, *teff* and wheat) appeared to receive close to the recommended rates, especially in the well-watered areas.

Mineral nutrient application in Ethiopia varied widely from crop-to-crop and among crop groups. Cereals accounted for 88% of the total mineral fertilizers used; of those, 29%, 27% and 22% were used on maize, *teff* and wheat, respectively, because these were the crops accorded the highest priority for food security by policy makers. Smallholder farmers use both mineral and organic fertilizers (the latter including farmyard manure, compost, household refuse and crop residue). Calculations using the Central Statistical Agency (CSA) data show that mineral fertilizer use in Ethiopia grew by 3.16% annually between 2006 and 2019, whereas organic fertilizers declined by 2.24% during the same period. The decline in organic fertilizer use is attributed partly to the shift towards mineral fertilizers and partly to the removal of crop residue and farmyard manure for other uses. Crop residue and farmyard manure are important sources of livestock fodder and firewood for smallholder farmers and rural households in Ethiopia. Cereals crop residue is used as animal feed, for construction and firewood; given the scarcity of fuel wood, dried cow dung has become an important source of energy for cooking. Grain legumes residue is also totally removed from the field for animal feed. All this means that there is heavy biomass removal and little or no biomass that goes back to the soil, leading to increasing soil mining and degradation across the highland regions.

Overall, smallholder farmers in Ethiopia selectively apply mineral fertilizers on crops with higher commercial value (such as major cereals, dry pepper, onion and potato) and organic fertilizers on traditional crops primarily grown for own consumption. Mineral and organic fertilizers covered about 62% and 12%, respectively, of the total cultivated area in 2017–2019. In general, smallholder farmers apply suboptimal amounts of mineral fertilizers – less than the extension package recommendations – mainly because they cannot afford the high price or because they are farming in climatically risky areas.

The use of microbial fertilizers (MF) has not received the due attention commensurate with its economic, environmental and other benefits. This technology is mature to take off (even though research has yet to perfect the delivery mechanisms). Demonstrations on farmers' fields have shown that MF inoculation of grain legumes, with or without P, increased seed yield by 21–79%, depending on the crop type and geographical location. A good number of smallholder farmers are already buying from local small businesses in Addis Ababa and using MF on grain legumes in many parts of the country. However, there is a compelling case for the government to incorporate MF and micronutrients as an essential component of the extension package for increased productivity of grain legumes.

Agricultural pesticide use

As of August 2020, 184 AI used in synthetic agricultural pesticides, represented by 586 trade names, have been registered for use in Ethiopia; 22 of these AI are banned and another 21 are on the watch list in other countries. The reasons for their ban are international conventions and their acute and chronic toxicity to humans and the environment, including the honeybee. Herbicides, insecticides, fungicides-bactericides, and others constituted roughly 75%, 15%, 9% and <1%, respectively, of agricultural chemicals registered for use. The largest number of individual chemicals by active ingredient (n=50), are registered for use against cut flower pests.

Ethiopia imported close to 18,000 MT (of agrichemical products) valued at US$108 million each year, during 2016–2020. The total pesticide consumption for this country was about 4100 MT of active ingredient per annum (2016–2020 average). This translates to an application rate of 0.2568 kg AI per ha of arable land. Comparing the world's top 10 countries with the largest area of arable land, the rates for Brazil, Argentina, France, Türkiye, Mexico, China and USA, respectively, were 25-fold, 23-fold, 14-fold, 9-fold, 9-fold, 9-fold, 9-fold higher than that for Ethiopia (Fig. 1.4). Among African countries, South Africa, Morocco, Zambia, Cameroon, Malawi and Zimbabwe apply 8-fold, 7-fold,

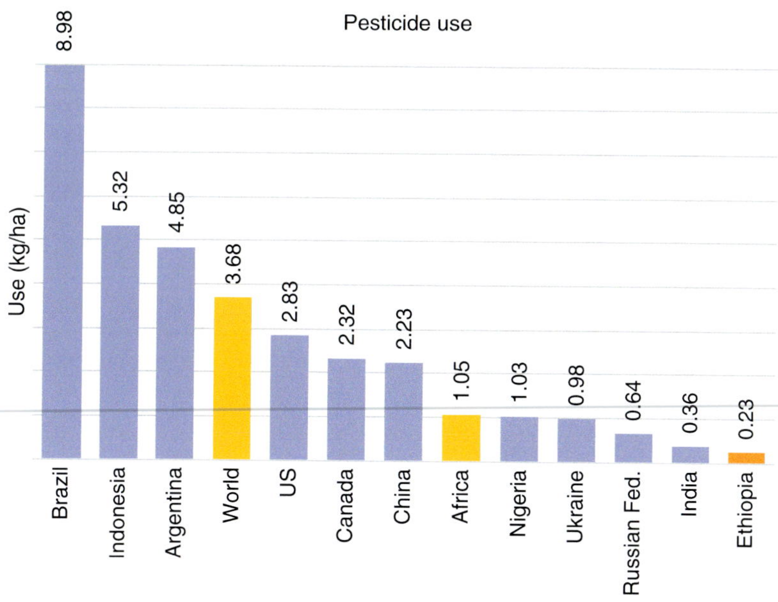

Fig. 1.4. Pesticide use (kg active ingredient per ha of cropland) of synthetic pesticides in Ethiopia, compared with the top ten countries with the largest cropland area in the world, 2016–2020. Note: Countries with the largest cropland area are India, US, China, Russian Federation, Brazil, Indonesia, Nigeria, Argentina and Ukraine; Ethiopia is 19th out of 223 countries and territories. Source: calculated and constructed by the author from FAO (2023).

4-fold, 4-fold, 2-fold and 2-fold, respectively, more synthetic pesticide than Ethiopia.

There are two plausible explanations to the relatively low level of pesticide use in Ethiopia: (i) the research system has developed viable IPM approaches, using cultural practices and host plant resistance as major components (especially for commercial agriculture); and (ii) Ethiopian agriculture is smallholder-dominated whereby crop production is carried out under complex, diverse cropping systems rather than monoculture and, therefore, regular pest outbreaks are uncommon. Historically, the largest amounts of pesticides in Ethiopia were targeted against migratory pests – the desert locust (*Schistocerca gregaria*), the African armyworm (*Spodoptera exempta*) and the weaver bird (*Quelea quelea*) – invasions and outbreaks.

Synthetic pesticide consumption in Ethiopia showed accelerated growth from 2006 but seems to have remained virtually unchanged since 2010. Pesticides covered about 25% of the total cultivated area, and roughly 11% of smallholders use them in Ethiopia (2016–2020, average). Two factors dictate smallholder farmers' choice to use pesticides on their crops: value of the crop as source of cash and availability of alternative options for controlling the pest. Smallholder farmers selectively apply pesticides to those crops that they produce for sale rather than those they produce for their own consumption. These findings suggest that there is no widespread contamination of the environment in Ethiopian agriculture.

Climate change and irrigation water use

Ethiopia is highly vulnerable to climate variability and change, which has significant implications for the realization of its development objectives. Rainfed agriculture, which accounts for more than 95% of cultivated area, is most vulnerable to climate shocks. The threat of climate change is already affecting productivity in many countries, reflected in terms of frequent droughts, pest and disease incidence and other extreme events that reduce land and labour productivity. Climate change will reduce national food security and affect export earnings from major export commodities such as coffee and tea. Cross-sectoral efforts, investments and policies are needed to expand the adoption of climate smart solutions and reduce exposure and sensitivity of agriculture and food systems to climate change. Investments in sustainable irrigation and improved management of natural resources, including reduced soil degradation and deforestation and protection of watersheds, would increase climate resilience. The review suggests the risk of climate change is well recognized in Ethiopia and significant progress has been made in formulating several climate-related policies, strategies and plans to reduce vulnerability and build resilience.

Despite this remarkable progress, however, there is a deficit in government-led planned adaptation actions at the local level. Most of the reviewed adaptation projects and programmes are financed by international organizations, and hence these financial sources may not be sustainable and adequate in the long run. To translate its national adaptation policies and strategies into concrete adaptation actions at the local level, the government is facing financial, institutional, technical and informational barriers. On the other hand, smallholder farmers are adopting numerous agricultural adaptation measures to combat the risk of climate change and improve their livelihoods, though various barriers impede these efforts. Together, Ethiopia needs to address the issues connected to barriers to autonomous and planned adaptation actions if it is to effectively tackle future impacts of climate change.

Ethiopia has huge water potential based on its available groundwater and surface-water resources. Increased irrigation could substantially increase crop productivity and production if properly planned and managed. Surface water, groundwater and rainwater are the sources of irrigation in this country. Ethiopia possesses 12 basins, with a total annual flow of 122 billion cubic metre (BCM) surface water and 2.65 BCM groundwater. Estimates for land equipped for irrigation vary, depending on the source. The FAO estimates suggest that about 858,000 ha of the total cropland (or 4.80%) could currently be irrigated (2016–2020 average), whereas the total cropland area actually irrigated was 554,000 ha (or 3.10%; average for 2017, 2018 and 2020). However, recent studies in Ethiopia estimate the total physical irrigation potential to be about 7.50 million ha.

The total suitable land of all the 12 basins for both surface and pressurized irrigation is estimated at about 37 million ha. This figure reflects only the land suitability, regardless of water availability. Irrigation potential should be necessarily based on both land and water availability. It was therefore said that about 8%, 58% and 34%, respectively, of the estimated land area here is considered highly suitable, moderately suitable and marginally suitable for irrigation.

Currently, there are 269 irrigation schemes implemented by the MOA. About 90% of those are medium and small scale and 10% are large scale. Of those, 41% are fully functional, 50% are partially functional and 9% are non-functional. This shows that the efficiency of irrigation schemes in Ethiopia is low – about 45% overall average (40% large scale and 60% for medium scale). An additional 59 irrigation schemes, consisting of 22 medium and small scale and 37 large scale are under study or construction (as of 2021). The GTP II under the MOA planned to bring more than 3 million ha (or about 17%) of cropland under irrigation (about 14% under small scale and 3% under medium and large scale) by 2019/20, but there is no indication how much of this has been achieved. However, this remains critical for building resilience and buffering smallholder production from climate shocks and ensuring food security under the changing climate.

Agricultural mechanization

The importance of agricultural mechanization, defined here as the use of any kind of tools powered by humans, animals or engine/motor in the process of agricultural production, has long been recognized in Ethiopian agriculture, but its uptake has been low until recent years. Less than 1% of the arable land is prepared using tractors. The exception to this has been wheat, where 25% is harvested using combiners. Continued efforts have been made to develop and introduce improved farm tools such as the *enset* processor, the *maresha* (single-tine plough), row planter, weeder, tied-ridge maker and donkey cart. Those efforts have yet to demonstrate substantial success.

Agricultural extension

Depending on the political system of the time, Ethiopia has experimented with various agricultural technology transfer models and approaches over the past decades. Several actors have been involved. These included various bilateral projects and innovations by the national research system; each has left behind varying levels of legacies. The country has the largest public extension system in Africa. The government has increased the number of FTCs to 15,000 as of 2016; there were more than 72,000 DAs serving about 16.7 million smallholders. This means a DA-to-farmer ratio of 43:10,000 (or 1:230), the densest in the world.

The number of smallholders and crop area receiving extension packages grew rapidly between 2006 and 2019. Each of them has at least doubled during this period. The overall average proportion of smallholders and crop area under extension grew from 7% to 26% and from 13% to 36%, respectively. Cereals were by far the largest recipients. The proportion of cereal growers receiving an extension package rose from 19% in the base year to 54% in the final year, while the area coverage was 17% in the base year versus 46% in the final year. Other crop groups also showed varying levels of growth.

The proportion of smallholders receiving extension packages among individual cereals varied from 55%, 50%, 47%, 45% and 42%, respectively, for wheat, maize, rice, *teff* and millet, compared to 3% for yam, and 5% each for taro and papaya, and 4% each for avocado and sugarcane. In a similar fashion, maize, wheat, millet, potato and rice area coverage under extension package was 64%, 59%, 50% and 49% each, respectively, whereas yam, taro, coffee and *noug* had 4% each; the coverage for avocado was 3%. Data for *enset* and cotton are conspicuously absent from CSA records of extension packages.

The priorities of the smallholders are not necessarily always aligned with government policy. For example, smallholders invest more in *ch'at* (*Catha edulis*) than coffee; proportions of area covered by extension packages for potato are not very far from those for major cereals; the level of farmers' investment on rapeseed, tomato, onion, dry pepper and garlic is comparable to that of *teff*. The smallholders' decision to invest in a crop is predicated by its market value rather than by government priority.

Socioeconomic benefits of investment in agriculture

Two important socioeconomic benefits from investment in AR&D are discussed – one is improvements in the overall nutrition and the other in the growth of agricultural exports. Ethiopia's dietary energy intake was estimated at about 1600 kcal/person/yr, 1700 kcal/person/yr and 2500 kcal/person/yr in 1975–1979, 1995–1999 and 2015–2019, respectively, indicating that the country has met the World Health Organization (WHO) minimum requirement of 2100 kcal/person/day by 2010. However, the diets are low in fat, protein and micronutrients compared to many African peers and models in Asia and Latin America. Plant sources accounted for 96% of dietary energy, 91% of dietary protein and 71% of dietary fat in the 2015–2019 period. This study has demonstrated that complex carbohydrates, protein and fat constituted 78%, 14% and 4%, respectively, of the total dietary energy versus the WHO recommendation of 55–70% carbohydrates and 15–30% fat for a healthy diet. Diversification of food production towards more sustainable, nutritious and healthy options to increase access and diversify dietary intake is key to addressing hunger and malnutrition in Ethiopia, and elsewhere in Africa.

Even though there was modest growth in per capita consumption of fruits (~9 kg/person/yr, 2015–2019 average) and vegetables (~14 kg/person/yr), Ethiopia compared unfavourably with most of the countries reviewed as part of this study (Fig. 1.5). The per capita consumption for red meat (5.40 kg/person/yr), chicken (0.72 kg/person/yr) and egg (0.40 kg/person/yr) has progressively declined over the years and is among the lowest in Africa. Milk consumption (~28 kg/person/yr) showed slight improvements starting in 2002 and was about the same as the continental average. Fish consumption (0.51 kg/person/yr) also showed some growth but was the lowest among all African countries.

Maize has displaced *teff* and sorghum as the top staple food crop, while *enset* and other root crops have shown the fastest growth rate over the last four decades despite the fact that they have not been given high priority by policy makers – i.e. their production and usage has expanded as part of farmers' innovation and increased demand by the general public who cannot afford high-priced cereals like *teff*. Another major shift was from traditional oilseeds to imported processed oil (dominated by palm oil). The per capita consumption of oilseeds declined from 3.32 kg/person/yr in 1975–1979 to 1.44 kg/person/yr in 2015–2019; conversely, manufactured oil consumption grew from 0.85 kg/person/yr in 1975–1979 to 3.65 kg/person/yr in 2015–2019.

In conclusion, this study has brought to light the need for Ethiopia to reduce heavy reliance on limited cereals and upgrade its investment in objectively identified native staples and oilseeds, as well as other potential under-utilized crops, to harness their full potential for achieving a sustainable food and nutrition security in the foreseeable future.

Ethiopia relies heavily on agricultural commodity export as a major source of foreign currency earnings. It also imports a large number of agricultural items to compensate for the shortfall in food and nutrition security and raw material supply for the local industry. The country has made substantial growth in the quantity of agricultural products exported and therefore the amount of earnings obtained over the last two decades. Export earnings showed fast growth starting from 2009 and peaked at ~US$3.61 billion in 2014; they took a downward trend thereafter but showed signs of recovery in 2019 and 2020 (see Chapter 25). The average export earnings for 2016–2020 were ~US$1.45 billion, compared to ~US$382 million in the 1996–2000 and ~US$313 million in the 1976–1980 periods. The number of export items also grew from 76 in 1976–1980 to 195 in 2016–2020; the top ten export items are coffee, oilseeds, grain legumes, meat, vegetables, spices, crude materials, prepared food, live animals and fruits. The top ten fastest growing export items are fresh vegetables, faba bean, honey, alcoholic beverages, pastry, goat meat, beer of barley, chickpea, common bean and mutton; by contrast, earnings from ginger, mustard, frozen vegetables and non-alcoholic beverages have declined.

In addition to the yield gaps for several agricultural products, the major challenge for Ethiopia's export market competitiveness is that

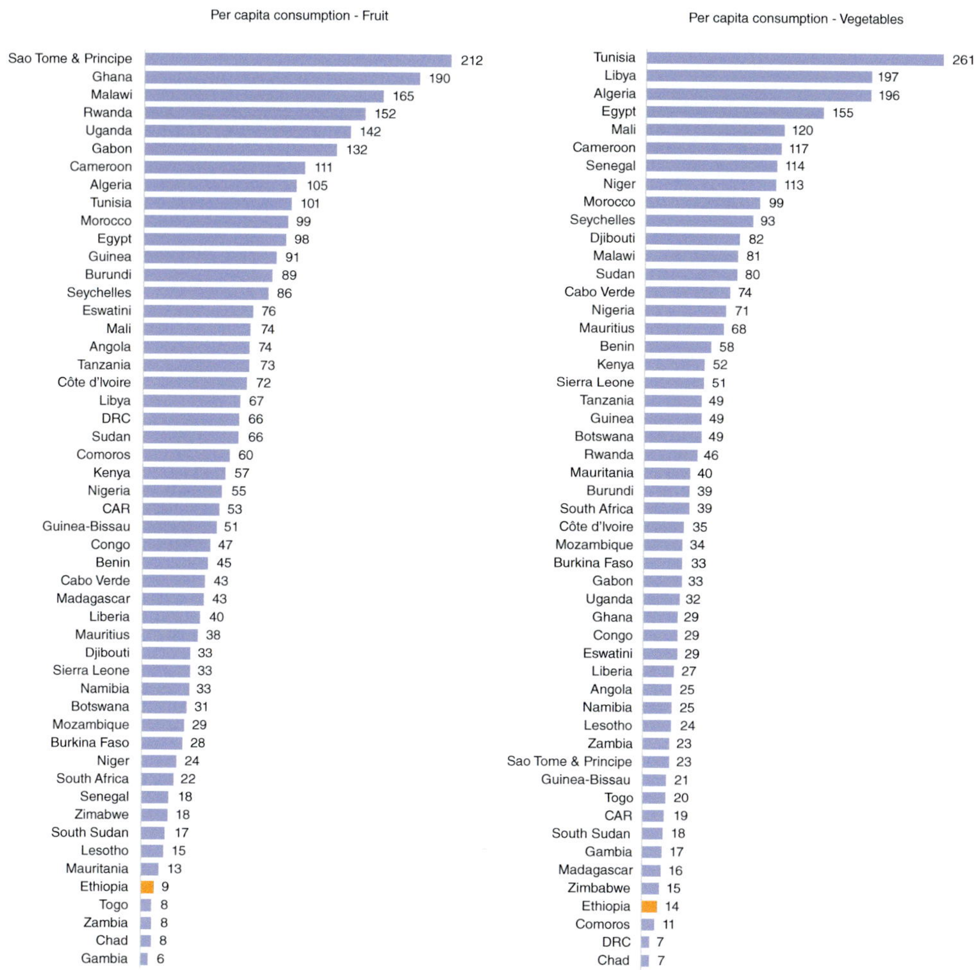

Fig. 1.5. Fruit and vegetable per capita consumption in Ethiopia compared with other African countries. CAR = Central African Republic; DRC = Democratic Republic of the Congo. Source: calculated and constructed by the author from FAO (2022a).

approximately 89% of all agricultural products are exported without value addition. For example, this country is the world's fifth largest producer of coffee (after Brazil, Viet Nam, Colombia and Indonesia), but nearly 99% of it is sold as green coffee. By contrast, Italy, France, the Netherlands and Germany produce no coffee, but roasted coffee accounted for approximately 93%, 88%, 76% and 41%, respectively, of their total coffee export; the USA produces less than 1% of its total export, of which 69% is roasted.

That said, Ethiopia had good examples of value addition with its hides and skins and goat meat export. For example, the country exported 4341 MT of dry or wet salted sheep skins and sheep skins with wool at an estimated value of ~US$20.3 million in 1976–1980; similarly, it exported 1077 MT for an estimated value of nearly ~US$16.9 million in 1996–2000. There was a negligible number of unprocessed hides and skins exported in 2016–2020; the hides and skins were either processed or semi-processed locally and then exported for increased earnings, or they were used by the local leather industry to manufacture leather goods for export or for the local market. By contrast, the country earned approximately US$78.4 million annually from the export of 14,000 MT of goat meat in 2016–2020;

the earning in 1996–2000 was about US$2.9 million from 1400 MT, whereas there was no export in 1976–1980.

Ethiopia has also made substantial progress with diversification of its export commodities. For example, the number of export commodities grew to 195 in 2016–2020 compared to 81 and 76 in the 1996–2000 and 1976–1980 periods, respectively. Coffee accounted for 77% and 82% of the country's total export earnings in 1976–1980 and 1996–2000, respectively, whereas its share in 2016–2020 was roughly 45%. By contrast, the contribution of oilseeds grew from about 2% in 1976–1980 and 6% in 1996–2000, compared to about 27% in 2016–2020. Vegetables and meat also showed substantial growth in recent years.

There is one good example of import substitution in Ethiopia. The country transitioned from a net importer to a net exporter of tea in 2001. Ethiopia imported 1001 MT of tea in 1976–1980, compared to 522 and 47 MT, respectively, in 1996–2000 and 2016–2020. On the other hand, it exported 1254 MT of tea for an estimated value of >US$2.2 million in 2016–2020; export quantity in 1996–2000 was 316 MT valued at US$617,000; there was no history of tea export from Ethiopia in 1976–1980. This achievement was made possible by the introduction of tea research and development initiatives in the late 1970s and increased production from area expansion. Ethiopia's experience with tea is a good example of import substitution and export diversification.

Import values showed continued growth trends from 2009 to 2018; they peaked at US$2.90 billion in 2015 and started to decline in 2019 and 2020. The average for 2016–2020 was US$2.09 billion (compared to the export revenue of US$1.45 billion). Ethiopia has become a net importer of agricultural commodities since 2016. The top ten commodities in descending order of value are wheat, cooking oil, rice, sugar, prepared food, sorghum, industrial goods (such as malt, raw tobacco, hops, natural rubber and molasses), dry peas, fruits and manufactured goods (fatty acid, cigarettes, glucose/dextrose, essential oils and lactose) (see Chapter 25). The number of import items grew from 120 in 1976–1980 to 184 in 1996–2000 and 275 in 2016–2020. This means that the government policy to improve import substitution should address the issue as a system rather than focusing on a limited commodity. Whether the current effort for import substitution, including increases in domestic wheat production and processing of edible oils, would reverse the recent trends and make the country less dependent on imported food is yet to be seen.

Factors for success

Ethiopia's successes (with cereals and grain legumes in particular) are attributed to the 3Is – institutions, investment and inputs. The country took a homegrown systems approach (Abate *et al.*, 2011, 2015) to the research and development of its crops. Ethiopia has one of the strongest national research systems on the African continent. It is unique among its peers in that it was built from scratch (not inherited from colonial past), run entirely by Ethiopians and, above all, has enjoyed sustained government budgetary support and has not been donor-dependent for funding. Large numbers of crop varieties (and hybrids) and their management practices were already available by the late 1990s. The government also built the strongest public extension system in Africa. This has enabled aggressive demonstration and promotion of the improved technologies and mineral fertilizer use to smallholder farmers. Ethiopia has also been fortunate in recent years not to have faced devastating droughts on the scale similar to that of 1984 or one a decade earlier than that (Abate *et al.*, 2015).

Cereals (mainly maize, wheat and *teff*) and grain legumes (common bean and chickpea) were successful in Ethiopia because they were accorded the highest priority by policy makers – the cereals for food security and the grain legumes for export earnings. Consequently, they received preferential treatment in terms of input use and extension services. It should be noted that Uganda, with its maize and common bean, and Ghana and Tanzania, with their common bean production, have achieved successes somewhat similar to those of Ethiopia.

Partnerships

Ethiopia has had a long-term partnership with the international development community in its AR&D endeavours. Notable among the partners are the United Nations (UN) agencies (e.g. the United Nations Development Programme and the Food and Agriculture Organization), bilateral agencies (such as the US Agency for International Development, the Canadian International Development Agency and the Swedish International Development Agency [SIDA]), the CGIAR (formerly the Consultative Group for International Agricultural Research) centres, the World Bank, and others such as Sasakawa Global 2000 and Sasakawa Africa Association. Each one of them has contributed to different kinds of legacy. The initial intentions and outcomes were positive and had far-reaching implications. For example, the US government supported the establishment of agricultural education system in the 1950s; the UN system played a key role in building the research system; SIDA introduced the first model of an extension system; the CGIAR centres' major achievement was support with the germplasm for the development of high-yielding crop varieties/hybrids and seed sector support; and the World Bank supported infrastructure and human capacity development through different phases.

On the other hand, the research system was under heavy pressure from the World Bank to devolve into RARIs (regional agricultural research institutes) in the early 2000s. EIAR was supposed to hand over almost all of its mandates (with the exception of coffee research in Jimma) to different regions. My management team at the time had to make extraordinary efforts to come up with a convincing argument that there was no way the government could achieve its aim of, for example, a fivefold increase in its export earnings from grain legumes within 5 years if the research programmes were transferred to the RARIs. This is because the latter did not have the technical capacity at the time. The move was thus put on hold and a possible disaster was averted.

At times there was direct competition between the CGIAR centres and the national system for donor funding. Shortly before my EIAR management team arrived in December 2003, the International Livestock Research Institute (ILRI) secured about US$20 million funding from the Canadian government for a project on technology delivery, named Improvement of Productivity and Market Success. The project was renewed for another phase in spite of EIAR's continued official request to avoid undesirable competition that crowds out the national AR&D system in delivering the technologies it developed. What is so unjust and ironic about this project was the fact that all of the technologies were developed by the national system and ILRI did not have the competence in these fields. Under my management team, the EIAR was the first to introduce and implement the concept of scaling up/scaling out technologies long before the CGIAR centres (cf. Abate, 2006). As their supposed 'guiding principle', the CGIAR centres are supposed to complement and help 'crowd-in' the national system through strategic upstream research and technology development to address major constrains such as climate change, rather than engaging in downstream activities that 'crowd-out' the national system (see McIntire, 2023).

There is significant quantitative and anecdotal evidence that the partnerships with the CGIAR centres and other organizations are not based on mutual respect and have not been equitable (McIntire, 2023). With the reduced funding for the CGIAR system, competition with national systems became a common occurrence. This approach seems also to benefit from the general perception that national AR&D systems in Africa are dysfunctional, which was not the case in Ethiopia (even in situations where the national system is weak, the sustainable solution is to invest in building domestic and regional capacity rather than bypassing it altogether). Knowingly or unknowingly, the national systems have always been junior partners. The research and development agenda of the international community may not necessarily align with the national agenda. As such, priority has been given to CGIAR-led commodities such as selected cereals and grain legumes. Consequently, the more sustainable indigenous crop systems such as *enset* have been neglected and not been accorded the attention they deserve.

Missed opportunities and unfulfilled promises

There have been missed opportunities and unfulfilled promises both at the national and continental levels. Ethiopia missed a big opportunity to further strengthen the transformation agenda the national programme spearheaded in the early 2000s and transform its technology delivery system, and hence its agricultural productivity as a whole, when it introduced the very high-profile Agricultural Transformation Agency (ATA) in 2011. The ATA failed to deliver on its goals, primarily because: (i) its implementation followed a top-down approach from the very beginning, i.e. the leadership position was more of a political appointment handpicked by the highest government body of the country at the time rather than merit or competence based; (ii) it did not have institutional capacity; and (iii) it lacked clarity of purpose. The institution therefore ended up doing field experimentation, in direct competition with the national AR&D system, whereas its main focus should have been enabling the national programme staff in research and building bargaining power for them. There is no evidence linking the ATA with the major successes Ethiopia enjoyed with its cereals and grain legumes. One can only hope that policy makers have learnt the appropriate lessons and would use this experience when planning future strategies and priorities.

At the continental level, there was great optimism about transforming African agriculture when the Alliance for a Green Revolution in Africa (AGRA) was launched in 13 countries across the continent in 2006. Some authors (e.g. Blaustein, 2008) even declared soon after that the green revolution had arrived in Africa. No, it had not (Garbero *et al.*, 2018; Wise, 2020; Abate, 2021), and still has not even after nearly US$446 million has been expended and 15 years of effort committed to AGRA's goals. It is true that Ethiopia has more than doubled its maize yield within two decades; others like Uganda, Ghana and Malawi also showed positive growth; Rwanda also doubled its maize yield but started from a very low base and is still below 2000 kg/ha; Niger and Tanzania are worse off in 2021 versus 2000; yields changed little in Burkina Faso, Kenya and Mozambique (Fig. 1.6). Even if all countries showed significant positive change, AGRA could hardly claim the credit, because agricultural progress is not a one-shot affair, rather it builds upon past cumulative investments. Progress can be attributed only to domestic efforts of those countries who got their 3Is right. AGRA's role could only supplement the national effort at best and cannot be seen as the ultimate mover and driver of agricultural transformation in Africa.

Unfinished agenda

It is obvious from the foregoing discussion that Ethiopia has made substantial progress with its agriculture, especially cereals and grain legumes, but there are a multitude of issues that still need to be addressed as a matter of urgency in order to help the government's agenda for agricultural

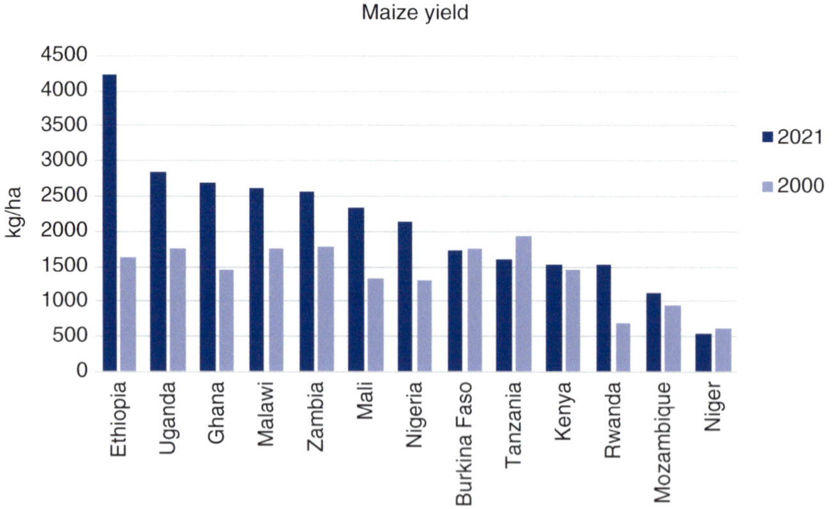

Fig. 1.6. Yield of maize during two periods in countries where the Alliance for a Green Revolution in Africa initiative was carried out. Source: calculated and constructed by the author from FAO (2023b).

transformation. Even the successes made so far are not going to last for much longer without continued strategic investment – it is only a matter of time before the yields start to plateau (Fig. 1.7) and eventually take a downward trend (Fischer *et al.*, 2014; Fischer, 2019). Grain legumes area is already showing sharp declines and might continue to do so in the future, owing to the fact that farmers are shifting to more profitable crops. Incentives should be put in place to keep farmers producing crops that are deemed high national priority, for both ensuring food security and domestic industry or exports.

Priority setting in the past has been heavily biased towards cereals – at the expense of indigenous sustainable systems like *enset* and other root crops. The area of traditional oilseeds such as linseed, safflower, *gomenzer* and *noug* has declined over the years, mainly due to the fact that research has not delivered higher yielding improved cultivars, as a consequence of which smallholder farmers are shifting to *teff* because the latter is more profitable. There is a strong need to revisit the current priorities on these and many other crops, as mentioned in the preceding paragraph.

The national average landholdings by smallholders are <1 ha and it would be unrealistic to expect Ethiopian agriculture to fulfil its functions of food and nutrition security, export earning, import substitution and feeding the local industry with raw material in a sustainable manner. Policy makers in this country should ask themselves as to how viable is smallholder agriculture. Obviously, the solution is long term, but a transition plan should start sooner than later.

National coordination in the EARS has been weakened since the hasty devolution of research into the RARIs in 2000. This has resulted in weak economies of scale, disparities in the development and diffusion of agricultural technologies and the loss of opportunities to mentor the younger generation of researchers, especially in the regions where capacity is lacking. Inevitably, there has also been duplication of efforts. There is a strong need to optimize coordination of AR&D in Ethiopia. The current Ethiopian Agricultural Research Council needs to be effectively led by an independent body composed of competent, meritorious professionals with adequate knowledge and experience rather than high government officials. Ethiopia should learn from its peers in

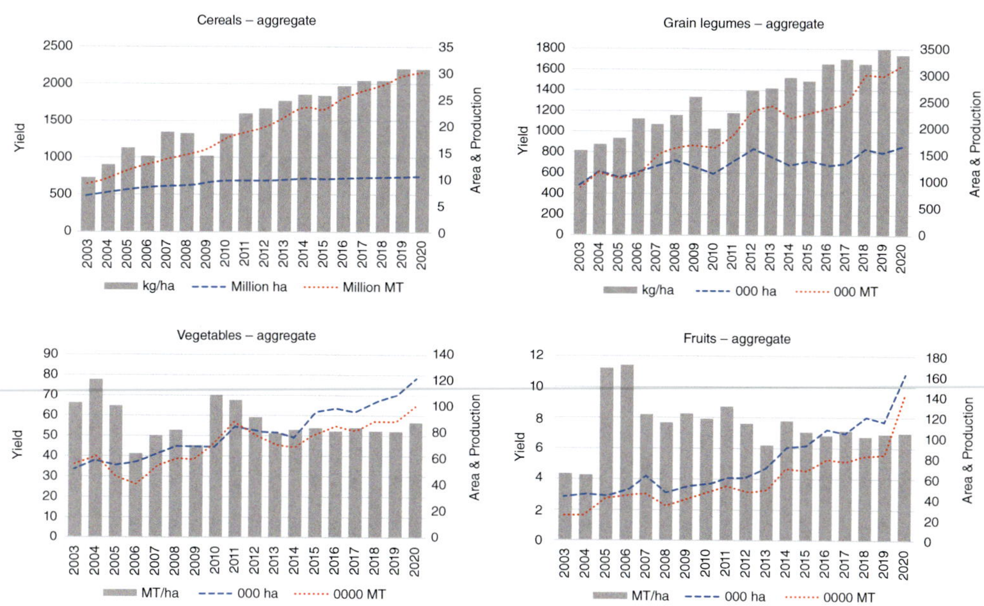

Fig. 1.7. Indicator trends for aggregate cereals, grain legumes, vegetables and fruit crops in Ethiopia. Note: Please note: the scale for production figures of the bottom two graphs are in 10,000s. Source: calculated and constructed by the author from CSA (www.csa.gov.et) data.

Asia (e.g. India and Bangladesh) in terms of organizing its AR&D system.

Many of the national programmes in Africa naïvely believed that international partnerships could substitute national efforts to accelerate their agricultural development but now realize that development cannot be outsourced. Such partnerships have not built a critical mass of African scholars and champions with solid knowledge of African development challenges. At present, there are few Africans who write and speak about African agriculture with authority. As stated in the introduction, all scholarships, policies and strategies dealing with the continent's agriculture and food security should be formulated by Africa's own best thinkers. Other development partners should support and complement Africa's own strategy and provide resources and incentives to help countries deal with climate change and improve productivity.

In 2020, African governments supplied more than 300 million MT of cereals, of which about 30% was imported at an estimated value of more than US$25.45 billion; the supply for roots and tubers was more than 336 million MT, with a negligible number of imports. By 2050, the continent needs to supply more than 588 million MT of cereals and close to 660 million MT of roots and tubers (Fig. 1.8) to ensure minimum food security requirements. In conclusion, Africa

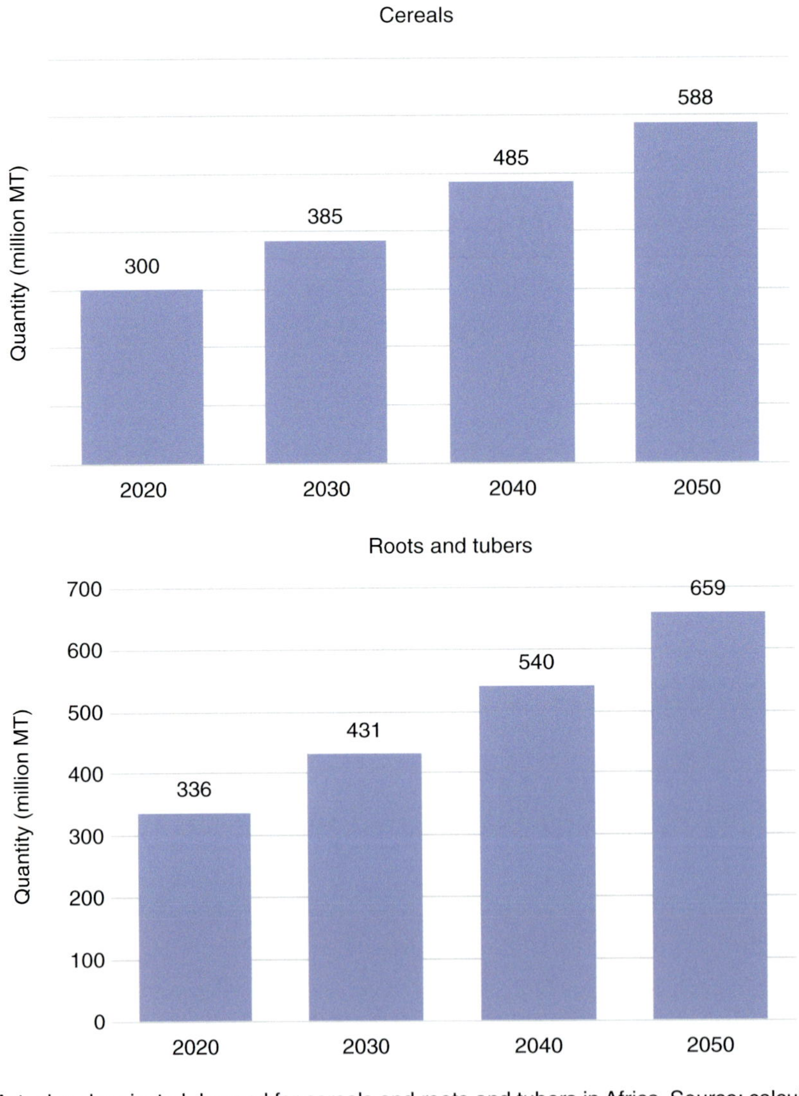

Fig. 1.8. Actual and projected demand for cereals and roots and tubers in Africa. Source: calculated and constructed by the author from FAO (2023d).

must overhaul the way future AR&D is conducted. It should follow a homegrown approach where qualified Africans lead their development agenda and be accountable for it at the same time if the continent is to achieve sustainable food sovereignty and economic independence.

References

Abate, T. (ed) (2006) *Successes with Value Chain: Proceedings of the International Conference on Scaling up and Scaling out Agricultural Technologies in Ethiopia, 9–11 May 2006*. EIAR, Addis Ababa, Ethiopia, 275 pp.

Abate, T. (2021) Would the green revolution ever come to Africa? *Academia Letters*, Article 1705. DOI: 10.20935/AL1705.

Abate, T., Shiferaw, B., Gebeyehu, S., Amsalu, B., Negash, K. *et al.* (2011) A systems and partnership approach to agricultural research for development: Lessons from Ethiopia. *Outlook on Agriculture* 40(3), 213–220.

Abate, T., Shiferaw, B., Menkir, A., Wegary, D., Kebede, Y. *et al.* (2015) Factors that transformed maize productivity in Ethiopia. *Food Security* 7, 965–981. DOI: 10.1007/s1257Fig015-0488-z.

Bindraban, P., Bulte, E., Giller, K., Meinke, H., Mol, A. *et al.* (2009) *Beyond Competition: Pathways for Africa's Agricultural Development*. Wageningen UR, the Netherlands, 82 pp.

Blaustein, R.J. (2008) The green revolution arrives in Africa. *BioScience* 58(1), 8–14. DOI: 10.1641/B580103.

Christiansen, L., Demery, L. (2018) *Agriculture in Africa: Telling Myths from Facts*. The World Bank, Washington, DC, USA.

Dixon, J., Garrity, D.P., Boffa, J.-M., Williams, T.O., Amede, T. *et al.* (eds) (2021) *Farming Systems and Food Security in Africa. Priorities for Science and Policy under Global Change*. Routledge Taylor and Francis Group, Milton Park, Oxfordshire, 674 pp.

FAO (Food and Agriculture Organization of the United Nations) (2022a) Statistical databases and data-sets of the Food and Agriculture Organization of the United Nations. Available at: http://www.fao.org/faostat/en/#data (accessed 14 August 2022).

FAO (2022b) Statistical databases and data-sets of the Food and Agriculture Organization of the United Nations. Available at: http://www.fao.org/faostat/en/#data (accessed 24 October 2022).

FAO (2023a) Statistical databases and data-sets of the Food and Agriculture Organization of the United Nations. Available at: http://www.fao.org/faostat/en/#data (accessed 3 January 2023).

FAO (2023b) Statistical databases and data-sets of the Food and Agriculture Organization of the United Nations. Available at: http://www.fao.org/faostat/en/#data (accessed 1 February 2023).

FAO (2023c) Statistical databases and data-sets of the Food and Agriculture Organization of the United Nations. Available at: http://www.fao.org/faostat/en/#data (accessed 19 February 2023).

FAO (2023d) Statistical databases and data-sets of the Food and Agriculture Organization of the United Nations. Available at: http://www.fao.org/faostat/en/#data (accessed 22 February 2023).

FAO (2023e) Statistical databases and data-sets of the Food and Agriculture Organization of the United Nations. Available at: http://www.fao.org/faostat/en/#data (accessed 23 October 2023).

Fischer, T. (2019) Closing crop yield gaps around the world. In: Robert Z. (ed.) *Sustaining Global Food Security, Chapter 14*. CSIRO Publishing, Melbourne, Australia. pp. 232–259.

Fischer, T., Byerlee, D., Edmeades, G. (2014) *Will Yield Increase Continue to Feed the World? Crop Yields and Global Food Security*. ACIAR, Canberra, Australia.

Garbero, A., Marion, P., Brailovskaya, V. (2018) *The Impact of the Adoption of CGIAR's Improved Varieties on Poverty and Welfare Outcomes: A Systematic Review*. IFAD, Rome, Italy. 87 pp.

Giller, K.E. (2020) The food security conundrum of sub-Saharan Africa. *Global Food Security Journal* 26, 100431. DOI: 10.1016/j.gfs.2020.100431

Giller, K.E., Delaune, T., Silva, J.V. *et al.* (2021) The future of farming: Who will produce our food? *Food Security* 13, 1073–1099. DOI: 10.1007/s12571-021-01184-6.

Haggblade, S., Hazell, P.B.R. (eds) (2010) *Successes in African Agriculture: Lessons for the Future*. IFPRI, Washington, DC, USA.

McIntire, J. (2023) *OneCGIAR is failing (and what to do about it).* DOI: 10.13140/RG.2.2.20565.09441.

MOALR (Ministry of Agriculture and Livestock Resource) (2017) *Plant Variety Register*, Issue No. 20, Addis Ababa, Ethiopia.

Schurman, R. (2018) Micro(soft) managing a 'green revolution' for Africa: The new donor culture and international agricultural development. *World Development* 112, 180–192.

Van Ittersum, M.K., Bussel, L.G.J., Wolfa, J., *et al.* (2016) Can sub-Saharan Africa feed itself? *Proceedings of the National Academy of Sciences* 113(52), 14964–14969. DOI: 10.1073/pnas.1610359113

Westphal, E. (1975) *Agricultural Systems in Ethiopia*. Joint publication of the College of Agriculture, Haile Sellassie I University, Addis Ababa, Ethiopia, and the Agricultural University, Wageningen, the Netherlands.

Wise, T.A. (2020) *Failing Africa's Farmers: An Impact Assessment of the Alliance for a Green Revolution in Africa*. Tufts University, Medford, USA. Available at: https://sites.tufts.edu/gdae/files/2020/07/20-01_Wise_FailureToYield.pdf (accessed 14 March 2024).

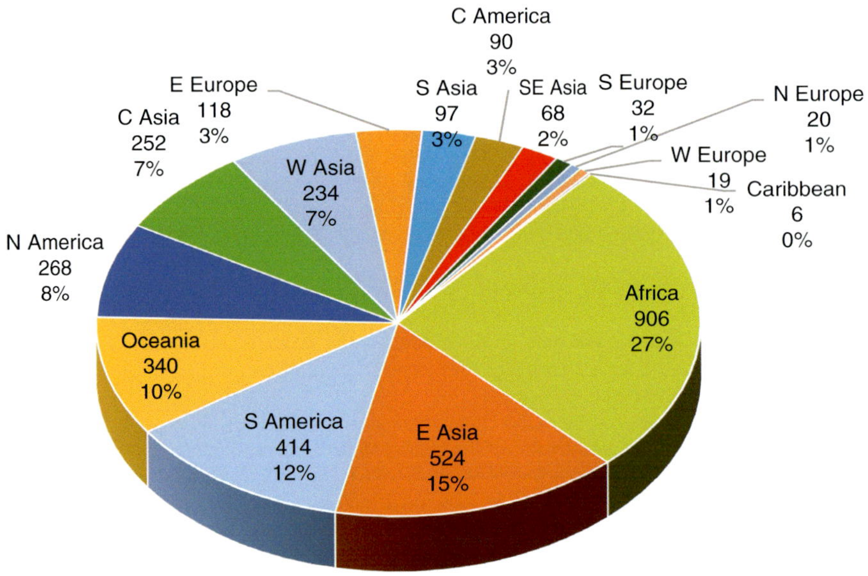

Annex Fig. 1.1. Unoccupied agricultural land area (million ha) and percentages of the world total in different regions, 2016–2020. C = Central, E = Eastern, N = Northern, S = Southern, SE = Southeastern, W = Western. Source: calculated and constructed by the author from FAO (2023e).

2 The Agricultural Resource Base

Tsedeke Abate,[1]* Regassa Feyissa[2] and Kindie Tesfaye[3]
[1]*Homegrown Vision, Addis Ababa, Ethiopia;* [2]*Organic Seed Association, Addis Ababa, Ethiopia;* [3]*CIMMYT-Ethiopia, Addis Ababa, Ethiopia*

Abstract

This chapter presents Ethiopia's agricultural resource base – land, water, agroecological/agroclimatic conditions, crop and livestock diversity, human population and institutional capacity. The country possesses approximately 38 million ha of agricultural land, of which about 16 million ha is arable, leaving approximately 22 million ha that is available for further development. Ethiopia has a dozen river basins and nearly 125 billion cubic metres of combined surface and underground water per annum. It is the land of wide-ranging agroecological/agroclimatic diversity where more than 70 plant species are cultivated and the largest head of livestock in Africa is raised each year. The country's population in 2022 of ~121 million comprises 28% between the age of 15 and 29 years and projected to reach ~205 million by 2050. Ethiopia has widely distributed research centres, >3000 researchers, 15,000 farmer training centres and a development agent to smallholder farmer ratio of 1:230, the densest in Africa.

Introduction

Understanding Ethiopia's available assets is the basis for planning its agricultural development. The assets include land and water resources, agroecological and climatic diversity, crop and livestock biodiversity, human population, and research and development capacity – including infrastructure and trained human resources.

Ethiopia is well known for its rich diversity in agroclimatic conditions and crop and livestock resources. This book deals with crops – livestock is mentioned here only in the context of its relevance to agricultural mechanization (including land preparation, cultivation, threshing and farm products transportation; Chapter 19) and human nutrition (Chapter 24). Below we provide more detailed descriptions of the various agricultural assets that Ethiopia possesses.

Land

With nearly 1.14 million km^2 of total area, Ethiopia is the tenth largest country on the African continent, after Algeria, the Democratic Republic of the Congo, Sudan, Libya, Chad, Niger, Angola, Mali and South Africa (FAO, 2022a). Figure 2.1A shows that nearly 38 million ha of the total land (or 34%) is considered to be agricultural area, while 17 million ha (15%) is forest area.

* Corresponding author: tsedeke@homegrownvision.org

DOI: 10.1079/9781800626386.0002

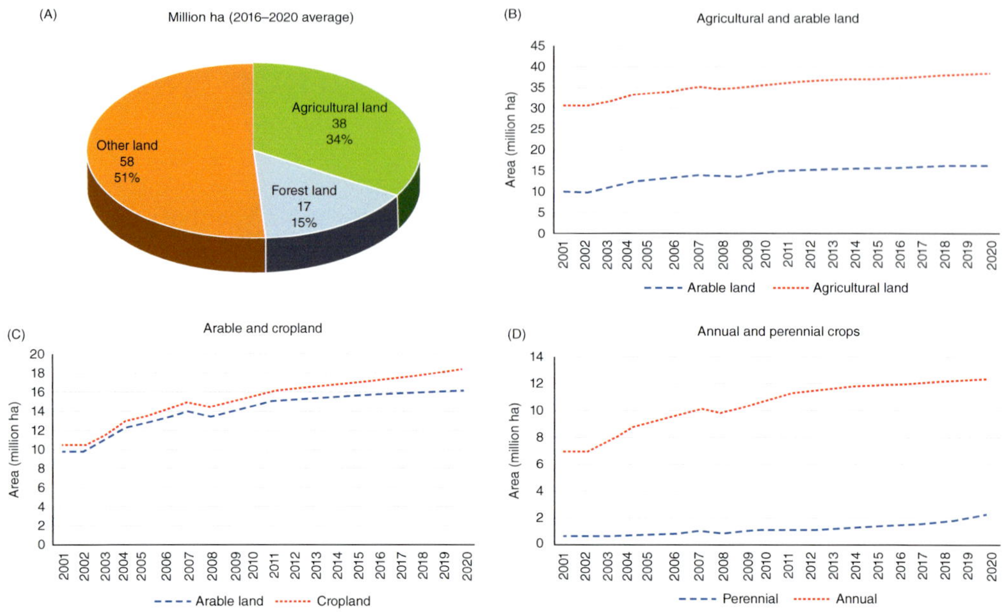

Fig. 2.1. A partial view of land use in Ethiopia. Source: calculated and constructed by the lead author from FAO (2022a).

Of the total agricultural area, a little over 16 million ha is counted as arable land (Fig. 2.1B), the fourth largest in Africa (after Nigeria, Sudan and Niger). This means that there is nearly 22 million ha of land in Ethiopia that can be further developed for agriculture. The total cropped area is slightly larger than the arable land (Fig. 2.1C) because some crops (annuals) are grown both in the *meher* and *belg* seasons. Annuals and perennials in Ethiopia account for roughly 87% and 13% of the total crop area (Fig. 2.1D).

Ethiopia has two crop growing seasons (at least in some parts of the country). These are the *meher* (also known as main crop season) and the *belg* (alternatively referred to as short rains or small rains season). According to the Central Statistics Agency (CSA; www.csa.gov.et), any crop harvested between September and February is considered *meher*, whereas any crop harvested between March and August is categorized as *belg*.

The major determinant of the two seasons is the amount of rainfall received and the length of crop growing period. In essence, *meher* covers the months of June, July, August and September, whereas *belg* rains include the months of February, March and April. According to the long-term data from the National Meteorological Agency, *meher* accounts for approximately 61% of the total annual precipitation while *belg* accounts for 15%.

Meher is the major crop growing season throughout the country, whereas *belg* production occurs in limited agroecological zones and with a narrow set of crops. Overall, the *belg* season accounts for approximately 13% of the total cereals area and 7% of production; among the three grain legumes for which data are available, the area and production for the *belg* season constitutes about 16% and 9%, respectively. However, the proportions for both groups of crops varied according to crops. Among the cereals, about 26% of emmer wheat, 24% of maize and 15% of barley area were grown in *belg*, whereas the corresponding proportion of production for each was 18%, 11% and 7%, respectively. Virtually all of the other cereals were produced in the *meher* season. Among grain legumes, the area and production for common bean in *belg* were approximately 39% and 23%, respectively, whereas virtually all of faba bean and fieldpea were grown in the *meher* season. The CSA records also show that about 71% of mungbean and 10% of lentil area are reported to be grown in *belg*. In general, the contribution of cereals grown in *belg* to the country's total production was only 5% and that for grain legumes was 6%.

Long-term data for *belg* production of oilseeds and roots and tubers are sketchy and it was not possible to make comparisons between seasons. The limited data available indicate that about 17% of safflower area and 22% of production came in *belg*. In a similar fashion, approximately 43% of the area and 44% of production of potato was in the *belg* season.

Farmers use early maturing, short-cycle varieties for *belg* production. Yields for crops grown in this season are very low in comparison to the main season. For example, the long-term (2003–2015) average yields of emmer, maize, sorghum, wheat, barley, lentil, *teff* and common bean were 852, 832, 681, 515, 384, 350 and 300 kg/ha, respectively.

Farm sizes are also small, thus resulting in low per capita production per smallholder. For example, our calculations showed that the average farm area per household for *teff*, maize, wheat, barley, sorghum, emmer, lentil and common bean was 0.2570, 0.2496, 0.2371, 0.1975, 0.1909, 0.1505, 0.1426 and 0.0672 ha, respectively, during the above period.

The per capita production per smallholder household for wheat, maize, barley, emmer, *teff*, sorghum and common bean (2007 and 2008 average) was estimated at 225, 193, 158, 134, 117, 96 and 30 kg, respectively. The *belg* season production is typically a subsistence mode of production but it is expected to change dramatically with the introduction of small-scale irrigation schemes that the government is promoting at present; this is particularly true for wheat.

Historically, Ethiopia enjoyed two good rainy seasons per year with abundant soil fertility (Pankhurst, 1966). This author remarked that 'The agricultural wealth of [Ethiopia] in former times impressed most foreign observers.' The rainfall situation has changed substantially in recent years (Komarek *et al.*, 2019) perhaps because of climate change, a result of which is a decline in the *belg* season cropping area.

Water

The latest available FAO data (FAO, 2022b) indicates that inland waters cover 768,000 of the total land area, the 19th largest in Africa. The top five African countries with the highest land area coverage by inland waters are the Democratic Republic of the Congo, Tanzania, Uganda, Chad and Malawi. Egypt, South Africa, Madagascar, Cote d'Ivoire and Senegal occupy the 20th to 24th ranks.

This means that 0.68% of the total land area of Ethiopia is covered by inland waters (which includes rivers and lakes). This is in sharp contrast to Guinea-Bissau, Malawi, Uganda, Liberia and Gambia, with 28.49%, 25.67%, 20.46%, 15.63% and 11.66%, respectively. The coverage for Egypt, Cameroon, Sudan, South Africa and Mauritius is 0.60%, 0.58%, 0.54%, 0.49% and 0.49%, respectively.

Agroecological Zones

Ethiopia has a wide-ranging agroecological diversity. The traditional land use systems categorize agroecosystems based on the major agroclimatic conditions, considering the relationship between elevation and temperature. Farming communities in Ethiopia recognize three major agroclimatic zones: *dega*, *weyna-dega* and *kolla*. As population pressure increased and agriculture activities extended, two additional zones at the higher and the lower temperature extremities have also been recognized as *woorch* and *bereha* (Hurni, 1998). These are depicted in Table 2.1.

This traditional classification has enabled farming and local communities to recognize broad vegetation types and environmental conditions, and to select crops and livestock breeds best suited for each specific agroecological condition. However, a more comprehensive and proper characterization of Ethiopia's agroecosystems should link the interrelated physical, abiotic and biotic parameters of farms and farm environments, soils, vegetation, animals and human activities with the climate in specific locations. The complexity of all these elements determines the diversity of strategies to be employed for sustainable management of the frequently diverse agroecologies and the agrobiodiversity resources the agroecologies harbour. Both the formal and the informal attributes as combined would be important to make agroecological differentiation.

Table 2.1. Traditional climatic zones of Ethiopia and their characteristics.

Zone	Climate	Altitude (m)	Average temperature (°C)	Rainfall (mm)
Woorch	Cold, moist	>3200	<11	>2200
Dega	Cool, humid	2400–3200	11–16	1200–2200
Weyna-Dega	Cool, semi-humid	1800–2400	16–20	800–1200
Kola	Warm, semi-arid	500–1800	20–27	200–800
Bereha	Hot, arid	<500	>27	<200

Actual agroecological characterization for promoting ecological agriculture should consider attributes such as farming and production systems, farm landscape suitability for growing specific crops/plants, and for raising livestock. It should also be characterized by specific conditions of farms and farm environments. Precipitation, temperature, soil types and its fertility, farm slopes and farm environment and its biomass density, livestock components of the system, local culture and practices together with farmers' objectives and needs are essential elements to be considered for the management of agrobiodiversity resources in all habitats. A range of complexity of all these elements determines the practices of agrobiodiversity resource management within the frequently diverse agroecosystems of Ethiopia.

Westphal (1975) recognized four agricultural systems in Ethiopia: (i) the seed farming complex (mainly consisting of cereals, grain legumes and oilseeds); (ii) the *enset* planting complex; (iii) shifting cultivation; and (iv) the pastoral complex. According to Getahun (1978), the mixed agriculture of the highlands (where both the crop and livestock production are integrated) and pastoralism in the lowlands were the two dominant agricultural systems of the country. He further divided these into four agricultural systems of: (i) the highland mixed farming system; (ii) the low plateau and valley mixed agriculture; (iii) the pastoral livestock production of the semi-arid and arid zone; and (iv) commercial agriculture.

Schmidt and Thomas (2018) divide Ethiopia's agricultural production areas merely into highlands (that account for 92% of cereals area and production) and lowlands. Viste *et al.* (2013) divide the country into 14 precipitation zones. Based mainly on varying altitudinal and precipitation regimes, MOA (2005) divides Ethiopia into 32 major agroecological zones (Fig. 2.2). The major agroecological zones can be aggregated into three broad categories: the lowlands (71.9%), the mid-highlands (27.8%) and the sub-afro-alpine or afro-alpine highlands (0.3%). The lowlands can be further subdivided into two major categories of sub-humid/humid and sub-moist/moist areas, accounting for about 38%, and arid/semi-arid areas occupying about 34% of the total area of the country. It should be noted that approximately 35% of Ethiopia is classified as arid or semi-arid, of which about 34% is in the lowlands and 1% is the tepid arid mid-highlands.

In the mid-highlands and highlands, there is little chance of further expansion as they are all occupied under smallholder farming – both crops and crop-livestock mixture. By contrast, the lowlands offer the opportunity for opening up new frontiers for irrigated large-scale agricultural development schemes, as stated in the preceding section.

The wide agroecological diversity has its own perks and penalties. On the one hand, it allows the country to grow a wide range of crops intended for domestic consumption, local industry, import substitution or export. On the other, it poses the challenge of tackling the myriads of abiotic and biotic stresses – such as drought, climate change, insect pests, plant diseases, and local and invasive weed species. The wide agroecological diversity is accompanied by a wide range of rainfall and temperature regimes in Ethiopia.

The farming systems in Ethiopia are also diverse, mainly traditional and based on small-scale production, which is managed with simple production technology. Farmers' strategies are based on combined crop and livestock production. There are three major categories of farming and production systems in Ethiopia. These are the highland mixed farming and production system; the lowland plateau and valley mixed

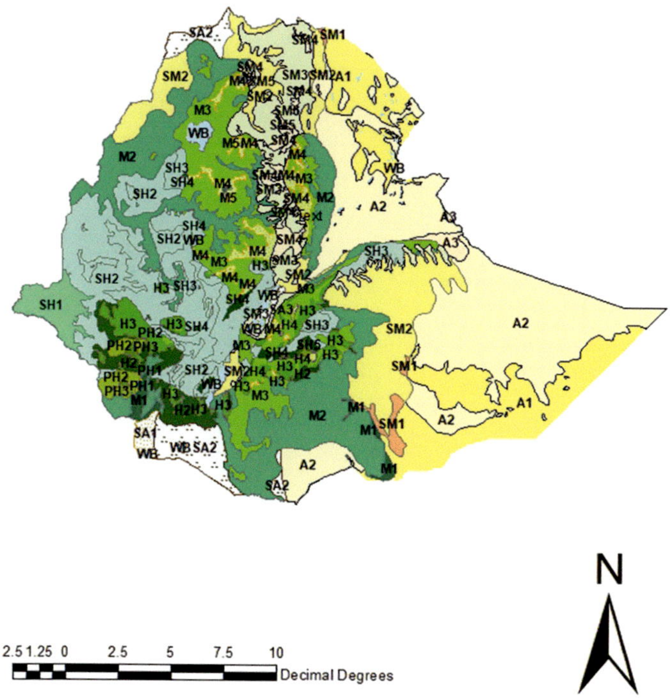

Fig. 2.2. Agroecological map of Ethiopia (see Annex Table 2.1 for details). Source: courtesy of Ato Friew Kelemu.

farming and production system; and the pastoral and agro-pastoral farming and production system of the arid and semi-arid zones (Westphal, 1975). However, there are varying sub-systems that are interspersed in each major farming system. In the highlands, continuous cropping is practiced through crop rotation where cereal production alternates with legumes as a means of maintaining soil fertility. Both crop and livestock production are economically important in the intermediate or lower highlands mountain foothills and upper valleys. Pastoral agriculture dominates at elevations below 1500 m with annual rainfall of below 450 mm, with limited semi-sedentary practices. See Annex Tables 2.1 and 2.2 for further details on agroecological diversity and land use in Ethiopia.

Crop Biodiversity

Ethiopia is one of Vavilov's eight gene centres (Zeven and de Wet, 1982), with the rest including South Mexico and Central America, Middle Eastern, Central Asiatic, Indian, China, Mediterranean and South America. Vavilov (1951) designated Ethiopia as a primary gene centre for seven crop species, including coffee, *enset*, *teff*, *noug*, buckthorn, garden cress and *ch'at*, and a secondary gene centre for an additional 27 crop plant species, as depicted in Table 2.2. Westphal (1975) added a variety of cotton (*Gossypium herbaceum* var. *acerifolium*) to the list.

The wide range of agroecological conditions influence the occurrence of genetic diversity in the cultivated crops and their wild relatives. The Ethiopian tetraploid wheat (*Triticum durum)* is one of the important cultivated crops with wide genetic diversity for which Ethiopia is a centre of diversity (Vavilov 1951; Porceddu 1976; Negassa 1986). Barley (*Hordeum vulgare*) is considered as one of the crops with high genetic diversity cultivated in Ethiopia for over 5000 years (Purseglove, 1976). There are several forms of Ethiopian barley that are sources of genes for yellow dwarf virus resistance, for resistance to diseases such as powdery mildew, leaf rust and net blotch. Ethiopian barley is also

Table 2.2. Crop species connected with Ethiopia as primary or secondary gene centre. Source: adapted from Vavilov (1951), as reported by Westphal (1975).

Local (common) name	Scientific name	Local (common) name	Scientific name
Primary gene centre			
Teff	*Eragrostis tef*	Feto (garden cress)	*Lepidium sativum*
Enset	*Ensete ventricosum*	*Ch'at*	*Catha edulis*
Coffee	*Coffea arabica*	*Gesho* (buckthorn)	*Rhamnus prinoides*
Noug (Niger seed)	*Guizotia abyssinica*		
Secondary gene centre			
Durum wheat	*Triticum durum*	Grasspea	*Lathyrus sativus*
Poulard wheat	*T. turgidum*	Cowpea	*Vigna unguiculata* cv.-group *Unguiculata*
Emmer	*T. dicoccum*	Catjang	*Vigna unguiculata* cv.-group *Biflora*
Polish wheat	*T. polonicum* var. *abyssinicum*	Hyacinth bean	*Dolichos lablab*
Barley	*Hordeum vulgare*	Lupin	*Lupinus albus* cv.-group *Albus*
Sorghum	*Sorghum bicolor*	Linseed	*Linum usitatissimum*
Finger millet	*Eleusine coracana*	Safflower	*Carthamus tinctorius*
Pearl millet	*Pennisetum typhoides*	Sesame	*Sesamum indicum*
Chickpea	*Cicer arietinum*	Castor	*Ricinus communis*
Lentil	*Lens culinaris*	Coriander	*Coriandrum sativum*
Pea	*Pisum sativum*	Onions	*Allium* spp.
Faba bean	*Vicia faba*	Okra	*Hibiscus esculentus*
Fenugreek	*Trigonella foenum-graecum*	Cotton[a]	*Gossypium herbaceum* var. *acerifolium*

[a]Not mentioned by Vavilov (1951).

considered to be an important source of the high lysine gene (Qualset, 1975).

Teff, as an Ethiopian domesticate and with its wide adaptation to different agroecologies, has *Eragrostis pilosa* as its wild relative (Harlan, 1976). Sorghum (*Sorghum bicolor*) has four out of its five races (except race kafir) being grown widely with different sub-races distributed across the country. There are also many intermediate, wild and weedy forms that occur over diverse agroecologies (Snowden, 1955; Stemler *et al.*, 1975). Maize (*Zea mays*) is believed to have been grown in Ethiopia for centuries and has a range of phenotypic and seed colour diversity. Finger millet is frequently mentioned in the literature to be of an Ethiopian origin (de Wet *et al.*, 1984).

Chickpea is an ancient crop in Ethiopia. Archaeological evidence in the caves of Lalibela has shown an age of 500 BCE (Dombrowski, 1969). Faba bean is the most important pulse in Ethiopia occurring mainly as small seed type (Lafiandra *et al.*, 1981). Fieldpea is also an old crop in Ethiopia; it is believed that subsp. *abyssinicum* has developed over a period of thousands of years (Vavilov, 1957). Lentil and grasspea (also known as vetch), having wild relatives in Ethiopia (Edwards, 1991), are all essential dietary legumes used along with the other staples. The dietary values of the legumes mean they are staple crops on farms, while they also serve as components of crop rotation for soil fertility improvement.

The major oil crops such as Ethiopian kale (*Brassica carinata*), believed to be of Ethiopian origin, are a result of the natural crossing of *Brassica nigra* and *Brassica oleracea* (Tacenko *et al.*, 1985). Sesame is a widely dominant oil crop in the lowlands. Although a number of wild species are abundant in Africa, Ethiopia is known to be a centre of origin for sesame (Vavilov, 1951; Nayar, 1976). *Noug* is believed to have originated from *Guizotia scabra*, which is widely spread in the wild. Linseed occurs on the intermediate and high elevations (Vavilov, 1951).

Coffee is widely distributed in the eastern, southern and southwestern parts of Ethiopia, having enormous phenotypic diversity and qualitative characteristics. Different types of spices such as *Aframomun corarima*, *Capsicum* spp. and *Piper longum* occur alongside the coffee plantations, particularly in the humid southwestern part of Ethiopia. Industrial crops such as cotton (*Gossypium* spp.) with its indigenous wild genome species (Saunders, 1961) and kenaf (*Hibiscus* spp.) mainly occur at elevations below 1500 m above sea level and serve as sources of cash, particularly for the lowland small-scale farmers.

Enset, *anchote* (*Coccinia abyssinica*), yam species and some other indigenous root and tuber crops, many of which occur in the high and mid-altitudes, are used as important staples. Oromo dinich (*Plectranthus edulis* [syn. *Coleus edulis*]), for example, occurs both as cultivated and as wild crops, having about 30 species. The cultivated species of this crop are grown in the humid south and southwestern parts of the country, while the wild types are distributed throughout the country (Edwards, 1991). Okra has *Abelmoschus esculentus* as cultivated vegetable crop and two wild forms, *A. manihot* and *A. moschatus* as wild (Vavilov, 1951).

There are also a number of wild plant species that are used as food, particularly during food shortage in times between planting and harvesting. The majority of such plants are those that are used as leafy vegetables and those with edible fruits, tubers or roots. The wild relatives of indigenous crops/plants or the wild and weedy species of non-cultivated plants could have high potential value as sources of important traits for crop improvement traits such as earliness, disease and insect pest resistance, and resistance to stresses are characters that are useful, especially in low-input agriculture. Reports show that scientists from many parts of the world have identified highly desirable genetic characters in a few Ethiopian germplasm collections of various crop species. These characters are intensively utilized in a number of international breeding programmes.

Livestock Diversity

As stated in the introduction above, livestock is beyond the scope of this book, but we mention it here as it relates to draft and pack farm power, a source of nutrition, a mechanism for pollination, and as an important source of socioeconomic benefits. Ethiopia's livestock resources include cattle, goats, sheep, donkeys, bees, horses, camels, mules and hinnies, chickens and swine (Table 2.3).

With close to 66 million head, this country has the largest number of cattle in Africa and fourth largest in the world. This is followed by goats (~47 million), sheep (~39 million), donkeys (~34 million), bees (~7 million), horses (~2 million), camels (~2 million), mules and hinnies

Table 2.3. Major livestock classes of Ethiopia and their characteristics (2016–2020 average). Source: constructed by the lead author from FAO (2022b).

			Rank		
Class	Total population	Unit	Africa	World	Top ten countries in Africa
Cattle	65,718,708	Head	1	4	ET, CH, SU, TA, KY, NI, NG, UG, SS, ML
Goats	47,309,695	Head	2	6	NI, ET, CH, KY, SU, M, TA, NG, BF, UG
Sheep	38,609,884	Head	4	9	NI, CH, SU, ET, AL, KY, MR, SA, ML, NG
Donkeys	33,803,854	Head	1	1	ET, SU, CH, NG, BF, ML, EG, MR, ZI
Bees	7,105,876	Number	1	5	ET, TA, AL, MR, SN, CA, KY, AN, EG, TN
Horses	2,192,443	Head	1	8	ET, CH, SU, ML, SN, SL, SA, NG, MR, NI
Camels	1,697,823	Head	6	6	CH, SO, SU, KY, NG, ET, MU, ML, AL, ER
Mules	399,192	Head	1	4	ET, MR, TN, SO, AL, SA, NM, BT, EG, CV
Chickens	59,289	1000 head	10	50	MR, EG, SA, NI, AL, TN, GH, SN, IC, ET
Swine	37,368	Head	35	130	NI, MW, AN, BF, UG, CR, MZ, RW, SA, ZM

AL = Algeria; AN = Angola; BF = Burkina Faso; BT = Botswana; CA = Central African Republic; CH = Chad; CR = Cameroon; CV = Cape Verde; EG = Egypt; ER = Eritrea; ET = Ethiopia; GH = Ghana; IC = Cote d'Ivoire; KY = Kenya; ML = Mali; MR = Morocco; MU = Mauritania; MW = Malawi; MZ = Mozambique; NG = Niger; NI = Nigeria; NM = Namibia; RW = Rwanda; SA = South Africa; SL = Sierra Leone; SN = Senegal; SO = Somalia; SU = Sudan; SS = South Sudan; TA = Tanzania; TN = Tunisia; UG = Uganda; ZI = Zimbabwe; ZM = Zambia.

(~400,000), chickens (~59 million) and swine (~37,000), according to the FAO (2022b) data for 2021. This means that Ethiopia has the:

- largest number of cattle in Africa (fourth largest in the world);
- second largest number of goats in Africa (sixth largest in the world);
- fourth largest number of sheep in Africa (ninth largest in the world);
- largest head of donkeys in Africa (largest in the world);
- largest number of bee colonies in Africa (fifth largest in the world);
- largest number of horses in Africa (eighth largest in the world);
- sixth largest number of camels in Africa (sixth largest in the world);
- largest head of mules and hinnies in Africa (fourth largest in the world);
- tenth largest number of chickens in Africa (50th largest in the world); and
- 35th largest head of swine in Africa (130th largest in the world).

Of the ten classes of livestock mentioned here, cattle, donkeys, horses, camels, and mules and hinnies provide farm power. Bees are the most important pollinators, in addition to providing honey and beeswax. Goats, sheep, chickens and swine are important sources of nutrition security and/or export earnings or import substitution.

Ethiopia has a long tradition of beekeeping, with three known geographical races of bees that occur in county. *Apis mellifera esculenta* occurs at altitudes ranging from 500–2400 m, while *Apis mellifera monticola* is of afro-alpine vegetation at altitudinal range from 2400–3400 m. *Apis mellifera yemenitica* is a lowland race in southern escarpments. There is another stingless bee race that commonly occurs in the range of medium altitudes that produces special honey of medicinal value known as *tazma* (Fitchtl and Adi, 1994).

Honeybees face enormous threats, mainly due to vegetation removal. Shrubs and herbs are usually considered as weeds and are mechanically removed or killed with herbicides. However, many such plant species are essential for the occurrence and maintenance of honeybee populations and other pollinators. Kiros *et al.* (2004) reported that species diversity of trees and shrubs and of the diversity of landraces were positively correlated with the number of weeds and insects but negatively correlated with the status of soil erosion. Threats to the flowering plants can, therefore, be considered as equal to threats to genetic resources of honeybee and other pollinator insects and may result in the total destruction of many generations of important insects.

Rangeland agrobiodiversity resources

The rangelands of Ethiopia are unique ecosystems inhabited by various plants that are progenitors of many food, feed, fibre and medicinal plant species. The ecological conditions of the rangelands are characterized by low and erratic precipitation, rough and rocky topography, intractable soils and poor drainage. Most of the non-arable lowlands of Ethiopia found within the altitudinal range of below 1500 m are classified as rangelands unsuitable for rain-fed agriculture. Pastoralism is the predominant land use system and crop production is insignificant due to the semi-arid climatic conditions of such agroecosystems.

Pastoralists of the rangeland agroecosystem totally depend on livestock to support their livelihoods. The main source of livestock feed is natural forage rather than cultivated fodder and pastures (Abate *et al.*, 2012). Scarcely distributed water points, soil degradation, breakdown of the traditional institutions and systems of management, as well as loss of traditional knowledge of the people, are among the challenges of the rangeland agrobiodiversity resources. The rangeland agrobiodiversity resources of Ethiopia are under the pressure of adverse impacts exerted by an increased human population and by encroachment into the agroecosystems for crop cultivation (Abate and Angassa, 2016).

Importance of genetic diversity for smallholder agriculture

Smallholder farming in Ethiopia is diversity based, where diversity is a common denominator for all farming systems and is critical for promoting stable production. A range of diversity within and among crop/plant species is utilized, enabling farmers' access to diverse sources of livelihood options. Failure of one crop is compensated by the yields of other crops, and for this reason smallholder farmers usually grow several genetically distinct varieties of crops in a single field as a risk aversion strategy. Complexity of diversity therefore sustains farm productivity by balancing out yield variability within and among crop species. This is how smallholder farmers always spread the risk of crop failure due to pest and disease attack or adverse environmental conditions.

Diversity is also a source of options for satisfying changing food preferences and for supporting a constant supply of food at the household level. In cases where wild relatives occur together with their cultivated forms, this coexistence provides a wider option of food sources and supports a constant supply of food and generation of new genetic variation. This farmers' strategy also enables intensification of production with limited resources and access to a variety of food and income sources (Feyissa, 1999). It is for this reason that smallholder farmers grow diverse sources of their food in their home gardens and main fields, where yield volume at household level is sustained by cumulative outputs of the total crop diversity used.

The complexity of diversity they manage within and among crops usually declines as it extends to fields further away from residential areas. At the same time, the plot size of the types/varieties increases as they extend further away from their residence area. The types/varieties that are grown on larger plots further from the gene pool are initially tested through farmers' own variety development strategy, which is based on close observation and examination of genotype–environmental interactions with consideration of the range of variables across their plots. The selection of types, in some cases, goes down to the level of a single line or genotype for various crops. This is how farmers use the decentralized varietal development strategy to meet diverse agroecological requirements within their fields. Thus, farmers have tested strategies in specializing production of crop varieties while at the same time closely managing the gene pools. As a matter of fact, the tremendous crop/plant genetic diversity we see in farmers' fields today is not simply a random, accidental or chance process, but rather highly developed and refined over time based on farmers' observations and practices that are shaped by their intelligence and wisdom.

Farmers always develop and refine strategies that enable them to meet new challenges throughout their practices, spatially and temporally. It is in this way that they practice conservation strategies, but in its dynamic form where co-evolutionary or gene–environment interactions are constantly maintained. Genotypes are distributed across locations with diverse conditions under which the activities of genes may be amplified differently, and genes of some genotypes may be suppressed, thus new diversity emerges as natural selection plays its role. In the process, they strictly evaluate the dysfunctional effects of genotypes over space and under different biotic and abiotic conditions, where they retain strong varieties and discard the weak. Herein lies the essence of the strategy of on-farm conservation or genetic diversity management on farms, which at times remains abstract to the formal way of thinking about in situ conservation for agricultural crops.

Women's role in managing a wider range of crop genetic diversity around the household is greater, since the larger gene pool occurs near residence areas. The gene pool is always there to backstop the gene flow within fields, serving as a reserve for genotype replacement and to minimize selective genotype elimination.

Human Population

The human population of Ethiopia was estimated to be 19 million in 1950, with only 5% of the population living in urban areas. In 2020, the population increased to approximately 115 million, with 21% classed as urban; by 2050, the population is projected to be approximately 205 million (Fig. 2.3) with 36% defined as urban dwellers. This country's population is second only to Nigeria in Africa. This means is that the country will have an additional 90 million people to feed and shelter just in less than 30 years. This would require more than 57 million MT of cereal grains in 2050 to maintain the current level of supply and consumption, based on cereal supply (local production and imports) estimates for 2019 where Ethiopia's grain consumption was approximately 31 million MT and the population for the same period was estimated at nearly 112 million).

Ethiopia witnessed the fastest annual rate of growth (ROG) of its population in the 1980s. The trend continued through the first decade of the current century, but a significant downward trend in the ROG has been projected from the 2030s onward. In general, the urban population grew much faster than the rural and the total population (Fig. 2.3). Overall, the urban population is projected to grow at the rate of 4.48% per annum, compared to 2.07% for the rural population and 2.49% for the total population in the 100 years between 1951 and 2050.

It is projected that 36% of Ethiopians would be urban dwellers by 2050. This urban population figure is below the African (excluding North Africa) average of 55% (e.g. Nigeria, 74%; Ghana, 64%; Zambia, 60%; Senegal, 55%; Tanzania, 50%; Sudan, 48%; Kenya and Mozambique, 45% each; and Uganda, 33%), but this has significant implications for Ethiopian agriculture and therefore for the socioeconomic stability of the country. As stated earlier, the challenge for policy makers is how to ensure food security and economic wellbeing while maintaining the integrity of the natural resources base. This is particularly important in light of the fast-dwindling agricultural area that can be cultivated (Lowder *et al.*, 2015; Schmidt and Thomas, 2018) using the existing capacity of smallholder farmers in Ethiopia (see later section for details).

Even though providing adequate food and shelter for such a rapidly growing population poses a great challenge, policy makers in Ethiopia can turn this to their advantage by harnessing this asset into a skilled workforce for modern agricultural, industrial, scientific and technological advancement.

Institutional Capacity

The history of agricultural research and development in Ethiopia goes back to the establishment of the Jimma Agricultural and Technical School that made some efforts to improve coffee and some cereals in the 1950s and 1960s. The establishment of the Imperial Ethiopian College of Agriculture and Mechanical Arts (now Haramaya University) introduced the three functions of teaching, research and extension (modelled after the US Land Grant Universities system) in the 1960s. However, a nationally coordinated agricultural research system only began when the Institute of Agricultural Research (IAR), with the support from the United Nations Development Programme/Special Fund (UNDP/SF), was established and became operational in

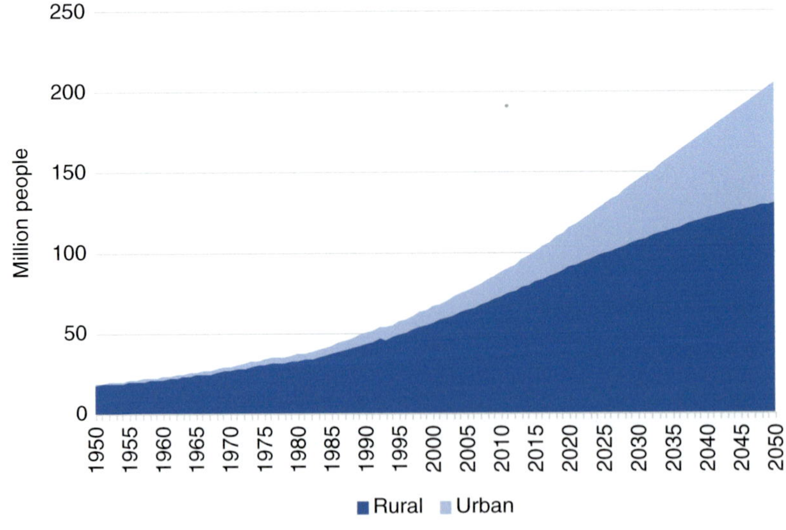

Fig. 2.3. Actual and projected population of Ethiopia. Source: constructed by the lead author from FAO (2020).

February 1966. This was part of the government's aim to achieve agricultural self-sufficiency through an increased volume and variety of agricultural products in the country.

Infrastructure

At the beginning, there were only three main centres located at Holetta, Jimma and Were. Sustained efforts were made to strengthen the research system and improve the coverage of wider agroecologies over the years. The UNDP/SF was extended in 1971, 1974 and 1979 (van der Graaff, 1981) and was instrumental in improving both infrastructure and research capacity. The government of Ethiopia also secured funding from the World Bank in the 1990s and 2000s for the same purpose.

Currently, there are a total of 21 research centres under the Ethiopian Institute of Agricultural Research (EIAR, formerly the Institute of Agricultural Research), consisting of 14 main research centres (Tepi, Jimma, Kulumsa, Melkassa, Debre-Zeit, Sebeta [fisheries], Ambo, Chiro, Holetta, Here, Asosa, Pawe, Fogera and Mehoni); and seven sub-centres (Mieso, Agaro, Gera, Metu, Haru, Mugi and Awada).

The combined regional agricultural research institutes, run a total of 50 research centres and sub-centres. These include:

- 17 in Oromia: Yabello, Bore, Sinana, Jimma (mechanization), Adami-Tulu, Ziway (soils), Ziway (fisheries), Asella (mechanization), Bedele, Mechara, Haro Sabu, Bako, Holetta (apiculture), Nekemte, Bako (mechanization), Fedis and Fiche;
- eight in Amhara: Debre-Berhan, Adet, Andassa, Bahr Dar (mechanization), Bahr Dar (fisheries), Sirinka, Gonder and Sekota;
- seven in Tigray: Alamata, Abergele, Mekele, Mekele (mechanization), Shire-Maitsebri, Axum and Humera;
- six in Southern Nations, Nationalities and Peoples Region: Jinka, Arba-Minch, Hawassa, Areka, Bonga and Worabe;
- five in Afar: Dalifagie, Semera, Dubti, Awra and Abala;
- four in Somali: Dolo Ado, Gode (mechanization), Fafen and Jigjiga; and
- three in Gambella: Goderer, Abobo and Lare.

The higher learning institutions have a total of 18 institutes or research centres. These are Pathobiology Institute (under Addis Ababa University), Haramaya University, Bahr Dar University, Hawassa College of Agriculture, Wondo-Genet College of Forestry, Ambo College of Agriculture, Jimma College of Agriculture, Mekelle University, Arba-Minch University, Addis Ababa University, Jimma University, Mada-Wolabu University, Wollo University, Gonder University, Semera University, Welkite University, Worabe University and Debre-Zeit Veterinary Medicine.

Overall, there are a total of 82 agricultural research centres and sub-centres administered by the federal, regional and university research institutions. Figure 2.4 shows the geographical spread of these centres (specific names not shown).

Research capacity

Expatriate staff almost exclusively ran the technical departments in the early years of the IAR. Major efforts were made to train researchers under UNDP/SF in the 1970s, and the Agricultural Research and Training Project and others before it, through World Bank loans and grants in the 1990s to early 2000s. Three unique features of the Ethiopian Agricultural Research System (EARS) as compared to most other African research systems have been the fact that: (i) it was established by Ethiopians (not inherited from the colonial era); (ii) it has been run almost exclusively by Ethiopian nationals since the early 1980s; and (iii) it has sustained government budgetary support through years. These have been the primary reason for the sustained progress of the research system in this country.

Agricultural research in Ethiopia is conducted under the auspices of government research institutions and higher learning institutions (universities). With the exception of Tendaho Cotton Plantation and the Middle Awash sugar estates, until the advent of the military government in the mid-1970s, Ethiopia has had little or no history of agricultural research done by the private sector. For example, ASTI (2018) data for 2010–2014 show that 90% of the total number of researchers belonged to government research institutions, whereas higher learning institutions accounted for 10% and none for the private sector.

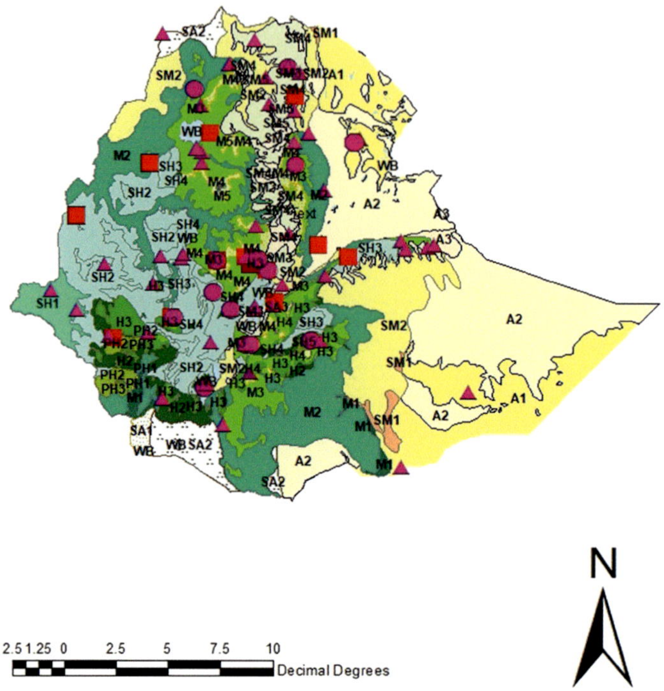

Fig. 2.4. Distribution of agricultural research centres covering various agroecological zones in Ethiopia as of 2015. Squares, triangles and circles represent centres under federal, regional and university systems, respectively. Source: courtesy of Ato Friew Kelemu.

Babu *et al.* (2011) reported a rapid growth in the number of researchers at EIAR (and EARS as a whole), from a humble beginning of a little more than 56 full-time employees (FTEs) in 1971 to 1318 FTEs in 2008. Beintema and Stads (2014) reported 1877 FTEs as the total capacity of public researchers in Ethiopia in 2011. In general, research capacity at EARS showed sustained rapid growth over the last three plus decades (Fig. 2.5). The total number of researchers in Ethiopia for the year 2016 was estimated at 3025 FTEs.

However, it should be emphasized here that the rapid growth was mostly in the number of junior staff with BSc degrees. As a whole, the EARS is dominated by BSc holders; for example, the average number of researchers for 2016 consisted of 7.8% PhD, 39.6% MSc and 52.6% BSc holders (ASTI, 2018).

According to the latest figures available at EIAR (as of October 2018), out of 1191 FTEs, PhD holders accounted for 9%, whereas MSc and BSc holders were 46% and 45%, respectively (Table 2.4). The proportion of PhD holders at EIAR declined by 34% between 2010 and 2014.

Women accounted for 13% of the total number of researchers at EIAR. Beintema and Stads (2014) reported that only 10% of the total researchers in 2014 in Ethiopia were women. The average age of researchers at EIAR was about 35 years. This consisted of 48 years (range: 31–65 years) for PhD holders, 36 years (range: 24–66 years) for MSc holders and 30 years (range: 22–61 years) for BSc holders. The report by Beintema and Stads (2014) indicated that nearly 83% of all researchers in Ethiopia were <41 years old, making it the youngest in Africa.

Overall, there are more than 3000 researchers working under the federal, regional and higher learning institutions, even though the majority have limited qualifications and experience. Over the five decades of its existence, the EARS has made substantial progress in

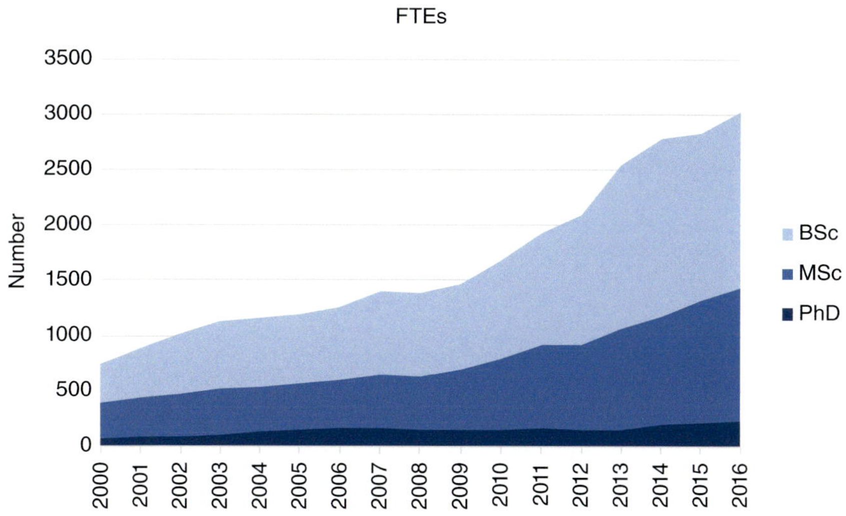

Fig. 2.5. Number of full-time agricultural researchers by qualification in Ethiopia. FTE = full-time employee. Source: calculated and constructed by the lead author from ASTI (http://www.asti.cgiar.org).

many fronts. It is fair to say that a good research culture exists in this country.

Extension capacity

Berhane *et al.* (2018) provide a comprehensive review of agricultural extension services (AES) in Ethiopia since its beginning in 1953–1963. AES delivery expanded rapidly with the establishment of agricultural technical and vocational education and training (ATVET) programmes starting in 2004 and expanding the number of farmer training centres (FTCs). A total of 25 ATVETs were built by 2016, resulting in rapid growth of the number of development agents (DAs) trained. By 2016, there were more than 15,000 FTCs.

The number of smallholder farmer households receiving AES was reported to be 15,000 in 2002, compared to >72,000 in 2016, a 4.8-fold increase. This means that more than 72,000 DAs were benefiting about 16.7 million smallholder farmer households. That is, the ratio of farmers to DAs was about 231:1, or 43 DAs serving 10,000 farmers. This is reported to be the densest farmer-to-DA ratio in the world. The government plan for 2019 was to raise the number of FTCs to 18,000, benefiting 18.2 million farmer households.

Reports for 2011 showed that 1 DA served about 476 farmers. In other words, the farmer-to-DA ratio for Ethiopia was 476:1, or 21 DAs to 10,000 farmers. In comparison, the number of DAs to 10,000 farmers in China, Indonesia, Tanzania, Nigeria and India was 16, 6, 4, 3 and 2, respectively.

Discussion and Conclusion

In this chapter we presented the agricultural resource base for Ethiopia, comprising its land and water, agroclimatic conditions, and crop and livestock diversity. We also covered the status of infrastructure, trained research personnel and agricultural extension capacity. Our study has revealed that the country has rich diversity and well-organized infrastructure and research and extension capacity. However, there are some issues that need immediate attention.

First, unplanned expansion of agricultural land is leading to land degradation, genetic erosion and habitat deprivation for honeybee and other important pollinators, and wildlife in general.

Second, there is a heavy and continued loss of plant genetic material either through material

Table 2.4. Research capacity at EIAR (as of October 2018). Source: constructed by the lead author from EIAR data.

	Qualification			
Discipline	PhD	MSc	BSc	Total
Plant sciences	5	39	207	251
Plant breeding	23	85	0	108
Horticulture	3	44	43	90
Human nutrition	6	33	44	83
Plant protection	9	59	7	75
Agricultural economics	9	31	31	71
Agronomy	19	43	5	67
Extension	2	23	37	62
Biotech	3	23	26	52
Natural resources	4	18	25	47
Soils	8	27	8	43
Livestock	6	12	21	39
Irrigation	0	16	21	37
Veterinary medicine	0	16	17	33
Engineering	1	13	16	30
Animal production	4	18	5	27
Climate	2	12	10	24
Seed science and technology	0	12	2	14
Fisheries	3	6	2	11
Animal nutrition	5	5	0	10
Range management	0	2	6	8
Animal breeding	1	4	0	5
Communications	0	2	1	3
Biometrics	0	2	1	3
Total	113	545	535	1193
Per cent	9	46	45	100

transfer agreements that cannot actually be enforced effectively, or under the guise of research collaboration for students' theses research, or through unlawful means. While the unlawful means is difficult to control, the Ethiopian Biodiversity Institute (EBI) and research institutions should ensure that DNA extracts rather than plant samples are transferred.

Third, Ethiopia is also losing its landraces of durum wheat to the introduced bread wheat varieties at a very fast rate. Durum wheat in Ethiopia accounted for about 60% of the total area in the early 1990s (Gebre-Mariam, 1991), compared to 0.9% of improved varieties and 20.2% of local varieties in 2011 (Yirga *et al.*, 2013), and about 4% in 2016 and 2017 (Hodson *et al.*, 2020).

Fourth, even though there are more than 70,000 accessions of more than 100 plant species being conserved at the EBI, its use in the national breeding programme is direly inadequate. Indigenous crops may possess important traits that may confer resistance or tolerance to biotic and abiotic stresses.

Fifth, there are diverging views on the size of available land for further development. The most recent FAO data suggest that Ethiopia has nearly 22 million ha of agricultural land that could be harnessed for expansion while other studies suggest that there is limited scope. In our opinion, there is actually a good amount of land (in the lowland areas) that could be developed under large-scale farming rather than under smallholder conditions. These are lands that require irrigation schemes and/or managing invasive plant species, or those that are found in the outlying areas and need infrastructure (particularly road nets) construction.

Finally, even though the research capacity has been growing steadily, the proportion of PhD holders with adequate skills and capacity is very low. Mechanisms need to be devised to retain trained personnel.

References

Abate, T., Angassa, A. (2016) *Conversion of Savanna Rangelands to Bush Dominated Landscape in Borana, Southern Ethiopia*. School of Animal and Range Sciences, Hawassa University, Hawassa, Ethiopia.

Abate, T., Ebro, A., Nigatu, L. (2012) Evaluation of rangeland in arid and semi-arid grazing land of South East Ethiopia. *International Journal of Agricultural Sciences* 2(7), 221–234.

ASTI (2018) Agricultural Science and Technology Indicators. ASTI Database, IFPRI. Available at: http://www.asti.cgiar.org/ (accessed 15 March 2024).

Babu, S.C., Annor-Frempong, I., Asenso-Okyere, K. (2011) *Enhancing Capacity for African Agricultural Research: Selected Models and Lessons*. Agricultural Science & Technology Indicators (ASTI), IFPRI/FARA, Washington DC, USA.

Beintema, N., Stads, G.-J. (2014) *Taking Stock of National Agricultural R&D Capacity in Africa South of the Sahara: ASTI Synthesis Report*. ASTI. IFPRI, Washington DC, USA.

Berhane, G., Ragasa, C., Abate, G.T., Assefa, T.W. (2018) *The State of Agricultural Extension Services in Ethiopia and their Contribution to Agricultural Productivity*. Strategy Support Program, Working Paper 118, May 2018. IFPRI/EDRI, Addis Ababa.

de Wet, J.M.J., Prasada Rao, K.E., Brink, D.E., Mengesha, M.H. (1984) Systematics and evolution of *Eleusine coracana*. *American Journal of Botany* 71, 550.

Dombrowski, J. (1969) Preliminary report on excavation in Lalibela and Natchabiet caves, Begemeder. *Annales d'Ethiopie* 8, 21–29.

Edwards, S. (1991) Crops with wild relatives found in Ethiopia. In: Engels, J.M.M., Hawkes, J.G., Melaku, W. (eds) *Plant Genetic Resources of Ethiopia*, pp. 42–60. Cambridge University Press, Cambridge, England.

FAO (Food and Agriculture Organization of the United Nations) (2020) Statistical databases and data-sets of the Food and Agriculture Organization of the United Nations. Available at: http://www.fao.org/faostat/en/#data (accessed 28 May 2020).

FAO (2022a) Statistical databases and data-sets of the Food and Agriculture Organization of the United Nations. Available at: http://www.fao.org/faostat/en/#data (accessed 24 October 2022).

FAO (2022b) Statistical databases and data-sets of the Food and Agriculture Organization of the United Nations. Available at: http://www.fao.org/faostat/en/#data (accessed 27 December 2022).

Feyissa, R. (1999) Mainstreaming biodiversity conservation towards sustainable agricultural development: An Ethiopian perspective. In: Taye, A. (ed.) *Food Security through Sustainable Land Use. Policy on Institutional, Land Tenure, and Extension Issues in Ethiopia*, pp. 255–276. NOVIB Partners Forum on Sustainable Land Use, Addis Ababa, Ethiopia.

Fitchtl, R., Adi, A. (1994) *Honeybee Floral of Ethiopia*. Margraf Verlag, Germany. 510 pp.

Gebre-Mariam, H. (1991) *Wheat Production and Research in Ethiopia*. EIAR, Addis Ababa.

Getahun, A. (1978) Agricultural Systems in Ethiopia. *Agricultural Systems* 3, 281–293.

Harlan, J.R. (1976) The origin of cereal agriculture in the old world. In: Reed, C.A. (ed.) *Origin of Agriculture*. Mouton Publishers, The Hague, pp. 357–383.

Hodson, D.P., Jaleta, M., Tesfaye, K., Yirga, C., Beyene, H. *et al.* (2020) Ethiopia's transforming wheat landscape: tracking variety use through DNA fingerprinting. *Scientific Reports* 10, 18532. DOI: 10.1038/s41598-020-75181-8

Hurni, H. (1998) Agro-ecological belts of Ethiopia. *Explanatory notes on three maps at a scale of 1:1,000,000*. Soil Conservation Research Program, Ethiopia. Research Report.

Kiros, H.M., van Bruggen, A.H.C., Epema, G.F. (2004) *Loss of agro-biodiversity and sustainability in relation to farming practices in Tigray, Northern Ethiopia*. Institute of Biodiversity Conservation/GTZ. Conference paper.

Lafiandra, D., Polignano, G.B., Filippet, A., Porcedddu, E. (1981) Genetic variability for protein content and s-aminoacids in broad-beans (*Vicia faba* L.). *Kulturpflanze* 29, 115–1127.

Komarek, A.M., Thurlow, J., Koo, J., De Pinto, A. (2019) Economywide effects of climate smart agriculture in Ethiopia. *Agricultural Economics* 50, 765–778. DOI: 10.1111/agec.12523.

Lowder, S.K., Skoet, J., Raney, T. (2015) The number, size, and distribution of farms, smallholder farms, and family farms worldwide. *World Development* 87, 16–29.

MOA (2005) *Agro-ecological Zones of Ethiopia*. Addis Ababa, Ethiopia.

Nayar, N.M. (1976) Sesame. In: Simmonds, N.W. (ed.), *Evolution of Crop Plants*. Longman, London, pp. 231–233.

Negassa, M. (1986) Estimates of phenotypic diversity and breeding potential of Ethiopian wheat. *Hereditas* 104, 41–48.

Pankhurst, R. (1966) The great Ethiopian famine of 1888-1892: A new assessment. *Journal of the History of Medicine and Allied Sciences* 21(2), 95–124.

Porceddu, E. (1976) Variation for agronomical traits in a world collection of durum wheat. *Zeitschrift fur Pflanzenzuchtung* 77, 314–329.

Purseglove (1976) The origin and migrations of crops in tropical Africa. In: Harlan, J.R., de Wet, J.M.J., Stemler, A.B.L. (eds) *Origin of African Plant Domestication*. Mouton Publishers, The Hague, pp. 291–309.

Qualset, C.O. (1975) Sampling germplasm in a center of diversity: an example of disease resistance in Ethiopian barley. In: Frankel, O.H., Hawkes, J.G. (eds) *Crop Genetic Resources for Today and Tomorrow*. Cambridge University Press, Cambridge, pp. 81–96.

Saunders, J.H. (1961) *The wild species of Gossypium and their Evolutionary History*. Empire Cotton Growing Corporation, London.

Schmidt, E., Thomas, T.S. (2018) Cropland expansion in Ethiopia: Economic and climatic conditions for highland agriculture. ESSP Working Paper 127. IFPRI, Washington, DC.
Snowden, J.D. (1955) The wild fodder sorghums of the section Eusorghum. *Journal of the Linnean Society* 55, 191–260.
Stemler, A.B.L., Harlan, J.R., de Wet, J.M.J. (1975) Evolutionary history of cultivated sorghum (*Sorghum bicolor* [Linn.] Moench) of Ethiopia. *Bulletin of the Torrey Botanical Club* 102, 325–33.
Tacenko, F.A., Ford-Lloyd, B.V., Astely, D. (1985) A numerical study of variation in a germplasm collection of *Brassica carinata* and allied species from Ethiopia. *Zeitschrift fur Pflanzenzuchtung* 94, 192–200.
van der Graaff, N.A. (1981) *Selection of Arabica Coffee Types for Resistance to Coffee Berry Disease in Ethiopia*. Mededelingen Landbouwhogeschool, Wageningen. pp. 8–11.
Vavilov, N.I. (1951) The origin, variation, immunity and breeding of cultivated plants. *Chronica Botanica* 33, 1–366.
Vavilov, N.I. (1957) *World Resources of Cereals, Leguminous Seed Crops and Flax, and their Utilization in Plant Breeding*. Israel Program for Scientific Translations, Jerusalem. (Published in English in 1960)
Viste, E., Korecha, D., Sorteberg, A. (2013) Recent drought and precipitation tendencies in Ethiopia. *Theoretical and Applied Climatology* 112 (3–5), 535–551.
Westphal, E. (1975) *Agricultural Systems in Ethiopia*. Joint publication of the College of Agriculture, Haile Sellassie I University, Ethiopia, and the Agricultural University, Wageningen, the Netherlands.
Yirga, C., Jaleta, M., Shiferaw, B., de Groote, H., Kassie, M. *et al.* (2013) Analysis of Adoption and Diffusion of Improved Wheat Technologies in Ethiopia. Research Report 101. EIAR & CIMMYT, Addis Ababa.
Zeven, A.C., de Wet, J.M.J. (1982) Dictionary of cultivated plants and their regions of diversity. *Excluding most Ornamentals, Forest Trees and Lower Plants*. Centre for Agricultural Publishing and Documentation, Wageningen.

Annex Table 2.1. Major agroecological zones of Ethiopia and their relative sizes. Source: calculated and constructed by the lead author from MOA (2005).

Agroecological zone	Symbol	Area	
		000 ha	Per cent
Warm arid lowland plains	A2	22,356	19.92
Warm moist lowlands	M2	17,110	15.24
Hot arid lowland plains	A1	12,202	10.87
Warm sub-moist lowlands	SM2	10,890	9.70
Tepid moist mid-highlands	M3	9,101	8.11
Warm sub-humid lowlands	SH2	8,047	7.17
Tepid sub-humid mid-highlands	SH3	7,504	6.68
Tepid sub-moist mid-highlands	SM3	5,850	5.21
Warm semi-arid lowlands	SA2	3,115	2.77
Tepid humid mid-highlands	H3	3,002	2.67
Warm humid lowlands	H2	2,593	2.31
Cool moist mid-highlands	M4	1,963	1.75
Hot sub-humid lowlands	SH1	1,893	1.69
Cool sub-moist mid-highlands	SM4	1,314	1.17
Cool humid mid-highlands	H4	926	0.83
Warm per-humid lowlands	PH2	765	0.68
Hot moist lowlands	M1	672	0.60
Hot sub-moist lowlands	SM1	637	0.57
Cool sub-humid mid-highlands	SH4	589	0.52
Tepid arid mid-highlands	A3	488	0.43
Hot semi-arid lowlands	SA1	450	0.40
Tepid semi-arid mid-highlands	SA3	219	0.19
Tepid per-humid mid-highland	PH3	152	0.14
Cold moist sub-afro-alpine to afro-alpine	M5	79	0.07

continued

Annex Table 2.1. Continued.

Agroecological zone	Symbol	Area	
		000 ha	Per cent
Cold sub-moist mid-highlands	SM5	77	0.07
Cold sub-humid sub-afro-alpine to afro-alpine	SH5	69	0.06
Cold humid sub-afro-alpine to afro-alpine	H5	63	0.06
Very cold humid sub-afro-alpine	H6	51	0.05
Very cold sub-humid sub-afro-alpine to afro-alpine	SH6	35	0.03
Very cold sub-moist mid-highlands	SM6	18	0.02
Very cold moist sub-afro-alpine to afro-alpine	M6	15	0.01
Hot per-humid lowlands	PH1	13	0.01
Total	All	112,258	100.00

Annex Table 2.2. Land use in Ethiopia (2016–2020 average). Source: calculated and constructed by the lead author from FAO (2022a).

Label	Area (000 ha)	Per cent	Per cent of
Country area (total)	113,624		
Land area (total)	112,856	99.32	Country area
Agricultural land	37,888	33.57	Total land
Permanent meadows & pastures	20,000	17.72	Total land
Cropland	17,888	47.21	Agricultural land
Forest land	17,215	15.25	Total land
Naturally regenerating forest	16,104	14.27	Forest land
Arable land	16,078	42.43	Agricultural
Annual crops	12,165	68.01	Crop land
Temporary fallow	2,057	1.82	Total land
Temporary meadows & pastures	1,856	1.64	Total land
Perennial crops	1,810	10.12	Crop land
Planted forest	1,111	6.45	Forest land
Land area equipped for irrigation	858	2.26	Agricultural land
Cropland area actually irrigated[a]	554	3.10	Crop land
Inland waters	768	0.68	Country area
Agriculture area under organic	201	0.53	Agricultural land
Agriculture area under certified organic	185	0.49	Agricultural land
Cropland area under organic[b]	186	1.04	Crop land
Cropland area under certified organic[b]	185	1.03	Crop land
Other land	57,754	50.83	Country area

[a]Average for 2017, 2018 and 2020; [b]2018 data only.

3 Factors that Transformed Cereals Productivity

Tsedeke Abate,[1]* Abebe Menkir,[2] Getachew Belay,[3] Regassa Feyissa,[4] Tolera Keno[5] and Olubunmi Bossey[2]

[1]*Homegrown Vision, Addis Ababa, Ethiopia;* [2]*IITA, Ibadan, Nigeria;* [3]*Formerly Ethiopian Institute of Agricultural Research, Bishoftu, Ethiopia;* [4]*Ethiopian Organic Seed Association, Addis Ababa, Ethiopia;* [5]*University of Queensland, Queensland, Australia*

Abstract

This chapter provides a critical assessment of productivity gains made so far and the opportunities for further improvement ahead for cereals in Ethiopia. Overall, the country achieved a very highly significant yield gain of 92 kg/ha/yr ($P<0.01$, $R^2=0.95$) between 2000 and 2020 – the fifth highest after Ukraine, Brazil, the USA and Argentina, among the world's top ten cereal-producing countries. This means that Ethiopia doubled its major cereals productivity during this period. Increases in yield accounted for ~70% of the change in production, compared to ~28% in area expansion. In all of Africa, only South Africa has achieved this level of cereals productivity under rainfed agriculture. The presence of well-functioning institutions, sustained government investment and increased use of mineral fertilizers were some of the factors that contributed to the success.

Introduction

The most recent data indicate that an estimated 15 million Ethiopians were in need of food aid in 2020, and the advent of COVID-19 was estimated to have doubled that number (Damene Darota, pers. com.). Ethiopia depends on cereals for 56% of its dietary energy, 54% of dietary protein and 20% of dietary fat (FAO, 2022d). The pre-COVID-19 demand for cereals in this country was estimated at 32 million MT in 2020, with projected demands of upwards of 40 million and 57 million MT in 2030 and 2050, respectively.

There are eight widely cultivated cereals in Ethiopia – in descending order on the basis of the number of households growing the crop: maize, *teff*, sorghum, wheat (both bread wheat and durum), barley (both food and malt), finger millet, emmer (erroneously reported as oats) and rice. Except for maize and rice, all the major cereals have their centre of origin or centre of diversity in Ethiopia (Westphal, 1975; Zeven and de Wet, 1982). Maize has been in cultivation in Ethiopia since its introduction towards the late 17th century (Huffnagel, 1961). Rice was introduced to Ethiopia in the early 1970s (Alemu *et al.*, 2018).

According to the Central Statistical Agency (CSA), cereals in Ethiopia employ 16 million smallholder households and occupy more than 10 million ha (or ~68%) of the total cultivated area in this country (Table 3.1). Their cultivation area has expanded at a very rapid rate, especially following the devastating droughts of the early 1970s and 1980s and the famines that resulted

* Corresponding author: tsedeke@homegrownvision.org

DOI: 10.1079/9781800626386.0003

Table 3.1. Indicators for major cereals that smallholders cultivated in the *meher* season in Ethiopia (2016–2020 average).[a] Source: number of holders, area, yield and production are constructed by the lead author from the CSA data (www.csa.gov.et).

Crop	000 Holders	000 ha	kg/ha	000 MT	ha/holder	kg/holder/yr	Number of varieties
Maize	10,694	2,287	4,005	9,186	0.21	859	119
Teff	6,927	3,030	1,780	5,391	0.44	778	70
Sorghum	5,284	1,823	2,717	4,946	0.35	936	65
Wheat	4,712	1,766	2,838	5,023	0.37	1,066	51
Barley	3,826	920	2,294	2,113	0.24	552	42
Finger millet	1,743	459	2,356	1,082	0.26	621	21
Emmer	184	20	2,046	42	0.11	226	3
Rice	169	62	2,895	180	0.37	1,065	36
Aggregate	15,999	10,366	2617	27,961	0.29	763	407

[a]Holders are 2016–2019 average; farm size (ha/holder) and per capita production (kg/holder/yr) are the lead author's own calculations; number of varieties are from MOALR (Ministry of Agriculture and Livestock Resources) (2017). Please note that the numbers of holders do not add up because each household grows more than one crop.

in the deaths of hundreds of thousands of people (Kumar, 1987).

Successive governments in Ethiopia over the last 50 years have accorded cereals research and development (R&D) the highest priority because of their significance in ensuring food security for the country. However, there are few comprehensive studies measuring the progress in their productivity resulting from these R&D efforts, the factors that contributed to the improvement, and their socioeconomic impacts and lessons learned so far.

Abate *et al.* (2015) studied the factors that transformed maize productivity in Ethiopia from 1961 to 2013. Past efforts by the Consultative Group on International Agricultural Research (CGIAR) centres as well as national programmes have relied on adoption rates as a metric for measuring successes with cereals and other crops in Ethiopia (Asfaw *et al.*, 2011; Shiferaw *et al.*, 2011; Jaleta *et al.*, 2013; Garbero *et al.*, 2018; Assefa *et al.*, 2019) or elsewhere in other African countries (Beyene and Kassie, 2015; Khonje *et al.*, 2015; Manda *et al.*, 2015, 2017). However, recent experience shows that adoption rates do not offer adequate measure of success, as they are not necessarily always correlated with productivity (Abate *et al.*, 2017). The use of yield gain and annual rate of growth (ROG) provide a better alternative to the use of adoption rates. For example, Abate *et al.* (2011) used these parameters to measure changes in grain legumes productivity in Ethiopia between 2003 and 2008. Abate *et al.* (2012) also used this method for measuring changes in tropical legumes productivity across Africa and India for the period 1985/87 to 2005/07. In a similar fashion, Abate *et al.* (2015) measured changes in the productivity of maize between 1990 and 2013.

In this chapter, we present detailed assessments of the changes in cereals productivity brought about by technological and policy interventions that have taken place in Ethiopia over a period of 21 years between 2000 and 2020, and lessons learned. We hope our findings will inform policy decisions on a comprehensive agricultural transformation agenda for the country and beyond.

Approach

To calculate the rate of yield gains (i.e. changes in productivity increase) over the 21-year period mentioned above (2000 to 2020), we computed the linear regression using the annual crop mean yield as a dependent variable and production year as an independent variable. The resulting regression coefficient (b-value) was taken as an annual rate of yield gain. We did this for each crop in the world's top ten growing countries for comparison with Ethiopia. Even though there are eight major cereals cultivated in Ethiopia, we present here the detailed analyses of maize, *teff*, sorghum and wheat as examples representing cereals.

In order to see the effect of technological change and yield growth on production, we used economic methodology to isolate the contribution of area and yield growth on national production for each of the crops included in our assessment, as employed by Abate *et al.* (2011, 2012, 2015). The contribution of area A and yield Y increases in production Q are calculated as follows:

$$Q = A * Y$$

If Q is growing at the rate of r per year,

$$Q_t = Q_0(1+r)^t = A_0(1+a)^t Y_0(1+v)^t$$

where a and v are growth rates for area and yield over time. Taking a simple logarithm of this equation, simplifying and taking anti-log, the growth rate in production can be given as:

$$r = a + v + av$$

This means that the ROG of production equals the ROG of area plus the ROG of yield, plus an interaction factor. The growth rates for production (and by implication area and yield) are computed as follows:

$$r = (Q_t / Q_0)^{(1/t)} - 1$$

Where, Q_0 is the value during the base year and Q_t the value after t years.

Accordingly, we have computed the value of a and v as indicators on the contribution of area and yield of growth rates for the total change in crop production. We also make graphical presentations of indicator trends for overall cereals as well as individual crops.

Results

Cereals in general

Table 3.2 summarizes the results of our analyses for all the cereals cultivated in Ethiopia. The aggregate yield gain was 92 kg/ha/yr and was very highly significant ($P<0.01$, $R^2=0.95$). Yield gain ranged from 56 kg/ha/yr for emmer to 127 kg/ha/yr for maize, and all were very highly significant. The aggregate ROG for area, yield and production was 2.10%, 5.16% and 7.37%, respectively. This meant that changes in yield explained about 70% of the change in production, while the area expansion was responsible for just over 28% and the interaction effect covered the balance. There were variations among the crops in terms of ROGs for area, yield and production. Details will be discussed under individual crops later.

We made comparisons of Ethiopia's cereal yield gain and indicators (area, yield and production) with the top ten largest cultivating countries of the world in 2016–2020, where this country was ranked 17th out of 180 countries, and territories for which data were available (Table 3.3). China, India, the USA, the Russian Federation and Brazil hold the top positions in the area harvested.

The yield gain for Ethiopia (92 kg/ha/yr) was the fifth highest after Ukraine, Brazil, the USA and Argentina, and ahead of many major growing countries, including Bangladesh, Canada, China, Indonesia, Türkiye, the Russian Federation and India. The yield gain for three

Table 3.2. Yield gain (b-value, kg/ha/yr) and annual rate of growth (ROG) for major cereals in Ethiopia, 2000–2020. Source: calculated and constructed by the lead author from FAO (2022a).

	Yield gain			ROG (%)		
Crop	b-value	P > |t|	R^2	Area	Yield	Production
Maize	127	<0.01	0.82	2.76	4.95	7.85
Wheat	96	<0.01	0.96	2.69	4.93	7.75
Sorghum	94	<0.01	0.98	2.71	5.07	7.92
Millet	88	<0.01	0.95	2.26	5.76	8.15
Rice	76	<0.01	0.64	16.69	3.38	20.63
Barley	68	<0.01	0.88	–0.05	4.40	4.35
Teff	64	<0.01	0.94	1.68	5.33	7.10
Emmer	56	<0.01	0.86	–4.11	3.83	–0.44
Aggregate	92	<0.01	0.95	2.10	5.16	7.37

other African countries of Sudan, Niger and Nigeria among the top ten was 1, 6 and 9 kg/ha/yr, respectively (Table 3.3).

Figure 3.1 illustrates the long-term trends of cereal production in Ethiopia. The mode of cereal production in this country through the 1990s can be broadly categorized as subsistence – i.e. yields were unstable and increases in production were achieved via area expansion rather than from increased productivity. Starting from the first half of the 2000s, however, it took the form of what could be termed as 'smallholder commercial agriculture (SCA)'. This mode is characterized by steady increases of production deriving from increases in productivity rather than area expansion.

The ROG for yield in Ethiopia was the highest among the top ten major cereal-cultivating countries. This was followed by Ukraine, Brazil, Argentina, the Russian Federation and Canada. In spite of the significant progress Ethiopia has made in terms of cereals productivity and

Table 3.3. Yield gain (b-value, kg/ha/yr) and annual rate of growth (ROG) for aggregate cereals in Ethiopia compared with the top ten cultivating countries of the world, 2000–2020. Source: calculated and constructed by the lead author from FAO (2022a).

	Yield gain			ROG (%)		
Country[a]	b-value	P > ltl	R^2	Area	Yield	Production
Brazil	133	<0.01	0.90	1.12	3.46	4.62
USA	129	<0.01	0.78	–0.19	1.86	1.67
Ethiopia	92	<0.01	0.95	2.10	5.16	7.37
Canada	83	<0.01	0.85	–0.86	2.53	1.65
China	79	<0.01	0.98	1.29	1.45	2.76
Indonesia	70	<0.01	0.93	0.17	1.51	1.68
Russian Federation	56	<0.01	0.76	0.33	2.54	2.88
India	56	<0.01	0.93	–0.01	2.04	2.03
Australia	16	0.29	0.06	–0.41	0.99	0.58
Kazakhstan	13	0.07	0.16	0.75	1.17	1.93
Nigeria	6	0.11	0.13	0.61	0.49	1.10

[a]Descending order of cultivated area: China, India, USA, Russian Federation, Brazil, Nigeria, Australia, Indonesia, Kazakhstan and Canada; Ethiopia has 17th largest cultivated area in the world.

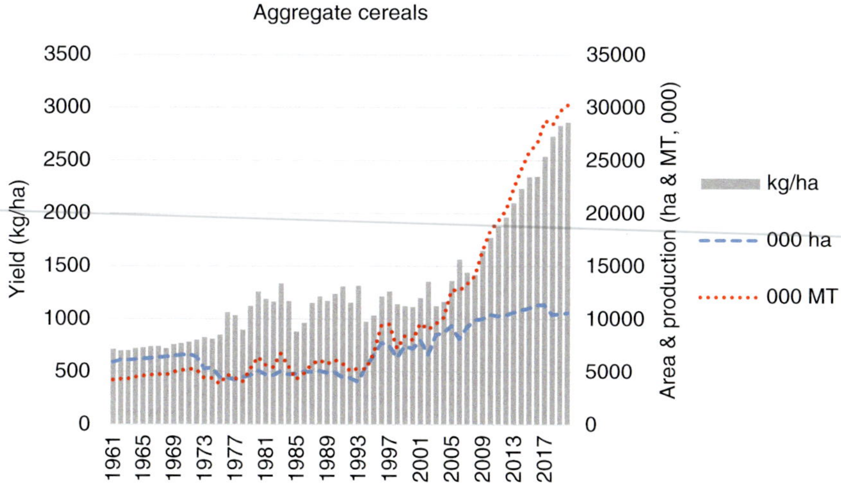

Fig. 3.1. Indicator trends for aggregate cereals in Ethiopia, 1961–2020. Source: constructed by the lead author from FAO (2022a).

production, it still remains a net importer. For example, this country imported >2 million MT of cereals valued at nearly US$810 million annually in 2016–2020. Wheat, rice and sorghum accounted for 61%, 23% and 12%, respectively, of the total import bill.

Performance of individual crops

It is obvious that not all crops would perform in the same way because of their intrinsic differences, the circumstances they are grown under and many other contributing factors thereof. Below we provide brief descriptions of progress for maize, *teff*, sorghum and wheat as examples of selected cereals grown in Ethiopia, prioritized according to the number of smallholder households growing them in the *meher* (main crop) season. Barley, finger millet, emmer and rice are also mentioned, but full details are not given for the sake of saving space.

Maize

Maize is known to have been introduced to Ethiopia towards the late 17th century (Huffnagel, 1961), but remained a subsistence crop through the 1970s (Abate *et al.*, 2015). The ROG for maize area declined following the drought of 1974. Maize production and its status in determining food security in the country received a major focus in the mid-1980s, particularly spurred by the 1984 devastating drought and the famine that followed. Thus, between 1975 and 1995, maize area expanded at an annual rate of 2.78%, compared to 0.94% for wheat, 0.43% for *teff* and −1.51% for sorghum. The introduction of drought-tolerant varieties and hybrids further helped the expansion of maize production.

At present, maize is the only cereal that is known to be cultivated in all regions of the country. Oromia, Amhara, Southern Nations, Nationalities and Peoples Region and Tigray accounted for 56%, 24%, 14% and 3%, respectively, in the 2018 *meher* season (CSA, 2019). According to the CSA (2017, 2018, 2019, 2020, 2021), close to 11 million smallholder households, more than for any other crop in the country, grew maize in the *meher* season each year (cf. Table 3.1). Ethiopia ranked 12th both in area and production out of 165 countries that grew maize in the 2016–2020 seasons (Table 3.4).

Maize is grown primarily for food security, but it is also important as a raw material for the local starch and milling industries. The 2015–2019 Food and Agriculture Organization of the United Nations (FAO) data show that maize accounted for 19.8% of the total dietary energy, 15.5% of total dietary protein and 3.4% of total dietary fat in Ethiopia (FAO, 2022c).

Table 3.4. Yield gain (b-value, kg/ha/yr) and annual rate of growth (ROG) for maize in Ethiopia compared with the top ten cultivating countries of the world, 2000–2020. Source: calculated and constructed by the lead author from FAO (2022a).

	Yield gain			ROG (%)		
Country[a]	b-value	P > |t|	R^2	Area	Yield	Production
Ukraine	199	<0.01	0.79	8.23	4.18	12.75
Indonesia	151	<0.01	0.98	1.85	3.75	5.66
Brazil	146	<0.01	0.85	2.05	3.54	5.67
Ethiopia	127	<0.01	0.82	2.76	4.95	7.85
USA	124	<0.01	0.54	0.97	1.27	2.26
China	84	<0.01	0.95	3.56	1.54	5.15
Argentina	73	0.01	0.31	6.16	1.12	7.35
Mexico	67	<0.01	0.88	−0.21	2.13	1.91
India	64	<0.01	0.88	1.99	2.76	4.81
Nigeria	4	0.61	0.01	5.16	0.29	5.47
Tanzania	−15	0.42	0.04	5.18	−0.21	4.97

[a]Descending order of cultivated area: China, USA, Brazil, India, Nigeria, Mexico, Argentina, Indonesia, Ukraine and Tanzania; Ethiopia has the 12th largest cultivated area.

Figure 3.2 presents the long-term trends of maize indicators in Ethiopia. The maize production system in this country can be characterized as subsistence through the early 1990s and SCA starting from the early 2000s. However, steady growth rates were obtained beginning in the second half of the 2000s. With >4 MT/ha, Ethiopia has the second highest yield for all of Africa (after South Africa) where maize is grown under rainfed conditions.

The area, yield and production for maize in Ethiopia increased at an annual rate of 2.76%, 4.95% and 7.85%, respectively, between 2000 and 2020 (Table 3.2). This means that the area expansion and increases in yield contributed about 35% and 63%, respectively (while the interaction effects covered the balance), to the growth of maize production in Ethiopia. These data show that Ethiopia has doubled its maize yield in less than two decades.

Ethiopia made a very highly significant ($P<0.01$, $R^2=0.82$) yield gain of 127 kg/ha/yr in the two decades between 2000 and 2020, the 7th highest after the Ukraine, the Russian Federation, Romania, Indonesia, South Africa and Brazil. This is similar to the USA, but superior to all of the top ten producing countries in the world (see Table 3.4).

The country's maize yield increased at 4.95% per annum, followed by the Russian Federation, Ukraine, Romania and South Africa; its average yield of >4 MT/ha is comparable to Mexico, but superior to the Philippines, India and all other African countries producing maize under rainfed conditions. Neither of the two African countries listed in the top 10 maize-producing countries in the world (Nigeria and Tanzania) achieved yields exceeding 2 MT/ha. Details of the factors that transformed maize productivity in Ethiopia can be found in Abate *et al.* (2015).

Teff

Teff is a unique crop to Ethiopia. With the exception of Eritrea (and very small pockets of Yemen), this crop is widely grown as a staple only in Ethiopia. This country is the primary centre of diversity for *teff* (Westphal, 1975). An estimated 6.9 million smallholder households grew *teff* on >3 million ha of land (approximately 29% of cereals or 20% of all crops grown in the country) and produced close to 5.4 million MT of grain annually in the 2016–2020 *meher* seasons (cf. Table 3.1). The average yield was about 1.7 MT/ha. The calculated average farm size for *teff* is about 0.44 ha/household (largest for all crops grown in Ethiopia), with the per capita production of about 778 kg/household/yr. According to the 2015/16 data from the CSA, *teff* is produced almost exclusively by smallholders.

The military/socialist government tried to downgrade *teff* R&D in the late 1970s and early

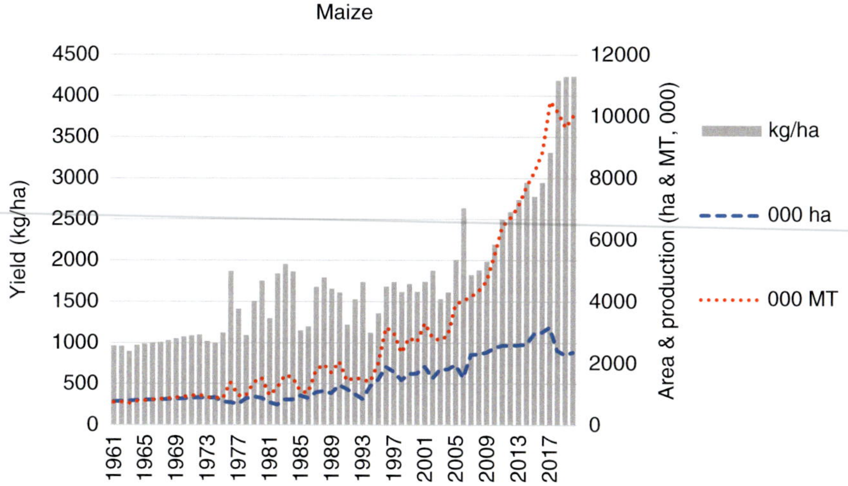

Fig. 3.2. Indicator trends for maize in Ethiopia, 1961–2020. Source: constructed by the lead author from FAO (2022a).

1980s on the grounds that it has limited genetic potential to increase productivity and production, but that has since changed. *Teff* has been one of the 'big three' cereals over the last three decades.

Teff is grown mostly for its grain, the most valued of all staples in Ethiopia, but farmers also highly value its biomass (hay) as an important source of animal feed and for construction purposes. Ethiopians obtain about 20.0% of their total dietary energy, 14.0% dietary protein and 3.8% of dietary fat from *teff*.

According to the 2018 CSA data, Oromia, Amhara, SNNPR and Tigray regional states accounted for 47%, 39%, 8% and 6%, respectively, of the total area occupied by *teff*.

Teff is grown traditionally for domestic consumption, but it has become an international commodity over the last two decades. There is a high demand because of the large population of Ethiopians in the diaspora and the gluten-free nature of the grain that has increased its use among European and American consumers. According to the Ethiopian Revenue and Customs Authority (ERCA), Ethiopia exported close to 7,000 MT of *teff* to 41 countries across the globe in the form of fresh *injera*, grain or flour at an estimated value of close to US$12 million each year in 2013–2017. The demand for *teff* products appears to continue to grow.

Figure 3.3 shows the long-term trends of *teff* indicators. The area, yield and production increased at annual rate of 1.68%, 5.33% and 7.10%, respectively. This means that the increases in yield and area accounted for roughly 75% and 24%, respectively, while the interaction effect covered the remaining 1% or so. The production system for *teff* was subsistence through the late 2000s. From the early 2010s it took the form of SCA. All indicators show continued upward trends for this crop in Ethiopia. Ethiopia has registered a very highly significant ($P<0.01$, $R^2=0.94$) yield gain of about 64 kg/ha/yr between 2000 and 2020. *Teff* yield doubled within two decades.

Because of its high demand, *teff* is one of the three major cereals that is accorded preferential treatment in terms of external input use and extension services. The 2019 data from the CSA indicate that this crop received 27% of all the fertilizers used in all crops grown in the country; nearly 71% of the area is fertilized; and 41% is treated with pesticides (mainly herbicides for weed control).

Sorghum

Sorghum is a traditional staple crop widely grown in the mid and lower altitudes of Ethiopia. According to Westphal (1975), Ethiopia is the

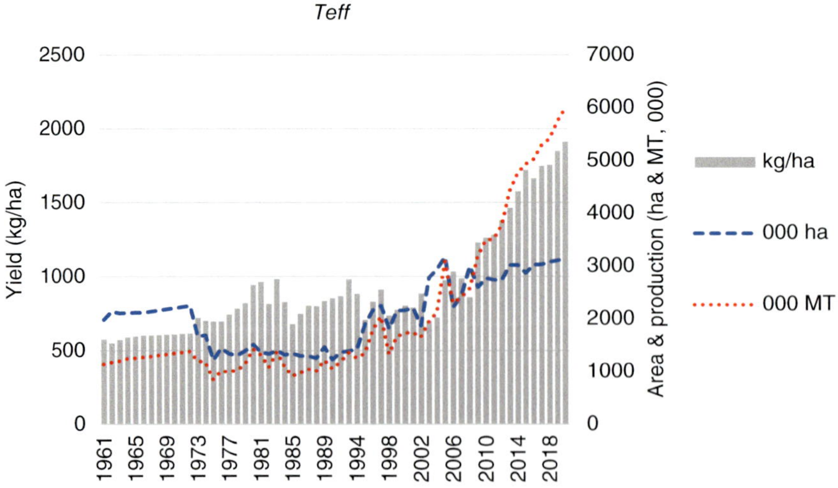

Fig. 3.3. Indicator trends for *teff* in Ethiopia, 1961–2020. Source: constructed by the lead author from FAO (2022a).

secondary gene centre for this crop. Nearly 5.3 million smallholder households grew sorghum annually on approximately 1.8 million ha and produced >4.9 million MT of grain in the 2016–2020 *meher* seasons; the average yield was a little over 2.7 MT/ha (cf. Table 3.1). The average area for a smallholder household under sorghum is estimated at 0.35 ha and the per capita production is 936 kg/household/year. Ethiopia is ranked sixth (after Sudan, Nigeria, India, Niger and the USA) in the area harvested, but second in production after the USA out of 111 countries in the world in 2016–2020 (FAO, 2022a).

Oromia, Amhara, Tigray and Southern Nations, Nationalities and Peoples Region (SNNPR) regional states accounted for 39%, 35%, 13% and 6%, respectively, of the total sorghum area in the 2018 *meher* season (CSA, 2019). Ethiopians obtain about 9.6% of their total dietary energy, 9.6% dietary protein and <0.5% dietary fat from sorghum.

Figure 3.4 summarizes the long-term trends of sorghum indicators in Ethiopia. The sorghum production system in Ethiopia was at subsistence level through the late 1990s; however, it has been in the SCA mode starting from the early 2000s. The yields are still continuing to grow.

Ethiopia achieved a very highly significant ($P<0.01$, $R^2=0.98$) yield gain of 94 kg/ha/yr between 2000 and 2020 (Table 3.5). The area, yield and production increased at an annual rate of 2.71%, 5.07% and 7.92%, respectively. This means that approximately 64% of the increase in production was due to increases in yield, whereas 34% was due to area expansion; the remaining 2% was attributed to the interaction factors.

Ethiopia's yield gain was superior to all the top ten sorghum-producing countries of the world – including Sudan, China and Brazil – and superior to many countries, including Australia, Bolivia and all African countries (Table 3.5). Its average yield of about 2.7 MT/ha was comparable to Brazil and Bolivia but lower than seven of the top ten countries that included China, the USA, Argentina, Mexico and Australia.

Sorghum in Ethiopia does not receive the share of inputs commensurate with its importance in the farming systems and its socioeconomic value for the country. Here are the facts (calculated from the CSA data for 2010–2014 and the 2016 *meher* seasons):

- share of fertilizer (of the total urea + DAP [diammonium phosphate]): 3%, compared to 31% for *teff*, 24% for maize and 23% for wheat;
- area covered under fertilizer: 13%, compared to 75% for wheat, 71% for *teff*, 51% for maize;
- fertilizer application rate: 11 kg/ha, compared to 98 kg/ha for wheat, 77 kg/ha for maize and 72 kg/ha for *teff*;

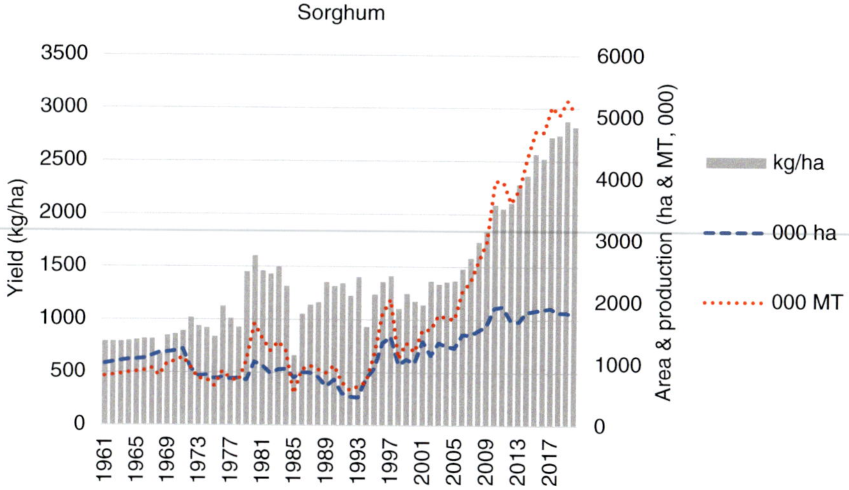

Fig. 3.4. Indicator trends for sorghum in Ethiopia, 1961–2020. Source: constructed by the lead author from FAO (2022a).

Table 3.5. Yield gain (b-value, kg/ha/yr) and annual rate of growth (ROG) for sorghum in the top ten cultivating countries of the world, 2000–2020. Source: calculated and constructed by the lead author from FAO (2022a).

	Yield gain			ROG (%)		
Country[a]	b-value	P > |t|	R^2	Area	Yield	Production
Ethiopia	94	<0.01	0.98	2.71	5.07	7.92
USA	48	0.01	0.29	−1.84	1.19	−0.67
Niger	16	<0.01	0.77	3.16	4.55	7.85
Chad	11	0.01	0.41	3.12	1.41	4.57
Mali	9	0.20	0.09	4.93	1.09	6.06
Mexico	7	0.55	0.02	−1.41	0.20	−1.22
India	5	0.10	0.14	−3.78	0.66	−3.14
Nigeria	4	0.32	0.05	−1.64	0.31	−1.33
Burkina Faso	2	0.50	0.02	1.36	0.22	1.59
Sudan	−4	0.23	0.08	NA	NA	NA

[a]Descending order of cultivated area: Sudan, Niger, India, Nigeria, USA, Ethiopia, Burkina Faso, Mali, Mexico and Chad.
NA = not available.

- area covered under improved seed: 0.15%, compared to 38% for maize and 7% for wheat; and
- extension package coverage: 10%, compared to 43% for wheat, 38% for maize, 33% for finger millet.

Ethiopia is a net importer of sorghum in spite of the progress made in productivity and production in recent times. The country imported close to 165,000 MT of sorghum grain, valued at >US$97 million annually in 2016–2020. By contrast, the ERCA data show that Ethiopia imported just over 9000 MT grain worth US$6.3 million versus its export of a little over 630 MT, valued at about US$198,000 annually in 2013–2017 (ERCA, 2013–2017).

Wheat

Ethiopia grows both bread and durum wheat and is the secondary centre of diversity for wheat (Westphal, 1975). It is a traditional crop widely grown in the mid and high altitudes (1900–2700 m above sea level). According to the CSA, roughly 4.7 million smallholder farmers grew wheat on close to 1.8 million ha of land and produced approximately 5 million MT of wheat grain annually in 2016–2020; the average yield was about 2.8 MT/ha (cf. Table 3.1). The farm size is estimated at 0.37 ha/household and each household produces close to 1100 kg/year. A total of 119 wheat varieties, the largest of all crops, have been officially registered as of 2017. Ethiopia is the 23rd largest wheat-growing country out of 124 countries for which data were available in 2016–2020, and the third largest in Africa, after Morocco and Algeria.

Oromia, Amhara, Tigray and SNNPR regional states accounted for roughly 48, 31, 12 and 8%, respectively, in the 2018 *meher* season (CSA, 2019). Wheat in Ethiopia is grown for food security and as import substitution for the local food industry. It provides about 13% each of dietary energy and dietary protein; the 2015–2019 FAO (2022c) records show that wheat has little or no contribution to fat supply in Ethiopian diets.

Ethiopia has registered a very highly significant ($P<0.01$, $R^2=0.96$) yield gain of 96 kg/ha/yr of wheat between 2000 and 2020. This is the third highest yield gain after Romania and China out of the top ten wheat-growing countries of the world (Table 3.6). The national average yield of 2.8 MT/ha in this country is a good sign of progress made in wheat productivity, but it has still a long way to go to catch up at least with the world average of nearly 3.3 MT/ha, or among the highest yielding countries of Germany, France and China, with 7.4, 6.7 and 5.5 MT/ha, respectively, among the top ten.

Figure 3.5 presents the long-term indicator trends for wheat in Ethiopia. Wheat production in this country remained in a subsistence mode through the early 1980s. Its transition to SCA started in the early 2000s. All indicators showed steady growth through 2020. The area, yield

Table 3.6. Yield gain (b-value, kg/ha/yr) and annual rate of growth (ROG) for wheat in Ethiopia, compared to the top ten cultivating countries of the world, 2000–2020. Source: calculated and constructed by the authors from FAO (2022a).

	Yield gain			ROG (%)		
Country[a]	b-value	P > |t|	R^2	Area	Yield	Production
China	101	<0.01	0.97	0.07	2.17	2.25
Ethiopia	96	<0.01	0.96	2.69	4.93	7.75
Canada	72	<0.01	0.78	−0.19	2.66	2.47
Russian Federation	55	<0.01	0.68	1.11	2.42	3.56
Türkiye	41	<0.01	0.80	−1.41	1.67	0.23
India	39	<0.01	0.72	0.97	1.30	2.28
USA	35	<0.01	0.61	−1.47	1.19	−0.30
Pakistan	29	<0.01	0.77	0.51	1.11	1.63
Australia	16	0.31	0.05	−0.52	1.06	0.53
Kazakhstan	9	0.25	0.07	0.34	0.80	1.14
Iran	2	0.86	0.00	0.77	0.11	0.88

[a]Descending order of cultivated area: India, Russian Federation, China, USA, Kazakhstan, Australia, Canada, Pakistan, Türkiye and Iran; Ethiopia is 23rd.

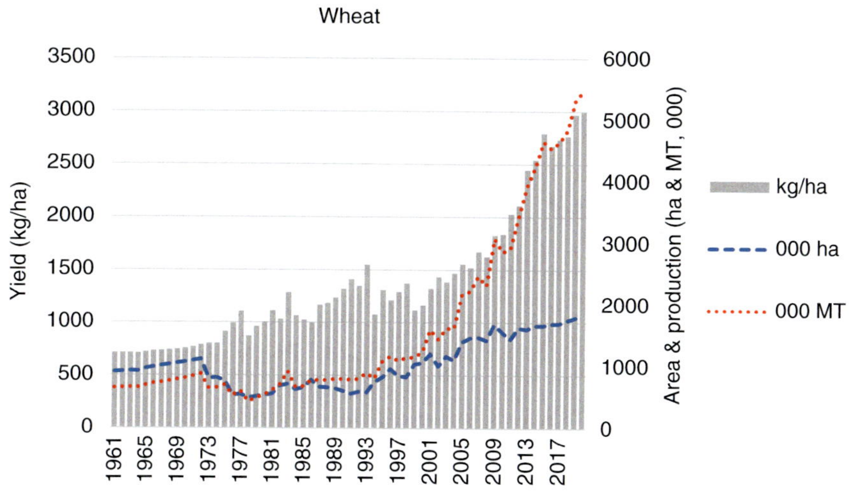

Fig. 3.5. Indicator trends for wheat in Ethiopia, 1961–2020. Source: constructed by the lead author from FAO (2022a).

and production increased at an annual rate of 2.69%, 4.93% and 7.75%, respectively. In other words, increases in yield and area contributed approximately 64% and 35%, respectively, to increases in production; the interaction effects (of area and yield) covered the remaining 1% or so.

Even though Ethiopia has made substantial progress in wheat productivity and production, the country remains a major importer. Imports started in the early 1970s and have shown increasing trends ever since. There was a shortfall of more than 25% of the total annual production in 2016–2020. Accordingly, the government imported nearly 1.5 million MT of wheat valued at close to US$496 million annually in 2016–2020.

By contrast, the ERCA data show that Ethiopia imported nearly 630,000 MT of wheat valued at nearly US$115 million annually during the same period. The huge difference between the two data sources may be that the FAO data include food aid whereas the ERCA reports show only the amount imported through trade.

At present, there is a government initiative to expand wheat production under irrigation in the Awash Valley, as well as double cropping in the traditional wheat areas. However, there is debate whether the country should be employing the scheme for producing wheat or other high-value crops such as cotton and horticultural crops that can be used as import substitution or export commodities. The economic advantages and disadvantages need to be weighed very carefully. At any rate, adequate technical management skills are essential to run such a complex scheme.

Barley

Ethiopia is a secondary centre of biodiversity for barley (Westphal, 1975) and it is a traditional crop widely grown in the high altitudes of this country. Close to 4 million smallholder households grew barley on 920,000 ha of land and produced more than 2.1 million MT of grain in Ethiopia in 2016–2020; the average yield was about 2.2 MT/ha (cf. Table 3.1). This country ranked 16th out of 103 countries that grew barley in 2016–2020, with the Russian Federation, Australia, Türkiye, Spain and Ukraine topping the list. In Africa, Ethiopia is the second largest grower of barley after Algeria.

Oromia, Amhara, SNNPR and Tigray regional states accounted for roughly 48, 29, 12 and 11%, respectively, in the 2018 *meher* season; the Benshangul-Gumuz regional state accounted for less than 1% of the total cultivated area (CSA, 2019). The *meher* season accounts for 85% of area and 93% of production. Barley in Ethiopia is grown for food security and as raw material supply for the local breweries. It provides about 5% each of dietary energy and dietary protein, and 1% of dietary fat (cf. Chapter 24 in this publication).

The long-term barley indicator trends are shown in Fig. 3.6. Like many of the other cereals, barley production in Ethiopia was in subsistence mode through the early 1970s. Area, yield and production showed heavy fluctuations in the 1980s and 1990s; it took SCA form starting in mid-2000s, during which period the productivity and production showed steady growth, even though the area declined. It appeared that wheat was replacing barley, especially in the lower high-altitude areas, and the latter is being pushed to further upper higher altitudes of Ethiopia.

The area declined (Fig. 3.6) by 0.05% while the yield and production increased at 4.40% per annum, respectively, with the production growth rate of 4.35%. This means that all of the changes in production resulted from increases in

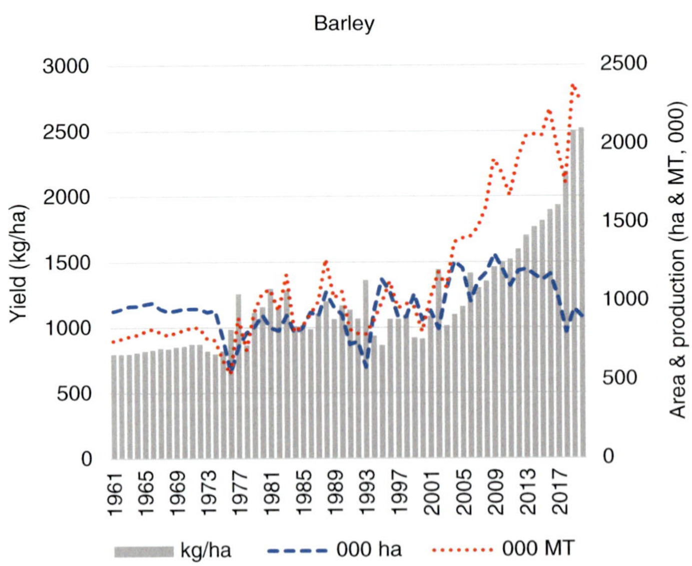

Fig. 3.6. Indicator trends for barley in Ethiopia, 1961–2020. Source: constructed by the lead author from FAO (2022a).

productivity. Yields in major cultivating countries range between 6.0 and 3.8 MT/ha. An average Ethiopian smallholder household produces about 552 kg per year.

Ethiopia registered a very highly significant yield gain of barley in the two decades between 2000 and 2020 (Table 3.7). With yield gain of 68 kg/ha/yr ($P<0.01$, $R^2=0.88$), Ethiopia was ahead of all of the top ten barley-growing countries of the world. It was followed by Ukraine, Canada and Germany. However, the national average yield of about 2.1 MT/ha is way below those for many of the top producing countries.

The ROG for area harvested and production of barley declined in six and five, respectively, of the top ten growing countries in the world during the 2000–2020 period. Despite the good progress Ethiopia has made in barley productivity, it is still a net importer. For example, the country imported 16,528 MT of barley (mainly malt), valued at US$5.86 million annually in 2016–2020.

Finger millet

Ethiopia is the secondary centre of diversity for finger millet (Westphal, 1975), even though Zeven and de Wet (1982) suggest that it is a primary centre. According to the CSA, >1.7 million smallholder households grew finger millet on close to half a million ha of land and produced a little over 1 million MT of grain annually in 2016–2020 (cf. Table 3.1). The average area per smallholder household for this crop is about 0.26 ha, and each household produces an average of >600 kg/yr. There are 21 finger millet varieties registered in this country as of 2017.

Ethiopia is ranked tenth in the total area harvested and third in production, out of 87 countries in the world for which data were available in 2016–2020. Eight of the top ten millet-growing countries are found in Africa, with Niger, Sudan, Mali, Nigeria and Burkina Faso leading the list, after India.

In Ethiopia, millet is grown mainly for domestic consumption, supplying about 2.7% of dietary energy, 2.6% dietary protein and about 2.0% of dietary fat. Amhara, Tigray, Oromia, Benshangul-Gumuz and SNNPR regional states accounted roughly for 56, 19, 18, 5 and 2%, respectively, in the 2018/19 *meher* season (CSA, 2019).

Ethiopia registered a very highly significant ($P<0.01$, $R^2=0.95$) yield gain of 88 kg/ha/yr between 2000 and 2020, the highest among the top 20 millet-growing countries of the world, followed by China, the USA, Tanzania and India (Table 3.8). Only China surpassed the national average yield for Ethiopia (2.4 MT/ha) among the top 20 cultivating countries of the world. The good progress made with finger millet in Ethiopia was due to concerted improved technology promotions, especially in the Rift Valley area, in the early 2000s (Anchala *et al.*, 2006).

Figure 3.7 presents the long-term indicator trends for finger millet in Ethiopia. Its

Table 3.7. Yield gain (b-value, kg/ha/yr) and annual rate of growth (ROG) for barley in the top ten cultivating countries of the world, 2000-2020. Source: calculated and constructed by the lead author from FAO (2022a).

	Yield gain			ROG (%)		
Country[a]	b-value	P > |t|	R^2	Area	Yield	Production
Ethiopia	68	<0.01	0.88	–0.05	4.40	4.35
Ukraine	66	0.01	0.49	–3.21	2.70	–0.59
Canada	59	<0.01	0.66	–3.19	1.86	–1.39
Germany	51	0.01	0.28	–1.50	0.82	–0.69
Spain	32	0.19	0.09	–1.22	1.13	–0.11
Russian Federation	30	0.01	0.44	–0.79	1.44	0.64
Australia	28	0.08	0.16	0.71	1.60	2.32
Türkiye	24	0.01	0.41	–1.67	0.98	–0.71
Kazakhstan	20	0.02	0.26	2.15	1.60	3.78
Iran	1	0.90	0.00	1.66	0.07	1.73
France	0	0.98	0.00	0.82	–0.03	0.79

[a]Top-ranking countries are Russian Federation, Australia, Türkiye, Spain, Ukraine, Canada, Kazakhstan, Iran, France and Germany; Ethiopia is 16th out of 103 countries in the world.

Table 3.8. Yield gain (b-value, kg/ha/yr) and annual rate of growth (ROG) for millet in the top ten cultivating countries of the world, 2000–2020. Source: calculated and constructed by the lead author from FAO (2022a).

	Yield gain			ROG (%)		
Country[a]	b-value	P > ltl	R^2	Area	Yield	Production
Ethiopia	88	<0.01	0.95	–1.80	2.63	0.78
China	58	<0.01	0.62	–1.80	2.63	0.78
India	26	<0.01	0.69	–1.92	2.58	0.61
Senegal	18	0.01	0.45	0.52	2.37	2.90
Mali	12	0.01	0.28	2.92	1.60	4.57
Sudan	9	0.02	0.24	NA	NA	NA
Niger	6	0.01	0.34	1.51	1.33	2.86
Chad	6	0.11	0.13	2.72	1.25	4.01
Burkina Faso	1	0.60	0.01	–0.34	0.22	–0.12
Nigeria	–30	0.01	0.30	–6.65	–2.66	–9.14

[a]Countries with the largest millet area are India, Niger, Sudan, Mali, Nigeria, Burkina Faso, Chad, Senegal, China and Ethiopia.
NA = not available.

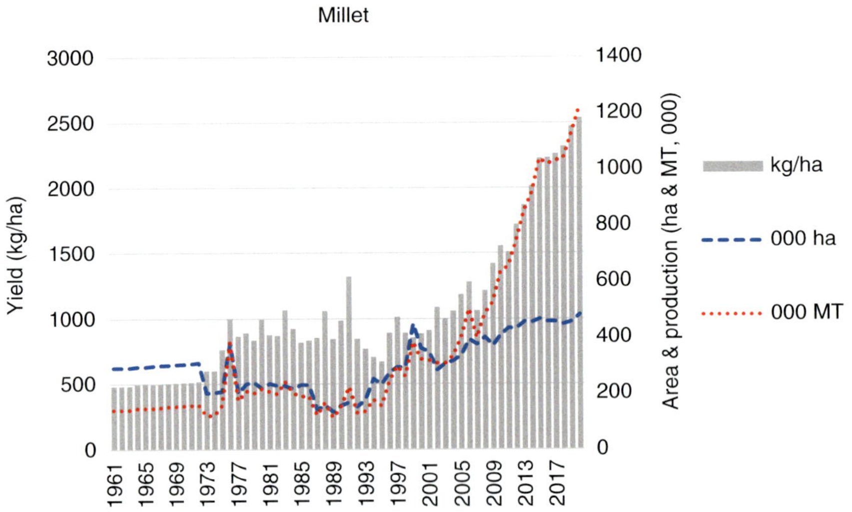

Fig. 3.7. Indicator trends for finger millet in Ethiopia, 1961–2020. Source: constructed by the lead author from FAO (2022a).

production system remained subsistence through the early 2000s and transitioned to SCA by the late 2000s. All indicators showed continued growth over the last two decades. The annual ROG for area, yield and production was 2.26, 5.76 and 8.15%, respectively. This meant that rises in yield accounted for nearly 71% of the change in production whereas the contribution of area expansion was nearly 28%, with the interaction factor covering the balance.

As stated above, this crop is grown mainly for local consumption, but records show that some small amounts are exported, making Ethiopia a net exporter. This country exported a little less than 4000 MT of millet grain, valued at about US$1.3 million annually in 2016–2020.

Emmer

Emmer, or emmer wheat (erroneously reported as oats by CSA) is a traditional crop to Ethiopia.

An estimated 184,000 smallholder households grew this crop on approximately 20,000 ha of land and produced 42,000 MT of grain annually with the average yield of about 2 MT/ha in 2016–2020 (cf. Table 3.2). The average farm size is estimated at 0.11 ha per holder, the lowest among all cereals, and the per capita production is 226 kg/household/yr (cf. Table 3.1). Only three improved varieties have been released through 2017 (MOALR, 2017). Oromia, Amhara, SNNPR and Tigray regional states accounted for 76%, 21%, 2% and 1%, respectively, of the total area planted to emmer in Ethiopia in the 2018/19 *meher* season (CSA, 2019).

Figure 3.8 presents the emmer indicator trends in Ethiopia. The crop area has been declining since the mid-2000s, while the yield increased at a good pace. According to the CSA, the area declined at an annual rate of 3.44 whereas the yield increased 3.51% per year; the ROG for production was 0.47% between 2000 and 2020.

In a similar fashion, the number of smallholders cultivating emmer declined at an annual rate of 3.58%. For example, there were close to 257,000 smallholders growing this crop in 2003 compared to ~146,000 in 2019, a 43% decline. By contrast, there was a very highly significant yield gain of 55 kg/ha/yr ($P<0.01$, $R^2=0.86$,) between 2000 and 2020 (cf. Table 3.2).

Rice

Rice is a relatively new crop to Ethiopia. Its R&D was emphasized by the military government following the devastating drought of 1984 and the famine that ensued that claimed thousands of human lives. The CSA records for rice in Ethiopia are available from 2003; the first record by FAO started in 1993.

Nearly 170,000 smallholder households grew rice on 62,000 ha and produced about 180,000 MT of grain annually in 2016–2020 (cf. Table 3.1). The average yield was about 2.9 MT/ha. Each rice-growing household has about 0.37 ha and the per capita production is estimated at about 1065 kg/household/yr. There are 19 rice varieties registered in Ethiopia as of 2017 (MOALR, 2017). The Amhara Regional State has about 96% share of the total area under rice in Ethiopia, with Tigray and Gambella covering nearly 3% and 1%, respectively (CSA, 2019).

We present rice trend indicators in Ethiopia using the latest available data from FAOSTAT as shown in Fig. 3.9. Analysis for 2007–2020 (data for earlier years did not seem to be reliable and therefore discarded) shows that the area, yield and production increased at an annual rate of 16.69%, 3.38% and 20.63%, respectively. In other words, 81% of the contribution to changes in production was due to the area expansion

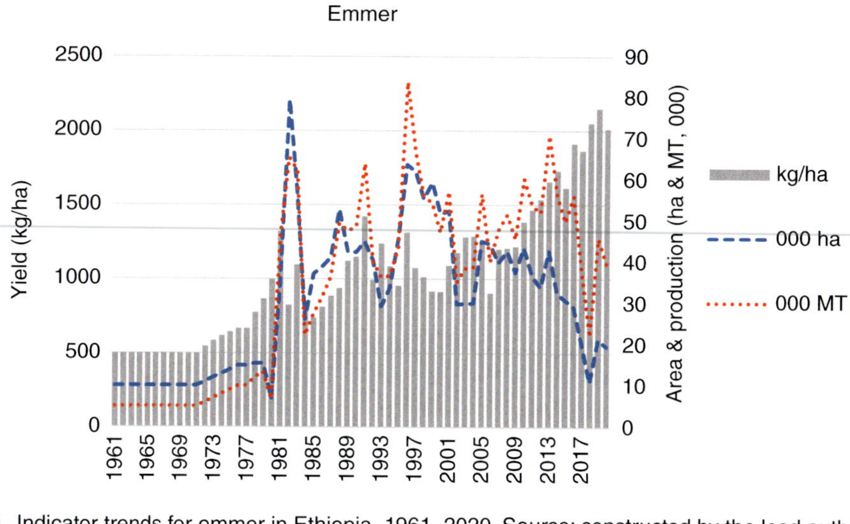

Fig. 3.8. Indicator trends for emmer in Ethiopia, 1961–2020. Source: constructed by the lead author from FAO (2022a).

whereas increases in yield accounted for 16%; the remaining <3% was due to the interaction factor.

Rice has become the fastest growing crop in area in Ethiopia. According to the CSA, a little over 25,000 smallholder households grew rice in 2003 compared to >178,000 in 2019. In a similar fashion, this crop occupied just over 24,000 ha in 2007 in comparison to >85,000 ha in 2020.

As presented in Table 3.9, the yield gain for Ethiopia was very highly significant and among the highest of the top ten largest cultivating countries of the world.

Ethiopia is a net importer of rice, and it is the country's second largest grain import (both in quantity and value) after wheat. According to the FAO, the country imported about 447,000 MT of rice valued at nearly US$186.5 million annually during 2016–2020. The ERCA data show import of about 262,000 MT valued at >US$144 million annually, in 2013–2017.

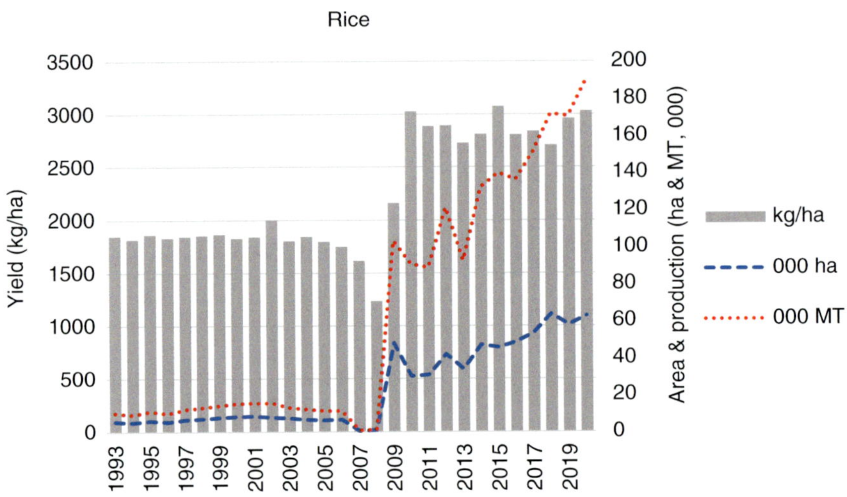

Fig. 3.9. Indicator trends for rice in Ethiopia, 1993–2020. Source: constructed by the lead author from FAO (2022a).

Table 3.9. Yield gain (b-value, kg/ha/yr) and annual rate of growth (ROG) for rice in the top ten cultivating countries of the world and Ethiopia, 2000–2020. Source: calculated and constructed by the lead author from FAO (2022a).

	Yield gain			ROG (%)		
Country[a]	b-value	P > |t|	R^2	Area	Yield	Production
Cambodia	95	<0.01	0.97	2.26	3.48	5.81
Viet Nam	80	<0.01	0.92	0.11	1.58	1.70
Ethiopia	76	<0.01	0.64	16.69	3.38	20.63
Bangladesh	75	<0.01	0.95	0.48	1.85	2.34
India	60	<0.01	0.91	0.09	1.79	1.88
China	50	<0.01	0.95	0.40	0.76	1.17
Indonesia	45	<0.01	0.83	−0.42	0.94	0.52
Philippines	43	<0.01	0.86	1.00	1.20	2.21
Myanmar	21	0.01	0.43	0.08	0.59	0.68
Nigeria	14	0.13	0.12	5.15	0.87	6.07
Thailand	8	0.08	0.15	5.30	2.19	7.60

[a]The top-ranking countries are India, China, Bangladesh, Indonesia, Thailand, Viet Nam, Myanmar, the Philippines, Nigeria and Cambodia; Ethiopia is 69th out of 119 countries and territories in the world.

Discussion and Conclusions

Cereals are the key to food security in Ethiopia, as they are across the globe, and their agricultural R&D has been accorded the highest priority. Ethiopia has made substantial improvements in the productivity of its major cereal crops, but comprehensive literature measuring the gains and the underlying causes for the productivity improvements are limited. These aspects are critical as the government is now in the process of revising its agricultural policy.

Using long-term data from the FAO and national sources (mainly CSA and ERCA), our study provides a critical assessment of the gains made so far and the opportunities for further improvement ahead. Overall, Ethiopia achieved a very highly significant yield gain of 92 kg/ha/yr ($P<0.01$, $R^2=0.95$) in its aggregate cereals in the two decades between 2000 and 2020. Increases in yield (productivity per unit area) contributed to about 70% of the change in production compared to 28% from area expansion, with the interaction factor accounting for the remaining <2%.

No other African country (except for South Africa) has achieved this level of cereals yield grown under rainfed conditions. Niger and Sudan have the second and third largest area of cereals in Africa and 18th and 19th, respectively, in the world, but both of them showed less than 1 MT/ha in 2016–2020. Nigeria has the largest area of cereals in Africa and sixth largest in the world, but its average yield was <2 MT/ha. The yield gain for Niger, Nigeria and Sudan was 9 kg/ha/yr, 6 kg/ha/yr and 1 kg/ha/yr, respectively. In general, all cereals (maize, wheat, sorghum, millet, rice, barley, *teff* and emmer) in Ethiopia attained highly significant ($P<0.01$) yield gains over the last two decades.

The successes with cereals in Ethiopia are attributed to the convergence of several factors. These crops received preferential treatment over all others in terms of support for technology development, external input (mineral fertilizer and pesticide) use and extension package coverage. As an example, approximately 88% of all mineral fertilizers (mainly urea and DAP) used in the country is consumed by cereals (cf. sections on inputs use and extension coverage in the later sections of this book).

As shown below, there were substantial variations among cereals in terms of their input usage and therefore their productivity gains. Overall, maize, wheat and *teff* were the cereals that received maximum attention. The absence of devastating droughts similar to those in the early 1970s and 1980s (Viste *et al.*, 2013; Abate *et al.*, 2015) has also contributed to the success of cereals in Ethiopia.

It is not at all surprising that cereals occupy the largest area (68%) of the total cultivated crop area (not including *enset*) in Ethiopia, as they do in many major agricultural countries across the globe. The 2016–2020 global average was 51%.

There are three critical issues surrounding cereals expansion in this country. First, cereals area expansion is based on the premise that Ethiopia is land abundant. But this notion has long been challenged – i.e. Africa is actually land scarce (Binswanger and Pingali, 1988). Ethiopia has so far cultivated ~16 million ha of its ~38 million ha of agricultural land (FAO, 2022d), indicating that there is ~22 million ha for further development. However, such land is found in remote lowland areas (Schmidt and Thomas, 2018; Degife *et al.*, 2019) and faces major challenges like invasive plant species such as mesquite (*Prosopis juliflora*) and cactus (*Opuntia* spp.). This country has almost reached the limits of agricultural area expansion in the highlands, as is the case in many countries across Africa (Chamberlin *et al.*, 2014). Farmers are cultivating the maximum land area they could cultivate with their existing economic and technological capacities. Areas that had once been natural forests, woodlands and meadows have been converted to farmlands. The negative impact of such activities on biodiversity of Ethiopia's agricultural systems is obvious and of great concern to the future of Ethiopian agriculture (De Beenhouwer *et al.*, 2015).

Second, unlike many of the top ten cereal-growing countries of the world (where Ethiopia is ranked 17th in area and 20th in production), it did very little to diversify its cropping system over the last four decades. By contrast, 11 countries (Canada, Argentina, Indonesia, Australia, Brazil, Niger, China, India, Thailand, the USA and Türkiye) reduced the proportion of their cereals area by up to 35% in 2016–2020 vs 1976–1980 and diversified their agricultural systems. The diversification took different forms in different countries. Several (e.g. Türkiye, Mexico, China and India) expanded and intensified fruit crops.

China increased its fruit crops area 9.24-fold within four decades (FAO, 2022b). Eighteen of its 33 major fruit crops were introduced and intensified in the mid-1980s, early 1990s, mid-1990s or early 2000s. The annual ROG in the area planted for cherries, papaya, quince, chestnut, apricot, walnuts, avocado and kiwi fruit ranged from 8.42–5.45%. Argentina and Australia intensified their soybean, rapeseed and chickpea; Brazil its soybean, oil palm and sunflower; Canada its lentil, dry peas and soybean; and India its soybean.

The overall world average for cereal proportions also continued to decline (and helped towards more diversification). For example, cereals constituted 65% of the total area harvested in 1976–1980 while their composition was 58% in 1996–2000 and 51% in 2016–2020.

Third, the average farm size of about 0.29 ha per household for overall cereals in Ethiopia is too small to lift smallholder farmers from the underdevelopment line (Dorosh and Schmidt, 2010; Harris and Orr, 2014) even though government interventions with improved agricultural technologies can substantially increase net returns per ha per cropping season (Thirtle *et al.*, 2003; Abate *et al.*, 2015; Manda *et al.*, 2015, 2017). Policy makers, therefore, should be asking themselves how far they intend to continue with smallholder agriculture. Furthermore, what would be the role of large-scale commercial farming in Ethiopia (Bekele, 2019)?

In spite of its demonstrated successes with cereals in the past four decades, Ethiopia still has a huge deficit of demand for cereal grains and their products. For example, the country imported an estimated 2.16 million MT of cereals, valued at nearly US$810 million each year in 2016–2020. The rapidly growing population and small farm sizes make it difficult to cater for the increasing demands (Harris and Orr, 2014) on food security, import substitution and supplying raw material for the local food industry. Unchecked expansion of cereals is becoming a major contributor to environmental degradation and a potential threat to Ethiopia's biodiversity (Kebede *et al.*, 2018; De Beenhouwer *et al.*, 2019), which is already the largest part of the Eastern Afro-Montane Biodiversity Hotspot (BirdLife International, 2012).

There are several key questions that need to be asked and solutions sought for cereals agriculture in Ethiopia. One such question would be: could the Ethiopian cereals system be diversified – are there other crops whose R&D investment could be elevated so that they can complement the food security requirements in a more sustainable manner?

References

Abate, T., Shiferaw, B., Gebeyehu, S., Amsalu, B., Negash, K. *et al.* (2011) A systems and partnership approach to agricultural research for development: Lessons from Ethiopia. *Outlook on Agriculture* 40(3), 213–220.

Abate, T., Alene, A.D., Bergvinson, D., Shiferaw, B., Silim, S. *et al.* (2012) *Tropical Grain Legumes in Africa and South Asia: Knowledge and Opportunities.* International Crops Research Institute for the Semi-Arid Tropics. 112 pp. ISBN: 978-92-9066-544-1. Order Code: BOE 056. ICRISAT, Patancheru, India.

Abate, T., Shiferaw, B., Menkir, A., Wegary, D., Kebede, Y. *et al.* (2015) Factors that transformed maize productivity in Ethiopia. *Food Security* 7, 965–981. DOI: 10.1007/s12571-015-0488-z.

Abate, T., Fisher, M., Abdoulaye, T., Kassie, G.T., Lunduka, R. *et al.* (2017) Characteristics of maize cultivars in Africa: How modern are they and how many do smallholder farmers grow? *Agriculture & Food Security* 6, 30. DOI: 10.1186/s40066-017-0108-6.

Alemu, D., Tesfaye, A., Assaye, A., Addis, D., Tadesse, T. *et al.* (2018) *A Historical Analysis of Rice Commercialisation in Ethiopia: The Case of the Fogera Plain.* Working Paper 18, Agricultural Policy Research in Africa. Future Agriculture Consortium, Nairobi, Kenya.

Anchala, C., Kidane, H.S., Mulatu, T. (2006) The impact of improved finger millet technology promotion in the central Rift Valley of Ethiopia. In: Abate, T. (ed) *Successes with Value Chain. Proceedings of Scaling up and Scaling out Agricultural Technologies in Ethiopia: An International Conference, 9–11 May 2006, Addis Ababa.* EIAR, Addis Ababa, Ethiopia, 275 pp.

Asfaw, S., Shiferaw, B., Simtowe, F., Haile, M.G. (2011) Agricultural technology adoption, seed access constraints and commercialization in Ethiopia. *Journal of Development and Agricultural Economics* 3(9), 436–447.

Bekele, M.S. (2019) Economic and agricultural transformation through large-scale farming: Impacts of large-scale farming on local economic development, household food security and the environment in Ethiopia. PhD dissertation, Leiden University, Netherlands.

Beyene, A.D., Kassie, M. (2015) Speed of adoption of improved maize varieties in Tanzania: An application of duration analysis. *Technological Forecasting & Social Change* 96, 298–307.

Binswanger, H.P., Pingali, P. (1988) Technological priorities for farming in Sub-Saharan Africa. *World Bank Research Observer* 3, 81–98.

BirdLife International (2012) *Eastern Afro-Montane Biodiversity Hotspot*. Critical Ecosystem Partnership Fund. Cambridge, England.

CSA (Central Statistical Agency) (2017) Agricultural Sample Survey 2016/17 (2009 E.C.), Report on Area and Production of Major Crops (Private Peasant Holdings, *Meher* Season), Volume I, Statistical Bulletin. CSA, Addis Ababa, Ethiopia.

CSA (2018) Agricultural Sample Survey 2017/18 (2010 E.C.), Report on Area and Production of Major Crops (Private Peasant Holdings, *Meher* Season), Volume I, Statistical Bulletin 586 CSA, Addis Ababa, Ethiopia.

CSA (2019) Agricultural Sample Survey 2018/19 (2011 E.C.), Report on Area and Production of Major Crops (Private Peasant Holdings, *Meher* Season), Volume I, Statistical Bulletin 589. CSA, Addis Ababa, Ethiopia.

CSA (2020) Agricultural Sample Survey 2019/20 (2012 E.C.), Report on Area and Production of Major Crops (Private Peasant Holdings, *Meher* Season), Volume I, Statistical Bulletin 587. CSA, Addis Ababa, Ethiopia.

CSA (2021) Agricultural Sample Survey 2020/21 (2013 E.C.), Report on Area and Production of Major Crops (Private Peasant Holdings, *Meher* Season), Volume I, Statistical Bulletin. CSA, Addis Ababa, Ethiopia.

Chamberlin, J., Jayne, T.S., Headey, D. (2014) Scarcity amidst abundance? Reassessing the potential for cropland expansion in Africa. *Food Policy* 48, 51–65.

De Beenhouwer, M., Aerts, R., Hundera, K., Van Overtveld, K., Honnay, O. (2015) Management intensification in Ethiopian forests is associated with crown habitat contraction and loss of specialized epiphytic orchid species. *Basic and Applied Ecology* 16(7), 592–600.

Degife, A.W., Zabel, F., Mauser, W. (2019) Land use scenarios and their effect on potential crop production: The case of Gambella Region. *Agriculture* 9(5), 105. DOI: 10.3390/agriculture9050105

Dorosh, P., Schmidt, E. (2010) *The Rural-Urban Migration in Ethiopia*. ESSP2 Discussion Paper 13. IFPRI & EDRI. 56 pp.

ERCA (Ethiopian Revenue and Customs Authority) (2013–2017) Available at: http://www.erca.gov.et/index.php/component/hscode/?view=importexport&iccaldate=2014-5-1 (Agricultural commodities export and import data) (accessed 29 December 2018).

FAO (Food and Agriculture Organization of the United Nations) (2022a) Statistical databases and data-sets of the Food and Agriculture Organization of the United Nations. Available at: http://www.fao.org/faostat/en/#data (accessed 8 March 2022).

FAO (2022b) Statistical databases and data-sets of the Food and Agriculture Organization of the United Nations. Available at: http://www.fao.org/faostat/en/#data (accessed 14 August 2022).

FAO (2022c) Statistical databases and data-sets of the Food and Agriculture Organization of the United Nations. Available at: http://www.fao.org/faostat/en/#data ((accessed 8 September 2022).

FAO (2022d) Statistical databases and data-sets of the Food and Agriculture Organization of the United Nations. Available at: http://www.fao.org/faostat/en/#data (accessed 24 October 2022).

Garbero, A., Marion, P., Brailovskaya, V. (2018) The impact of the adoption of CGIAR's improved varieties on poverty and welfare outcomes: A systematic review. IFAD, Rome, Italy. 87 pp.

Harris, D., Orr, A. (2014) Is rainfed agriculture really a pathway from poverty? *Agricultural Systems* 123, 84–96.

Huffnagel, H.P. (1961) *Agriculture in Ethiopia*. FAO, Rome.

Jaleta, M., Yirga, C., Kassie, M., de Groote, H., Shiferaw, B. (2013) Knowledge, adoption and use intensity of improved maize technologies in Ethiopia. In: *Invited Paper Presented at the 4th International Conference of the African Association of Agricultural Economists, 22–25 September 2013, Hammamet, Tunisia*. African Association of Agricultural Economists, Nairobi, Kenya.

Kebede, Y., Bianchi, F., Baudron, F., Abraham, K., de Valença, A. *et al.* (2018) Implications of changes in land cover and landscape structure for the biocontrol potential of stem borers in Ethiopia. *Biological Control* 122, 1–10.

Khonje, M., Manda, J., Alene, A.D., Kassie, M. (2015) Analysis of adoption and impacts of improved maize varieties in eastern Zambia. *World Development* 66, 695–706.

Kumar, G. (1987) *Ethiopian famines 1973–1985: A Case Study*. Wider Working Papers. World Institute for Development Economic Research of the United Nations University. Helsinki, Finland. 82 pp.

Manda, J., Alene, A.D., Gardebroek, C., Kassie, M., Tembo, G. (2015) Adoption and impacts of sustainable agricultural practices on maize yields and incomes: Evidence from rural Zambia. *Journal of Agricultural Economics* 67, 130–153. DOI: 10.1111/1477-9552.12127.

Manda, J., Khonje, M.G., Alene, A.D., Gondwe, T. (2017) Welfare impacts of improved groundnut varieties in eastern Zambia: A heterogeneous treatment effects approach. *Agrekon* 56, 313–329. DOI: 10.1080/03031853.2017.1400447.

MOALR (2017) *Crop Variety Register Issue No. 20*. Plant Variety Release, Protection and Seed Quality Control Directorate, MOALR, Addis Ababa, Ethiopia.

Schmidt, E., Thomas, T.S. (2018) *Cropland Expansion in Ethiopia: Economic and Climatic Conditions for Highland Agriculture*. ESSP Working Paper 127. IFPRI, Washington.

Shiferaw, B., Prasanna, B.M., Hellin, J., Bänziger, M. (2011). Crops that feed the world 6. Past successes and future challenges to the role played by maize in global food security. *Food Security* 2011 (3), 307–327. DOI: 10.1007/s12571-011-0140-5.

Thirtle, C., Lin, L., Piesse, J. (2003) The impact of research-led agricultural productivity growth on poverty reduction in Africa, Asia and Latin America. *World Development* 31(12), 1959–1975.

Viste, E., Korecha, D., Sorteberg, A. (2013) Recent drought and precipitation tendencies in Ethiopia. *Theoretical and Applied Climatology* 112(3–5), 535–551.

Westphal, E. (1975) *Agricultural Systems in Ethiopia*. Joint publication of the College of Agriculture, Haile Sellassie I University, Ethiopia, and the Agricultural University, Wageningen, the Netherlands.

Zeven, A.C., de Wet, J.M.J. (1982) *Dictionary of Cultivated Plants and Their Regions of Diversity. Excluding most Ornamentals, Forest Trees and Lower Plants*. Centre for Agricultural Publishing and Documentation, Wageningen, the Netherlands.

4 A Compelling Case for Increased Investment in *Enset* and other Root and Tuber Crops Research and Development

Tsedeke Abate,[1]* Berga Lemaga,[2] Terefe Belehu[3] and Zerihun Yemataw[4]
[1]*Homegrown Vision, Addis Ababa, Ethiopia;* [2]*Formerly CIP-Ethiopia, Addis Ababa, Ethiopia;* [3]*Formerly Agricultural Transformation Agency, Addis Ababa, Ethiopia;* [4]*South Agricultural Research Institute, Hawassa, Ethiopia*

Abstract

This chapter provides a critical assessment of the status of root and tuber crops and suggests the way forward with their research and development agenda. Ethiopia grows at least nine species of roots and tubers, including *enset (Ensete ventricosum)*, taro, potato, sweet potato, yam, anchote (*Coccinia abyssinica*), aerial yam (*Dioscorea bulbifera*), cassava and Ethiopian potato (*Plectranthus edulis*). Policy makers here give high priority to food security, but as their focus has been to limited cereals, roots and tubers have not received the attention commensurate with their socioeconomic significance for the country and their full potential has not been realized. For example, *enset* has many positive attributes and affects the economy of 39 million people in southern and southwestern parts of the country, and yet there are no decent vital statistics available for the crop.

Introduction

Ethiopia's efforts to ensure food security started in the mid-1980s, following the devastating drought of 1984. Cereals (maize in particular) were given the highest priority and have received the largest share of investment. Productivity and production of those crops saw substantial improvements over the past two decades. According to the Food and Agriculture Organization (FAO, 2022a), yields of aggregated cereals more than doubled in the 21–year period between 2000 and 2020.

In spite of the progress made so far, Ethiopia has not achieved food self-sufficiency perhaps because of, among others, the fast population growth that is further exacerbated by climate change. The country has become a net importer of food in recent years. It imported agricultural products at an estimated annual value of US$2.09 billion in 2016–2020, compared to its export value of just over US$1.45 billion, according to the FAO (2022b). Food imports accounted for about 95% of the total agricultural products imports.

Area expansion for aggregated cereals in Ethiopia increased at an annual rate of 1.77% between 2000 and 2020 (cf. Chapter 3). However, this rate of crop land expansion cannot be sustained because of limited agricultural area for further expansion (Schmidt and Thomas, 2018), especially in the highlands and mid-altitudes. The logical available option for the country to

* Corresponding author: tsedeke@homegrownvision.org

DOI: 10.1079/9781800626386.0004

feed more than 205 million people by 2050 is to look for alternatives to heavy reliance on cereals. Roots and tubers (RTs) offer such an opportunity. They have many desirable attributes – including higher productivity per unit area and input, environmental protection, resilience and sustainability, among others.

In this chapter we attempt to provide evidence on the status of RTs research and development (R&D), the critical knowledge gap, and suggest the path forward on these important food and nutrition security crops for Ethiopia and beyond.

Approach

We reviewed available literature to determine the composition and socioeconomic importance of RTs and the status of their R&D in Ethiopia. Using secondary data from the Central Statistical Agency (CSA) and FAO, we made extensive analysis to measure yield gains and the annual rate of growth (ROG), using a similar method to Abate *et al.* (2015) and as presented in Chapter 3. CSA data records (CSA, 2005–2021) for RTs are similar in content to those of the FAO, but we employed the latter because it is uninterrupted and complete.

Results

The crops in general

There are four cultivated RTs for which some sort of data are available in Ethiopia: taro, potato, sweet potato and yams (Table 4.1). No reliable data are available for *enset*, which is the most important RT in the country. In addition, there are four species of cultivated RTs that are considered minor and no statistical reports are available on them. These are *Coccinia abyssinica* (known as *anchote* in Oromifa, *usukia* or *ushushia* in Dawrotso); *Dioscorea bulbifera* (aerial yam); *Plectranthus edulis* (syn. *Coleus edulis*) – variously referred to as Ethiopian *dinich*, Habesha *dinich*, Awi *dinich*, Gurage *dinich*, Oromo *dinich*, Wolaita *dinich* and the like; and cassava (*Manihot esculenta*). At present, there are no national data on these crops in Ethiopia; only limited R&D activities are currently being carried out by a few researchers (Mekbib and Deressa, 2016; Melaku and Duguma, 2016). The CSA classifies beetroot, carrot, garlic and onion under RTs, but their more appropriate place is under vegetables and therefore discussed in Chapter 7.

According to the CSA, more than 6 million smallholders cultivated RTs on about 191,000 ha of land and produced >4 million MT of food annually in the 2016–2020 *meher* seasons; the average yield was about 20 MT/ha (however, we believe that the data for area under RTs and number of people growing them are grossly underestimated). Each holder had 0.0468 ha of land covered under RTs and the per capita production was a little over 1,100 kg/holder/season (Table 4.1).

Smallholders grow about 98% of all RTs in Ethiopia. According to the CSA data for 2013/14 and 2015/16, the *meher* (main rain) season accounts for 46% of the total area and 78% of total production of the RTs that are mentioned in this study. Conversely, approximately 54% of area and 22% of production comes from the *belg* season. Obviously, there are variations among crops (Annex Table 4.1). They are grown as staples and co-staples in southern, southwestern and eastern parts of Ethiopia and play a crucial

Table 4.1. Major root and tuber crops smallholders cultivated in the *meher* season in Ethiopia (2016–2020). Source: constructed by the authors from CSA (www.csa.gov.et).

Crop	Holders (000)	Area (000 ha)	Yield (MT/ha)	Production (000 MT)	Area per holder (ha)	Prod. per holder (kg)	Number of varieties released
Taro	2,066	60	25	1,528	0.0291	740	3
Potato	1,162	73	11	1,000	0.0631	860	36
Sweet potato	645	53	33	1,727	0.0817	2,678	26
Yams	365	5	9	46	0.0138	127	3
Total/average	6,203,939[a]	191	20	4,302	0.0469	1,101	68

[a]The total number of holders is the aggregate for all roots and tubers.

role in its food security; potato is grown in all parts of the country, especially in the highlands.

RTs in Ethiopia are produced primarily for domestic consumption, perhaps with the exception of potato with a significant number of exports to neighbouring countries such as Somalia and Djibouti. Over the past four decades, the contribution of RTs to food security in this country has grown faster than any of the major cereals. For example, RTs provided a dietary energy supply of 254 kcal/person/day in 2009–2013, compared to 94 kcal/person/day in 1969–1973 (FAO, 2020). Their contribution grew at an annual rate of 2.35% between 1970 and 2013, whereas the rates for wheat, maize, sorghum, unspecified pulses, *teff* and barley were 1.22%, 1.95%, –0.23%, –0.62%, –0.90% and –1.04%, respectively.

The aggregate area, yield and production of RTs in Ethiopia increased at 1.77%, 4.48% and 6.33%, respectively, between 2000 and 2020. The overall yield gain was estimated at 730 kg/ha/yr ($P<0.01$, $R^2=0.63$).

Figure 4.1 depicts the trends of aggregate RTs indicators in Ethiopia for the period from 2000 to 2020. All indicators remained almost constant through 2011. Yield and (therefore) production showed fast growth from 2012, peaked in 2014, and somewhat levelled off after that. Area and production started to take upward trends in 2019 and 2020 while the yield tended to decline during the same period.

This figure may not give a very clear picture because of the large differences among the indicators for the different crops. For example, the average yield was roughly 32, 25, 14 and 9 MT/ha, respectively, for sweet potato, taro, potato and yams. Furthermore, there is no way to determine whether the rapid increase in yield from 2011 to 2012 was due to an adjustment of data by the CSA (or FAO) made on the basis of new information, or an actual surge.

The next section provides brief accounts for *enset*, potato, sweet potato, taro and yam. Potato, sweet potato and taro data are relatively better represented because they have received better research attention than the other RTs.

Performance of individual crops

Enset (Ensete ventricosum)

The work of Ehret (1979) suggests that the Omotic people of southern and southwestern Ethiopia domesticated *enset* some 8 millennia ago. This unique, indigenous crop aptly nicknamed 'werkie' ወርቄ (literally meaning 'my gold') because of its resilience during drought conditions (Pankhurst, 1966), by none other than the Emperor Menelik II himself, accounts for the largest proportion of the total RTs cultivated in Ethiopia and plays a

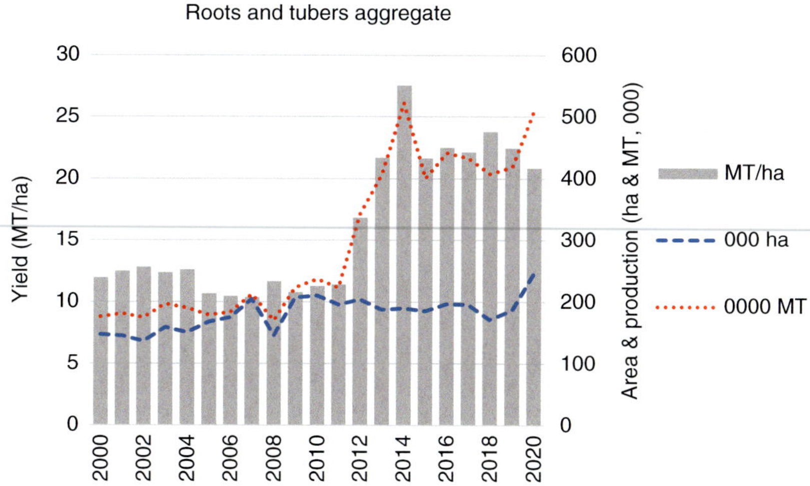

Fig. 4.1. Indicators for aggregate root and tuber crops in Ethiopia. Source: constructed by the lead author from FAO (2022a).

key role in the food and nutrition security of the country. In spite of its huge socioeconomic and environmental significance, *enset* is among the least understood crops in this country. The work of Sahle *et al.* (2018), attempting to determine the demand and supply of *kocho* in Gurage Zone, is perhaps the only quantitative survey available to date.

The statistics on this crop are either lacking or grossly misrepresented. The CSA sample surveys provide indicators (number of holders, area, yield and production) on *enset* only for the 2008/09 to 2011/12 *meher* seasons. According to data for the latter, more than 4.3 million smallholders grew *enset* on about 312,000 ha and produced approximately 729,000 MT of *kocho*; the yield was estimated at 2.3 MT/ha.

Sample surveys for the *meher* seasons between 2012/13 and 2020/21 give number of trees harvested (and yields per plant for *amicho*, *kocho* and *bulla*) instead of the standard indicators that are useful for planning purposes – i.e. number of holders, area, production and yield per ha. The proportion of *amicho* is highly exaggerated in these sample surveys.

Our calculations based on these data show that *kocho*, *amicho* and *bulla* account for roughly 52%, 46% and 2%, respectively, of the total production over the nine years between 2012/13 and 2020/21. As a matter of fact, *amicho* should not even count as a major portion of *enset* product because people eat it only if they are desperately poor and only under extremely desperate situations such as extended drought. In some Ethiopian cultures, it is eaten only during special commemorations. We conclude that the values for *amicho* are inflated.

With about 9% of the total dietary calorie supply in the Ethiopian diet, *enset* is the sixth largest source after maize, wheat, RTs dry equivalent, *teff* and sorghum in Ethiopia (out of a total of 69 items that constituted the total calorie intake). For example, a comparison of the 2009–2013 versus 1969–1973 averages show a 3.66-fold increase for *enset*, in contrast with 2.19-fold for maize, 1.56-fold for wheat, 1.12-fold for sorghum and 0.80-fold decrease for *teff*.

Historically, *enset* was associated with the Southern Nations, Nationalities and Peoples Region (SNNPR; including southwestern Ethiopia and Sidama regional states), but it is currently widespread in many parts of the Oromia Regional State and, to some extent, in Gambella and Benshangul-Gumuz as a major staple or co-staple. Our experience indicates that *enset* expansion in Oromia started some six decades ago in the Kofele area of West Arsi (Savard, 1973). *Enset*-growing areas in these regions and zones cover more than 39.6 million people (Annex Table 4.2), which translates to about 42% of the total population of Ethiopia. Borrell *et al.* (2019a,b) and many authors before them suggest that *enset* provides the staple diet for approximately 20 million people. This figure grossly underestimates the number of people who depend on the crop as a staple or co-staple.

The agricultural sample surveys of the CSA provide the standard indicators (area, yield and production) for *enset* only for four years – 2008/09 to 2011/12. FAO (2020) does not mention *enset* as such; it gives 'Roots and tubers, nes' (representing *enset*) through 1992 where the total area was estimated at 370,000 ha.

Considering what is known about *enset* yield in Ethiopia from the literature, the CSA report of 2.3 MT/ha is a gross underestimation. For example, Bezuneh (1996) reported that *enset* yield ranged between 37 MT/ha and 67 MT/ha, depending on the variety. Hibsch (1996) reviewed the available data on *enset* production in Ethiopia and reported that the average fresh weight yield was 49 MT/ha while the dry weight was 21.6 MT/ha (which translated into 4.4 MT/ha/yr) or 4.4 MT/ha/yr. Tabogie *et al.* (1994) reported a mean fresh *kocho* yield of about 57 MT/ha, taking 52.8% moisture for fermented *kocho*. Tsegaye and Struik (2001) demonstrated that (fresh) *kocho* yields were 19, 33 and 26 MT/ha per season, respectively, for *enset* plants transplanted once, twice and three times. The FAO (2022a) does not provide data from 1993 onwards; the estimates for 1992 were: area, 370,000 ha; yield, 7.03 MT/ha; and production, 2.6 million MT.

Circumstantial evidence suggests that *enset* yield has been declining over the past two decades. We attribute this decline to constraints associated with improper crop management, mainly the decline in organic matter content of the soil and crop diversity, even though much blame has been assigned to diseases, particularly the bacterial wilt of *enset*, caused by *Xanthomonas campestris* pv. *musacearum* (Merga *et al.*, 2019), or many insect pests (Abate, 1988), including the *enset* mealybug *Cataenococcus ensete* (Addis *et al.*, 2008; Azerefegne *et al.*, 2009).

The fact remains that, in general, *enset* gives higher dry matter yield than any other staples in Ethiopia; its energy yield is higher than all cereals, potato, sweet potato and banana; similarly, its protein yield is also higher than all these crops and cassava (Tsegaye and Struik, 2001; Tsegaye, 2002). *Enset* is deficient only in vitamin A (Atlabachew and Chandravanshi, 2008; Bosha *et al.*, 2016; Daba and Shigeta, 2016). According to Yemataw (2018), *enset* is the best source of carbohydrates, ash and fibre, but has less proteins and fat content. Minerals such as calcium and potassium are abundant in *enset* products. Although the protein content is limited, *enset*-based diets possess good concentrations of essential amino acids such as lysine, histidine and isoleucine. *Enset* products have moderate antioxidant properties compared to other staples. Data related to vitamin profiles of *enset*-based foods are limited. The work of Tamrat *et al.* (2020) demonstrates that *enset* is superior to all staples of Ethiopia in terms of its Ca, Fe, K, Mg and Zn content – all essential nutrients for human metabolism.

Enset has other advantages too: it can be stored for a long period (many years); it is readily available throughout the year; it has many by-products (e.g. Tabogie *et al.* [1994] reported 443 kg/ha yield of *enset* fibre); and the prolonged presence of its closed canopy has an ecological advantage similar to that of forest, including carbon dioxide sequestration to reduce global climate change. These attributes make *enset* the crop of choice for sustainable food security, especially where farm size is diminishing because of increasing population – like the mid and highland areas of Ethiopia.

In spite of its huge importance and potential (Demeke, 1986), *enset* R&D has not been given the commensurate attention it deserves. For example, the first variety (clone) release ever of *enset* was three clones – Gewada, Yanbule and Endale in 2009. Another set of three clones – Kelisa, Mesena and Zerietta – were released in 2010. Even though these have been reported as being under production (MOALR, 2017), it is not clear how many of those clones are being grown by farmers and where. An annotated bibliography on *enset* can be found in Yemataw *et al.* (2022).

Potato

Potato was first introduced to Ethiopia in 1858 by a German botanist named Schimper (Pankhurst, 1964). It has since become a well-known crop and widely grown in many parts of this country. According to a potato factsheet for policy makers (Lemaga, 2018), potato is an important food security and cash crop in Ethiopia. It is a nutritious, high-value, short production cycle crop and a major food in the highlands during the lean months before the harvest of grain crops. Potato produces more food per unit time, area and other resources than most major crops. In addition to carbohydrates, potato is rich in several micronutrients, particularly vitamins C, B1, B3 and B6 and minerals such as K, P and Mg. For every cubic metre of water applied, the potato produces 5600 calories of dietary energy, which is significant compared to 3860, 2300 and 2000 produced by maize, wheat and rice, respectively. It also yields 150 g of protein, twice that of wheat and maize for the same cubic metre of water. Moreover, the biological quality of protein in potato is excellent, making it a good weaning food for infants of low-income people.

Since potato is a short-cycle crop, its production fits in a wide range of farming systems, including double cropping and intercropping with crops that are planted in rows such as *enset*, coffee and others. It gives more than 25 MT/ha in just 90 to 120 days and is best suited to addressing gender equality and job creation for the youth along the production to consumption continuum.

More than 1 million smallholders grew potato annually in Ethiopia in the *meher* season (2016–2020 average). According to the FAO (2022a), the area, yield and production were estimated at ~73,000 ha, nearly 11 MT/ha and 1 million MT, respectively, during the 2016–2020 period. According to the CSA, the *belg* season accounted for 71% of the total area and 50% of production (average, for 2013/14 and 2015/16). This means that the average area harvested and production for these two seasons were 238,000 ha and 1,743,000 MT (Annex Table 4.1); the yield here was estimated at about 13 MT/ha. Potato in Ethiopia is mainly produced in Oromia (50%), SNNPR (25%), Amhara (24%) and Tigray (1%).

Ethiopia is ranked 37th (in area) out of 155 countries that grew potato in 2016–2020 (considering only *meher* season) whereas it would rank approximately 12th if the *belg* season area is included. China, India, Ukraine, the Russian Federation, Bangladesh, the USA, Peru, Nigeria, Poland and Belarus are the top ten potato-growing countries in the world.

Figure 4.2 presents indicator trends for potato in Ethiopia. The yield (and therefore production) declined between 2005 and 2011, but increased starting from 2012, reaching a peak in 2015 and showing slight decline thereafter. The area, yield and production increased at an annual rate of 2.47%, 3.17% and 5.72%, respectively, between 2000 and 2020. The increases in yield accounted for roughly 55% of the change in production whereas area expansion contributed approximately 43%, with the interaction effect between the two covering the balance (<2%). As a nationally coordinated and high priority crop, potato benefited from extensive promotion through working with farmers and conducting field days (Woldegiorgis *et al.*, 2006). The yield growth rate for Ethiopia is the fastest among the top ten countries.

Moreover, the research system has released over 35 high-yielding varieties with resistance to late blight and suitable to different agroecologies and to different processed products. For example, Sun Chips, the company that produces the most famous processed product (potato crisps), mainly uses the variety Gudenie that was released in Ethiopia in 2006. At least three other companies, including Midroc-Ethiopia, have now started producing potato crisps. Several of the released varieties are also suitable for boiling and *wot* (stew) making.

The yield gain was roughly 439 kg/ha/yr and highly significant ($P<0.01$, $R^2=0.69$). This was the sixth highest gain (following Poland, Bangladesh, the USA, Belarus and the Russian Federation) among the top ten potato-growing countries of the world (Table 4.2). Countries with high yields among the top ten include the USA and Poland.

Ethiopia's average yield for potato (~13 MT/ha) is much lower than the world average of nearly 21 MT/ha. However, this country has the potential to reach the world average yield and surpass it if clean seed potatoes of high-yielding varieties are grown under optimal agronomic practices (Lemaga *et al.*, 1994; Woldegiorgis *et al.*, 2015), major potato diseases are controlled (Kassa and Bekele, 2008), and farmers' participation in all these is ensured. Among the potato diseases, potato late blight and bacterial wilt are the major constraints (Damtew *et al.*, 2018), with the latter caused by *Ralstonia solanacearum* being the single most important threat to the potato industry in the country (Gorfu *et al.*, 2013; Tafesse *et al.*, 2018). In Ethiopia, the bacterium has spread to the highland areas of 3000 m above sea level with the movement of latently infected seed (Abdurahman *et al.*, 2017) infesting the fields and seriously reducing potato production.

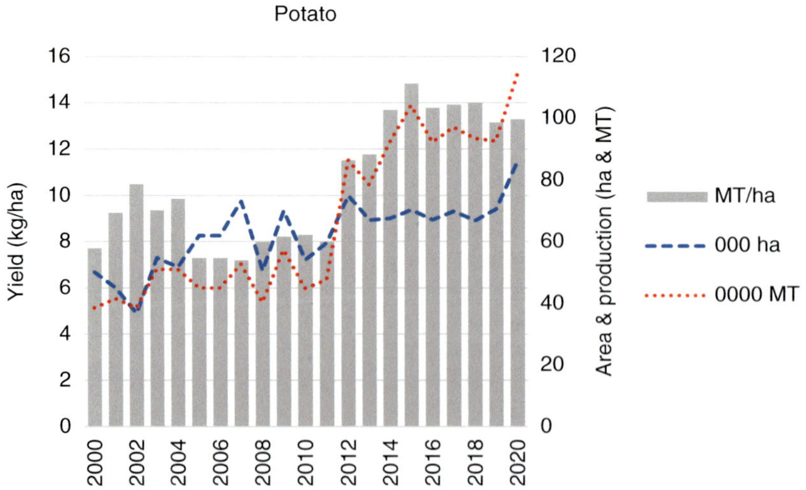

Fig. 4.2. Indicator trends for potato in Ethiopia. Source: constructed by the lead author from FAO (2022a).

Table 4.2. Yield gain (b-value, kg/ha/yr) and annual rate of growth (ROG) for potato in Ethiopia compared with the top ten cultivating countries of the world (2000–2020). Source: calculated and constructed by the lead author from FAO (2022a).

	Yield gain			ROG (%)		
Country[a]	b-value	P > ltl	R^2	Area	Yield	Production
Poland	590	<0.01	0.60	–7.91	2.61	–4.76
Bangladesh	475	<0.01	0.91	3.94	2.92	6.98
USA	473	<0.01	0.91	–1.34	1.04	–0.31
Belarus	384	<0.01	0.59	–4.52	2.23	–2.39
Russian Federation	343	<0.01	0.82	–3.71	2.54	–1.27
Ethiopia	336	<0.01	0.59	2.47	3.17	5.72
India	314	<0.01	0.71	3.02	1.53	4.60
Ukraine	306	<0.01	0.76	–1.15	2.23	1.05
China	246	<0.01	0.87	–0.06	1.52	1.46
Peru	246	<0.01	0.97	1.47	1.80	3.30
Nigeria	57	0.01	0.53	2.23	1.77	4.04

[a]Countries with largest areas (descending order) are China, India, Ukraine, Russian Federation, Bangladesh, USA, Peru, Nigeria, Poland and Belarus; Ethiopia is 37th.

The effect of the bacterium in Ethiopian highlands is further exacerbated by soil acidification. It is estimated that soil acidity affects about 40% of the area in Ethiopian highlands where potato is produced, making lime application essential to increase potato yield (Tafesse *et al.*, 2020). To reduce the inoculum or eliminate it from infested soils, it is important to practice long-term crop rotation with non-host crops such as wheat and maize (Lemaga *et al.*, 2001a) or *Crotalaria falcata* (Kakuhenzire *et al.*, 2013), and the use of soil amendments (Lemaga *et al.*, 2001b). With these practices, it is possible to get acceptable yields of ware potatoes even in the presence of potato bacterial wilt, but the tubers should not be used for seed as they maybe latently infected. According to Kakuhenzire *et al.* (2013), the bacterium can be completely eliminated with inclusion of *C. falcata* in long-term crop rotation. It is, however, recommended to use clean seed, fallowing, crop rotation with non-host crops and applying soil amendments for best results.

One of the most important constraints to potato production in Ethiopia is a critical shortage of quality seed of improved varieties. Attempts have been made to produce quality seed potatoes through a mix of formal and informal systems. The formal system produces clean early generation seed (EGS) using different rapid multiplication techniques in tissue culture, aeroponics, green houses and screen houses at research centres. Through the informal system, EGS is supplied to producer cooperatives and a few commercial seed producers for large-scale multiplication of quality-declared seed (QDS), which is sold to final ware potato producers. The production of EGS, however, lacks regular testing for diseases and production of QDS does not follow the field inspection guidelines issued by the Ministry of Agriculture in 2016. As a result, the QDS delivered to the producer does not possess the desired quality. Although the approach has enabled an increase in potato yields for farmers, its impact has been far below expected. In order for the crop to considerably improve the lives of potato farmers and contribute to the growth of the national economy, research centres and selected private partners have to be supported to produce large quantities of disease-free EGS. Similarly, QDS production fields have to be regularly inspected by trained inspectors, and producers need to be educated on seed potato production and management practices. Furthermore, a mechanism should be put in place to randomly test QDS for freedom from bacterial wilt and important virus diseases to avoid transmission of devastating diseases with seed. Finally, it is important to forge sustainable market linkages among seed and ware potato producers to make the system viable.

Sweet potato

Sweet potato is widely grown in southern and eastern parts of Ethiopia as a co-staple with cereals, such as maize and sorghum. Approximately

645,000 smallholder households cultivated 53,000 ha of land and produced >1.7 million MT annually in 2016–2020 (cf. Table 4.1). The area covered by sweet potato per holder was roughly 0.0817 ha and each holder produced nearly 2700 kg per season; the research system released 26 improved varieties as of 2017. Of these, nine are orange-fleshed sweet potato (OFSP) varieties, while the rest are white-fleshed sweet potato (WFSP) varieties. According to the latest FAO data, the country cultivated 57,000 ha sweet potato and produced nearly 1.8 million MT of tubers annually in 2016–2020.

Of the nine OFSP varieties, four (Kulfo, Alamura, Dilla and Kabode [a variety released in Uganda and registered in Ethiopia]) are more widely grown, mainly in SNNPR, with a rapid expansion to Amhara and Oromia regions. While WFSP and OFSP produce comparable yields, the latter is more nutritious. In addition to providing carbohydrates and comparatively high levels of protein (~3% on a fresh weight basis), calcium and antioxidants, OFSP provides high levels of vitamins A, B (thiamine, niacin, and pyridoxine), C, E and K. Both WFSP and OFSP produce more edible energy per hectare per day than other common sources of carbohydrates such as rice and maize. Moreover, OFSP leaves and vines are consumed as a vegetable, especially in SNNPR. The leaves of OFSP are an expedient substitute to *habesha gomen* (Ethiopian kale), which is a very common vegetable because it is usually available during the dry season where the former is heavily attacked by the cabbage aphid (*Brevicoryne brassicae*).

The *meher* season occupies 76% of the total area but accounts for 93% of production (cf. Annex Table 4.1). The SNNPR (60%), Oromia (37%) and Benshangul-Gumuz (2%) are the major growing regions of sweet potato in Ethiopia, with Harari, Gambella and Dire Dawa covering the balance.

Ethiopia is ranked 21st in area (China, Nigeria, Tanzania, Uganda and Malawi are the top five) and 6th in production among 111 countries and territories in the world that cultivated sweet potato in 2016–2020, according to the FAO (2022a).

Figure 4.3 depicts indicator trends for sweet potato in this country. Yield and production did not change much between 2000 and 2011, but it increased starting in 2012, with a peak in 2014. The area has increased slowly throughout the 21 seasons.

The yield gain for sweet potato was estimated at 1536 kg/ha/yr and was highly significant ($P<0.01$, $R^2=0.65$), as presented in Table 4.3. Poland, Bangladesh, the USA, Belarus and the Russian Federation were a distant second to sixth. Nigeria experienced the highest yield decline, which was highly significant; Uganda also showed a negative but non-significant yield gain.

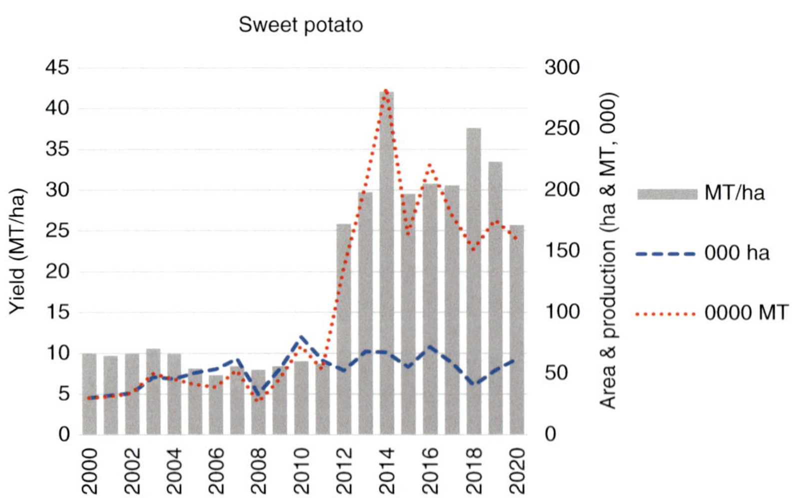

Fig. 4.3. Indicator trends for sweet potato in Ethiopia. Source: constructed by the lead author from FAO (2022a).

The area, yield and production for Ethiopia increased at an annual rate of 2.64%, 8.82% and 11.69%, respectively. The ROG for yield here was the highest among the top ten countries and was followed by Tanzania, Angola and Madagascar (Table 4.3). Similarly, the national average yield for Ethiopia was the highest among the top 20, followed by the USA, China, Malawi and Indonesia. This is a great improvement over what was reported by Belehu and Legesse (1994).

The single most important constraint to the production of sweet potato in Ethiopia is probably the critical shortage of quality planting material. Research centres and the private sector have to be encouraged to produce clean planting material in large quantities using different rapid multiplication techniques in tissue culture, net houses, greenhouses and screen houses to supply massive amounts of quality-declared planting material (QDPM) to producer cooperatives and commercial producers following the field inspection guidelines issued by the Ministry of Agriculture in 2016. Moreover, it is recommended to produce QDPM close to the production areas, as planting material is damaged during long-distance transportation.

Taro

Smallholders in the humid lower and mid-altitudes (1200–2000 m above sea level) of the southern and southwestern parts of Ethiopia have grown taro since ancient times. This country is reported to have high cultivar diversity of taro (Fujimoto, 2009). There are two major types of taro cultivated in Ethiopia. One grows in well-drained soils and the other is a marsh type grown under wetter conditions. The latter is minor in area and production but has larger corms and higher yields. It is more popular among consumers in the supermarkets in Addis Ababa and fetches a higher price. Both can be grown either as garden or field crops. Annex Table 4.1 shows that the *meher* season accounts for 90% of the area and 98% of production.

Among the most important characteristics of taro are its high yield per unit area and ease of processing (Fanta and Neela, 2019) and consumption, earning the nickname የምድር ቅቤ 'butter from the ground' because it can be boiled and eaten without the need for relish or supplements such as milk and milk products required for *kocho* or cereal products. Taro in Ethiopia is cultivated exclusively for domestic consumption.

It is estimated that the SNNPR and Oromia regions accounted for approximately 86% and 14%, respectively, of the total area planted to this crop in the 2018/19 *meher* season. The Gambella regional state grows very small amounts. Fujimoto (2009) observed high cultivar diversity in SNNPR and expressed concerns

Table 4.3. Yield gain (b-value, kg/ha/yr) and annual rate of growth (ROG) for sweet potato in Ethiopia compared with the top 20 cultivating countries of the world (2000–2020). Source: calculated and constructed by the authors from FAO (2022a).

	Yield gain			ROG (%)		
Country[a]	b-value	P > ltl	R^2	Area	Yield	Production
Ethiopia	1563	<0.01	0.65	2.64	8.82	11.69
Malawi	381	<0.01	0.57	NA	NA	NA
Angola	369	<0.01	0.76	3.59	5.72	9.52
Tanzania	269	<0.01	0.74	3.75	6.93	10.94
Madagascar	167	<0.01	0.81	1.91	2.51	4.47
India	136	<0.01	0.64	–0.39	1.38	0.99
Rwanda	72	0.07	0.16	–0.31	1.08	0.77
China	45	0.03	0.22	–4.96	0.21	–4.76
Papua New Guinea	17	<0.01	0.75	1.73	0.34	2.07
Uganda	–12	0.05	0.19	–2.48	–0.29	–2.76
Nigeria	–33	0.01	0.45	3.59	–1.16	2.39

[a]Countries with the largest areas (descending order) are China, Nigeria, Tanzania, Uganda, Malawi, Angola, Rwanda, Papua New Guinea, Madagascar and India; Ethiopia is 21st.
NA = not available.

about the possibility of biodiversity loss due to replacement by early maturing new varieties.

More than 2 million smallholders cultivated 65,000 ha and produced >1.5 million MT of taro annually in the 2016–2020 *meher* seasons (cf. Table 4.1). The national average yield was estimated at 25 MT/ha. The average landholding per family is about 0.0291 ha and the per capita production is approximately 740 kg. Only three improved clones ('varieties') have been released as of 2017, but there is no information on how widely these clones have been adopted by farmers.

According to the FAO, Ethiopia ranked sixth (after Nigeria, Cameroon, Ghana, China and Cote d'Ivoire) in area and fifth (following Nigeria, China, Cameroon and Ghana) in production out of 49 countries in the world that cultivated taro in 2016–2020.

Yams

Half a dozen species of yams are known to be cultivated in Ethiopia. These are: *D. alata, D. abyssinica, D. bulbifera, D. cayensis, D. rotundata* and *D. schimperiana*. Yams are almost exclusively grown in the SNNPR, where there is a high degree of genetic diversity (Oli *et al.*, 2008; Tilahun, 2009; Akassa *et al.*, 2015; Garedew *et al.*, 2017; Ousmael *et al.*, 2019), with only insignificant amounts in Gambella. Each household plants one to six landraces (average = 2.9) and the mean area ranges from 0.08 to 0.18 ha.

An estimated 365,000 smallholders in Ethiopia grew yams on 5000 ha and produced roughly 46,000 MT of tubers annually in 2016–2020; each holder planted about 0.0138 ha and produced 127 kg each season, with the national average yield of about 9 MT/ha (cf. Table 4.1). The research system has released only three modern yam varieties between 1970 and 2017. The area and production declined fast from 2013 onwards. Overall, the area and production declined at an annual rate of 11.79% and 11.37%, respectively, while the yield increased at a rate of 0.48%.

Ethiopia ranked 24th in area and 26th in production out of 56 countries in the world that grew yams in 2016–2020 (FAO, 2022a). The largest producers include Nigeria, Ghana, Cote d'Ivoire, Benin and Togo. Even though the area for yams declined so dramatically in Ethiopia, it expanded globally. For example, the total area harvested increased from about 1.2 million ha in 1961 to nearly 8.8 million in 2020; the area increased at an annual rate of 4.81% between 2000 and 2020.

The yield gain for yams in Ethiopia was 63 kg/ha/yr, but it was non-significant. Ghana, Benin, Cameroon, the Central African Republic and Haiti achieved highly significant yield gains. By contrast, Cuba, Cote d'Ivoire, Nigeria, Togo, Guinea, Gabon and Sudan suffered significant yield losses; the yield gain for South Sudan was also negative but non-significant.

The national average yield of 9 MT/ha for Ethiopia is below the world average of about 10 MT/ha. Countries that obtain higher yields include Japan, Ghana, Papua New Guinea, Jamaica and Benin – averaging about 23, 18, 17, 17 and 14 MT/ha, respectively (FAO, 2022a).

Discussion and Conclusions

Roots and tubers are among the least understood crops in Ethiopia. In this chapter, we have tried to assemble as much information as possible, conducted rigorous analyses and presented the data with the main purpose of helping priority setting and policy decisions. At least nine species of cultivated RTs are grown in Ethiopia, but the latest available reports from the CSA and FAO provide production data (or just mention) only for five of them – *enset*, potato, taro, sweet potato and yam. The rest (including aerial yam, *anchote*, cassava and Ethiopian potato) are considered minor and no production data are available on them.

At present, the data on *enset* are either totally missing or grossly misrepresented. For example, the CSA sample survey (2012/12) reported the area at a little over 312,000 ha, yield at 2.3 MT/ha and production at ~729,000 MT. The most recent data from FAO (2022a) does not give *enset* data beyond 1992; the estimates for that year were: area (>312,000 ha), yield (7 MT/ha) and production (2.6 million MT).

Most of the data from the CSA and all from FAO are based on the *meher* season production, leaving out the *belg* season data, which is especially crucial for potato. The 2013/14 and 2015/16 average show that about 71% of area was grown and roughly 50% of the total production was in the latter season.

Sweet potato and potato experienced highly significant yield gains, whereas yams and taro were non-significant. This shows that progress can be made only when crops are taken as high priority and investment is made. Potato is a national priority crop, whereas sweet potato is given a high priority only in the SNNPR. Taro and yams are not considered high priority crops and minimal investment (if any) is made in them.

As a primary or secondary centre of genetic diversity for many of the RTs (Westphal, 1975), Ethiopia possesses a wide range of genetic diversity but so far it has not made adequate efforts to explore and exploit these resources to make significant contributions towards its food and nutrition security goals. We call upon policy makers to revisit the country's crop priorities and take immediate actions to substantially increase investment in *enset* and other traditional RT crops.

The immediate task after that would be to develop survey methods and then launch extensive surveys covering all *enset*-growing zones and special *woredas* of the country. This should be followed up with developing models for determining suitability maps for expansion of the crop across the nation.

References

Abate, T. (1988) Insect and mite pests of horticultural and miscellaneous plants in Ethiopia. *IAR Handbook No. 1*. IAR, Addis Ababa, Ethiopia.115 pp.

Abate, T., Shiferaw, B., Menkir, A., Wegary, D., Kebede, Y. *et al.* (2015) Factors that transformed maize productivity in Ethiopia. *Food Security* 7, 965–981. DOI: 10.1007/s12571-015-0488-z.

Abdurahman, A., Griffin, D., Elphinstone, J., Struik, P.C., Schulz, S. *et al.* (2017) Molecular characterization of *Ralstonia solanacearum* strains from Ethiopia and tracing potential source of bacterial wilt disease outbreak in seed potatoes. *Plant Pathology* 66, 826–834. DOI: 10.1111/ppa.12661.

Addis, T., Azerefegne, F., Blomme, G. (2008) Density and distribution of enset root mealybugs on enset. *African Crop Science Journal* 16(1), 67–74.

Akassa, B., Negera, A., Chemeda, G., Wandimu, T., Teshome, N. *et al.* (2015) Registration of 'Bulcha' and 'Lalo' yam varieties. *East African Journal of Sciences* 9(1), 65–68.

Atlabachew, M., Chandravanshi, B.S. (2008) Levels of major, minor and trace elements in commercially available enset (*Ensete ventricosum* (Welw.), Cheesman) food products (*kocho* and *bulla*) in Ethiopia. *Journal of Food Composition and Analysis* 21, 545–552.

Azerefegne, A., Addis, T., Lemawork, S., Tafesse, E., Gemu, M. *et al.* (2009) *An IPM Guide for Enset Root Mealybug (Cataenococcus ensete) in Enset Production*. Bioversity International, Rome, Italy.

Belehu, T., Legesse G. (1994) Agronomic studies on sweet potato. In: Edward Herath, Lemma Desalegne (eds) Horticulture Research and Development in Ethiopia, Proceedings of the Second National Horticultural Workshop, 1-3 December 1992, EIAR, Addis Ababa, Ethiopia.

Bezuneh, T. (1996) An overview of enset research and future technological needs for enhancing its production and utilization. In: Abate, T., Steven, A.B., Clifton H., Seifu G. (eds.) *Enset-Based Sustainable Agriculture in Ethiopia*. Institute of Agricultural Research, Addis Ababa, Ethiopia. 1–14.

Borrell, J.S., Biswas, M.K., Goodwin, M., Blomme, G., Schwarzacher, T. (2019a) Enset in Ethiopia: A poorly characterized but resilient starch staple. *Annals of Botany* 123(5), 747–766. DOI: 10.1093/aob/mcy214.

Borrell, J.S., Goodwin, M., Blomme, G., Jacobson, K., Wendawek, A.M. *et al.* (2019b) Enset-based agricultural systems in Ethiopia: A systematic review of production trends, agronomy, processing and the wider food security applications of a neglected banana relative. *Plants People Planet PPP Open Access* 2, 212–228. DOI: 10.1002/ppp3.10084.

Bosha, A., Lagibo, A., Tana, T., Mohammed, W., Tesfaye, B. *et al.* (2016) Nutritional and chemical properties of fermented food of wild and cultivated genotypes of *enset* (*Ensete ventricosum*). *Food Research International* 89, 806–811. DOI: 10.1016/j.foodres.2016.10.016.

CSA (Central Statistical Agency) (2005) Agricultural Sample Survey 2004/2005 (1997 E.C.), Report on Area and Production of Crops (Private Peasant Holdings, *Meher* Season), Volume I, Statistical Bulletin 331. CSA, Addis Ababa, Ethiopia.

CSA (2006) Agricultural Sample Survey 2005/2006 (1998 E.C.), Report on Area and Production of Crops (Private Peasant Holdings, *Meher* Season), Volume I, Statistical Bulletin 361. CSA, Addis Ababa, Ethiopia.

CSA (2007) Agricultural Sample Survey 2006/2007 (1999 E.C.), Report on Area and Production of Crops (Private Peasant Holdings, *Meher* Season), Volume I, Statistical Bulletin.388. CSA, Addis Ababa, Ethiopia.

CSA (2008) Agricultural Sample Survey 2007/2008 (2000 E.C.), Report on Area and Production of Crops (Private Peasant Holdings, *Meher* Season), Volume I, Statistical Bulletin 417. CSA, Addis Ababa, Ethiopia.

CSA (2009) Agricultural Sample Survey 2008/2009 (2001 E.C.), Report on Area and Production of Crops (Private Peasant Holdings, *Meher* Season), Volume I, Statistical Bulletin 446. CSA, Addis Ababa, Ethiopia.

CSA (2010) Agricultural Sample Survey 2009/2010 (2002 E.C.), Report on Area and Production of Major Crops (Private Peasant Holdings, *Meher* Season), Volume I, Statistical Bulletin 446. CSA, Addis Ababa, Ethiopia.

CSA (2011) Agricultural Sample Survey 2010/2011 (2003 E.C.), Report on Area and Production of Major Crops (Private Peasant Holdings, *Meher* Season), Volume I, Statistical Bulletin. CSA, Addis Ababa, Ethiopia.

CSA (2012) Agricultural Sample Survey 2011/2012 (2004 E.C.), Report on Area and Production of Major Crops (Private Peasant Holdings, *Meher* Season), Volume I, Statistical Bulletin. CSA, Addis Ababa, Ethiopia.

CSA (2013) Agricultural Sample Survey 2012/2013 (2005 E.C.), Report on Area and Production of Major Crops (Private Peasant Holdings, *Meher* Season), Volume I, Statistical Bulletin 532. CSA, Addis Ababa, Ethiopia.

CSA (2014) Agricultural Sample Survey 2013/2014 (2006 E.C.), Report on Area and Production of Major Crops (Private Peasant Holdings, *Meher* Season), Volume I, Statistical Bulletin. CSA, Addis Ababa, Ethiopia.

CSA (2015) Agricultural Sample Survey 2014/2015 (2007 E.C.), Report on Area and Production of Major Crops (Private Peasant Holdings, *Meher* Season), Volume I, Statistical Bulletin 578. CSA, Addis Ababa, Ethiopia.

CSA (2016) Agricultural Sample Survey 2015/2016 (2008 E.C.), Report on Area and Production of Major Crops (Private Peasant Holdings, *Meher* Season), Volume I, Statistical Bulletin. CSA, Addis Ababa, Ethiopia.

CSA (2017) Agricultural Sample Survey 2016/17 (2009 E.C.), Report on Area and Production of Major Crops (Private Peasant Holdings, *Meher* Season), Volume I, Statistical Bulletin. CSA, Addis Ababa, Ethiopia.

CSA (2018) Agricultural Sample Survey 2017/18 (2010 E.C.), Report on Area and Production of Major Crops (Private Peasant Holdings, *Meher* Season), Volume I, Statistical Bulletin 586 CSA, Addis Ababa, Ethiopia.

CSA (2019) Agricultural Sample Survey 2018/19 (2011 E.C.), Report on Area and Production of Major Crops (Private Peasant Holdings, *Meher* Season), Volume I, Statistical Bulletin 589. CSA, Addis Ababa, Ethiopia.

CSA (2020) Agricultural Sample Survey 2019/20 (2012 E.C.), Report on Area and Production of Major Crops (Private Peasant Holdings, *Meher* Season), Volume I, Statistical Bulletin 587. CSA, Addis Ababa, Ethiopia

CSA (2021) Agricultural Sample Survey 2020/21 (2013 E.C.), Report on Area and Production of Major Crops (Private Peasant Holdings, *Meher* Season), Volume I, Statistical Bulletin. CSA, Addis Ababa, Ethiopia.

Daba, T., Shigeta, M. (2016) Enset (*Ensete ventricosum*) production in Ethiopia: its nutritional and socio-cultural values. *Agriculture and Food Sciences Research* 3(2), 66–74, Available at: http://www.asianonlinejournals.com/index.php/AESR.

Damtew, E., Tafesse, S., Lie, R., van Mierlo Lemaga, B., Sharma, B.K. *et al.* (2018) Diagnosis of management of bacterial wilt and late blight in potato in Ethiopia: A systems thinking perspective. *NJAS-Wageningen Journal of Life Sciences* 86, 12–24. DOI: 10.1016/J.NJAS.2018.03.003.

Demeke, T. (1986) Is Ethiopia's *Ensete ventricosum* crop her greatest potential food? *Agriculture International* 38, 362–365.

Ehret, C. (1979) On the antiquity of agriculture in Ethiopia. *Journal of African History* 20, 161–177.

FAO (Food and Agriculture Organization of the UN) (2020) Available at: http://www.fao.org/faostat/en/#home (accessed 30 July 2020).

FAO (2022a) Statistical databases and data-sets of the Food and Agriculture Organization of the United Nations. Available at: http://www.fao.org/faostat/en/#data (accessed 19-28 February 2022).

FAO. (2022b) Statistical databases and data-sets of the Food and Agriculture Organization of the United Nations. Available at: http://www.fao.org/faostat/en/#data (accessed 4 August 2022).

Fanta, S.W., Neela, S. (2019) A review on nutritional profile of the food from enset: A staple diet for more than 25 per cent population in Ethiopia. *Nutrition & Food Science* 40, 824–843. DOI: 10.1108/NFS-11-2018-0306.

Fujimoto, T. (2009) Taro (*Colocasia esculenta* [L.] Schott) cultivation in vertical wet-dry environments: Farmers' techniques and cultivar diversity in Southwestern Ethiopia. *Economic Botany* 63(2), 152–166.

Garedew, B., Haile, B., Ayiza, A. (2017) Distribution, diversity and potential production of yams (*Dioscorea* spp.) in Sheko district, southwest Ethiopia. *American Journal of Life Sciences* 5(3), 86–92. doi: 10.11648/j.ajls.20170503.12.

Gebre, E., Woldegiorigis, G.M., Lemaga, B. (2008) Potato seed management. In: Gebremedhin, W., Endale, G., Berga, L. (eds) *Root and Tuber Crops: The Untapped Resources*. Ethiopian Institute of Agricultural Research, Addis Ababa, Ethiopia, pp. 53–78.

Gorfu, D., Woldegiorgis, G.M., Kassa, B. (2013) Sustainable seed potato production in Ethiopia: From farm-saved to quality declared seed. In: Woldegiorigis, G.M., Schulz, S., Berihun, B. (eds.). *Seed Potato Tuber Production and Dissemination*. International Potato Center, Addis Ababa, Ethiopia, pp. 211–222. ISBN: 978-99944-53-87-x.

Hibsch, C. (1996) Yield of *Ensete ventricosum*—a concept. In: Tsedeke, A., Clifton, H., Steven, B., Seifu, G. (eds) *Enset-Based Sustainable Agriculture in Ethiopia*. IAR, Addis Ababa.

Kakuhenzire, R., Lemaga, B., Kashaija, I., Ortiz, O., Mateeka, B. (2013) Effect of *Crotalaria falcata* in crop rotation and fallowing on potato bacterial wilt incidence, disease severity and latent infection in tubers and field soil. *Biopesticides International* 9(2), 182–194.

Kassa, B., Bekele, E. (2008) Potato disease management. In: Gebremedhin, W., Endale, G., Berga, L. (eds) *Root and Tuber Crops: The Untapped Resources*. Ethiopian Institute of Agricultural Research, Addis Ababa, Ethiopia, pp. 79-85.

Lemaga, B. (2018) *Potato Factsheet for Policy Makers*. International Potato Center, Addis Ababa, Ethiopia. 2 pp.

Lemaga, B., Woldegiorgis, G.M., Jaleta, T., Tuku, B.T. (1994) Potato agronomy research in Ethiopia. In: Edward, H., Lemma, D. (eds) *Horticulture Research and Development in Ethiopia. Proceedings of the 2nd National Horticulture Workshop*. 1-3 December 1992, EIAR, Addis Ababa, Ethiopia, pp. 101–119.

Lemaga, B., Siriri, D., Ebanyat, P. (2001a) Effect of soil amendments on bacterial wilt incidence and yield of potatoes in southwestern Uganda. *African Crop Science Journal* 9, 267–278.

Lemaga, B., Kanzikwera, R., Kakuhenzire, R., Hakiza, J.J., Manzi, G. (2001) The effect of crop rotation on bacterial wilt incidence and potato tuber yield. *African Crop Science Journal* 9, 257–266.

Mekbib, Y., Deressa, T. (2016) Exploration and collection of root and tuber crops in East Wollega and Ilu Ababor. *Indian Journal of Traditional Knowledge* 15(1), 86–92.

Melaku, Y., Duguma, T. (2016) Proximate composition, phytochemical screening and antioxidant activities of the tubers of *Plecthrans edulis*. *International Research Journal of Biological Sciences* 5(12), 13–17.

Merga, I.F., Tripathi, L., Hvoslef-Eide, Endale, G. (2019) Application of genetic engineering for control of bacterial wilt disease of enset, Ethiopia's sustainability crop. *Frontiers in Plant Science* 10, 133. doi: 10.3389/fpls.2019.00133.

MOALR (Ministry of Agriculture and Livestock Resources) (2017) Crop Variety Register Issue No. 20. *Plant Variety Release, Protection and Seed Quality Control Directorate*. MOALR, Addis Ababa.

Oli, M.T., Becker, H.C., Maass, B.L. (2008) Diversity, distribution and management of yam landraces (*Dioscorea* spp.) in southern Ethiopia. *Genetic Resources and Crop Evolution* 55, 115–131. DOI 10.1007/s10722-007-9219-4.

Ousmael, K.M., Tesfaye, K., Hailesilassie, T. (2019) Genetic diversity assessment of yams (*Dioscorea* spp.) from Ethiopia using inter simple sequence repeat (ISSR) markers. *African Journal of Biotechnology* 18(30), 970–977. DOI: 10.5897/AJB2018.16446.

Pankhurst, R. (1964) Notes on the history of Ethiopian Agriculture. *Ethiopian Observer* 7, 210–240.

Pankhurst, R. (1966) The Great Ethiopian Famine of 1888-1892: A New Assessment: Part Two. *Journal of the History of Medicine and Allied Sciences* 21(3), 271–294. Oxford University Press.

Sahle, M., Yeshitela, K., Saito, O. (2018) Mapping the supply and demand of enset crop to improve food security in southern Ethiopia. *Agronomy for Sustainable Development* 38, 7.

Savard, G.C. (1973) The people of Ethiopia: Draft of text for the freshmen of Haile Sellassie I University. PhD thesis, Columbia University, New York, USA.

Schmidt, E., Thomas, T.S. (2018) *Cropland Expansion in Ethiopia: Economic and Climatic Conditions for Highland Agriculture*. ESSP Working Paper 127. IFPRI, Washington, DC.

Tafesse, S., Damtew, E., van Mierlo, B., Lie, R., Lemaga, B. *et al.* (2018) Farmers' knowledge and practices of potato disease management in Ethiopia. *NJAS - Wageningen Journal of Life Sciences* 86–87, 25–38. DOI: 10.1016/j.njas.2018.03.004.

Tabogie, E., Belehu, T., Diro, M., Legesse, G. (1994) Improvement studies on enset and sweet potato. In: Edward, H., Lemma, D. (eds) *Horticulture Research and Development in Ethiopia, Proceedings of the Second National Horticultural Workshop, 1–3 December 1992.* EIAR, Addis Ababa, Ethiopia.

Tafesse, S., Braam, C., Lemaga, B., van Mierlo, B., Struik, P.C. (2020) Association between soil acidity and bacterial wilt occurrence and genetic diversity of *Ralstonia solanacearum* strains in Ethiopia. *Agronomy* 2021 11(8), 1541. doi: 10.3390/agronomy11081541

Tamrat, S., Borrell, J.S., Biswas, M.K., Gashu, D., Wondimu, T. *et al.* (2020) Micronutrient composition and microbial community analysis across diverse landraces of the Ethiopian orphan crop enset. *Food Research International* 137, 109636.

Tilahun, A. (2009) Effect of processing on some physicochemical and anti-nutritional factors of taro (*Colocasia esculenta* (L.) Schott.) grown in Ethiopia. M.Sc thesis, School of Graduate Studies, Addis Ababa University, Addis Ababa.

Tsegaye, A. (2002) On indigenous production, genetic diversity and crop ecology of enset (*Ensete ventricosum* (Welw.) Cheesman). Ph.D. thesis, Wageningen University, The Netherlands, 105 pp. ISBN 90-5808-628-3.

Tsegaye, A., Struik, P.C. (2001) Enset (*Ensete ventricosum* (Welw.) Cheesman) *kocho* yield under different crop establishment methods as compared to yields of other carbohydrate-rich food crops. *Netherlands Journal of Agricultural Science* 49, 81–94.

Westphal, E. (1975) *Agricultural Systems in Ethiopia*. Agricultural University, Wageningen. Center for Agriculture Publishing and Documentation, Wageningen, the Netherlands, 278 pp.

Woldegiorgis, G., Solomon, A., Gebre, E., Tesfaye, A., Kassa, B. *et al.* (2006) Transforming the traditional potato production through improved technologies in the Central Highlands of Ethiopia. In: Tsedeke, A. (ed.) *Successes with Value Chain, Proceeding of Scaling up and Scaling out Agricultural Technologies in Ethiopia: An International Conference, 9-11 May 2006*. EIAR, Addis Ababa, Ethiopia.

Woldegiorgis, G.M., Gebre, E., Lemaga, B., Abebe, T. (2008) Potato agronomy. In: Gebremedhin, W., Endale, G., Berga, L. (eds) *Root and Tuber Crops: The Untapped Resources*. Ethiopian Institute of Agricultural Research, Addis Ababa, Ethiopia, pp. 33–51.

Woldegiorgis, G.M., Hailemariam, G., Lemaga, B., Schultz, S. (2015) Quality seed potato production: experiences from the highlands of Ethiopia. In: Jan, L., Moses, N., Sara, Q., Monica, P. (eds) *Potato and Sweet Potato in Africa. Transforming the Value Chains for Food and Nutrition Security*. CAB International, pp. 186–198. ISBN: 978-1-78064-4020-2.

Yemata, G. (2020) *Ensete ventricosum*: A multipurpose crop against hunger in Ethiopia. *Hindawi* 2020, 6431849, 10 pages. DOI:10.1155/2020/6431849.

Yemataw, Z. (2018) Agro-morphological and Molecular Characterization of Enset (*Ensete ventricosum* (Welw.) Cheesman) Accessions from Ethiopia. PhD thesis, Addis Ababa University, Addis Ababa, Ethiopia. 283 pp.

Yemataw, Z., Fikre, H., Muzemil, S., Borrell, J., Magule, T. (2022) Annotated bibliography of enset (*Ensete ventricosum* (Welw.) Cheesman) 1790 to 2021. Eds Getinet, A., Agdew, B., Fekadu, G. Southern Agricultural Research Institute, Hawassa, Ethiopia. 92 pp.

Annex Table 4.1. Area occupied and production of some root and tuber crops smallholders cultivated in the *meher* and *belg* seasons in Ethiopia[a]. Source: constructed by the authors from CSA (www.csa.gov.et).

	Area (000 ha)					Production (000 MT)				
Crop	*Meher*	*Belg*	Total	% *meher*	% *belg*	*Meher*	*Belg*	Total	% *meher*	% *belg*
Potato	68	169	238	29	71	864	879	1743	50	50
Taro	46	5	51	90	10	1202	23	1226	98	2
Sweet potato	47	15	62	76	24	1578	127	1705	93	7
Yam	4	2	7	66	34	52	13	65	80	20
Total	166	191	357	46	54	3696	1042	4739	78	22

[a]Data are averages for 2013/14 and 2015/16.

Annex Table 4.2. Area, production, yield, holders, total population and area per holder in the *enset*-growing regions, zones and special *weredas* in Ethiopia (2011/12 *meher* season).

Region	Zone	Area (ha)	Production (MT)	Yield (MT/ha)	000 Holders	000 People	ha/holder
SNNPR	Sidama	76,305	288,851	3.785	797	3,668	0.0958
SNNPR	Gurage	24,390	10,232	0.420	245	1,635	0.0998
SNNPR	Keffa	20,670	83,289	4.029	187	1,102	0.1103
SNNPR	Hadiya	18,219	28,949	1.589	245	1,591	0.0742
SNNPR	Gedeo	17,388	33,321	1.916	185	1,109	0.0941
SNNPR	Gamo-Gofa	11,589	14,750	1.273	285	2,044	0.0406
SNNPR	Kembata-Tembaro	8,147	10,703	1.314	121	902	0.0671
SNNPR	Dawro	8,031	14,545	1.811	70	615	0.1153
SNNPR	Wolayita	7,643	10,042	1.314	269	1,948	0.0284
SNNPR	Siltie	5,587	2,344	0.420	85	937	0.0660
SNNPR	South Omo	4,211	6,749	1.603	61	723	0.0690
SNNPR	Amaro SW	4,035	8,387	2.078	38	186	0.1069
SNNPR	Sheka	3,975	16,016	4.030	37	269	0.1075
SNNPR	Bench-Maji	2,322	5,206	2.242	78	847	0.0298
SNNPR	Yem SW	1,755	1,853	1.056	19	104	0.0929
SNNPR	Konta SW	1,594	3,313	2.078	28	116	0.0578
SNNPR	Basketo SW	861	1,558	1.811	16	72	0.0528
SNNPR	Burji SW	580	494	0.852	4	72	0.1522
SNNPR	Derashe SW	351	635	1.811	10	182	0.0341
SNNPR	Alaba SW	73	96	1.314	12	302	0.0061
Others	Others	1,690	4,526	NA	35	NA	0.0479
SNNPR total/avg.		219,414	54,586	1.837	3	18	0.0737
Oromiya	Guji	32,949	69,354	2.105	155	1,812	0.2123
Oromiya	West Arsi	20,701	53,975	2.607	186	2,591	0.1116
Oromiya	Jimma	15,692	32,369	2.063	485	3,209	0.0323
Oromiya	South West Shewa	10,081	7,673	0.761	163	1,452	0.0620
Oromiya	Borana	6,202	13,055	2.105	58	1,253	0.1066
Oromiya	W Shewa	3,660	3,076	0.841	174	2,701	0.0210
Oromiya	Illubabor	2,276	1,916	0.842	156	1,660	0.0146
Oromiya	Kelem	536	408	0.761	56	1,041	0.0096
Oromiya	Arsi	162	421	2.607	17	3,459	0.0095

Region	Zone	Area (ha)	Production (MT)	Yield (MT/ha)	000 Holders	000 People	ha/holder
Oromiya	West Welega	109	83	0.761	19	1,768	0.0059
Oromia total/avg.		92,368	18,233	1.545	1	21	0.0585
Gambella	Mezhenger	383	677	1.769	9	79	0.0434
BSG	Mao	7	NA	NA	1	65	0.0046
Ethiopia total/avg.		312,172	728,869	1.741	4,305	39	0.0725

BSG = Benshangul-Gumuz; NA = not available; SNNPR, Southern Nations, Nationalities and Peoples Region; SW = special *wereda*.

5 Increased Productivity and Socioeconomic Benefits of Grain Legumes

Tsedeke Abate,[1]* Girma Tesfahun Kassie,[2] Seid Ahmed Kemal,[2] and Asnake Fikre[3]

[1]*Homegrown Vision, Addis Ababa, Ethiopia;* [2]*ICARDA-Morocco, Rabat, Morocco;* [3]*Ethiopian Institute of Agricultural Research, Debre-Zeit Research Center, Bishoftu, Ethiopia*

Abstract

This chapter provides a critical assessment of productivity gains of the major grain legumes cultivated in Ethiopia and their contribution to export earnings for the country. Ethiopian farmers grow over 20 species of cultivated grain legumes for different purposes in different agroecologies. Approximately 8.5 million smallholders grew grain legumes on ~1.5 million ha and produced ~2.8 million MT of grain annually during 2016–2020. Ethiopia doubled its aggregate grain legumes productivity within two decades between 2000 and 2020. The yield gain for aggregate grain legumes was 56 kg/ha/yr and was highly significant ($P<0.01$, $R^2=0.97$), the highest among the top ten producing countries of the world. Ethiopia's high-priority grain legumes (chickpea and common bean, in particular) production system is transitioning from subsistence to small-scale commercial agriculture. The successes are attributed to several factors, including the research system following a systems approach, suitable technologies, aggressive promotion and availability of market access.

Introduction

Ethiopia is well known for its diversity of grain legumes, otherwise known as pulses. Available literature suggests that there are at least 19 species of cultivated grain legumes (groundnut and soybean treated under oilseeds, and fenugreek under spices, more appropriately) in this country (Westphal, 1974, 1975; Edwards *et al.*, 2019). The Ministry of Agriculture and Livestock Resources (MOALR, 2017) shows a dozen species on which research is conducted and several cultivars have been released.

The Central Statistical Agency (CSA) collects and publishes production data for smallholder producers on a yearly basis for eight of the known grain legume species mentioned by the above two sources. These are faba bean (*Vicia faba*), common bean (*Phaseolus vulgaris*), fieldpea (*Pisum sativum*) [here, *dekeko* (*Pisum sativum* var. *abyssinicum*) is considered as a separate species, which it should not be], chickpea (*Cicer arietinum*), lentil (*Lens culinaris*), grasspea (*Lathyrus sativus*), mungbean (*Vigna radiata*) and lupin (*Lupinus albus* and *L. mutabilis*). Lupin area is increasing in locations where soil acidity is a problem.

The CSA data indicate that ~8.5 million smallholders cultivated over 1.5 million ha of land under grain legumes and produced more than 2.8 million MT of grain annually in the

* Corresponding author: tsedeke@homegrownvision.org

DOI: 10.1079/9781800626386.0005

2016–2020 *meher* seasons (Table 5.1). Our calculations, based on the CSA data, show that an average smallholder household cultivates 0.18 ha and produces approximately 288 kg of grain per year.

Grain legumes in Ethiopia are grown for domestic consumption as well as foreign export earnings. In addition, they are an important source of soil nutrition, as they fix nitrogen from the air, and are therefore widely used in crop rotation with cereals. Ethiopians obtain about 17% of their daily dietary protein requirement, ~7% of dietary energy and ~4% of dietary fat from grain legumes. That is perhaps why they are affectionately referred to as a 'poor man's meat'. Low-income people can hardly afford to buy meat on regular basis in Ethiopia; the vast majority of people in this country purchase meat during major religious and traditional holidays.

Successive political systems in Ethiopia have made significant investment in grain legumes research and development (R&D) over the past five decades. However, even though there are many reports and scholarly articles published on the progress of these crops, technological advances and their impacts on productivity and their socioeconomic benefits have not been comprehensively measured. We provide an analysis of such information below and offer insights into the path ahead for grain legumes R&D in this country and beyond.

Approach

This chapter employs similar methods to those presented in Chapter 3 to measure the benefits accruing from investments in grain legumes R&D in Ethiopia. That is, to calculate the rate of yield gains over the years (i.e. changes in productivity increase) over a period of the 21 years mentioned above (2000 to 2020), the linear regression was computed using the annual crop mean yield as a dependent variable and production year as an independent variable. The resulting regression coefficient (b-value) was taken as an annual rate of yield gain. This was done for each crop and comparisons were made with the world's top ten producing countries.

Results

Grain legumes in general

The research system has released 183 cultivars between 1970 and 2017. The release rate increased exponentially from the early 2000s (Fig. 5.1). For example, the total number of cultivars released in the 27 years between 1973[1] and 1999 was 51 (averaging 1.89 cultivars per year), compared to 132 in the 18 years (average 7.33 cultivars per year) between 2000 and 2017.

Table 5.1. Indicators for major grain legumes smallholders cultivated in the *meher* season in Ethiopia (2016–2020 average). Source: number of holders, area, yield and production are constructed by the authors from CSA sample surveys (www.csa.gov.et); area and production per holder are the lead author's own calculations from CSA data; number of varieties are from MOALR (2017).

Crop	Holders (000)	Area (000 ha)	Yield (kg/ha)	Production (000 MT)	Area per holder (ha)	Production per holder (kg)	Number of varieties released
Faba bean	3826	466	2112	984	0.12	257	34
Common bean	3375	296	1712	506	0.09	150	65
Fieldpea	1699	219	1726	369	0.13	217	42[a]
Chickpea	882	228	1982	459	0.26	521	26
Lentil	731	101	1410	143	0.14	196	11
Grasspea	632	143	2025	289	0.23	458	1
Mungbean	235	45	1155	52	0.19	221	4
Lupin	100	18	1574	28	0.18	282	0
Aggregate	8466	1508	1712	2813	0.18	288	183

[a]Number of varieties for fieldpea includes the two varieties registered for *dekeko* (*Pisum sativum* var. *abyssinicum*).

Many of those cultivars possess desirable traits, including high yield potential, disease resistance and market demand, among others.

Overall, grain legumes showed sustained stable yield and production growth trends starting in the mid-2000s (Fig. 5.2). Ethiopia succeeded with its grain legumes mainly because it followed a knowledge-based approach (Abate and Orr, 2011). Yields started to increase from the early 2000s and continued through to 2018, after which time they levelled off. The area peaked in 2013, levelled off until 2017 and declined sharply after that. Overall, the area, yield and production in Ethiopia increased at 2.22%, 4.44% and 6.75%, respectively (Table 5.2).

The average yield gain for major legumes in Ethiopia was ~56 kg/ha/yr ($P<0.01$, $R^2=0.97$) for the period 2000–2020, the highest for the top ten producing countries of the world (Table 5.3). This was followed by Canada, Brazil, China, Niger, Australia, Nigeria, Myanmar, India and the Russian Federation; however, the gains were highly

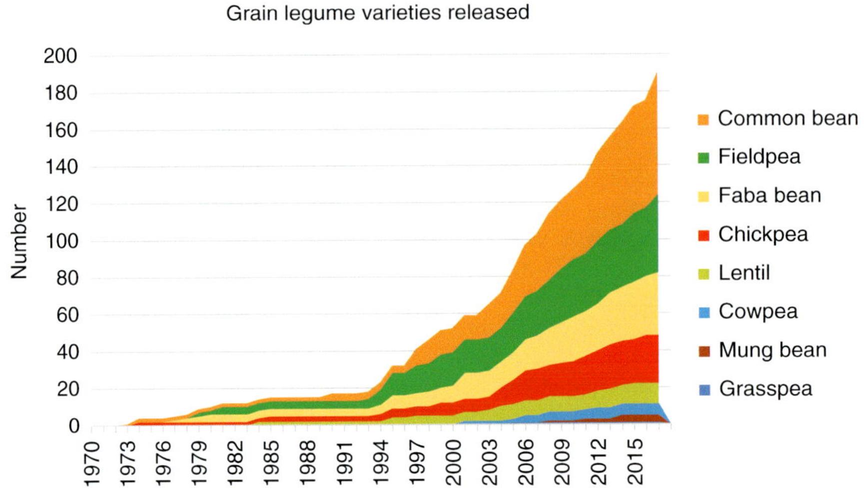

Fig. 5.1. Running totals of grain legume varieties released in Ethiopia, 1973–2017. Source: constructed by the lead author from MOALR (2017).

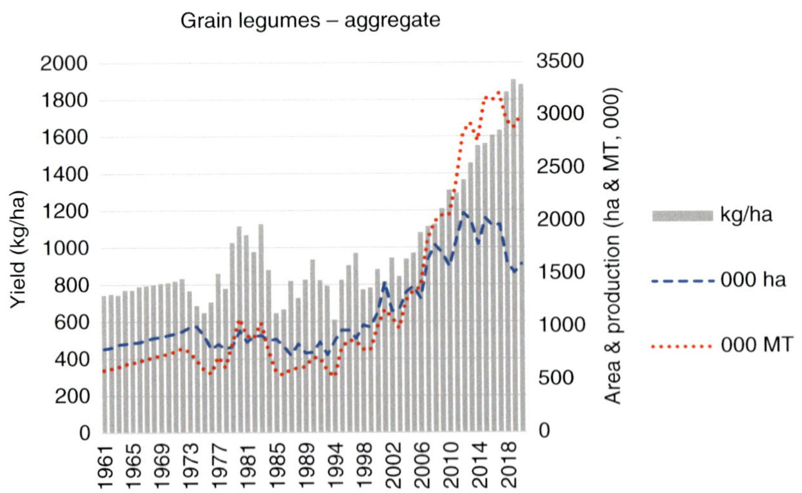

Fig. 5.2. Indicator trends for aggregate grain legumes in Ethiopia. Source: calculated and constructed by the lead author from FAO (2022a).

Table 5.2. Yield gain and annual rate of growth (ROG) for selected grain legumes in Ethiopia (2000–2020). Source: calculated and constructed by the lead author from FAO (2022a).

	Yield gain			ROG (%)		
Crop[a]	b-value	P > |t|	R^2	Area	Yield	Production
Chickpea	74	<0.01	0.95	1.32	5.41	6.80
Grasspea	67	<0.01	0.94	2.61	4.74	7.47
Faba bean	64	<0.01	0.92	0.76	4.24	5.03
Fieldpea	55	<0.01	0.93	0.59	4.87	5.48
Common bean	53	<0.01	0.83	4.54	5.32	10.11
Lentil	48	<0.01	0.86	2.37	4.83	7.31
Aggregate	56	<0.01	0.97	2.22	4.44	6.75

[a]Note: FAO data do not include mungbean and lupin for Ethiopia.

Table 5.3. Yield gain (b-value, kg/ha/yr) and annual rate of growth (ROG) for aggregate grain legumes in the top ten cultivating countries of the world (2000–2020). Source: calculated and constructed by the lead author from FAO (2022a).

	Yield gain			ROG (%)		
Country[a]	b-value	P > |t|	R^2	Area	Yield	Production
Ethiopia	56	<0.01	0.97	2.22	4.44	6.75
Canada	32	0.01	0.37	2.90	1.93	4.88
Brazil	21	<0.01	0.94	–2.49	2.41	–0.15
China	20	<0.01	0.72	–1.90	1.24	–0.69
Niger	17	<0.01	0.90	2.81	7.33	10.34
Australia	12	0.25	0.07	0.70	1.27	1.98
Nigeria	11	0.20	0.09	0.91	1.45	2.37
Myanmar	6	0.33	0.05	2.70	0.68	3.40
India	4	0.02	0.26	2.81	0.65	3.48
Russian Federation	–1	0.95	0.00	4.90	–0.07	4.82

[a]Descending order of the largest cultivating countries are India, Niger, Nigeria, Myanmar, Canada, Brazil, China, Australia, Russian Federation and Ethiopia.

significant only for the first four. The world average was 21 kg/ha/yr ($P<0.01$, $R^2=0.75$).

The ROG (annual rate of growth) for area and production were negative, while Niger and Ethiopia showed the fastest yield growth rate.

The average yield for Ethiopia in 2016–2020 was estimated at nearly 1800 kg/ha, third highest among the top ten countries, after Canada and China, but superior to the Russian Federation, Australia, Brazil, Myanmar, Nigeria, India and Niger. The world average yield is ~1430 kg/ha and the average for the top ten countries of the world by area cultivated is ~1240 kg/ha. The increased productivity and production of grain legumes in Ethiopia resulted in substantial increases in export volume and value (see the details in Table 5.10 later in this chapter).

This shows that Ethiopia's grain legumes agriculture is transitioning from subsistence to small-scale commercial agriculture, as shown in Fig. 5.2. It should be said that no other country on the African continent has registered such achievements with aggregate grain legumes under rainfed agriculture.

We present brief accounts of the progress made for each of the eight grain legumes (ranked on the basis of number of households growing) in the sections that follow.

Performance of individual crops

Faba bean

Faba bean (also known as broad bean or horse bean) is a traditional crop widely grown in the Ethiopian highlands, even though its centre of origin or centre of diversity is known to be the Central Asiatic Center (Zeven and de Wet,

1982). About 3.8 million smallholders grew this crop on about 466,000 ha and produced ~984,000 MT of grain annually in 2016–2020 (cf. Table 5.1). The average farm size for faba bean is estimated at 0.12 ha per holder. The Ethiopian National Research System has released 34 improved varieties between 1977 and 2017 (cf. Fig. 5.1).

According to the CSA, the regional share of faba bean area planted in the 2018 *meher* (main rainy season) was 43% Oromia, 39% Amhara, 15% SNNPR (Southern Nations, Nationalities and Peoples Region) and 3% Tigray. This crop is grown mostly for domestic consumption. Only a little over 2% of the total production is exported, largely to Sudan.

Ethiopia ranked second (in the area harvested) after China out of 68 countries in the world that grew faba bean during 2016–2020; they were ahead of Australia, the UK, Morocco, France, Italy, Tunisia, Peru and Germany in the top ten producing countries.

The yield gain for faba bean in Ethiopia was 64 kg/ha/yr ($P<0.01$, $R^2=0.92$) – the highest among the top ten countries, including Tunisia, China, Italy, Peru, Australia, Morocco, Germany, UK and France (Table 5.4). France had a highly significant ($P<0.01$, $R^2=0.36$) yield loss. UK also had yield loss but this was non-significant.

Faba bean production in Ethiopia showed stable increases starting from the early 2000s (Fig. 5.3). According to the latest Food and Agriculture Organization data, the annual ROG for area, yield and production was 0.76%, 4.24% and 5.03%, respectively, between 2000 and 2020 (cf. Table 5.2). This means that increases in yield contributed 84% to production compared to 15% for area and about 1% for the interaction factor.

The national average yield of faba bean in Ethiopia (2111 kg/ha), is close to the world average of 2160 kg/ha (FAO, 2022a). However, experiences suggest that Ethiopia has the potential to raise faba bean yields to more than 3000 kg/ha. The major threats for crop production in this country come from diseases such as the recently recorded faba bean gall, parasitic weeds (*Orobanche* spp.) and regular diseases such as rust and chocolate spot (Hailu *et al.*, 2014; Belay *et al.*, 2020).

Common bean

Common bean (also referred to as haricot bean or dry beans), especially the coloured, food type, was historically grown as an understory in maize and other crop fields mainly in southern and southwestern parts of Ethiopia. This changed with the introduction of white pea bean type commercial variety Mexican-142 by private investors producing the crop for the export market in the Central Rift Valley in the early 1970s.

Ethiopia ranked 15th in the area harvested and 10th in production out of 122 countries

Table 5.4. Yield gain (b-value, kg/ha/yr) and annual rate of growth (ROG) for faba bean in the top ten cultivating countries of the world (2000–2020). Source: calculated and constructed by the lead author from FAO (2022a).

	Yield gain			ROG (%)		
Country[a]	b-value	P > |t|	R^2	Area	Yield	Production
Ethiopia	64	<0.01	0.92	0.76	4.24	5.03
Tunisia	32	0.01	0.46	1.08	3.82	4.94
China	25	<0.01	0.68	–2.28	1.40	–0.91
Italy	23	0.01	0.55	1.15	1.32	2.48
Peru	17	<0.01	0.87	1.86	1.30	3.19
Australia	14	0.43	0.03	2.02	1.23	3.27
Morocco	6	0.03	0.46	–0.88	1.12	0.24
Germany	5	0.71	0.01	6.37	0.12	6.50
UK	–21	0.29	0.06	11.13	–0.59	10.47
France	–84	0.01	0.36	1.41	–2.63	–1.26

[a]Descending order of the top ten largest countries: China, Ethiopia, Australia, UK, Morocco, France, Italy, Tunisia, Peru and Germany.

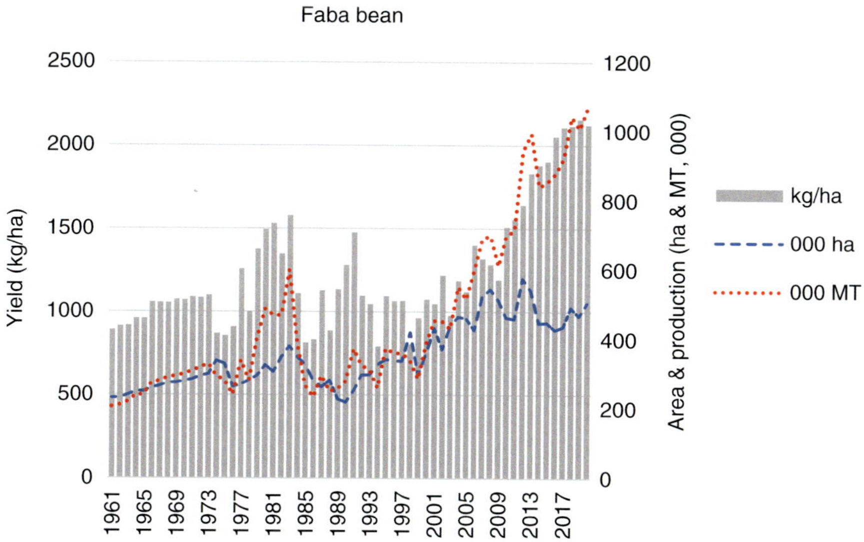

Fig. 5.3. Indicator trends for faba bean in Ethiopia. Source: constructed by the lead author from FAO (2022a).

and territories in the world that grew common bean in the 2016–2020 seasons (FAO, 2022a). Approximately 3.4 million smallholder households grew common bean on ~296,000 ha and produced ~506,000 MT of grain annually in the 2016–2020 *meher* seasons (cf. Table 5.1). The national average land area allocated to common bean is less than 0.1 ha and the per capita production per smallholder household is estimated at 150 kg per year. The research system has released a total of 65 improved cultivars, the highest for all grain legumes, through to 2017 (MOALR, 2017). The national average yield is estimated at ~1700 kg/ha.

Common bean in Ethiopia is produced both for domestic consumption and the export market. Like most other grain legumes, common bean is a major source of protein for people who cannot afford meat and other animal products. Roughly 27% of the country's total production was destined for the export market (2016–2020 average). According to the CSA data for 2013/14 and 2015/16, this crop is grown in Oromia (43%), Amhara (32%), SNNPR (21%), Benshangul-Gumuz (2%), Tigray (1%) and Dire Dawa (<1%). Approximately 71% of the white and 52% of coloured types are grown in the *meher* season. Assuming 46% of area and 34% of the total production for the *belg* season, the total area harvested and production each year could be about 560,000 ha and 624,000 MT, respectively. Smallholders account for 99% of the total common bean production in Ethiopia.

Ethiopia registered a yield gain of 53 kg/ha/yr ($P<0.01$, $R^2=0.83$) between 2000 and 2020 (Table 5.5). Yield gains were also highly significant ($P<0.01$) for Tanzania, Brazil, Mozambique and Kenya; China also registered significant ($P<0.05$) yield gain. Burundi was the only country with a highly significant ($P<0.01$) decline among the top ten countries of the world. There was little or no change in Myanmar, Angola, Mexico and India.

The area, yield and production in Ethiopia increased at an annual rate of 4.54%, 5.32% and 10.11%, respectively. This means that increases in the yield and area accounted for about 53% and 45%, respectively, for the change in production while the interaction factor between the two covered the balance (~2%). Angola, Burundi and Ethiopia showed fast growth in the area, whereas it shrank in China, Brazil and Mexico.

Figure 5.4 presents long-term indicator trends for common bean in Ethiopia. Yields fluctuated from the mid-70s to early 2000s but attained steady and sustained growth over the last decade. The national average yield for 2016–2020 was estimated at ~1500 kg/ha, second only to China (~1700 kg/ha) but ahead of

Table 5.5. Yield gain (b-value, kg/ha/yr) and annual rate of growth (ROG) for common bean in Ethiopia compared to the top ten cultivating countries of the world (2000–2020). Source: calculated and constructed by the lead author from FAO (2022a).

	Yield gain			ROG (%)		
Country[a]	b-value	P > |t|	R^2	Area	Yield	Production
Ethiopia	53	<0.01	0.83	4.54	5.32	10.11
Tanzania	34	<0.01	0.75	2.14	3.68	5.89
Brazil	22	<0.01	0.94	−2.50	2.43	−0.14
Mozambique	17	0.01	0.56	NA	NA	NA
Kenya	16	<0.01	0.70	1.97	3.07	5.10
China	16	0.02	0.52	−3.63	1.03	−2.63
Myanmar	8	0.31	0.05	3.15	0.87	4.05
Angola	5	0.11	0.13	8.30	1.32	9.73
Mexico	1	0.55	0.02	−0.64	0.24	−0.40
India	0	0.94	0.00	3.79	0.10	3.89
Burundi	−18	<0.01	0.71	6.41	−2.28	3.98

[a]Descending order of countries with the largest area are India, Myanmar, Brazil, Mexico, Kenya, Tanzania, Angola, China, Mozambique and Burundi; Ethiopia is the 15th.
NA = not available.

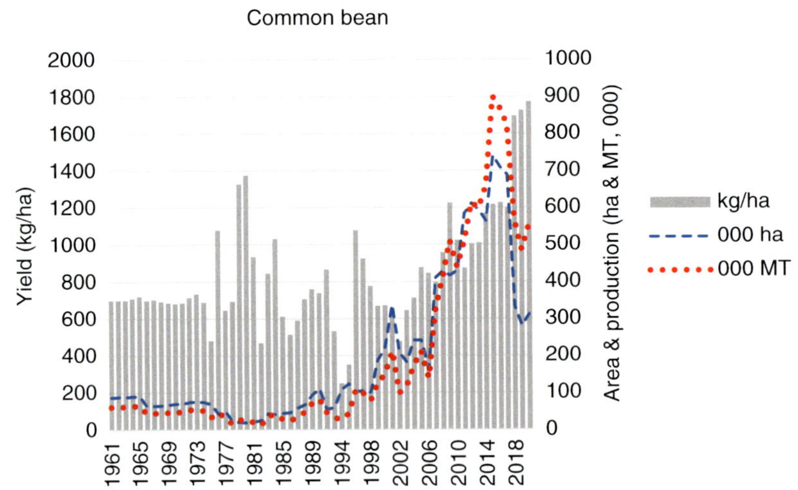

Fig. 5.4. Indicator trends for common bean in Ethiopia. Source: constructed by the lead author from FAO (2022a).

the rest of the top ten, including Tanzania, Brazil, Myanmar, Mexico, Kenya, Burundi, Mozambique, India and Angola (cf. Box 5.1). It is reported that, with ~1700 kg/ha, only Uganda has a superior yield of common bean compared to Ethiopia among the African countries (FAO, 2022a).

The annual export volume jumped from 34,000 MT in 2000–2004 to 179,000 MT in 2016–2020; similarly, earnings increased from US$10 million to US$110 million during the same period.

Fieldpea

Fieldpea (also known as dry peas) is traditionally grown in Ethiopia for a long period of time. The available data perhaps include both the regular pea and the Ethiopian pea (*Pisum sativum* var. *abyssincum*), locally known as '*dekeko*'. Approximately 1.7 million smallholder households grew fieldpea on >219,000 ha of land and produced ~369,000 MT of grain each year during the 2016–2020 *meher* seasons (cf. Table 5.1). The

Box 5.1: The story behind common bean successes in Ethiopia

Common bean production by smallholder farmers in Ethiopia, like many other grain legumes, used to follow a subsistence production mode until the early 2000s. That is, farmers in common bean-growing agroecologies invested their labour mostly in cereals (mainly maize, sorghum and *teff*) where they would plough the soil three to four times, sow the local seed at the beginning of the rainy season and weed their plots as many times as needed. By contrast, common bean was broadcast later in the season following just one ploughing; it was considered a secondary crop. No weeding was done. Farmers came and harvested the crop at the end of the season and obtained a yield of about 700 kg/ha or less. There was no source of improved seed and the market was poorly developed.

This was the time the government was entering into agricultural transformation initiative and demanded a lot from the research community in a short period of time. For example, the government plan in 2004 called for a 2.47-fold increase in the export volume of grain legumes within 5 years, and the pressure on research to provide good-quality seed varieties and the accompanying crop management technologies was very high.

Research released an export-type pea bean variety, 'Awash Melka', which exhibited all the desirable attributes that made it highly acceptable by the canning industry abroad. Optimum agronomic practices – including proper seedbed preparation and weeding up to two times – were also developed by research. The technology package was piloted in the 2005 *meher* season with 700 farmers on more than 600 ha of land in three *weredas* (districts) in East Shewa, where all stakeholders – farmers, extension, research, seed producers, community-based organizations and market actors – were involved (see Assefa *et al*., 2006 for further details). Participating farmers obtained an average of 2000 kg/ha.

This technology was adopted widely across the country and the national average yield more than doubled (cf. Fig. 5.4). The current national average yield of common bean in Ethiopia is comparable to that of China and superior to many of the major producing countries, including Brazil, Mexico and India (Fig. Box 5.1).

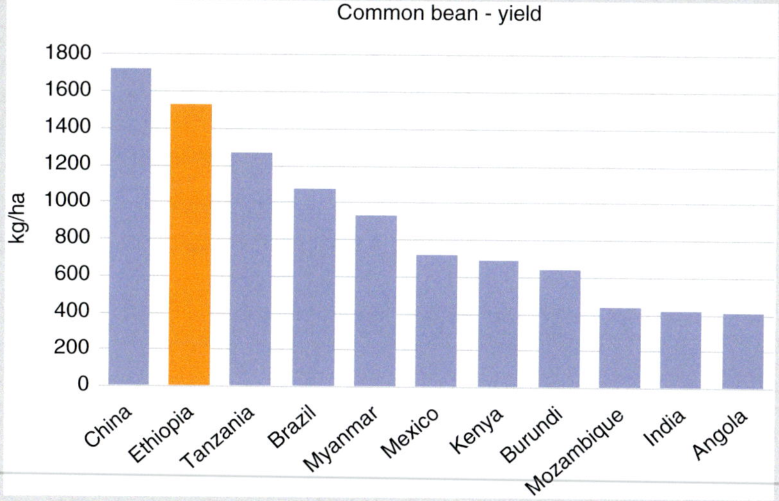

Fig. Box 5.1. Average yield (2016–2020) of common bean in Ethiopia compared to the top ten cultivating countries of the world. Source: constructed by the lead author from FAO (2022a); descending order of top ten countries with highest cultivated area are: India, Myanmar, Brazil, Mexico, Kenya, Tanzania, Angola, China, Mozambique and Burundi; Ethiopia is 15th of 122 countries in the world.

average plot size is about 0.13 ha/holder, whereas the per capita production is approximately 217 kg/holder/yr. According to the CSA (2016), the *meher* season accounts for 90% of the area harvested and 94% of production. There is little or no large-scale production of fieldpea in Ethiopia.

According to Ruelle *et al.* (2019), its adaptation agroecologies include mainly the tepid mid-highlands (1600–2400 m above sea level [masl]) and, to some extent, the cool mid-highlands (2400–3200 masl) and the humid warm lowlands (500–1600 masl). The Amhara and Oromia regional states account for about 39% each of the total area occupied by fieldpea in this country, with the south and Tigray accounting for 19% and 3%, respectively. Ethiopia ranks eighth (both in the area harvested and total production) out of a total of 95 countries that grew fieldpea in the world in 2016–2020 (FAO, 2022a).

Fieldpea is an essential staple in Ethiopian diets and a major source of protein for a large proportion of the population. It is mainly used in the form of *kik* (shelled and split grain) or *shiro* (shelled and ground). A smaller proportion is used as *kolo* (roasted snacks) and *nifro* (boiled whole grain).

Ethiopia ranked eighth both in area and production of 96 countries that produced fieldpea in the world in the 2016–2020 seasons (FAO, 2022a). It registered a yield gain of 55 kg/ha/yr ($P<0.01$, $R^2=0.93$), the largest of top ten producing countries of the world, followed by Canada, Ukraine, the Russian Federation, China, Spain, India, the USA, Australia and France (Table 5.6). The ROGs for area were negative in the Ukraine, Australia and France; for yield it was negative only in France, and for production it was negative in France and Australia.

Figure 5.5 presents long-term fieldpea indicators in Ethiopia. The yield showed steady growth from the early 2000s onwards. Production also grew in a similar fashion but started to decline or slowed down from 2013 onwards, owing to decline in the area.

The national average yield for Ethiopia, <1700 kg/ha, is below the world average of a little over 1800 kg/ha. The averages for France, Canada, Ukraine, the USA and the Russian Federation ranged between ~3300 kg/ha and ~2100 kg/ha, respectively. Ethiopia's yield is ahead of Spain, China, India and Australia. Despite the progress Ethiopia has made in fieldpea production, the country is still a net importer (cf. Table 5.10 below). This crop constitutes the largest proportion of Ethiopia's import of grain legumes.

The major constraint to fieldpea production in Ethiopia is the pea aphid, *Acyrthosiphon pisum* (Harris) (Ahmed, 2002; Ali *et al.*, 2005; Tesfaye *et al.*, 2013). Damage by this insect pest can cause up to 48% yield loss and total crop loss could occur when the main season rains stop suddenly and are followed by dry spells (Ali and Habtewold, 1994).

Even though attempts have been made to control the pea aphid with insecticides (Ali, 1997) their use is not a viable option where the crop is grown by smallholders and because of the hazards to human beings and natural enemies of the pest. Successful biological control programmes have been used against this pest in North America (Mackauer, 1982, 1983; Chow and Mackauer, 1984, 1985, 1986; Mackauer *et al.*, 1990; Boivin *et al.*, 2012), but there is no capacity for this option in Ethiopia at present.

Table 5.6. Yield gain (b-value, kg/ha/yr) and annual rate of growth (ROG) for fieldpea in the top ten cultivating countries of the world (2000–2020). Source: calculated and constructed by the lead author from FAO (2022a).

	Yield gain			ROG (%)		
Country[a]	b-value	P > |t|	R^2	Area	Yield	Production
Ethiopia	55	<0.01	0.93	0.59	4.87	5.48
Canada	49	0.01	0.43	1.32	2.61	3.96
Ukraine	28	0.11	0.13	–0.97	1.46	0.47
Russian Federation	21	0.04	0.21	4.07	1.14	5.26
China	17	0.01	0.29	0.14	1.27	1.41
Spain	14	0.22	0.08	3.99	0.86	4.88
India	12	0.02	0.24	0.73	1.14	1.88
USA	7	0.49	0.03	7.39	0.32	7.73
Australia	4	0.78	0.00	–2.74	0.77	–1.99
France	–69	0.01	0.49	–3.86	–1.84	–5.63

[a]Descending order of countries with the largest cultivated area: Canada, Russian Federation, China, India, USA, Ukraine, Australia, Ethiopia, France and Spain.

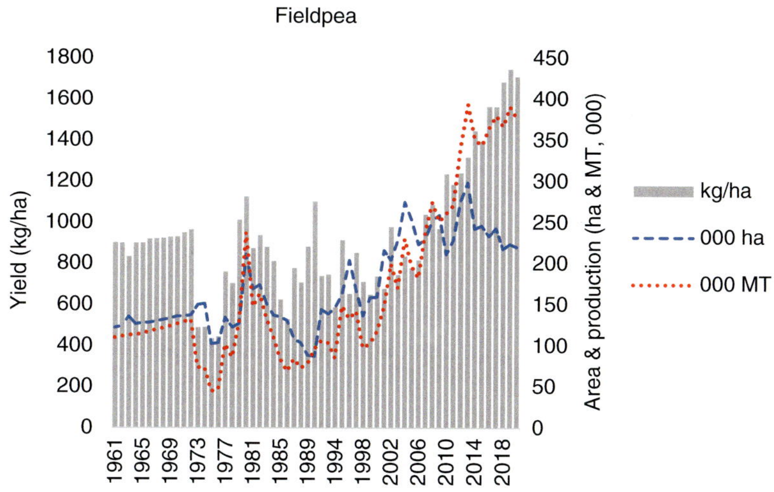

Fig. 5.5. Indicator trends for fieldpea in Ethiopia. Source: constructed by the lead author from FAO (2022a).

An integrated pest management approach based on the use of host plant resistance would be an appropriate method of pea aphid management in Ethiopia. There is evidence of genetic variability among fieldpea lines in their response to the pea aphid damage (Auclair, 1957; Cartier, 1963; Fernandez *et al.*, 2014; Khan *et al.*, 2015). Ali (2005) screened a limited number of pea accessions and identified some promising lines. However, a more robust and targeted approach using large numbers of accessions is needed to develop an effective and sustainable management system. Currently, there are more than 2000 local accessions of fieldpea at the Ethiopian Biodiversity Institute (Teshome *et al.*, 2015). Additional germplasm can be obtained through international cooperation and exchange from the International Center for Agricultural Research in the Dry Areas or other national research systems (Corinto, 2014).

Chickpea

Chickpea is a traditional cool-season legume widely grown on the black soils in the mid-altitude agroecologies in Ethiopia. Nearly 900,000 smallholder households planted ~228,000 ha of land to chickpea annually and produced ~459,000 MT of grain in Ethiopia in the 2016–2020 *meher* seasons (cf. Table 5.1). Ethiopia was ranked eighth in area, fifth in production and third in yield out of 56 countries in the world that grew chickpea during the above period (FAO, 2022a).

The *meher* season accounts for about 93% of the total area and 96% of total production. More than 99% of the total production comes from smallholder farmers, where the average holding is about 0.26 ha/holder, the average yield is ~2000 kg/ha and the per capita smallholder production is about 520 kg/yr. The research system has released a total of 26 varieties through to 2017.

The Amhara and Oromia regional states account for 51% and 42% of the total area, respectively, whereas the share of SNNPR and Tigray was 4% and 2%, respectively.

Chickpea in Ethiopia is produced both for local consumption as well as export market. This country exported about 10% of its total production annually in 2016–2020 (details in Table 5.10 below). The largest buyers are Pakistan, the United Arab Emirates and India. Iran, Sudan and Jordan also have a good share of this country's chickpea exports.

The yield gain for chickpea in Ethiopia was 74 kg/ha/yr, which is highly significant ($P<0.01$, $R^2=0.95$) and the highest among the top ten (Table 5.7). This was followed by Canada, Myanmar, Türkiye and India, among the top ten. Gains were negative in Pakistan and the Russian Federation but non-significant.

The area, yield and production in Ethiopia increased at an annual rate of 1.32%, 5.41%

Table 5.7. Yield gain (b-value, kg/ha/yr) and annual rate of growth (ROG) for chickpea in the top ten countries of the world (2000–2020). Source: calculated and constructed by the lead author from FAO (2022a).

	Yield gain			ROG (%)		
Country[a]	b-value	P > |t|	R^2	Area	Yield	Production
Ethiopia	74	<0.01	0.95	1.32	5.41	6.80
Canada	39	0.01	0.48	−2.86	2.73	−0.20
Myanmar	26	<0.01	0.62	4.32	2.24	6.66
USA	19	0.01	0.29	11.08	1.24	12.46
Türkiye	19	<0.01	0.75	−2.15	1.77	−0.42
Australia	13	0.23	0.07	7.50	1.25	8.83
India	13	<0.01	0.65	3.02	1.48	4.55
Iran	1	0.66	0.01	−1.93	0.36	−1.57
Russian Fed.	−5	0.85	0.01	NA	NA	NA
Pakistan	−10	0.08	0.16	−0.16	−2.07	−2.22

[a]Descending order of countries with largest area: India, Pakistan, Australia, Russian Federation, Iran, Türkiye, Myanmar, Ethiopia, USA and Canada.
NA = not available.

and 6.80%, respectively. In other words, increases in yield and area expansion contributed to increases in production of about 80% and 19%, respectively, while their interaction factor covered the remaining 1% or so.

The USA and Australia had the fastest growth rate in area (and therefore production). By contrast, the ROG for production declined in Pakistan, Iran, Türkiye and Canada. Pakistan was the only country among the top ten where the ROG for yield declined. The area declined in nine of the top ten countries – including Canada, Syria, Türkiye, Mexico, Iran, Yemen, Morocco, Pakistan and Tanzania (Table 5.7 and Box 5.2). The yield declined in Kazakhstan and Pakistan, while Canada, Mexico, Türkiye, Iran, Syria and Pakistan experienced a decline in production. The decline in production in the major producing countries would open up more opportunities for Ethiopia's chickpea export markets.

Chickpea yield showed substantial improvement from the mid-2000s; area and production followed similar trends but started to decline or slowdown from 2014 (Fig. 5.6). It can be seen that Ethiopia doubled its chickpea yield in less than 20 years (between 2005 and 2020).

Lentil

Lentil is one of the traditional cool-season grain legumes grown in Ethiopia. About 731,000 smallholder households planted 101,000 ha of land to lentil and produced about 143,000 MT of grain each year in 2016–2020 (cf. Table 5.1). Ethiopia ranked 12th in area and 10th in production, out of 51 countries in the world that grew lentil in 2016–2020 (FAO, 2022a). The Amhara, Oromia and Tigray regional states grow approximately 68%, 22% and 10% of lentil, respectively.

Smallholder households account for 99% of lentil production in Ethiopia, almost exclusively during the *meher* season (Eshete *et al.*, 2006). The national average yield is about 1400 kg/ha. The average area allocated to lentil is 0.14 ha and each household produces a little less than 200 kg/yr. Eleven improved cultivars of lentil are known to be under production in Ethiopia. This staple is highly valued for its high protein content and highly sought after in Ethiopian diets.

Ethiopia registered a highly significant yield gain of 48 kg/ha/yr ($P<0.01$, $R^2=0.86$) between 2000 and 2020 (Table 5.8), the second highest among the top ten producing countries in the world, after China and followed by Canada, Nepal, Bangladesh, Türkiye, Australia, India, the USA and Kazakhstan. Kazakhstan and the USA had non-significant yield losses.

The area, yield and production increased at 2.37%, 4.83% and 7.31%, respectively, per annum. That is to say, increases in yield and area expansion contributed about 66% and 32%, respectively, to changes in production, whereas their interaction factor covered the remaining 2%.

Box 5.2: Factors that transformed chickpea production in Ethiopia

Just like common bean discussed above, the government had a very ambitious plan of quadrupling chickpea production and export volume in the early 2000s. The research system developed a systems approach, where multi-stakeholders worked together (Abate *et al.*, 2011). One of the major constraints was waterlogging, which created conducive conditions for ascochyta blight (a fungal disease) and wilt/root rot complex, so that farmers planted chickpea at the end of the rainy season.

Release of ascochyta blight and wilt root rot disease-resistant varieties allowed farmers to plant in August using their traditional ridge and furrow technologies to drain excess water, and yield increased up to 4000 kg/ha. Traditionally, Ethiopian farmers grew the small-seeded, coloured *desi* types. In the late 1990s and early 2000s, research also released the white-coloured *kabuli* types such as Arerti (FLIP 89-84C), Shasho (ICCV-93512) and Habru (FLIP 88-42C) that have more export market demand. Farmers who adopted the innovation increased their incomes over non-adopters (Verkaart *et al.*, 2017). Chickpea yield in this country is among the highest in the world (Fig. Box 5.2).

Research also organized selected farmers to be seed producers and suppliers. Ethiopia has the advantage of more fertile soils in the chickpea-growing agroecologies than many of its competitors. Recently, many farmers have introduced chickpea in maize and cotton-based cropping systems as a double crop.

Chickpea yield in Ethiopia showed sustained growth since the early 2000s (cf. Fig. 5.6), whereas yields in other major producing countries have levelled off, stayed stagnant or declined. This opened up opportunities for Ethiopia to expand its export market.

This country's annual export volume grew from a little over 19,000 MT in 2000–2004 to nearly 48,000 MT in 2016–2020, with the earnings rising from less than US$6 million to more than US$38 million over the same period.

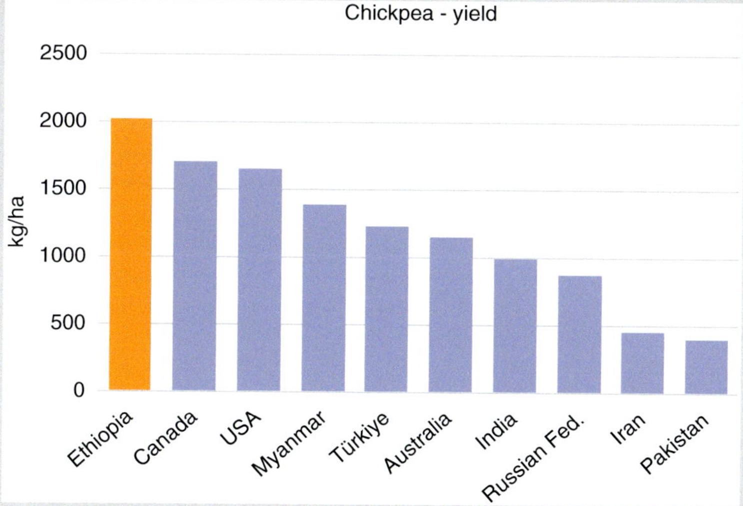

Fig. Box 5.2. Average yield (2016–2020) of chickpea in the top ten cultivating countries of the world. Descending order of top ten countries with the largest cultivated area are: India, Pakistan, Australia, the Russian Federation, Iran, Türkiye, Myanmar, Ethiopia, the USA and Canada. Source: constructed by the authors from FAO (2022a).

Lentil in Ethiopia showed appreciable yield growth starting from the mid-2000s, peaked between 2015 and 2017 and showed slight declines thereafter (Fig. 5.7). The area and production somewhat fluctuated among different years and have been in decline since 2017. With about 1400 kg/ha, Ethiopia's is the fourth highest yield among the top ten producers; it is superior to the world average of >1000 kg/ha and in the same group as Canada, Türkiye and the USA.

Canada, Australia and Kazakhstan, along with Ethiopia, showed the fastest growth in production. However, other than Ethiopia, the growth in the other countries was mainly driven by area expansion.

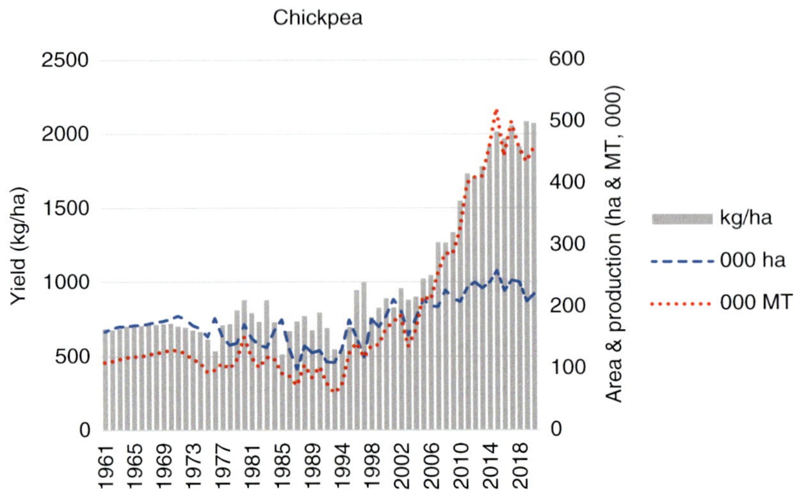

Fig. 5.6. Indicator trends for chickpea in Ethiopia. Source: constructed by the lead author from FAO (2022a).

Table 5.8. Yield gain (b-value, kg/ha/yr) and annual rate of growth (ROG) for lentil in the top ten cultivating countries of the world (2000–2020). Source: calculated and constructed by the lead author from FAO (2022a).

	Yield gain			ROG (%)		
Country[a]	b-value	Pr > |t|	R^2	Area	Yield	Production
China	66	<0.01	0.77	–2.21	3.74	1.45
Ethiopia	48	<0.01	0.86	2.37	4.83	7.31
Canada	30	0.01	0.46	7.12	2.37	9.66
Nepal	27	<0.01	0.86	0.87	2.70	3.60
Bangladesh	26	<0.01	0.85	–0.56	2.75	2.17
Türkiye	23	0.03	0.21	–3.60	1.95	–1.72
Australia	13	0.43	0.03	6.59	2.64	9.40
India	13	0.01	0.48	–0.17	1.65	1.48
USA	–1	0.92	0.00	5.60	–0.10	5.49
Kazakhstan	–5	0.21	0.08	8.45	–0.66	7.74

[a]Descending order of countries with largest area: Canada, India, Australia, USA, Türkiye, Nepal, Bangladesh, Kazakhstan, China and Ethiopia.

Ethiopia used to be a net exporter of lentil from the early 1960s to early 1980s; then there was little or no export till the early 2000s. After peaking in 2010, lentil export went down to nil in recent years and the country is now a net importer. This is perhaps because of the increased demand within the country (see Table 5.10 for further details).

These import figures are not high, for now, but importing grain legumes such as lentil (and fieldpea, discussed above) cannot be sustainable and is not justifiable for a country that has both comparative and competitive advantages to increase productivity. First, the country has suitable land and a long tradition of lentil production. Second, Ethiopia's lentil productivity has shown positive growth, whereas the annual production in many of the major producing countries including Türkiye, Iran, Syria, Pakistan and Mexico is declining because of declines either in the area harvested (Türkiye, Bangladesh, Iran, Syria, China, Morocco and Pakistan) or yield per unit area (as in the USA, Syria, Pakistan, Greece and Mexico). Third, there is already an established market in the Middle East and South Asia. These facts suggest good opportunities for Ethiopia's export markets for lentil.

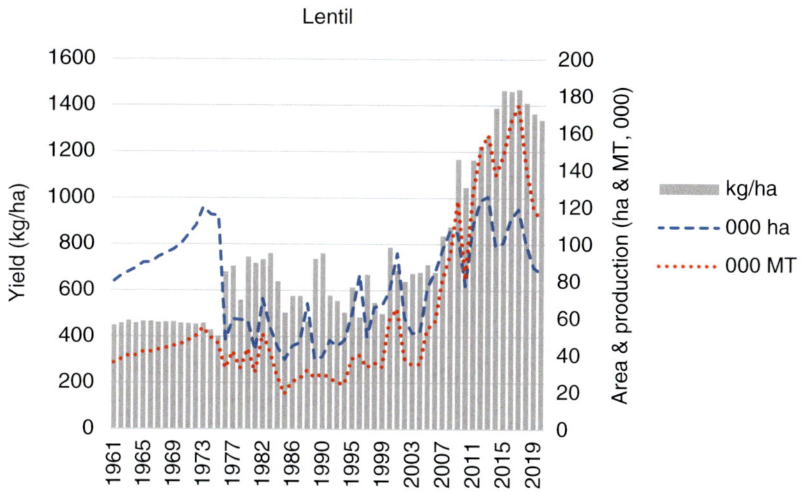

Fig. 5.7. Indicator trends for lentil in Ethiopia. Source: constructed by the author from FAO (2022a).

Grasspea

Grasspea (also known as vetch in Ethiopia and chickling vetch) is another cool-season grain legume widely grown in the central and north-central parts of Ethiopia. Approximately 632,000 smallholder households grew grasspea on ~143,000 ha and produced close to 289,000 MT of grain annually in the *meher* seasons of 2016–2020 (cf. Table 5.1). The national average yield was about 2000 kg/ha, while the average land area allocated to this crop is 0.23 ha per smallholder household that produces close to 460 kg/yr. Ethiopia has the largest area and production out of 35 countries that grew grasspea in 2016–2020 (FAO, 2022a).

About 66% of the total harvested area of grasspea in Ethiopia is found in the Amhara Regional State, with the Oromia and Tigray accounting for approximately 28% and 6%, respectively. Little or none of this crop is grown in SNNPR. Grasspea in this country is grown in the *meher* season mainly for domestic consumption.

Ethiopia registered a highly significant yield gain of 67 kg/ha/yr ($P<0.01$, $R^2=0.94$), the highest among the top ten producing countries ahead of Türkiye, Syria, Spain, Belarus, the Russian Federation, Mexico, Serbia, Greece and Australia (Table 5.9). Australia, Greece and Serbia showed non-significant yield losses.

The area, yield and production in Ethiopia increased at an annual rate of 0.26%, 4.74% and 7.49%, respectively. This country also showed the fastest growth rate for production mainly driven by increases in productivity rather than by area expansion. The ROG for production declined for Türkiye, the Russian Federation, Syria and Australia (Table 5.9).

The area planted to grasspea in Ethiopia showed upward trends from the mid-1980s to the early 2010s but has been in decline since (Fig. 5.8). The yield started to grow steadily from the early 2000s but somewhat levelled off from 2016 onwards.

As stated above, grasspea in Ethiopia is produced exclusively by smallholders. It is considered a minor grain legume and little resources are allocated for its research, at least at the federal level. Only one improved variety, 'Wasie' was released (in 2005) by the research system so far, but it is not clear how much of this variety has been adopted by smallholders as there do not seem to be any demonstrations and popularization carried out.

This crop is grown purely for local consumption in Ethiopia, but some is believed to be smuggled to Sudan and Eritrea. It is eaten in the form of *shiro* (flour for making traditional sauce), *kitta* (unleavened local bread), *kolo* (roasted snacks) or *nifro* (boiled grain). The major constraint with consumption of grasspea is its high content of a neurotoxin known as β-ODAP, which causes lathyrism, a crippling disease (Tekle-Haimanot *et al.*, 1993). There are no known varieties with a low level of β-ODAP; steeping the grain is known to remove about 30% of the neurotoxin. Grasspea is known to be low in two

Table 5.9. Yield gain (b-value, kg/ha/yr) and annual rate of growth (ROG) for grasspea in the top ten cultivating countries of the world (2000–2020). Source: calculated and constructed by the lead author from FAO (2022a).

	Yield gain			ROG (%)		
Country[a]	b-value	Pr > ItI	R^2	Area	Yield	Production
Ethiopia	67	<0.01	0.94	0.26	4.74	7.47
Türkiye	28	<0.01	0.33	−1.79	3.37	−15.11
Syria	16	0.01	0.38	−0.66	2.12	−4.64
Spain	13	0.22	0.09	NA	NA	NA
Belarus	2	0.90	0.00	NA	NA	NA
Russian Fed.	2	0.84	0.00	−0.68	0.14	−6.62
Mexico	2	0.97	0.00	0.01	0.01	0.12
Serbia	−10	0.57	0.03	NA	NA	NA
Greece	−11	0.10	0.16	NA	NA	NA
Australia	−16	0.05	0.20	0.08	−3.08	−2.31

[a]Descending order of countries with the largest cultivated area: Ethiopia, Russian Federation, Spain, Australia, Türkiye, Belarus, Syria, Serbia, Mexico and Greece.
NA = not available.

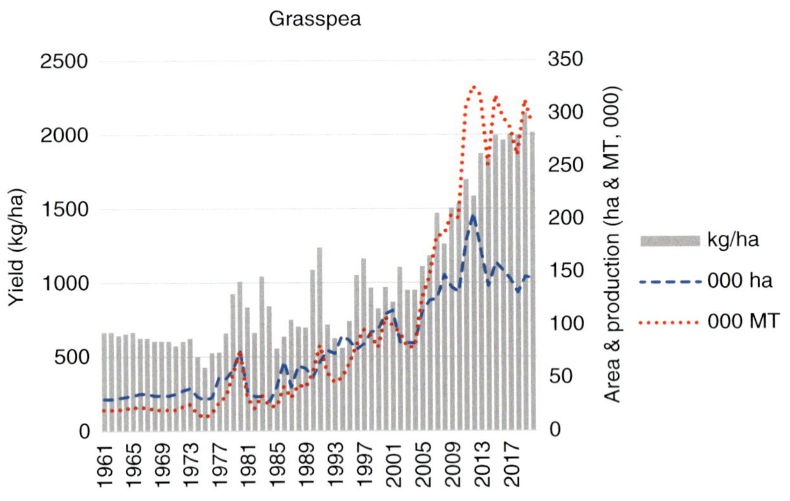

Fig. 5.8. Indicator trends for grasspea in Ethiopia. Source: constructed by the lead author from FAO (2022).

of the essential amino acids, i.e. methionine and tryptophan (Girma and Korbu, 2012). This crop has suffered from green aphid attack in recent years and farmers have started using insecticidal sprays to control this insect pest.

Farmers in Ethiopia still grow grasspea in spite of the underlying problems because of its ability to grow under adverse environmental conditions such as terminal drought, high salinity, waterlogging and marginal soils. In addition, the crop has a high protein content compared to the other grain legumes mentioned earlier. The research system has the responsibility either to develop cultivars with desirable traits such as low β-ODAP and high essential amino acids, or to find a better substitute legume that meets the farmers' demand.

Mungbean

Traditionally, mungbean has been known as a minor crop grown in the warmer parts of North Shewa (such as the Shewa-Robit areas) and South Welo (Kemisie areas) for food, fodder, income generation and improved soil fertility. According to the CSA (2019), the Amhara

Regional State accounted for 100% of this crop grown in Ethiopia in the 2018/19 *meher* season. However, recent studies (Asegie, 2015) show that it is also grown in West Shewa, Gamo Gofa, Segen and Gambella. There are also reports that the crop is produced in Benshangul-Gumuz Regional State. The CSA data for 2015/16 indicate that smallholder households account for 75% of the total production while commercial growers are responsible for the remaining 25%.

Records on the number of smallholder households, area, yield and production entered the CSA sample survey starting from the 2013/14 *meher* season. The FAO database does not include records for mungbean in Ethiopia or anywhere in the world, even though this is an important grain legume in South Asia, Southeast Asia (where India, Myanmar, China and Indonesia are among the major producers) and East Africa (Schreinemachers *et al.*, 2019; Sequeros *et al.*, 2021).

The surge in interest for mungbean in Ethiopia is because of its high international demand. For example, information from the Ethiopian Commodity Exchange suggests that the country exported 30,694 MT of the grain, valued at US$35.8 million in 2015/16 (www.ecx.com.et). This means that approximately 41% of the total production is exported.

Mungbean used to be primarily a subsistence crop, but this has changed in recent years. According to the CSA sample survey for 2015/16, this crop is grown in both seasons – with the *belg* and *meher* seasons shares being about 65% and 35%, respectively. It is grown as a sole crop or intercropped with other crops such as maize, sorghum, common bean and, to a lesser extent, barley (Asegie, 2015).

Approximately 235,000 smallholder households planted mungbean on approximately 45,000 ha of land and produced ~52,000 MT annually in the *meher* seasons of 2016–2020 (cf. Table 5.1). Considering 65% of area and 64% of production for the *belg* season,[2] we estimate that the total area under mungbean for both seasons is ~84,000 ha and the production is ~92,000 MT.

Some low-level research was started in the early 1980s but was stopped because the crop was considered minor and did not meet the criteria for high-priority grain legumes at that time. The devastating effect of the foliar disease, halo blight (caused by the bacterial pathogen *Pseudomonas syringae* pv. *phaseolicola*) (Assefa and Gorfu, 1986), also contributed to mungbean relegation from the research agenda.

The cultivated area, yield and production increased at an annual rate of 24.51%, 4.74% and 30.43%, respectively, between 2013 and 2020, making this crop the fastest growing grain legume in Ethiopia. The number of smallholders cultivating mungbean in Ethiopia increased at an annual rate of 29.48% between 2013 and 2019. Figure 5.9 presents trends for mungbean indicators (only in the *meher* seasons) in Ethiopia.

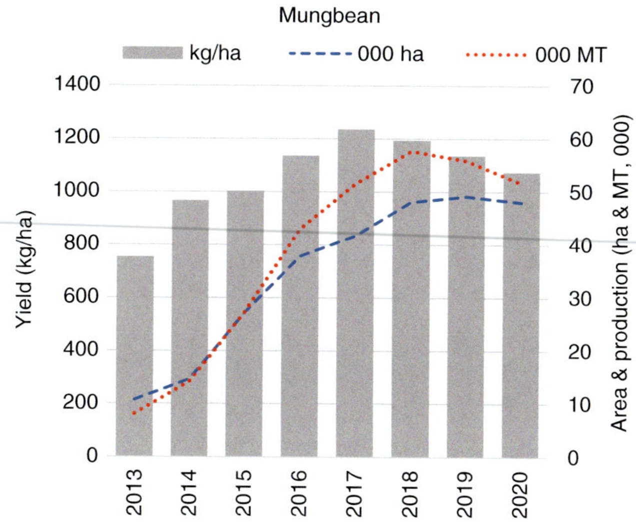

Fig. 5.9. Indicator trends for mungbean in Ethiopia. Source: constructed by the lead author from CSA data (www.csa.gov.et).

Halo blight is still the main constraint to mungbean production in Ethiopia. Research has released a total of four improved varieties between 2008 and 2014, but the extent of their use is not clear. It is possible that many farmers are still using their landraces. Even though there is variability in the Ethiopian landraces (Asegie, 2015; Yoseph *et al.*, 2020) and international collections elsewhere, no effort has been made to use this resource to develop resistant varieties in the integrated management of the disease.

Ethiopia would have both comparative and competitive advantage in the international market for mungbean as long as there is a concerted effort to improve management practices – including pest management (Burlyaeva *et al.*, 2019) and improved inputs (mainly microbial fertilizer) use to enhance productivity and production (Tenna *et al.*, 2016). First, the crop is grown twice per year. Second, Ethiopia's average yield (1155 kg/ha) is superior to the world average of approximately ~730 kg/ha (Schreinemachers *et al.*, 2019) and there is potential for further improvement. Third, there is already a well-established market, especially in India and Pakistan, and possibly in China.

Lupin

Lupin (*Lupinus albus*) is a traditional cool-season grain legume grown almost exclusively in Gojjam (mainly in the West Gojjam and Awi zones). Westphal (1974) reported *Lupinus albus* and *L. albus* cv.—group *Albus*, whereas Edwards *et al.* (2019) recognize two distinct species of lupin, viz. *L. albus* and *L. mutabilis*.

Data for lupin in Ethiopia are not included in the FAO database. The first report of lupin production in the CSA sample surveys started in the 2003/04 *meher* season. According to the latter, approximately 100,000 smallholder households grew lupin on ~18,000 ha of land and produced ~28,000 MT of grain each year in the 2016–2020 *meher* seasons (cf. Table 5.1). The average land area allocated to lupin is ~0.18 ha per household where the per capita production is ~282 kg/yr. The national average yield is estimated at 1574 kg/ha, which is superior to the world average of 1365 kg/ha (FAO, 2022a). There are no improved cultivars of lupin registered in Ethiopia through to 2017 (MOALR, 2017), but there are reports that three varieties (Sanabor, Vitabor and Welela) were released in recent years for food and feed uses. Because of its adaptation to acid and sandy soils, lupin deserves to be given priority to include as a rotation crop.

Ethiopia's lupin production is typically subsistence, where the indicators fluctuate from year to year (Fig. 5.10). The area, yield and production increased at the rate of 7.11%, 0.41% and 10.67%, respectively, between 2003 and 2020. The number of households growing lupin increased at the rate of 0.16% annually.

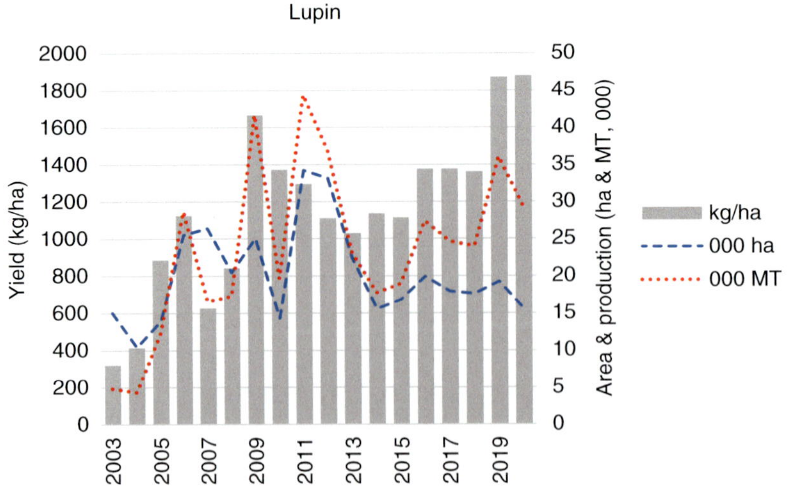

Fig. 5.10. Indicator trends for lupin in Ethiopia. Source: constructed by the lead author from CSA data (www.csa.gov.et).

Ethiopia is ranked 25th out of 27 countries that are known to produce lupin in the world. Australia, Poland, Morocco, the Russian Federation and Germany are among the top producing countries. There is approximately 900,000 ha of land planted to lupin in the world. However, different world regions have different species of this crop. For example, the Mediterranean region grows mainly *L. albus* whereas North Africa and Australia grow *L. angustifolius*, and *L. mutabilis* is grown in Latin America.

Grain legumes trade in Ethiopia

Ethiopia has made tremendous progress in its grain legumes export trade over the years (Fig. 5.11). Export value and quantity showed fast growth starting from the mid-2000s and peaked in 2014. Both tended to decline in 2015, but they trended upwards in 2020. At its peak in 2014, the country exported an estimated 249,000 MT of grain legumes, valued at US$333 million annually.

Each year in 2000–2004 the country exported an estimated 57,000 MT of grain, valued at US$17 million. By contrast, the quantity in 2016–2020 was 249,000 MT annually, for an estimated value of US$158 million – a 4.4-fold increase in quantity and a 9.3-fold increase in earnings.

Overall, Ethiopia is a net exporter of grain legumes. Common bean, chickpea and faba bean accounted for approximately 70%, 24% and 6%, respectively, of the total export earnings, as presented in Table 5.10. Meanwhile, the country's imports of grain legumes included >73,000 MT of grain for an estimated value of US$53 million. Fieldpea and lentil accounted for 94% and 5%, respectively, of the total import bills.

Discussion and Conclusions

In this chapter, we provide a critical review of grain legumes agriculture in Ethiopia. Our study shows that this country has a wide range of grain legumes diversity. There are close to 20 species of this group of crops grown in Ethiopia. Of those, only faba bean, common bean, fieldpea, chickpea, lentil, grasspea, mungbean and lupin are featured in the national statistics, even though the national research system recognizes additional species such as cowpea and adzuki bean (MOALR, 2017). While there is good potential to develop the value chain for cowpea (Alemu *et al*., 2016), the status of adzuki bean and others that are not mentioned here is uncertain.

The average annual ROG for yields of selected grain legumes in Ethiopia was 5.59% compared to 1.53% for the world average

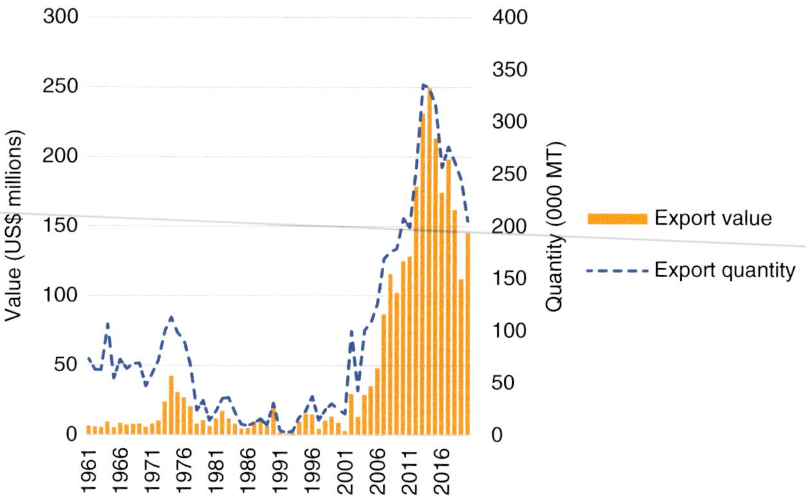

Fig. 5.11. Export value and quantity trends for grain legumes in Ethiopia, 1961–2020. Source: constructed by the lead authors from FAO (2022a).

Table 5.10. Grain legumes trade in Ethiopia (2016–2020 average). Source: constructed by the authors from FAO (2022b).

	Export			Import	
Commodity	US$ 000	MT	Commodity	US$ 000	MT
Common bean	110,067	179,030	Fieldpea	50,025	69,415
Chickpea	38,331	47,511	Lentil	2,884	3,389
Faba bean	9,063	21,525	Common bean	346	293
Pulse flour	593	299	Chickpea	66	46
Fieldpea	208	350	Faba bean	8	9
Lentil	2	5	Pulse flour	6	2
Total	158,264	248,720	Total	53,335	73,155

(Table 5.3). Similarly, the average ROG for production was 8.59% for Ethiopia versus 2.30% for the world average. The benefits accruing from investment in grain legumes in Ethiopia varied from crop to crop, perhaps reflecting the emphasis and efforts invested in each crop. The maximum change in productivity, and therefore in benefits, was seen for chickpea and common bean, two of the major export commodities for the country. Even though the productivity has increased for lentil and fieldpea, the country still needed additional quantities to make up for the shortfalls in domestic demand. The growth in the productivity of grasspea is the result of the farmers' own innovations rather than efforts by the research system. Lupin has remained a subsistence crop mainly because the R&D system has not invested in it.

Ethiopia's export quantity and value of total legumes increased 4.37- and 9.29-fold, respectively, in 2016–2020 compared to 2000–2004. Common bean and chickpea constituted the largest proportion of grain legumes export in this country. The largest buyers for common bean include Pakistan, India, Kenya, Yemen, Russian Federation, South Africa and Türkiye; Pakistan, the United Arab Emirates, India, Sudan, Jordan and Türkiye are among the biggest importers of chickpea from Ethiopia; Sudan is the largest buyer of Ethiopia's faba bean grains, while Vietnam, Indonesia, Malaysia and Taiwan also import from Ethiopia.

This country has huge potential to expand its export market for grain legumes. First, many countries in the developed and emerging economies are moving away from producing at least some of the grain legumes. For example, the area, yield or production for common bean in Brazil, Burundi, Mexico, China, Uganda and Argentina; for chickpea in Pakistan, Iran, Türkiye, Mexico, Canada, Australia, Morocco and Syria; lentil for Türkiye, Iran, Syria, China, Morocco, Pakistan, Greece and the USA; and for faba bean in China, Egypt, Brazil, France, Germany, Paraguay, Sudan, Syria and the UK have been declining over the years (FAO, 2020). Second, Ethiopia has a competitive edge on yield for many of the grain legumes, compared to many other major producers. Third, there is a growing demand for grain legumes in many of the countries mentioned above which can be important destinations for Ethiopia's export market. Fourth, Ethiopia is either the centre of origin or centre of diversity such that there is good opportunity to commercialize those crops that have not been in the system so far. Finally, this country has a well-established R&D system and a long tradition of export for these crops.

Notes

[1] The first grain legume variety, Mexican-142, was released in 1973.

[2] Based on the 2013/14 and 2015/16 averages.

References

Abate, T., Orr, A. (2011) A knowledge-based strategy towards tropical legumes research and development. *Journal of SAT Agricultural Research* 10, 1–12.

Abate, T., Shiferaw, B., Gebeyehu, S., Amsalu, B., Negash, K. *et al.* (2011) A systems and partnership approach to agricultural research for development: Lessons from Ethiopia. *Outlook on Agriculture* 40(3), 213–220.

Ahmed, K.A. (2002) An integrated approach to pest management in fieldpea, *Pisum sativum* (L.), with emphasis on pea aphid, *Acyrthosiphon pisum* (Harris). PhD thesis, Department of Zoology & Entomology, Faculty of Natural and Agricultural Sciences, University of the Free State, Bloemfontein, South Africa. 313 pp.

Alemu, M., Asfaw, Z., Woldu, Z., Fenta, B.A., Medvecky, B. (2016) Cowpea (*Vigna unguiculata* (L.) Walp.) (Fabaceae) landrace diversity in northern Ethiopia. *International Journal of Biodiversity and Conservation* 8(11), 297–309. DOI: 10.5897/IJBC2016.0946.

Ali, K. (1997) Economic threshold for pea aphid. *Pest Management Journal of Ethiopia* 1(1&2), 63–68.

Ali, K., Habtewold, T. (1994) Research on insect pests of cool season food legumes. In: Asfaw T. *et al.* (eds) *Cool-Season Food Legumes of Ethiopia*. IAR, Addis Ababa, Ethiopia, pp. 367–396.

Ali, K., Louw, S., van den, M., Swart, W.J. (2005) Components and mechanisms of resistance in selected fieldpea *Pisum sativum* lines to the pea aphid *Acyrthosiphon pisum* (Homoptera: Aphididae). *International Journal of Tropical Insect Science* 25(2), 114–121.

Asegie, T.A. (2015) Mungbean (*Vigna radiata* (L.) Wilczek) (Fabaceae) Landrace Diversity in Ethiopia. MSc thesis, Addis Ababa University, Addis Ababa, Ethiopia.

Assefa, H., Gorfu, D. (1986) A review of grain legume diseases research in Ethiopia. In: Abate, T. A. (ed) *A Review of Crop Protection Research in Ethiopia*. IAR, Addis Ababa, Ethiopia.

Assefa, T., Reda, F., Amsalu, B., Abate, T. (2006) Promotion of improved haricot bean production systems in East Shewa: An example of a successful partnership among stakeholders. In: Abate, T. (ed) *Successes with Value Chain. Proceedings of Scaling up and Scaling out Agricultural Technologies in Ethiopia: An International Conference, 9-11 May 2006.* EIAR, Addis Ababa, Ethiopia. 275 pp.

Auclair, J.L., Maltais, J.B., Cartier, J.J. (1957) Factors in resistance of peas to the pea aphid, *Acyrthosiphon pisum* (Harr.) (Homoptera: Aphididae). II. Amino acids. *The Canadian Entomologist* 89(10), 457–464.

Belay, G., Tesfaye, K., Hamwieh, A., Ahmed, S., Dejene, T. *et al.* (2020) Genetic diversity of *Orobanche crenata* populations in Ethiopia using microsatellite markers. *International Journal Genomics* 2020, 3202037. DOI: 10.1155/2020/3202037.

Boivin, G., Hance, T., Brodeur, J. (2012) Aphid parasitoids in biological control. *Canadian Journal of Plant Science* 92(1), 1–12. DOI: 10.4141/cjps2011-045.

Burlyaeva, M., Vishnyakova, M., Gurkina, M., Kozlov, K., Lee, C.-R. *et al.* (2019) Collections of mungbean [*Vigna radiata*) (L.) R. Wilczek] and urdbean [*V. mungo* (L.) Hepper] in Vavilov Institute (VIR): Traits diversity and trends in the breeding process over the last 100 years. *Genetic Resources and Crop Evolution* 66, 767–781. DOI: 10.1007/s10722-019-00760-2.

Cartier, J.J. (1963) Varietal resistance of peas to pea aphid biotypes under field and greenhouse conditions. *Journal of Economic Entomology* 56(2), 205–213. DOI: 10.1093/jee/56.2.205.

Chow, F.J., Mackauer, M. (1984) Inter- and intraspecific larval competition in *Aphidius smithi* and *Praon pequodorum* (Hymenoptera: Aphidiidae). *Canadian Entomologist* 116, 1097–1107.

Chow, F.J., Mackauer, M. (1985) Multiple parasitism of the pea aphid: Stage of development of parasite determines survival of *Aphidius smithi* and *Praon pequodorrum* (Hymenoptera: Aphidiidae). *Canadian Entomologist* 117, 133–134.

Chow, F.J., Mackauer, M. (1986) Host discrimination and larval competition in the aphid parasite *Ephedrus californicus*. *Entomologia Experimentalia & Applicata* 41, 243–254.

Corinto, G.L. (2014) Nikolai Vavilov's centers of origin of cultivated plants with a view to conserving agricultural biodiversity. *Human Evolution* 29(4), 285.

CSA (Central Statistical Agency) (2016) Agricultural Sample Survey 2015/16 (2008 E.C.), Report on Area and Production of Major Crops (Private Peasant Holdings, *Meher* Season), Volume I, Statistical Bulletin. CSA, Addis Ababa, Ethiopia.

CSA (2019) Agricultural Sample Survey 2018/19 (2011 E.C.), Report on Area and Production of Major Crops (Private Peasant Holdings, *Meher* Season), Volume I, Statistical Bulletin 589. CSA, Addis Ababa, Ethiopia.

Edwards, S.B. *et al.* (2019) *A Guide to Edible Legumes Found in Ethiopia: For Extension Officers and Researchers*. Mekelle University Press, Mekelle, Ethiopia. 109 pp. ISBN 978-99944-74-56-1.

Eshete, M., Aliye, S., Assefa, S., Ahmed, S., Tadesse, N. *et al.* (2006) Improved lentil technologies improve farmers' livelihoods in Gimbichu Wereda, East Shewa. In: Abate, T. (ed) *Successes with Value Chain, Proceedings of Scaling up and Scaling out Agricultural Technologies in Ethiopia: An International Conference, 9-11 May 2006*. EIAR, Addis Ababa, Ethiopia, pp. 47–55.

FAO (Food and Agriculture Organization) (2020) Statistical databases and data-sets of the Food and Agriculture Organization of the United Nations. Available at: http://www.fao.org/faostat/en/#data (accessed 9 May 2020).

FAO (2022a) Statistical databases and data-sets of the Food and Agriculture Organization of the United Nations. Available at: http://www.fao.org/faostat/en/#data (accessed 10 March 2022).

FAO (2022b) Statistical databases and data-sets of the Food and Agriculture Organization of the United Nations. Available at: http://www.fao.org/faostat/en/#data (accessed 16 April 2022).

Fernandez, T.A, Carillo, E., Valdez-Patto, M.C., Rubiales, D. (2014) Screening pea (*Pisum sativum*) for aphid (*Acyrthosiphon pisum*) and weevil (*Bruchus pisorum*) resistance under field and under controlled conditions. Institute for Sustainable Agriculture, CSIC, Apdo. 4084, 14080, Cordoba, Spain.

Girma, D., Korbu, L. (2012) Genetic improvement of grasspea (*Lathyrus sativus*) in Ethiopia: An unfulfilled promise. *Plant Breeding* 131, 231–236. DOI:10.1111/j.1439-0523.2011.01935.x 2012.

Hailu, E., Getaneh, G., Sefera, T., Tadesse, N., Bitew, B. *et al.* (2014) Faba bean gall; a new threat for faba bean (*Vicia faba*) production in Ethiopia. *Advances in Crop Science and Technology* 2, 144. DOI: 10.4172/2329-8863.1000144.

Mackauer, M. (1982) Fecundity and host utilization of the aphid parasite *Aphidius semiflavus* (Hymentoptera: Aphelinidae) at two host densities. *Canadian Entomologist* 114, 721–726.

Mackauer, M. (1983) Quantitative assessment of *Aphidius smithi* (Hymenoptera: Aphidiidae): Fecundity, intrinsic rate of increase, and functional response. *Canadian Entomologist* 115, 399–415.

Mackauer, M., Ehler, L.E., Roland, J. (eds) (1990) *Critical Issues in Biological Control*. Intercept, Andover, Hants.

MOALR (Ministry of Agriculture and Livestock Resources) (2017) *Plant Variety Release, Protection and Seed Quality Control Directorate*. Crop Variety Register Issue No. 20, Addis Ababa, Ethiopia.

Ruelle, M.L., Asfaw, Z., Dejen, A., Tewolde-Berhan, S., Nebiyu, A. *et al.* (2019) Inter- and intraspecific diversity of food legumes among households and communities in Ethiopia. *PLoS ONE* 14(12), e0227074. DOI: 10.1371/journal.pone.0227074.

Schreinemachers, P., Sequeros, T., Rani, S., Rashid, M.A., Gowdru, N.V. *et al.* (2019) Counting the beans: Quantifying the adoption of improved mungbean varieties in South Asia and Myanmar. *Food Security* 11, 623–634. DOI: 10.1007/s12571-019-00926-x.

Sequeros, P. *et al.* (2021) Mungbean in Southeast Asia and East Africa. *Agriculture and Food Security* 10, 2. DOI: 10.1186/s40066-020-00273-7.

Tekle-Haimanot, R., Abegaz, B.M., Wuhib, E., Kassegn, A., Kidane, Y. *et al.* (1993) Pattern of *Lathyrus sativus* (grasspea) consumption and beta-N-oxalyl-α-β-diaminoproprionic acid (β-ODAP) content of food samples in the lathyrism endemic region of northwest Ethiopia. *Nutrition Research* 13(10), 1113–1126. DOI: 10.1016/S0271-5317(05)80736-5.

Tena, W., Wolde-Meskel, E., Walley, F. (2016) Symbiotic efficiency of native and exotic Rhizobium strains nodulating lentil (*Lens culinaris* Medik.) in soils of southern Ethiopia. *Agronomy* 6(1), 11. DOI: 10.3390/agronomy6010011.

Tesfaye, A., Wale, M., Azerefegne, F. (2013) *Acyrthosiphon pisum* (Harris) (Homoptera: Aphididae) feeding preference and performance on cool-season food legumes in northwestern Ethiopia. *International Journal of Pest Management* 59(4), 319–328.

Teshome, T., Bryngelsson, T., Dagne, K., Geleta, M. (2015) Assessment of genetic diversity in Ethiopian fieldpea (*Pisum sativum* L.) accessions with newly developed EST-SSR markers. *BMC Genetics* 16, 102. DOI: 10.1186/s12863-015-0261-5.

Verkaart, S., Munyua, B.G., Mausch, K., Michler, J.D. (2017) Welfare impacts of improved chickpea adoption: A pathway for rural development in Ethiopia? *Food Policy* 66, 50–61.

Westphal, E. (1974) Pulses in Ethiopia, their taxonomy and agricultural significance. Doctoral thesis, Wageningen. ISBN 90 220 0501 1, (viii) + 276 p.

Westphal, E. (1975) *Agricultural Systems in Ethiopia*. Joint publication of the College of Agriculture, Haile Sellassie I University, Ethiopia, and the Agricultural University, Wageningen, the Netherlands.

Yoseph, T., Mekbib, F., Amsalu, B., Tadele, Z., Alemaw, G. (2020) Screening mung bean [*Vigna radiata* (L.) Wilczek] genotypes for drought tolerance. *Ethiopian Journal of Agricultural Sciences* 30(4), 223–236.

Zeven, A.C., de Wet, J.M.J. (1982) *Dictionary of Cultivated Plants and their Regions of Diversity. Excluding most Ornamentals, Forest Trees and Lower Plants*. Centre for Agricultural Publishing and Documentation, Wageningen, The Netherlands.

6 The Rise of Oilseeds in Export Diversification and Earnings

Tsedeke Abate[1]* and Getinet Alemaw[2]

[1]*Homegrown Vision, Addis Ababa, Ethiopia;* [2]*Ethiopian Institute of Agricultural Research, Addis Ababa, Ethiopia*

Abstract

This chapter presents an assessment of the performance of edible oilseeds in terms of their contributions to alleviating heavy reliance on coffee for foreign exchange earnings as well as nutrition diversity and security, and raw material supply for the local food industry. Edible oilseeds cultivated in Ethiopia include *noug* (*Guizotia abyssinica*), linseed, *gomenzer* (*Brassica carinata*), sesame, groundnut, safflower and soybean, together employing >3,000,000 smallholders on ~853,000 ha of land. The aggregate yield gain was 49 kg/ha/yr and highly significant ($P<0.01$, $R^2=0.94$), ranging from 110 kg/ha/yr for soybean to 7 kg/ha/yr for sesame. Export earnings rose from ~US$32.8 million (~10% of the total) in 2000 to ~US$355.8 million (~24%) in 2020. Oilseed export comprised 99.96% raw seed while the import was 99.88% processed. Current yields are still low compared to their potential, and research is needed to introduce modern technologies to increase productivity and reduce the burgeoning oil import costs.

Introduction

There are seven economically important edible oilseeds cultivated in Ethiopia, according to the records of the Central Statistical Agency (CSA) of the Ethiopian government. These include *noug* (*Guizotia abyssinica*), linseed (*Linum usitatissimum*), rapeseed (includes both *Brassica napus* and *B. carinata*), sesame (*Sesamum indicum*), groundnut (*Arachis hypogea*), soybean (*Glycine max*) and safflower (*Carthamus tinctorius*). According to the Food and Agriculture Organization of the United Nations (FAO, 2022a), this country also grows nearly 6000 ha of sunflower and more than 70,000 ha of seed cotton as sources of edible oil. Manore and Yohannes (2019) report unknown amounts of camelina (*Camelina sativa*) grown for edible oil purposes. With the exception of soybean, which was introduced in the 1950s, Ethiopia is either the primary gene centre (*noug*) or secondary gene centre for many of the crops (Westphal, 1975) which smallholder farmers have grown traditionally in many parts of the country over the centuries.

Oilseeds in Ethiopia are grown for food and nutrition security, export earnings and raw material for the local industry. According to the FAO (2020), Ethiopians obtain about 29% of their dietary fat from oilseeds. On average, they accounted for 24% of the export earnings in 2015–2019, second only to coffee.

Ethiopia has invested in the research and development of oilseeds over the last several decades and achieved varying levels of productivity gains. Wijnands *et al.* (2007) and Allaro (2011) have reviewed the export market for oilseeds in

* Corresponding author: tsedeke@homegrownvision.org

 Untold Stories of African Agriculture: Lessons from Ethiopia (T. Abate ed.)
DOI: 10.1079/9781800626386.0006

Ethiopia and Alemaw (2013) provides a detailed review of the challenges facing increased production of these crops in the country. This chapter presents an assessment of the current status of oilseeds and offers the path forward for their increased productivity and production.

Approach

To measure increases in productivity of the crops, we calculated the yield gain as well as the annual rate of growth (ROG) for area, yield and production, using the procedures used for cereals in Chapter 3.

CSA and FAOSTAT were the major sources of data presented here. The CSA data for the major oilseeds are available starting from the 1994/95 *meher* (main rainy season), but the more complete data (including number of smallholders growing the crops and production data) are given from 2004/05 for all but soybean, for which data are complete from 2010 onwards. The FAO data for linseed, sesame and groundnut are available starting in 1961, whereas data for rapeseed (*gomenzer*) and safflower are given starting in 1994.

Results

Oilseeds in general

According to the CSA, more than 3 million smallholder households grew oilseeds on ~853,000 ha of land and produced ~950,000 MT annually in 2016–2020 (Table 6.1). The average farm size is estimated at 0.22 ha per household (ranging from 0.03 ha for rapeseed to 0.50 ha for sesame). The per capita production per household per year is estimated at ~300 kg. The research system has released 116 cultivars of oilseeds as of 2017.

The yield gain for the aggregate oilseeds was 49 kg/ha/yr between 2000 and 2020 and highly significant ($P<0.01$, $R^2=0.94$,); meanwhile, the ROGs for area, yield and production increased at an annual rate of 3.12%, 5.03% and 8.51%, respectively (Table 6.2). However, the yield gains and ROGs varied from crop to crop.

With the exception of sesame, all crops showed highly significant yield gains. Even though the yield and production showed positive ROGs, there were differences among the crops in area. For example, soybean, sesame and groundnut showed fast growth in area, while it shrank for linseed, safflower, *gomenzer* and *noug* (Table 6.2).

In summary, Ethiopia has doubled the yields of all the oilseeds and their aggregate during the period mentioned above (Fig. 6.1). The only exception was sesame, whose yield increased by 75% over the 21 years between 2000 and 2020. According to FAO (2022b), Niger, Ethiopia (ranked 27th) and the Ukraine are the only major growing countries that have doubled their oilseeds yield during this period.

China, the USA, Argentina, Brazil, Canada, Paraguay, India, the Russian Federation, Nigeria and Sudan are the top ten oilseed-producing countries out of the 162 countries in the world that cultivate oilseeds. The aggregate oilseeds yield for Ethiopia was estimated at nearly 1000 kg/ha

Table 6.1. Indicators for major edible oilseeds smallholders cultivated in the *meher* season in Ethiopia (2016–2020 average). Source: number of holders, area, yield and production are constructed by the author from CSA sample surveys (www.csa.gov.et); area per holder and per capita production are calculated by the author from CSA data; number of varieties are from MOALR (2017).

Crop	Holders (000)	Area (000 ha)	Yield (kg/ha)	Production (000 MT)	Area per holder (ha)	Production per holder (kg)	Number of varieties released
Noug	835	256	1117	286	0.31	342	5
Linseed	769	78	1107	87	0.10	113	18
Gomenzer	679	19	1784	34	0.03	50	11
Sesame	628	350	714	250	0.56	398	29
Groundnut	473	88	1768	156	0.19	330	25
Safflower	184	7	1184	8	0.04	43	2
Soybean	158	56	2318	130	0.35	822	26
Total/average	3048	853	1427	950	0.22	300	116

Table 6.2. Yield gain (b-value, kg/ha/yr) and annual rate of growth (ROG) for major oilseeds cultivated in Ethiopia compared to the top ten cultivating countries of the world (2000–2020). Source: calculated and constructed by the lead author from FAO (2022a).

	Yield gain			ROG (%)		
Crop	b-value	P-value	R^2	Area	Yield	Production
Soybean	110	<0.01	0.81	19.87	9.49	31.25
Gomenzer	59	<0.01	0.80	–0.70	4.91	4.17
Groundnut	49	<0.01	0.81	10.68	3.85	14.94
Noug	45	<0.01	0.92	–0.66	6.83	6.12
Linseed	38	<0.01	0.79	–3.26	5.07	1.64
Safflower	36	<0.01	0.77	–1.02	4.10	3.04
Sesame	7	0.14	0.11	11.32	1.26	12.73
Total	49	<0.01	0.94	3.12	5.03	8.51

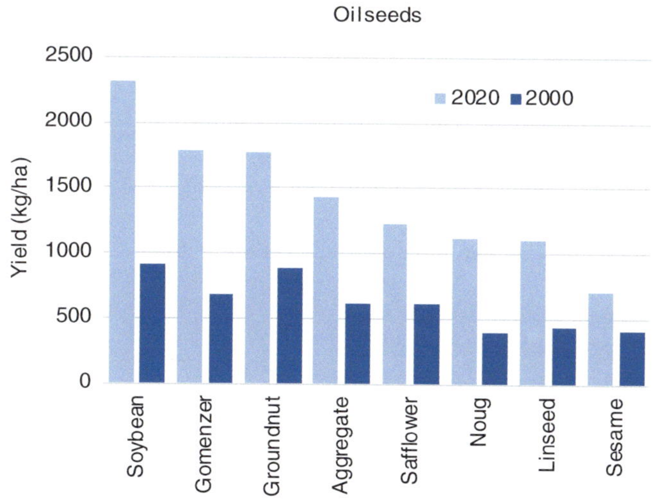

Fig. 6.1. Oilseeds yield during two periods in Ethiopia. Source: calculated and constructed by the lead author from FAO (2022a).

compared to about 1400, 1200, 1000 and 1000 kg/ha, respectively, for the first four countries and about 500 kg/ha for the world average. Average yields for Canada, Paraguay, India, the Russian Federation, Nigeria and Sudan were about 700, 600, 600, 500, 300 and 100 kg/ha, respectively (FAO, 2022b).

Performance of individual crops

Noug

Ethiopia is the primary gene centre for *noug* (variously reported as *neug*, *noog*, or *nug*). It occupies more than ~256,000 ha of land and employs ~835,000 smallholder households, the largest for all oilseeds in Ethiopia (cf. Table 6.1). The landholding per farmer is 0.31 ha and the national average yield is estimated at about 1100 kg/ha. Per capita production is ~342 kg/household/yr.

Noug is well adapted to altitudes ranging from about 1500 m to 2400 m above sea level. Oromia, Amhara, Benshangul-Gumuz (BSG) and Tigray account for approximately 71%, 22%, 5% and 2%, respectively, of the total area occupied by this crop in Ethiopia in the 2018 *meher* season (CSA, 2019). *Noug* in Ethiopia is produced both for domestic consumption where it is used for high-quality cooking oil extraction (and high protein cake for animal feed) as well as for export. Out of the 2016–2020 average annual production of ~286,000 MT, this country exported more than 26,000 MT (or approximately 9%).

The *noug* indicator trends for Ethiopia are shown in Fig. 6.2. The area occupied by this crop has declined at an annual rate of 0.66% even though the yield and production increased by 6.83% and 6.12%, respectively, between 2000 and 2020 (cf. Table 6.2). *Noug* area appeared to continue declining because the traditional *noug*-growing areas are now cultivating either hybrid maize as the case in West Gojjam, Wellega and Assossa, or rice in the Fogera plain. The CSA (2020) data show that the number of smallholders cultivating this crop in Ethiopia declined by 2.04% per annum between 2003 and 2020.

The number of smallholder households growing *noug* also declined at the rate of 2.04% between 2003 and 2019, according to the CSA figures. Farmers have shifted from *noug* to *teff* because the latter is more profitable. Research has not made breakthroughs to significantly increase the productivity of *noug*. The yield jumped from 398 kg/ha in 2000 to 1120 kg/ha in 2020. It was not possible to make yield comparisons with other countries because the FAO data do not list *noug* specifically, it is presented as "Oilseeds, nes" (meaning, oilseeds not elsewhere specified).

Linseed

Ethiopia is the secondary gene centre for linseed (Vavilov, 1951). It is a traditional crop grown for its cooking oil. Approximately 769,000 smallholder households grew linseed on ~78,000 ha of land and produced ~87,000 MT annually in the 2016–2020 *meher* seasons (cf. Table 6.1). The average land area allocated to linseed is estimated at 0.10 ha per household, with the per capita production of about 113 kg/household/yr.

Linseed in Ethiopia grows at altitudes of 1600–2700 m above sea level. The largest producer is Oromia (66%), followed by Amhara (26%), Tigray (6%), BSG (1%) and Southern Nations, Nationalities and Peoples Region (SNNPR) (1%). Ethiopia ranked seventh (after Kazakhstan, the Russian Federation, Canada, China, India and the USA) both in the area harvested and production, out of 55 countries in the world that produced linseed (FAO, 2022a). This country accounts for ~3% of the world's total production of over 3 million MT.

The yield gain was estimated at 38 kg/ha/yr (Table 6.3) and highly significant ($P<0.00$, $R^2=0.79$), the highest among the top ten producing countries of the world between 2000 and 2020. It was followed by the UK, China and Canada; the yields declined in Egypt, France and Belgium.

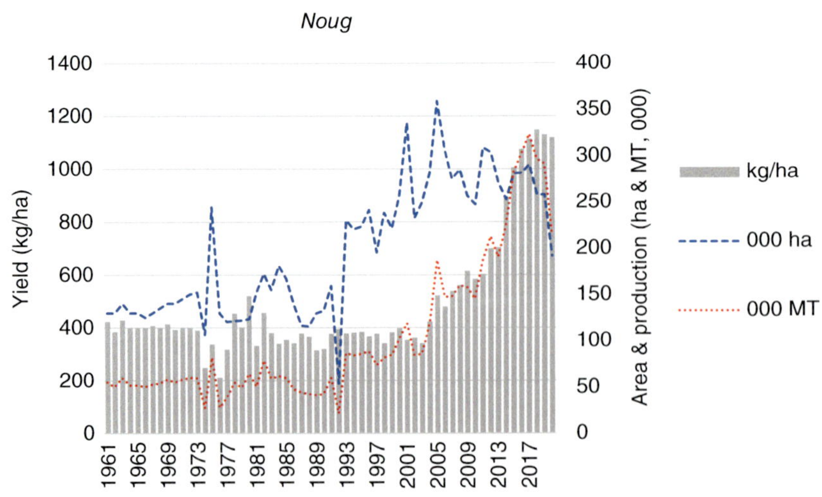

Fig. 6.2. Indicator trends for *noug* in Ethiopia, 1961–2020. Source: constructed by the lead author from FAO (2022c).

For Ethiopia, the area declined by 3.26%, whereas the yield and production increased at an annual rate of 5.07% and 1.64%, respectively. The area declined in eight and production in six of the top ten producing countries.

Figure 6.3 depicts linseed indicator trends for Ethiopia. The production system for linseed in this country is a subsistence mode – where the bulk of increases in production are attributed to increases in the area rather than increases in yield. According to the CSA, the number of smallholders growing linseed has also declined at the rate of 3.08% per year between 2003 and 2019.

Gomenzer

The CSA data do not make a distinction between rapeseed and Ethiopian mustard (and reports them as rapeseed), whereas the FAOSTAT treats the two crops separately. According to the former

Table 6.3. Yield gain (b-value, kg/ha/yr) and annual rate of growth for linseed in Ethiopia compared to the top ten cultivating countries of the world (2000–2020). Source: calculated and constructed by the lead author from FAO (2022a).

	Yield gain			ROG (%)		
Country[a]	b-value	P > ltl	R^2	Area	Yield	Production
Ethiopia	38	<0.01	0.79	–3.26	5.07	1.64
UK	34	<0.05	0.29	–1.81	2.65	0.79
China	32	<0.01	0.83	–3.32	3.21	–0.21
Canada	24	<0.05	0.56	–3.35	1.87	–1.54
Argentina	23	<0.01	0.61	–6.13	2.37	–3.91
Poland	23	<0.05	0.34	12.11	2.01	14.36
USA	7	0.45	0.03	–5.02	0.51	–4.53
Belgium	–5	0.44	0.04	–17.90	19.33	–2.03
France	–14	0.02	0.24	5.70	–0.73	4.93
Egypt	–18	<0.05	0.40	–6.02	–1.10	–7.06

[a]Countries with the largest cultivated area (descending order) are France, UK, Belgium, Egypt, Canada, Poland, China, USA, Argentina and Ethiopia.

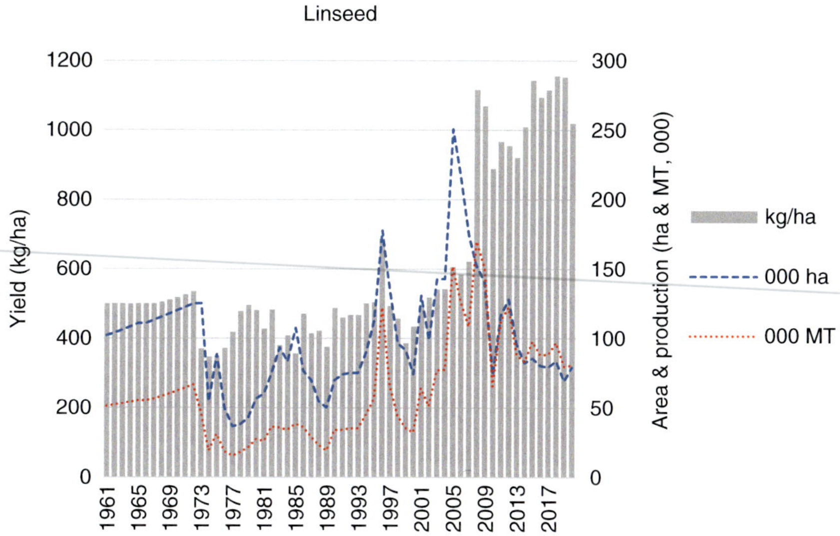

Fig. 6.3. Indicator trends for linseed in Ethiopia, 1961–2020. Source: constructed by the lead author from FAO (2022a).

Table 6.4. Yield gain (b-value, kg/ha/yr) and annual rate of growth (ROG) for *gomenzer* in Ethiopia compared to the top ten cultivating countries of the world (2000–2020). Source: calculated and constructed by the lead author from FAO (2022a).

	Yield gain			ROG (%)		
Country[a]	b-value	P > |t|	R^2	Area	Yield	Production
Ukraine	96	<0.01	0.86	13.68	5.80	20.27
Ethiopia	59	<0.01	0.80	–0.70	4.91	4.17
Canada	53	<0.01	0.81	4.49	2.93	7.55
USA	30	<0.01	0.58	2.87	1.71	4.63
Poland	29	0.03	0.24	4.25	1.16	5.46
China	25	<0.01	0.84	–0.21	1.41	1.20
Russian Federation	24	<0.05	0.48	13.45	1.97	15.69
India	20	<0.01	0.65	0.92	1.78	2.72
France	11	0.36	0.04	1.05	0.35	1.41
Australia	9	0.32	0.05	4.62	0.95	5.61
Germany	4	0.80	0.00	–0.58	0.12	–0.45

[a]Countries with the largest cultivated area (descending order) are Canada, China, India, Australia, France, Russian Federation, Germany, the Ukraine, Poland and USA; Ethiopia is 42nd.

source, ~679,000 smallholder households grew this crop on ~19,000 ha of land and produced ~34,000 MT each year, with the average yield of ~1800 kg/ha in 2016–2020 (cf. Table 6.2). The average area allocated to rapeseed in Ethiopia is 0.03 ha/household and the per capita production about 50 kg/household/yr. The national research system has released 11 cultivars as of 2017 (MOALR, 2017).

Ethiopia ranked 42nd in area, 43rd in yield and 44th in production out of 64 countries that cultivated *gomenzer* in 2016–2020 (according to FAO, 2022a). Canada, China, India, Australia, France, the Russian Federation, Germany, the Ukraine, Poland and the USA are the world leaders in the area cultivated of this crop. This country has registered a very highly significant ($P<0.00$, $R^2=0.80$) yield gain of 59 kg/ha/yr during the 2000 to 2020 period (Table 6.4). *Gomenzer* area in Ethiopia declined at an annual rate of 0.70%, whereas the yield and production increased at 4.91% and 4.17%, respectively.

Figure 6.4 presents rapeseed indicator trends for Ethiopia between 2000 and 2020. The area fluctuated from early 2000s and has been falling since 2014; production also followed a similar pattern. The yield also showed some growth through 2014 but was little changed through 2019, after which it showed a decline. The decline in area appeared to be related to the farmers shifting from rapeseed to bread wheat, perhaps because the latter is more profitable. This is particularly true in the Gojjam, Arsi and Bale highlands.

Sesame

Ethiopia is the secondary gene centre for sesame (Vavilov, 1951). It is a traditional crop grown in the lower altitudes (500–1500 m above sea level) of the country. It is also cultivated at altitudes as low as 300 m (e.g. Omorate, Kangaten) under irrigation. Close to 630,000 smallholder households grew sesame on ~350,000 ha of land and produced ~250,000 MT seed annually in the 2016–2020 *meher* seasons (cf. Table 6.1). The average farm size for sesame in Ethiopia is estimated at 0.56 ha, the largest for all crops. On average, a smallholder household produces ~400 kg sesame seed per year. The Amhara, Tigray, Oromia and BSG regional states had a share of roughly 55%, 33%, 8% and 5%, respectively, of the total area planted to sesame in the 2018/19 *meher* season (CSA, 2019). Sesame is the second largest earner of all agricultural exports in Ethiopia, after coffee.

Ethiopia ranked eighth in the area harvested and seventh in production among 74 countries in the world that cultivated sesame in 2016–2020 (FAO, 2022a). Sudan, India, Myanmar, Tanzania, Nigeria, South Sudan, Burkina Faso, Chad and China are the other top ten sesame-cultivating countries of the world.

The yield gain for Ethiopia was 7 kg/ha/yr and non-significant ($P=0.14$, $R^2=0.11$), as

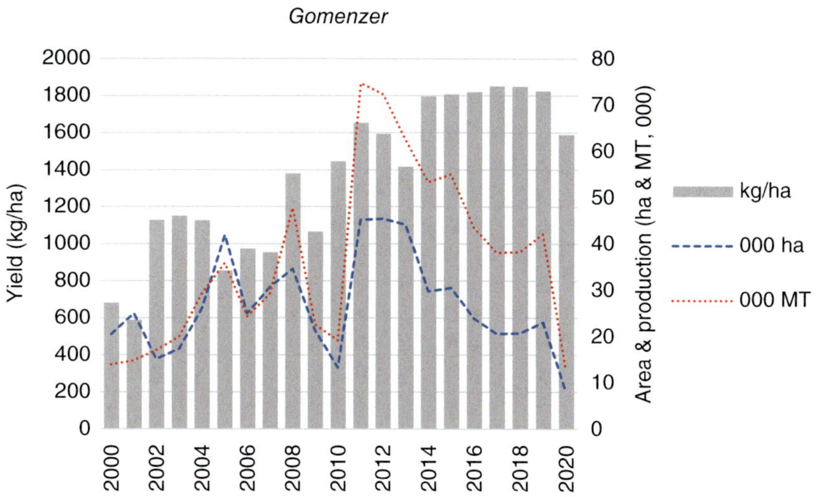

Fig. 6.4. Indicator trends for *gomenzer* in Ethiopia, 2000–2020. Source: constructed by the lead author from FAO (2022c).

Table 6.5. Yield gain (b-value, kg/ha/yr) and annual rate of growth (ROG) for sesame in in Ethiopia compared to the top ten cultivating countries of the world (2000–2020). Source: calculated and constructed by the lead author from FAO (2022a).

	Yield gain			ROG (%)		
Country[a]	b-value	P > |t|	R^2	Area	Yield	Production
China	32	<0.01	0.74	–5.61	2.48	–3.27
Nigeria	25	0.06	0.17	9.82	3.84	14.04
Myanmar	11	<0.01	0.57	1.14	2.83	4.00
Tanzania	11	0.15	0.11	18.04	2.23	20.67
Chad	10	<0.01	0.61	10.64	2.24	13.12
Burkina Faso	8	0.02	0.26	18.14	1.74	20.20
Ethiopia	7	0.14	0.11	11.32	1.26	12.73
Sudan	6	<0.01	0.68	NA	NA	NA
India	6	<0.05	0.39	–0.18	1.45	1.27
South Sudan	3	0.14	0.28	NA	NA	NA

[a]Countries with the largest cultivated area (descending order) are Sudan, India, Myanmar, Tanzania, Nigeria, South Sudan, Burkina Faso, Ethiopia, Chad and China
NA = not available.

presented in Table 6.5. By contrast, China has registered 32 kg/ha/yr, followed by Nigeria, Myanmar and Tanzania.

Ethiopia, along with Tanzania, Chad and Burkina Faso, was among the countries with double-digit ROG for both area and production. By contrast, both area and production declined for China; the area showed a slight negative change for India.

Figure 6.5 depicts indicator trends for sesame in Ethiopia. The area and production showed fast growth from the early 2000 but slowed down and fluctuated after 2010. The fast growth in area may be due to the increasing cultivation of the crop under irrigation in the Omo and Awash valleys. It should be noted that both the area and production are showing downward trends in recent years. Experience shows that the rainfall intensity in the western highlands has been increasing in recent years – creating favourable conditions for diseases and shattering. However, sesame production in irrigated lowlands (including Omo1, Omo2, Omo3 and Omo4) is increasing at a rapid rate.

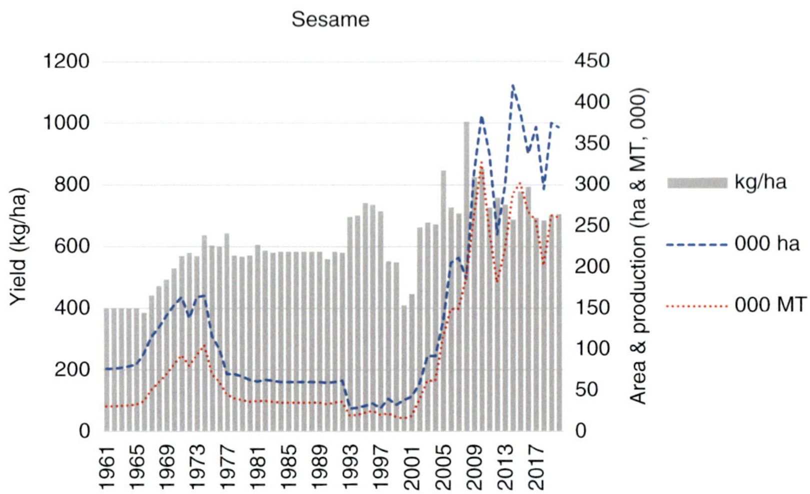

Fig. 6.5. Indicator trends for sesame in Ethiopia, 1961–2020. Source: constructed by the lead author from FAO (2022c).

The average yield for Ethiopia is a little over 700 kg/ha, the fourth largest among the top 20 producing countries, but slightly below the world average of about 770 kg/ha (FAO, 2022a). The yields for the top 20 producing countries range from nearly 1600 kg/ha for China to a little less than 300 kg/ha for Sudan (which is the highest producer in the world). Our experience shows that sesame yield in marginal areas (such as Kobo) is <500 kg/ha, whereas in areas with good rainfall and irrigation it is about 900 and 2100 kg/ha, respectively.

Groundnut

Groundnut in Ethiopia is grown largely for domestic consumption where it is used for oil extraction and snack food. There was no evidence of groundnut (or any of its products) export from Ethiopia in the 2016–2020 period. More than 470,000 smallholder households planted approximately 88,000 ha of land to this crop and produced 156,000 MT of seed annually in the 2016–2020 *meher* season (cf. Table 6.1). The estimated average farm size is about 0.19 ha/household and the per capita production about 300 kg/household/yr. The national research system has released 25 improved varieties as of 2017. This crop is adapted to 300–1600 m above sea level and is largely produced in Oromia, esp. East Hararge, (72%), BSG (25%) and Harari Region (3%).

Several studies have shown that groundnut cultivated in the lowlands under irrigation is the best choice for self-sufficiency, due to its high oil content (48%) and yield under irrigation. Again, this crop is cultivated under marginal rainfall such as Babile and Bisidimo, high rainfall area such as Gambella, and under irrigation in the Omo and Awash valleys. The average oil yield per hectare is the highest for groundnuts among all oilseeds.

When cultivated under irrigation, the crop is clean from rust as compared to high rainfall areas such as Gambella and Fincha. The major problem of ground nut is lifting and shelling, which can be performed using animal- or tractor-drawn implements.

Ethiopia ranked 33rd in area and 29th in production out of 114 countries in the world that produced groundnut in 2016–2020 (FAO, 2022a). China, India, Nigeria, the USA, Sudan and Senegal are the top producers.

Ethiopia has registered a highly significant ($P<0.00$, $R^2=0.81$) yield gain of 49 kg/ha/yr (Table 6.6). This is the second highest yield gain after China. By contrast, Guinea and Nigeria showed highly significant yield losses.

Ethiopia had the fastest ROG for groundnut yield (3.85% per annum) in comparison with the top ten cultivating countries of the world. Figure 6.6 presents groundnut indicator trends for Ethiopia. Both the area and yield (and therefore production) showed upward trends even

Table 6.6. Yield gain (b-value, kg/ha/yr) and annual rate of growth (ROG) for groundnut in Ethiopia compared to the top ten cultivating countries of the world (2000–2020). Source: calculated and constructed by the lead author from FAO (2022a).

	Yield gain			ROG (%)		
Country[a]	b-value	P > ltl	R^2	Area	Yield	Production
China	53	<0.01	0.91	–0.23	1.60	1.37
Ethiopia	49	<0.01	0.81	10.68	3.85	14.94
India	34	<0.05	0.50	–1.36	2.79	1.40
Senegal	24	0.01	0.32	2.20	2.74	5.00
Myanmar	15	<0.05	0.52	3.39	1.06	4.48
Chad	13	0.01	0.33	3.97	1.30	5.33
Niger	11	<0.01	0.62	6.91	2.32	9.39
Sudan	3	0.72	0.01	5.81	0.81	6.66
Tanzania	3	0.72	0.01	9.37	0.36	9.77
Nigeria	–24	<0.05	0.56	4.03	–1.77	2.19
Guinea	–26	<0.05	0.46	10.30	–2.32	7.74

[a]Countries with the largest area (in descending order) are India, China, Nigeria, Sudan, Senegal, Myanmar, Tanzania, Niger, Guinea and Chad; Ethiopia is 33rd.

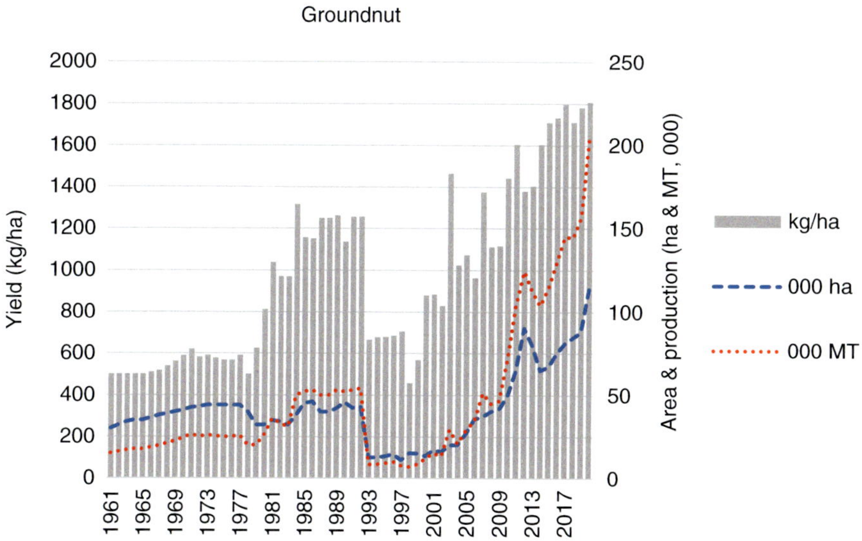

Fig. 6.6. Indicator trends for groundnut in Ethiopia, 1961–2020. Source: constructed by the lead author from FAO (2022c).

though the area increased faster than the yield. The CSA data also indicate that the number of smallholder farmers producing groundnut in Ethiopia increase at the rate of 7.49% per annum between 2003 and 2018. All indicators (area, production and yield) showed upward trends starting from the early 2000s. The area and production appeared to continue growing while the yield seemed to level off around 1800 kg/ha.

Safflower

Ethiopia is the secondary gene centre for safflower (Vavilov, 1951) and it is traditionally cultivated in the mid-altitudes (1600–2200 m above sea level) in central and north-central Ethiopia. The Amhara and Oromia regional states accounted for 83% and 17%, respectively, of the total safflower production in this country in the 2018/19 *meher* season (CSA, 2019).

An estimated 184,000 smallholder farmers grew safflower on ~7000 ha of land and produced ~8000 MT annually in 2016–2020 (cf. Table 6.1). The national average area allocated to safflower in Ethiopia is estimated at 0.04 ha/smallholder household, and the per capita production is estimated at 43 kg/household/yr. The national research system has released only two cultivars as of 2017, suggesting that it does not give priority to this crop. The bulk of safflower produced in Ethiopia is utilized for domestic consumption.

Countries with the largest area are Kazakhstan, the Russian Federation, India, the USA, Mexico, Argentina, Tanzania, Türkiye, China and Uzbekistan. Ethiopia ranked 12th both in the area and production out of 20 countries that produced safflower in the 2016–2020 period (FAO, 2022a). Kazakhstan, the Russian Federation, India, the USA and Mexico are the largest producers. The yield gain for Ethiopia was 36 kg/ha/yr, third highest after Türkiye and Mexico, and highly significant (R^2=0.77; P<0.00), as presented in Table 6.7.

Türkiye and Russian Federation showed the fastest area expansion. On the other hand, the area declined in India, Mexico, the USA and Argentina. Safflower area in Ethiopia declined at an annual rate of 1.02% per annum during the period, whereas the yield and production increased by 4.10% and 3.40%, respectively. All indicators showed declining trends for the USA and Argentina.

Figure 6.7 presents the indicator trends for safflower in Ethiopia. The cultivated area, yield and production all fluctuated from year to year – a typical subsistence condition. However, the yield jumped from just over 600 kg/ha in 2000 to well over 1200 kg/ha in 2020. The yield for all countries ranged from 1800 kg/ha for Mexico to just over 100 kg/ha for Palestine, with the world average of ~850 kg/ha (FAO, 2022a).

Soybean

Soybean is known to have been introduced to Ethiopia in the 1950s, but its research began in the 1970s under the grain legumes research group. According to the CSA, ~158,000 smallholder households grew soybean and produced ~130,000 MT of grain annually in 2016–2020 (cf. Table 6.1). The average soybean farm size for each household was about 0.35 ha and the per capita production 822 kg/holder/yr. The research system has released 26 improved varieties as of 2017.

According to the CSA, Benshangul-Gomuz and SNNPR accounted for 93% and 7%, respectively, of the total soybean area in the 2018/19 *meher* season. In Ethiopia, soybean is produced for export or as an import substitution for the local industries that use it for oil extraction or as a part of infant weaning formula.

Ethiopia ranked 34th in area and 35th in production, respectively, out of 100 countries that cultivated soybean each year in 2016–2020

Table 6.7. Yield gain (b-value, kg/ha/yr) and annual rate of growth (ROG) for safflower in Ethiopia compared to the top ten cultivating of the world (2000–2020). Source: calculated and constructed by the lead author from FAO (2022a).

	Yield gain			ROG (%)		
Country	b-value	P > ltl	R^2	Area	Yield	Production
Türkiye	48	<0.01	0.59	45.85	4.60	52.56
Mexico	39	<0.01	0.43	–4.21	2.74	–1.58
Ethiopia	36	<0.01	0.77	–1.02	4.10	3.04
Uzbekistan	23	<0.05	0.52	2.65	7.95	10.81
Russian Fed.	19	<0.05	0.47	40.03	4.52	46.36
Kazakhstan	12	<0.05	0.48	8.46	2.21	10.85
Tanzania	9	<0.01	0.92	2.18	1.79	4.01
India	3	0.44	0.03	–8.93	0.55	–8.43
China	0	0.97	0.00	0.24	0.02	0.27
USA	–4	0.49	0.03	–1.43	–0.26	–1.69
Argentina	–4	0.36	0.04	–0.39	–0.54	–0.93

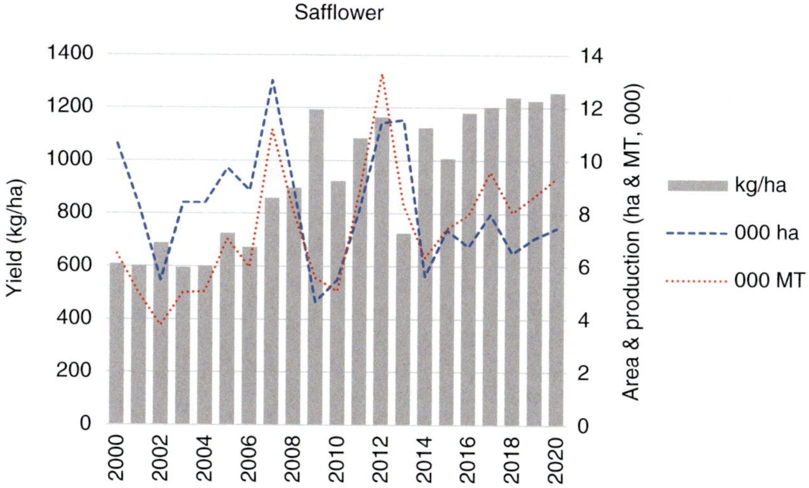

Fig. 6.7. Indicator trends for safflower in Ethiopia, 2000–2020. Source: constructed by the lead author from FAO (2022c).

(FAO, 2022c). The countries with the largest cultivated area (in descending order) are Brazil, the USA, Argentina, India, China, Paraguay, the Russian Federation, Canada, the Ukraine and Bolivia.

Both the area coverage and productivity (and therefore production) of soybean are growing at an increasing rate due to its use in poultry industry and baby food formula. There are 23 factories in Ethiopia that require soybean as raw material. However, most of the crop is exported while leaving the local factories without adequate supply. Soybean yield in the West Gonder Zone is very high, 4.5–6.0 MT/ha on experimental plots and possibly 3.2 MT/ha on farmers' fields.

Ethiopia registered a yield gain of 110 kg/ha/yr during the 2000 to 2020 period ($P<0.00$; $R^2=0.81$), as presented in Table 6.8. This is higher than any of the top ten soybean-producing countries of the world, and was followed by the Ukraine, the USA and Brazil.

Ethiopia, along with the Ukraine and the Russian Federation, showed double-digit growth rate for area, yield and production of soybean. Its average yield of ~2300 kg/ha is superior to the world average of ~1800 kg/ha (FAO, 2022c). The yield for the top ten countries with the highest area harvested ranged from ~3400 kg/ha in the USA to ~1100 kg/ha in India.

All indicators (area, yield and production) for soybean in Ethiopia showed fast growth starting in the early 2000s and peaked in the early 2010s (Fig. 6.8). The yield jumped from ~900 kg/ha in 2000 to ~2500 kg/ha in 2020. Our calculations show that increases in area accounted for ~64% of the change in production, whereas the contribution from increases in yield was ~30%, with their interaction factor covering the remaining ~6%. These results suggest that there is urgent need for research to develop a mechanism to increase productivity – through developing high-yielding cultivars and improved management practices, such as the scaling of microbial fertilizer use and good agronomy.

Trade

Oilseeds constitute an important part of Ethiopia's agricultural trade. This country exported close to 358,000 MT of oilseeds valued at US$392 million annually (second largest after coffee) between 2016 and 2020 (Table 6.9). At the same time, Ethiopia spent nearly US$427 million on cooking oil imports, 93% of which was palm oil.

As shown in Fig. 6.9, export earnings from oilseeds showed steady growth from 2000 to 2014 but took a downward trend from 2015 and levelled off in 2019 and 2020. Import value also increased continually and peaked in 2015, after which time it took downward trends.

Table 6.8. Yield gain (b-value, kg/ha/yr) and annual rate of growth (ROG) for soybean in Ethiopia compared to the top ten cultivating countries of the world (2000–2020). Source: calculated and constructed by the lead author from FAO (2022a).

	Yield gain			ROG (%)		
Country	b-value	P > ltl	R^2	Area	Yield	Production
Ethiopia	110	<0.01	0.81	19.87	9.49	31.25
Ukraine	67	<0.01	0.84	17.94	4.13	22.82
USA	44	<0.01	0.74	0.89	1.52	2.42
Brazil	44	<0.01	0.64	4.74	1.56	6.37
Canada	38	<0.05	0.43	5.28	1.60	6.97
Paraguay	31	0.09	0.15	5.66	1.24	6.97
Russian Fed.	31	<0.01	0.81	12.21	2.53	15.05
Argentina	24	0.06	0.17	2.80	0.86	3.68
Bolivia	18	0.06	0.18	4.45	0.89	5.38
China	13	<0.05	0.41	–0.92	0.75	–0.18
India	7	0.30	0.06	3.44	0.67	4.13

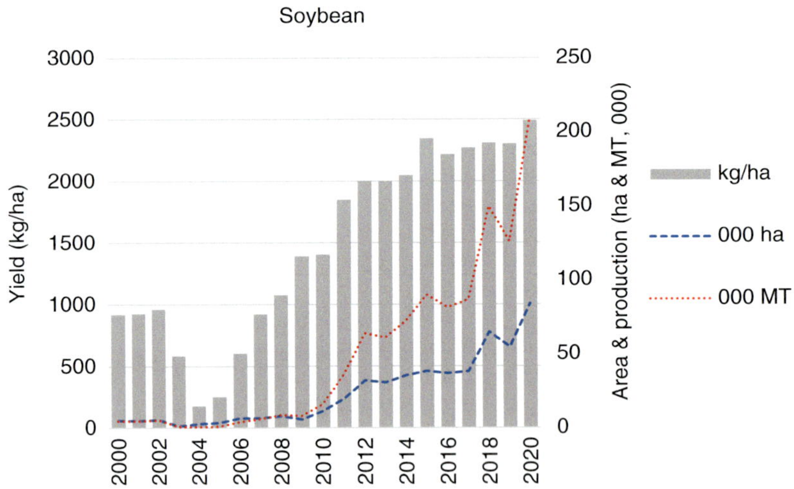

Fig. 6.8. Indicator trends for soybean in Ethiopia, 2000–2020. Source: constructed by the lead author from FAO (2022c).

Ethiopia's export of oilseeds and their products comprises 99.96% raw seed and 0.04% processed products. Imports, to the contrary, consist of 0.12% raw seed and 99.88% processed/semi-processed products (oil).

Discussion and Conclusions

Available records show that there are seven oilseeds grown in Ethiopia from which edible oil is extracted. Conventionally, soybean is categorized under grain legumes, but to our thinking oilseeds is its more appropriate place as it is produced primarily for oil extraction, in addition to its use in infant food formula.

With the exception of sesame, all oilseeds showed highly significant (P<0.01), yield gain, ranging from 110 kg/ha/yr for soybean to 36 kg/ha/yr for safflower, between 2000 and 2020 (cf. Table 6.2).

We observed the highest decline in area for linseed, safflower, rapeseed and *noug* whereas soybean, sesame and groundnut showed

Table 6.9. Value and quantity of international trade in oilseeds and cooking oil in Ethiopia (2016–2020 average). Source: constructed by the lead author from FAO (2022b).

	Import			Export	
Item	Value (US$ 000)	Quantity (MT)	Item	Value (US$ 000)	Quantity (MT)
Palm oil	372,155	428,191	Sesame	332,981	262,256
Sunflower oil	36,204	28,902	Soybean	27,781	58,994
Soybean oil	6,525	5,908	*Noug*	24,334	26,451
Vegetable origin oil	5,883	4,484	Cottonseed	4,301	4,201
Olive oil (virgin)	1,611	747	Soybean cake	921	1,660
Palm kernel oil	1,408	1,257	Rapeseed cake	637	3,954
Margarine (short)	1,067	864	Vegetable origin oil	508	88
Sesame	477	289	Palm oil	83	65
Olive oil (residue)	342	86	Rapeseed	47	218
Cottonseed oil	243	169	Linseed	35	37
Margarine (liquid)	222	155	Sunflower oil	34	<1
Sunflower	165	234	Sunflower	25	26
Rapeseed oil	152	115	Poppy	24	4
Cottonseed	53	13	Mustard	7	2
Coconut oil (copra)	48	28	Olive oil (residue)	3	<1
Noug oil	25	20	Maize oil	2	<1
Sesame oil	21	6	Linseed oil	1	<1
Mustard	18	13	Cottonseed oil	<1	<1
Maize oil	16	9	Olive oil (virgin)	<1	1
Groundnut oil	14	5	Sesame oil	<1	<1
Linseed oil	11	3			
Soybean	7	2			
Linseed	5	1			
Rapeseed	2	<1			
Poppy seed	<1	<1			
Total	426,675	471,502	Total	391,725	357,958

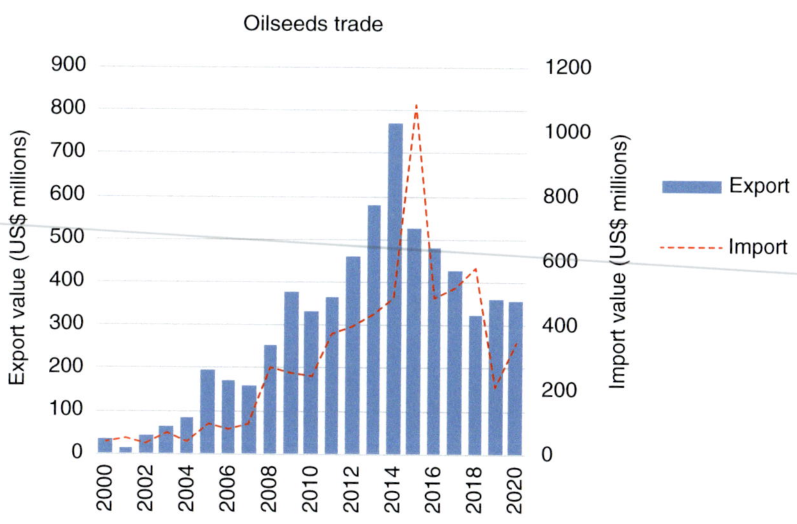

Fig. 6.9. Export and import value of oilseeds and their products in Ethiopia. Source: constructed by the lead author from FAO (2022a).

double-digit growth rates in the area cultivated (cf. Table 6.2). The declines in the area, and in some cases in the number of holders, appeared to be related to the farmers shifting from oilseeds to *teff* because the latter is more profitable. The crops showed varying levels of growth rates for yield – ranging from 1.26% for sesame to 9.49% for soybean.

Even though yield gains for most of the oilseeds in Ethiopia are high, the present level of oilseeds production does not meet the growing demand for raw material by the local industry and cooking oil. As a result, the country imports cooking oil (>90% of which is palm oil that has health concerns because of its high saturated fatty acid content), valued at nearly US$427 million annually. This is not sustainable and does not make sense for a country with high potential to produce beyond its domestic needs. We offer the following suggestions to be considered by way of ameliorating the current situation.

First, Ethiopia can intensify commercialization of *Rhizobium* fertilizer, for which proven technology is already available, for rapid commercialization and growth of soybean productivity and production. Second, as a primary or secondary gene centre for many of the oilseeds (Westphal, 1975), Ethiopia has the opportunity to unlock the potential of its traditional oilseeds through identification and development of desirable traits from its rich genetic pool (Allaro, 2011). Third, there are possibilities for developing safflower both for its high-quality oil production as well as export for its petals for the food dye market in Europe and elsewhere. Fourth, local processing and bottling high-quality oil (Wijnands *et al.*, 2007) should be incentivized for the country to extricate itself from heavy imports of palm oil that is high in saturated fatty acids. Finally, the shrinking area in several of the major oilseed-producing countries (e.g. India and Pakistan for *noug*; Indonesia, India and China for groundnut; Belgium, Belarus, Egypt, USA and India for linseed; China, Uganda and India for sesame; and India, Mexico, Palestine, Australia and the USA for safflower) may open up potential markets for Ethiopia, should it improve its productivity and production capacity.

References

Alemaw, G. (2013) A review on the challenges for the increased production of oilseeds in Ethiopia. *Ethiopian Journal of Agricultural Sciences* 24(1), 1–24.

Allaro, H.B. (2011) Export performance of oilseeds and its determinants in Ethiopia. *Journal of Cereals and Oilseeds* 2(1), 1–15.

CSA (Central Statistical Agency) (2019) Agricultural Sample Survey 2018/19 (2011 E.C.), Report on Area and Production of Major Crops (Private Peasant Holdings, *Meher* Season), Volume I, Statistical Bulletin 589. CSA, Addis Ababa, Ethiopia.

FAO (2020) Statistical databases and data-sets of the Food and Agriculture Organization of the United Nations. Available at: http://www.fao.org/faostat/en/#data (accessed 30 July 2020).

FAO (2022a) Statistical databases and data-sets of the Food and Agriculture Organization of the United Nations. Available at: http://www.fao.org/faostat/en/#data (accessed 17 March 2022).

FAO (2022b) Statistical databases and data-sets of the Food and Agriculture Organization of the United Nations. Available at: http://www.fao.org/faostat/en/#data (accessed 4 April 2022).

FAO (2022c) Statistical databases and data-sets of the Food and Agriculture Organization of the United Nations. Available at: http://www.fao.org/faostat/en/#data (accessed 23 April 2022).

Manore, D., Yohannes, A. (2019) Evaluating growth, seed yield and yield attributes of camelina (*Camelina sativa* L) in response to seeding rate and nitrogen fertilizer levels under irrigation condition, Southern Ethiopia. *Agriculture, Forestry and Fisheries* 8(2), 31–35. Available at: http://www.sciencepublishinggroup.com/j/aff. doi: 10.11648/j.aff.20190802.11

MOALR (Ministry of Agriculture and Livestock Resources) (2017) Crop Variety Register, Issue No. 20. *Plant Variety Release, Protection, and Seed Quality Control Directorate*. Addis Ababa, Ethiopia.

Vavilov, N.I. (1951) The origin, variation, immunity and breeding of cultivated plants. *Chronica Bot* 13 (1-6), 1–364.

Westphal, E. (1975) *Agricultural Systems in Ethiopia*. Joint publication of the College of Agriculture, Haile Sellassie I University, Addis Ababa, Ethiopia, and the Agricultural University, Wageningen, the Netherlands.

Wijnands, J., Biersteker, J., Hiel, R. (2007) *Oilseeds Business Opportunities in Ethiopia*. Commissioned by Ministry of Agriculture, Nature and Food Quality, the Netherlands.

7 The Rise and Decline of Vegetables in Export Diversification and Earnings

Tsedeke Abate*
Homegrown Vision, Addis Ababa, Ethiopia

Abstract

This chapter examines the performance and socioeconomic benefits of vegetable crops in Ethiopia. Successive government regimes in this country invested in vegetable crops research and development, with the aim of improving nutrition as well as export diversification. Annual average export earnings were US$4 million in 2000–2004, US$485 million in 2011–2015 and US$54 million in 2016–2020, meaning that the accelerated growth was not sustained. The national average yield estimates varied from 5.03–6.32 MT/ha, compared to the world average of 16.87 MT/ha. The aggregate yield gain was ~74 kg/ha/yr ($P<0.01$, $R^2 = 0.70$), compared to the top ten cultivating countries where it varied from 291 kg/ha/yr to 531 kg/ha/yr. The current per capita consumption of 14 kg/person/yr has changed little over the last four decades and is among the lowest in Africa. However, Ethiopia has great potential to enhance productivity and improve its nutrition security and export earnings.

Introduction

Vegetables had only marginal importance in traditional Ethiopian diets in the past. The only exception to this would be cabbages – especially the Ethiopian cabbage. Only those who could not afford the luxury of milk and meat would use cabbages as part of their regular diet. The significance of vegetables and other horticultural crops in Ethiopia received a significant boost with the introduction of horticultural research by the Ethiopian Institute of Agricultural Research (EIAR) at the Melkassa Research Center in 1969. National research capacity building was the major drive in the initial years. Technical assistance was received from various bilateral and multilateral sources in the form of backstopping and both short- and long-term training.

The research system introduced, tested and developed varieties of different kinds of vegetables and their production technologies (e.g. Abate and Ayalew, 1994, 1997; Ayalew and Abate, 1993) and popularized their use. The Horticulture Development Corporation of the then Ministry of State Farms (MSFD) played a key role in the large-scale commercialization (Agonafer, 1989, 1994; Desalegne *et al.*, 1994; Herath and Desalegne, 1994) as well as smallholder agriculture (K-Yesus, 1994) of vegetable crops in Ethiopia. At present, these crops have become essential components of the farming

* Corresponding author: tsedeke@homegrownvision.org

 Untold Stories of African Agriculture: Lessons from Ethiopia (T. Abate ed.)
DOI: 10.1079/9781800626386.0007

system, making significant contributions to the country's economy and its nutrition security.

However, in-depth reviews on how much of the aims and objectives of the various development efforts have been realized to date is not available. Ethiopia is currently revising its agricultural policy. This chapter provides a comprehensive review of the status of vegetable production in this country to help policy decisions and priority setting.

Approach

Data Sources

We used three major sources to conduct our analyses of vegetable crops in Ethiopia. In general, production data were from the Central Statistical Agency (CSA; http://www.csa.gov.et) and the Food and Agriculture Organization (FAO) of the United Nations (http://www.fao.org/faostat/en/#data). Seemingly comprehensive and complete data for vegetables in Ethiopia by CSA start in 2003. We also noted that both sources have similar data for the most part for most of the crops starting from this period. We gave priority to CSA data where they are available and complete (CSA, 2004–2021), because FAOSTAT does not provide information on the number of households growing the crops. For long-term production data, we relied on FAOSTAT. Our information for long-term agricultural trade data is also from the FAO, whereas we made reference to the Ethiopian Revenues and Customs Authority (ERCA) database for comparison and identifying export destinations and import sources (for 2013–2017). Information on variety releases was obtained from the Ministry of Agriculture and Livestock Resources (MOALR) (now Ministry of Agriculture [MOA]).

Measuring yield gains

In order to see the effect of technological change and yield growth on production, we used economic methodology to isolate the contribution of area and yield growth on national production for each of the crops included in our assessment, as employed by Abate *et al.* (2015) and described in Chapter 3.

Results

Vegetables in general

Ethiopia grows more than 20 species of cultivated vegetables. According to the CSA, an estimated 7.4 million smallholder households grew about a dozen species of vegetables on more than 123,000 ha of land and produced more than 1 million MT annually in 2016–2020 (Table 7.1). The estimated average yield was 6.32 MT/ha while the average holder produced roughly 100 kg per ha per year.

Smallholder vegetables in this country are cultivated in small gardens. An average smallholder cultivates approximately 0.0148 ha (or 148 m^2) and produces 100 kg per season.

According to the FAO (2022a), vegetables in Ethiopia occupied more than 312,000 ha, the average yield was 5.03 MT/ha and the annual production nearly 1.6 million MT in the 2016–2020 period. The difference between the CSA and FAO figures is that the former is concerned with only the smallholder horticulture, whereas the latter is inclusive of both the smallholder and commercial/private production.

Even though the CSA reported 115 cultivars of vegetables, reports from the National Variety Register of MOALR show that the national research system has released a total of 143 cultivars between 1980 and 2017 (Fig. 7.1). Tomato, onion/shallot and capsicum accounted for approximately two-thirds of the total varieties released. The first vegetable released in Ethiopia was the onion variety 'Adama Red', in 1980. The rate of variety release was slow initially but increased at a rapid pace starting from the second half of 2000s. Both the rate of release and number of vegetable crops introduced to the farming systems showed a steady increase over the latter period. For example, the total number of vegetable varieties released between 1980 and 1999 was seven (~0.35 per year), while it was 136 (~7.56 per year) between 2000 and 2017. It is interesting to note that 'Adama Red' is still the dominant onion variety grown in Ethiopia despite being more than 40 years old. The major factors for this accelerated growth rate in the number of varieties released are discussed in more detail in the discussion section later in this chapter.

Table 7.1. Indicators for major vegetables smallholders cultivated in the *meher* season in Ethiopia (2016–2020 average). Source: number of holders, area, yield and production are constructed by the author from the CSA sample surveys (www.csa.gov.et); area per holder and per capita production are the author's own calculations from CSA data; number of holders is average for 2016–2019; number of varieties are from MOALR (2017).

Crop	000 Holders	Area (000 ha)	Yield (MT/ha)	Production (000 MT)	Area per holder (ha)	Production per holder (kg)	Number of varieties released
Ethiopian cabbage	3709	41.13	9.89	405.96	0.0111	109	11
Garlic	1872	17.99	8.54	156.03	0.0096	83	7
Green pepper	1447	11.32	5.84	65.70	0.0078	45	6
Onion/shallot	789	33.76	8.92	300.75	0.0428	381	31
Head cabbage	447	5.53	6.72	37.47	0.0124	84	3
Beetroot	441	3.19	8.57	27.29	0.0072	62	1
Tomato	237	5.66	5.45	31.32	0.0239	132	42
Carrot	190	3.57	4.15	14.88	0.0187	78	2
Swiss chard	116	0.55	3.07	1.98	0.0047	17	NA
Lettuce	44	0.41	0.75	0.20	0.0095	5	12
Total	7418	123.11	6.32	1041.58	0.0148	100	115

NA=not available.

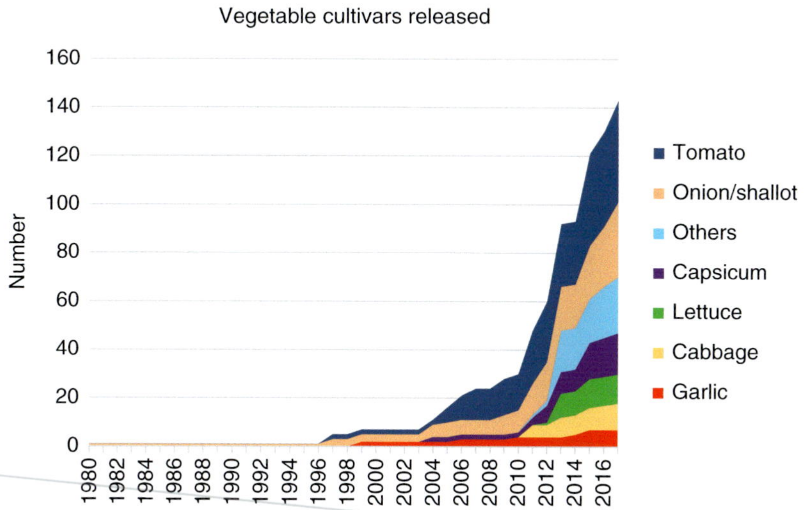

Fig. 7.1. Cumulative number of vegetable cultivars released in Ethiopia between 1980 and 2017. Source: constructed by the author from MOALR (2017).

Performance of aggregate vegetables

Overall, vegetables in Ethiopia showed upward trends, especially starting in the 2000s. However, performances varied from crop to crop. The area expanded steadily through the early 2010s but slowed down and levelled off after that. The yield remained little changed till the early 2000s (Fig. 7.2).

On average, the area, yield and production increased at a rate of 1.37%, 1.72% and 3.11%, respectively, in 2016–2020. This means that the yield and area expansion contributed to changes in production of roughly 55% and 44%, respectively,

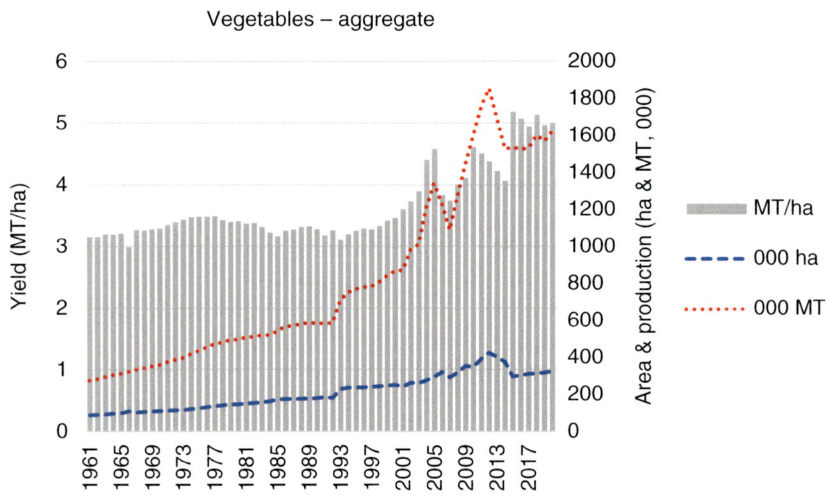

Fig. 7.2. Indicator trends for aggregate vegetables in Ethiopia, 1961–2020. Source: constructed by the author from FAO (2022c).

with the interaction factor covering the balance. The yield gain was estimated at 74 kg/ha/yr ($P<0.01$; $R^2 = 0.70$) as presented in Table 7.2.

Even though the yield gain was very highly significant, the country has a long way to go to catch up with countries with higher yields. The USA, Türkiye, Mexico, China and Viet Nam were among the countries that showed the highest levels of yield gains. By contrast, Nigeria had a significant decline.

Fresh vegetables

Fresh vegetables in Ethiopia occupy nearly 190,000 ha (~61% of the total vegetables area) of land, with an estimated yield of about 2.76 MT per ha and total production of approximately nearly 1.6 million MT. Ethiopia ranked 11th in area (after China, India, Viet Nam, Nigeria, the Philippines, Tanzania, Myanmar, Nepal, Brazil and Bangladesh) and 36th in production (with China, India, Viet Nam, Nigeria, the Philippines, Myanmar, Republic of Korea (ROK) and Brazil leading the list) out of 191 countries and territories in the world that grew fresh vegetables in the 2016–2020 period (FAO, 2022a).

With 2.76 MT/ha, Ethiopia ranked 182nd in yield among those countries. Examples of the highest yielding countries include Japan (22.63 MT/ha), Russian Federation (19.67 MT/ha), Viet Nam (17.39 MT/ha), China (16.62 MT/ha), Nepal (14.19 MT/ha) and India (14.00 MT/ha).

Available data indicate that, like all other vegetables in Ethiopia, the bulk (85%) of fresh vegetables is used for domestic consumption. Fresh vegetables in this country account for ~40% of all vegetable export values. According to ERCA, approximately 80%, 16% and 3% of Ethiopia's fresh vegetable export earnings come from Somalia, Djibouti and the UK, respectively. Other destinations include Israel, India, the United Arab Emirates and the Netherlands.

Figure 7.3 depicts trends of fresh vegetables production in Ethiopia. Their area, yield and production increased at an annual rate of 1.30%, 0.60% and 1.36%, respectively, between 2000 and 2020. This means that 95% of the change in production was due to area expansion, whereas increases in yield contributed 4%; the interaction factors accounted for the remaining 1%. In other words, there was no significant change in the yield level over the 21-year period. The national average yield of about 2.76 MT/ha compared unfavourably with the world average of 12.66 MT/ha.

Regression analysis showed that Ethiopia made an extremely low yield gain of 2 kg/ha/yr ($P<0.01$, $R^2=0.70$), as presented in Table 7.3. Compare this with 294, 226, 150 and 140 for Viet Nam, Nepal, Bangladesh and Nigeria, respectively. By contrast, China, Tanzania, Myanmar and the

Table 7.2. Yield gain (b-value, kg/ha/yr) and annual rate of growth (ROG) for aggregate vegetables in Ethiopia compared to the top ten cultivating countries of the world (2000–2020). Source: constructed by the author from FAO (2022c).

	Yield gain			ROG (%)		
Country[a]	b-value	P > |t|	R^2	Area	Yield	Production
USA	486	<0.01	0.86	–2.07	1.51	–0.59
Türkiye	434	<0.01	0.92	0.20	1.41	1.61
Mexico	365	<0.01	0.87	1.02	1.98	3.02
China	351	<0.01	0.97	1.88	1.60	3.51
Viet Nam	291	<0.01	0.92	3.49	2.12	5.68
India	168	<0.01	0.81	2.79	1.24	4.06
Indonesia	144	<0.01	0.93	1.83	1.60	3.46
Ethiopia	74	<0.01	0.70	1.37	1.72	3.11
Cameroon	26	0.05	0.19	4.83	0.69	5.55
Philippines	15	<0.05	0.52	1.48	0.17	1.65
Nigeria	–118	<0.01	0.63	6.02	–2.47	3.40

[a]The world's top ten cultivating countries (in descending order) are China, India, Nigeria, Indonesia, Viet Nam, USA, Philippines, Cameroon, Türkiye and Mexico; Ethiopia is 24th out of 194 countries in the world.

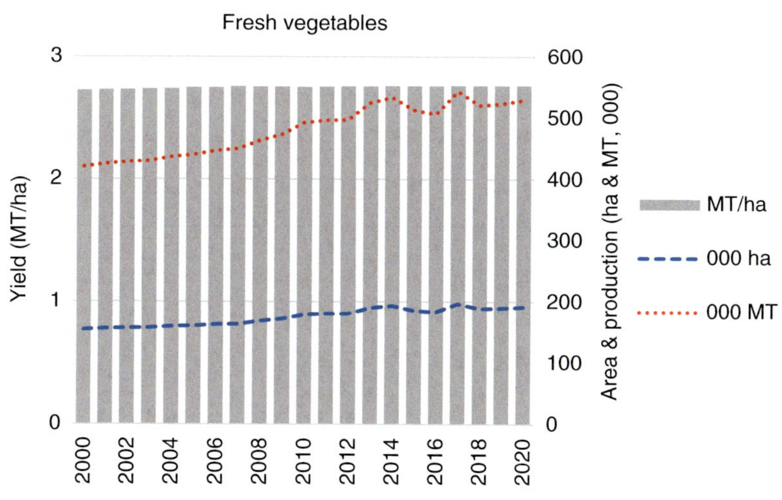

Fig. 7.3. Indicator trends for fresh vegetables in Ethiopia. Source: Constructed by the author from FAO (2022b).

Philippines all showed negative (but non-significant) yield gains.

Performance of individual crops

Cabbages

Cabbages, especially the Ethiopian cabbage, are the most traditional and most widely cultivated of all vegetables in Ethiopia. According to the CSA, ~3.7 million smallholder households (the largest for all vegetables cultivated in Ethiopia) grew Ethiopian cabbage on ~41,000 ha of land (~33% of all vegetables) and produced ~406,000 MT (39% of the total) each year during the *meher* (refers to any annual crop harvested between the months of September and February) period of 2016–2020 (cf. Table 7.1). The national average yield is estimated at 10 MT/ha. According to the FAO (2022a), cabbages occupied ~42,000 ha of land, with an average

Table 7.3. Yield gain (b-value, kg/ha/yr) and annual rate of growth (ROG) for fresh vegetables in Ethiopia compared to the top ten cultivating countries of the world, 2000–2020. Source: constructed by the author from FAO (2022a).

	Yield gain			ROG (%)		
Country[a]	b-value	P > |t|	R^2	Area	Yield	Production
Viet Nam	294	<0.01	0.84	3.81	2.00	5.89
Nepal	226	<0.01	0.96	3.34	1.85	5.25
Bangladesh	150	<0.01	0.91	2.21	2.07	4.32
Nigeria	140	<0.01	0.74	1.04	1.84	2.89
India	108	0.01	0.31	1.30	0.86	2.18
Brazil	84	<0.01	0.76	1.16	0.69	1.86
Ethiopia	2	<0.01	0.79	1.30	0.06	1.36
Philippines	–10	0.03	0.22	1.72	–0.12	1.60
Myanmar	–36	<0.01	0.61	1.54	–0.27	1.27
Tanzania	–40	<0.05	0.38	5.48	–0.65	4.80
China	–92	<0.01	0.66	2.13	–0.53	1.59

[a]The world's top ten cultivating countries (in descending order) are China, India, Viet Nam, Nigeria, the Philippines, Tanzania, Myanmar, Nepal, Brazil and Bangladesh; Ethiopia is 11th out of 191.

yield of 9.78 MT/ha and total production of ~414,000 MT during the same period.

Ethiopia ranked 7th in area (after China, India, Angola, Russian Federation, Indonesia and the Ukraine) and 20th in production (with China, India, the Russian Federation, the ROK, the Ukraine, Indonesia, Japan, Romania, Poland and the USA being the top ten) out of 147 countries that cultivated cabbages in the 2016–2020 period. This means that Ethiopia's share of the world area was about 1.72% of ~2.4 million ha and 0.59% of production of ~70 million MT. This country ranked 133rd in yield in the world. The highest yielding countries out of the top ten included the ROK and China, with yields ranging between 68.28 MT/ha and 34.75 MT/ha (FAO, 2022a).

Figure 7.4 depicts indicator trends for cabbages in Ethiopia. It can be observed that the yields since 1993 have been lower than they were in previous years. Cabbages area increased at an annual rate of 5.95% between 2000 and 2020, fastest of all vegetables in the country. Meanwhile, the yield and production increased by 0.56% and 6.55%, respectively. This means that the area expansion accounted for ~91% of the change in production and the yield contributed ~9%.

Cabbages yield gain in Ethiopia was 50 kg/ha/yr between 2000 and 2020 and non-significant (Table 7.4). In contrast, highly significant (P<0.01) yield gains were recorded in the Russian Federation, ROK, Ukraine, Viet Nam and China. Angola had a yield loss of 952 kg/ha/yr (P<0.03) among the top ten cultivating countries of the world. There was little change for India and Indonesia. The annual rate of growth (ROG) for area increased at an annual rate of 5.96% for Ethiopia while the yield increased by 0.56%. Cabbages area declined in five of the top ten countries while production declined in two of the top ten countries (Table 7.4).

Dry onion/shallot

According to the CSA, ~789,000 smallholders grew onions/shallots on ~34,000 ha of land and produced ~300,000 MT in the 2016–2020 *meher* season period (cf. Table 7.1). The average land area allocated to onions/shallots is 0.0428 ha per smallholder household and the per capita production is about 381 kg/household/yr. The national average yield was ~8.92 MT/ha. A total of 31 cultivars have been released as of 2017. There was little or no difference between the data reported by the CSA and FAO.

Figure 7.5 presents dry onions/shallots indicator trends. The figure shows that there have been increases in production but those came purely from the area expansion. The area and production for dry onions/shallots in Ethiopia increased at an annual rate of 5.67% and 4.46%, respectively, but the yield decreased by –1.15%.

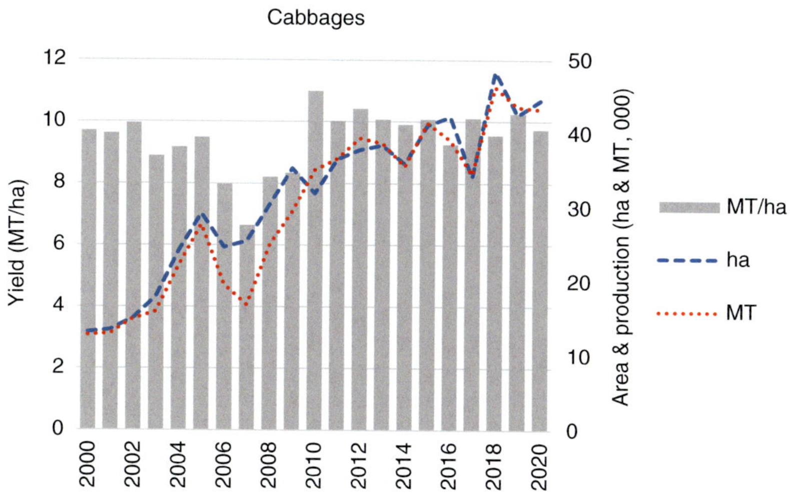

Fig. 7.4. Indicator trends for cabbages in Ethiopia. Source: constructed by the author from FAO (2022b).

Table 7.4. Yield gain (b-value) and annual rate of growth (ROG) for cabbages in Ethiopia compared to the top ten cultivating countries of the world (2000–2020). Source: constructed by the author from FAO (2022a).

	Yield gain			ROG (%)		
Country[a]	b-value	P > ltl	R2	Area	Yield	Production
Russian Federation	727	<0.01	0.68	–2.97	2.63	–0.41
Republic of Korea	613	<0.05	0.53	–2.27	0.96	–1.33
Ukraine	580	<0.01	0.79	–0.95	2.86	1.89
Viet Nam	540	<0.01	0.87	1.35	2.49	3.88
Romania	111	0.07	0.16	–1.08	0.54	–0.54
China	72	<0.01	0.69	–0.46	0.21	–0.25
Ethiopia	50	0.16	0.10	5.95	0.56	6.55
India	19	0.42	0.03	3.15	0.09	3.24
Indonesia	12	0.83	0.00	0.69	0.02	0.71
Angola	–952	0.03	0.34	NA	NA	NA

[a]The world's top ten cultivating countries (in descending order) are China, India, Angola, the Russian Federation, Indonesia, Ukraine, Ethiopia, Viet Nam, Republic of Korea and Romania.
NA = not available.

The data for yield gains for onion/shallot in Ethiopia compared to the top ten producing countries of the world are presented in Table 7.5. Ethiopia's yield declined by 117 kg/ha/yr (P=0.01, R^2 = 0.29). By contrast, Egypt, India, China and Viet Nam obtained yield gains ranging from 411 to 38 kg/ha/yr.

The national average yield of dry onions for Ethiopia is estimated at 8.95 MT/ha. The world average is 22.95 MT/ha. Among the top ten producing countries, Egypt, China, Sudan, India and Pakistan obtain about 335.78 MT/ha, 31.85 MT/ha, 18.16 MT/ha, 17.70 MT/ha and 13.71 MT/ha, respectively; on the other hand, the yields for Bangladesh, Indonesia, Uganda, Viet Nam and Nigeria are estimated at 10.11 MT/ha, 9.64 MT/ha, 3.95 MT/ha, 3.55 MT/ha and 2.42 MT/ha, respectively.

Onion thrips (*Thrips tabaci*) and purple blotch (caused by *Alternaria porii*) are the major constraints in onion production under irrigation and wet season, respectively, in Ethiopia.

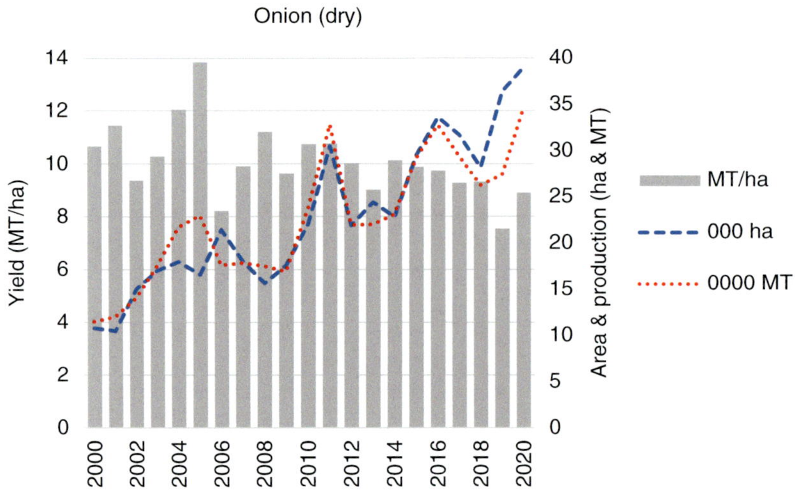

Fig. 7.5. Indicator trends for dry onion in Ethiopia. Source: constructed by the author from FAO (2022b).

Table 7.5. Yield gain (b-value, kg/ha/yr) and annual rate of growth (ROG) for dry onion in Ethiopia compared to the top ten countries of the world (2000–2020). Source: constructed by the author from FAO (2022a).

	Yield gain			ROG (%)		
Country[a]	b-value	P > |t|	R^2	Area	Yield	Production
Egypt	411	0.01	0.34	7.14	1.26	8.49
India	387	<0.01	0.81	6.09	2.79	9.05
Bangladesh	359	<0.01	0.95	9.18	5.41	15.09
China	44	0.01	0.29	2.15	0.21	2.36
Viet Nam	38	<0.01	0.82	1.73	1.17	2.92
Indonesia	38	0.06	0.18	3.96	0.41	4.39
Sudan	27	0.77	0.00	–1.06	3.05	1.95
Uganda	0	0.65	0.01	4.45	–0.01	4.44
Pakistan	–27	0.23	0.07	1.49	–0.19	1.30
Ethiopia	–117	0.01	0.29	5.67	–1.15	4.46
Nigeria	–169	0.01	0.45	8.59	–4.92	3.24

[a]The world's top ten cultivating countries (in descending order) are: India, China, Nigeria, Bangladesh, Indonesia, Pakistan, Viet Nam, Sudan, Egypt and Uganda, respectively; Ethiopia is 22nd out of 139.

Garlic

Garlic is a well-known traditional vegetable in Ethiopia. Nearly 1.9 million households cultivated garlic on ~18,000 ha of land and produced ~156,000 MT in the 2016–2020 period (cf. Table 7.1). The yield was estimated at 8.54 MT/ha and an average household produced about 83 kg per year. The CSA and FAO estimates provide almost identical reports on garlic production in Ethiopia.

The country ranked ninth in the area cultivated out of 102 countries for which data were available for the above period. China, India, Bangladesh, Myanmar and Spain are among the largest cultivating countries.

Figure 7.6 presents garlic indicators for Ethiopia. The ROG for area and yield were estimated at 6.13% and –1.74%, respectively, between 2000 and 2020. The production fluctuated very much.

The yield gain for garlic in Ethiopia declined by 190 kg/ha/yr between 2000 and 2020 (Table 7.6). Uzbekistan, China, Algeria, Brazil, Bangladesh and Peru were among the top countries with significant yield gains in the world. Average yields for Uzbekistan, China,

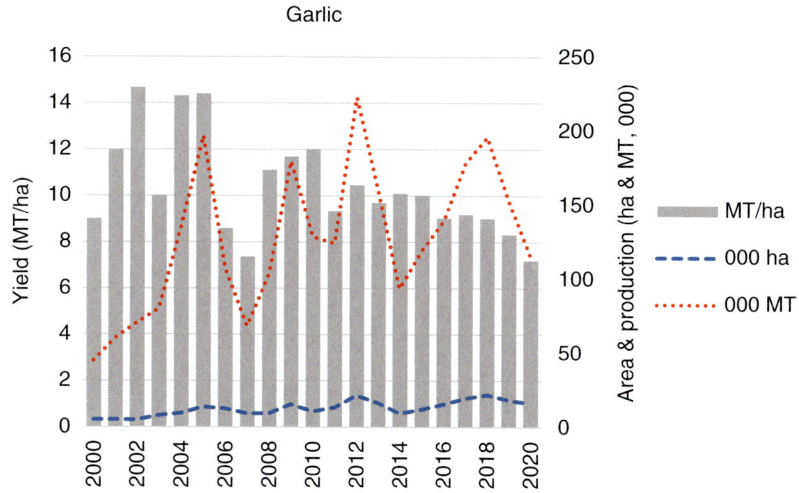

Fig. 7.6. Indicator trends for garlic in Ethiopia. Source: constructed by the author from FAO (2022b).

Table 7.6. Yield gain (b-value, kg/ha/yr) and annual rate of growth (ROG) for garlic in Ethiopia compared to the top ten cultivating countries of the world (2000–2020). Source: constructed by the author from FAO (2022a).

	Yield gain			ROG (%)		
Country[a]	b-value	P > |t|	R^2	Area	Yield	Production
China	665	<0.01	0.87	1.96	3.44	5.47
Bangladesh	219	<0.01	0.96	9.52	5.02	15.02
Ukraine	166	<0.01	0.78	0.63	2.33	2.98
Myanmar	134	<0.01	0.81	1.90	2.18	4.12
India	132	<0.01	0.60	6.34	2.40	8.89
Republic of Korea	121	0.01	0.47	–2.11	0.98	–1.15
Russian Federation	113	**<0.01**	0.82	–1.27	1.36	0.07
Spain	107	<0.01	0.63	1.20	1.25	2.47
Argentina	–6	0.62	0.01	0.55	–0.05	0.50
Ethiopia	–190	0.01	0.30	6.13	–1.74	4.29

[a]The world's top ten cultivating countries (in descending order) are: China, India, Bangladesh, Myanmar, Spain, Republic of Korea, Ukraine, Russian Federation, Ethiopia and Argentina; Ethiopia is 9th out of 102.

Egypt, the USA and Algeria ranged between 30.57 MT/ha and 14.07 MT/ha, compared to Ethiopia's 8.54 MT/ha.

Lack of high-yielding improved varieties, diseases (mainly garlic rust caused by *Puccinia porii*) and poor agronomic management are the main factors limiting garlic production in Ethiopia.

Green pepper

Green pepper is a traditional vegetable used as a relish and is widely grown both in smallholder and commercial agriculture in Ethiopia. According to the CSA, ~1.4 million smallholder households grew green pepper annually on ~11,000 ha of land and produced ~66,000 MT, with an average yield of ~5.84 MT/ha during the 2016–2020 *meher* season period (cf. Table 7.1). The average green pepper farm size was 0.0078 ha (or 78 m^2) per household and each household produced 45 kg/yr. Six cultivars of green pepper have been registered in Ethiopia as of 2017. According to the CSA data, 78% of the green pepper smallholders cultivate is grown in the *meher* season, with the remaining 22% grown in the *belg* (refers to any annual crop harvested between the months of March and August) season. The CSA and FAO data for green pepper are identical.

Ethiopia ranked 23rd in area out of 127 countries in the world that cultivated green pepper in the 2016–2020 period. The largest producers were China, Indonesia, Mexico, Nigeria and Türkiye (FAO, 2022a).

Figure 7.7 presents green pepper indicator trends for Ethiopia. All indicators showed upward trends starting in the early 2000s, but the area and production sharply declined starting in 2013. The ROGs for area, yield and production (2000–2020) were –8.54%, 7.92% and –1.30%, respectively. The average yield for Ethiopia was 7.92 MT/ha. Spain, the USA, Algeria, Türkiye and China (with yields ranging from 63.45 to 22.63 MT/ha) were the highest yielders among the top 20 producing countries in the world.

Ethiopia has made a yield gain of 251 kg/ha/yr ($P<0.01$, $R^2 = 0.70$) between 2000 and 2020, but several other countries (notably Spain, Algeria, Tunisia, the USA and Türkiye) have made much higher gains (Table 7.7). Ethiopia's

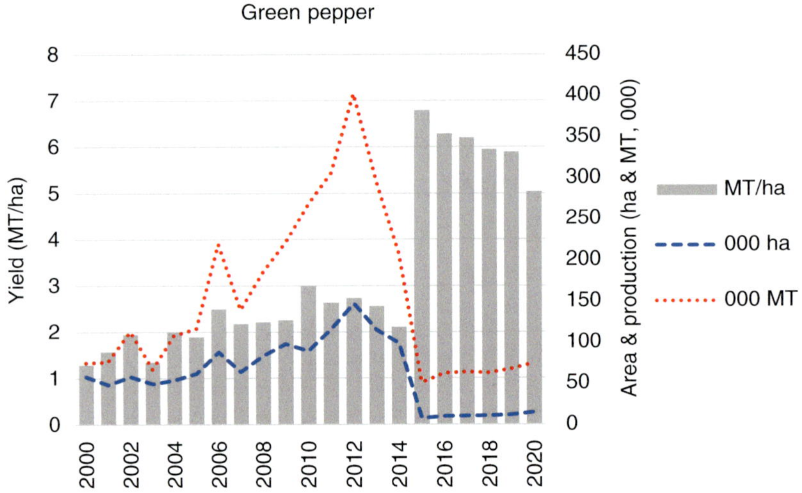

Fig. 7.7. Indicator trends for green pepper in Ethiopia. Source: constructed by the author from FAO (2022b).

Table 7.7. Yield gain (b-value, kg/ha/yr) and annual rate of growth (ROG) for green pepper in Ethiopia compared to the top ten countries of the world (2000–2020). Source: constructed by the author from FAO (2022a).

	Yield gain			ROG (%)		
Country[a]	b-value	P > ltl	R^2	Area	Yield	Production
Türkiye	410	<0.01	0.97	1.06	1.66	2.73
Sierra Leone	339	0.23	0.88	NA	NA	NA
Mexico	301	0.01	0.40	1.66	1.90	3.59
Ethiopia	251	<0.01	0.70	–8.54	7.92	–1.30
China	223	<0.01	0.83	1.76	1.08	2.86
Indonesia	216	<0.01	0.88	3.59	3.54	7.26
Egypt	111	0.01	0.28	2.69	0.65	3.36
ROK	102	<0.01	0.61	–4.41	1.66	–2.83
Benin	91	<0.01	0.67	1.18	4.26	5.49
Cameroon	15	0.63	0.01	11.39	1.68	13.27
Nigeria	–26	<0.01	0.96	0.69	–0.33	0.35

[a]The world's top ten cultivating countries (in descending order) are: China, Indonesia, Mexico, Nigeria, Türkiye, Egypt, Republic of Korea, Sierra Leone, Benin and Cameroon; Ethiopia is 23rd out of 127.
NA = Not available

ROG for green pepper was higher than any of the top 20 countries, followed by Algeria, Ghana, Benin, Sri Lanka and Indonesia.

Tomato

The CSA data show that ~237,000 smallholder households grew tomato on ~6,000 ha of land and produced ~31,000 MT in the 2016–2020 *meher* season period, with an average yield of ~5.45 MT/ha (cf. Table 7.1). The average tomato area per household was estimated at 0.0239 ha and the per capita production at 132 kg/household/season. A total of 42 cultivars, the largest for all vegetables, have been released between 1996 and 2017. There was not much difference between the CSA and FAO data for tomato in Ethiopia.

Ethiopia ranked 72nd in area (out of 166 countries and territories). China, India, Nigeria, Türkiye, Egypt and the USA are among the top-ranking countries in the world. Figure 7.8 presents long-term trends for tomato indicators in Ethiopia. Tomato area and production showed a slow but steady rise through the late 1990s. All indicators fluctuated heavily starting early 2000s. Overall, the ROG for area, yield and production was 1.83%, –2.06% and –0.26% per annum, respectively.

The performance of tomato in Ethiopia (5.52 MT/ha) is in sharp contrast to what is observed for the world trend. There is a huge gap between this country's yield and many of the top ten producing countries of the world. The average yields for the USA, Türkiye, Italy, China, Mexico, Iran, Egypt, India, Cameroon and Nigeria were 99.43 MT/ha, 69.63 MT/ha, 60.59 MT/ha, 47.35 MT/ha, 44.17 MT/ha, 39.45 MT/ha, 24.98 MT/ha, 12.52 MT/ha and 4.21 MT/ha, respectively. The world average was 57.84 MT/ha for the 2016–2020 period (FAO, 2022a).

Tomato in Ethiopia lost 178 kg/ha/yr between 2000 and 2020 (Table 7.8). This compares poorly with the USA, Türkiye, Mexico, China and Iran, where they ranged from 1760 kg/ha/yr to 1482 kg/ha/yr. Yield gains were also negative in Nigeria and Cameroon, among the top ten countries.

Carrot

The CSA reports carrot as a part of roots and tubers. This crop is considered minor in Ethiopia. According to the CSA, ~179,000 smallholders grew carrot on <4000 ha of land and produced <15,000 MT in the 2016–2020 *meher* season period (cf. Table 7.1). The average area allocated to carrot is 0.0196 ha and the per capita production is ~75 kg/household/yr. The average yield is less than 4 MT/ha and there are two cultivars registered as of 2017. Ethiopia ranked 47th in area and 84th out of 128 countries and territories in the world for which data were available in the 2016–2020 period (FAO, 2022a). The top

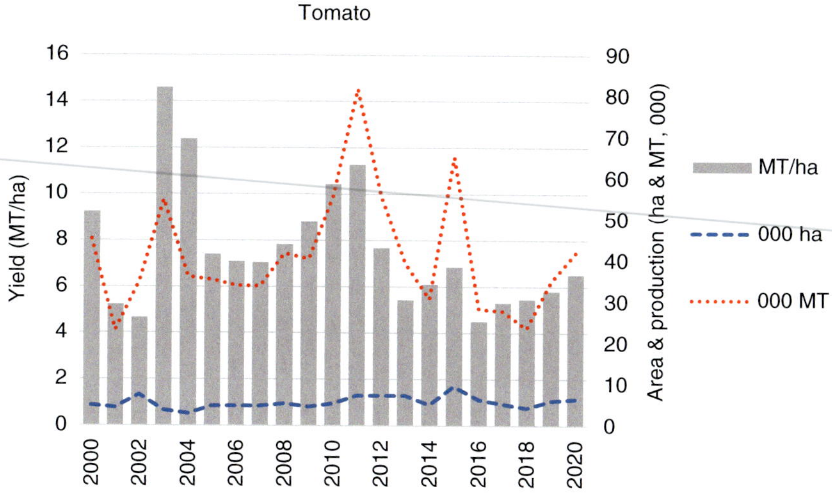

Fig. 7.8. Indicator trends for tomato in Ethiopia. Source: constructed by the author from FAO (2022b).

Table 7.8. Yield gain (b-value, kg/ha/yr) and annual rate of growth (ROG) for tomato in Ethiopia compared to the top ten cultivating countries of the world, 2000–2020. Source: constructed by the author from FAO (2022a).

	Yield gain			ROG (%)		
Country[a]	b-value	P > Itl	R^2	Area	Yield	Production
USA	1760	<0.01	0.81	–1.85	2.06	0.18
Türkiye	1567	<0.01	0.99	–0.73	2.83	2.07
Mexico	1547	<0.01	0.92	–2.03	4.77	2.64
China	1482	<0.01	0.93	2.01	3.40	5.48
Iran	854	<0.01	0.77	0.00	2.38	2.38
Italy	558	<0.01	0.68	–1.82	1.01	–0.83
India	529	<0.01	0.92	3.63	2.67	6.39
Egypt	186	0.01	0.46	–0.58	0.50	–0.08
Cameroon	–46	0.65	0.01	7.53	–0.15	7.37
Nigeria	–134	0.01	0.31	8.44	–2.48	5.76
Ethiopia	–178	0.06	0.17	1.83	–2.06	–0.26

[a]The world's top ten cultivating countries (in descending order) are China, India, Nigeria, Türkiye, Egypt, the USA, Iran, Italy, Cameroon and Mexico; Ethiopia is 72nd out of 166.

producers include China, the Russian Federation, Iran, Indonesia and India.

Figure 7.9 presents carrot indicator trends for Ethiopia. The area increased at the rate of 9.17% per annum, whereas the yield declined at the rate of 5.70%; production increased at 2.73%. In other words, the area increased relatively fast, and the yield declined whereas production fluctuated from time to time.

The world average yield was estimated at 27.37 MT/ha. Top yielding countries included Uzbekistan, the UK, Germany, the USA and China, where the yields ranged between 70.42 MT/ha and 44.27 MT/ha. The ROGs for the world area, yield and production increased at an annual rate of 0.43%, 0.21% and 3.41% (FAO, 2022a).

Regression analysis results show that Ethiopia's yield gain for carrots declined at the rate of ~54 kg/ha/yr between 2000 and 2022 (Table 7.9). Other countries with negative yield gain included Iran, the UK, Pakistan and Italy. By contrast, Uzbekistan, China, the Russian Federation, the USA, Poland and the Ukraine showed highly significant yield gains ranging between 258 kg/ha/yr and 41 kg/ha/yr.

Green bean

Green bean is a relatively new addition to Ethiopian horticulture. The CSA does not include this vegetable in its sample surveys, perhaps because the crop is not widely known among smallholders. It is grown by commercial farmers for export. The FAO data records start in 1998. This country grows ~1800 ha of green beans and produces ~7000 MT of green bean each year; the average yield is ~4 MT/ha. Ethiopia ranked 43rd (out of 110) in area and 53rd (out of 111) in production compared to the total number of countries and territories in the world that produced green bean.

Figure 7.10 presents the green bean indicator trends in Ethiopia. Even though the area increased by 4.32% per annum, the yield stayed stagnant. Green bean yield in this country is among the lowest of the top ten producing countries of the world. The top yielding countries included China, India, Thailand, Indonesia, Türkiye, Egypt, Bangladesh, France, Italy and Serbia, where the yields ranged between 26.50 MT/ha and 15.03 MT/ha. Ethiopia's national average yield was 4.12 MT, while that for the world was 7.84 MT/ha.

Ethiopia registered a yield gain of 4 kg/ha/yr, which was non-significant (Table 7.10). By contrast, China, Türkiye, France, Indonesia and Bangladesh obtained highly significant yield gain ranging between 628 kg/ha/yr and 119 kg/ha/yr (Table 7.10).

Lettuce

The CSA sample surveys started reporting lettuce from 2003. According to these reports, ~44,000 smallholder households grew lettuce

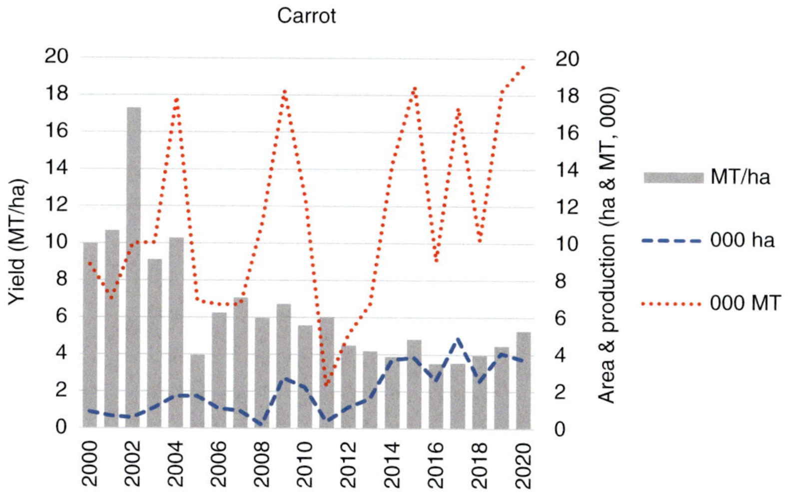

Fig. 7.9. Indicator trends for carrot in Ethiopia. Source: constructed by the author from FAO (2022b).

Table 7.9. Yield gain (b-value, kg/ha/yr) and annual rate of growth (ROG) for carrot in Ethiopia compared to the top ten cultivating countries of the world, 2000–2020. Source: constructed by the author from FAO (2022a).

	Yield gain			ROG (%)		
Country[a]	b-value	P > |t|	R^2	Area	Yield	Production
Uzbekistan	258	<0.01	0.97	5.68	5.57	11.57
China	165	<0.01	0.94	0.55	5.61	6.19
Russian Federation	66	<0.01	0.90	–1.88	2.83	0.90
USA	63	0.01	0.53	–1.95	1.38	–0.60
Poland	52	0.01	0.46	–2.74	1.66	–1.12
Ukraine	41	<0.01	0.85	0.74	2.56	3.32
India	4	<0.01	0.92	2.04	0.27	2.31
Indonesia	3	0.47	0.03	3.78	0.18	3.97
Nigeria	2	<0.01	0.95	–0.17	0.23	0.06
Pakistan	–8	<0.01	0.83	0.07	–0.43	–0.36
Ethiopia	–54	0.23	0.08	9.17	–5.90	2.73

[a]The world's top ten cultivating countries (in descending order) are China, the Russian Federation, the Ukraine, Indonesia, India, Uzbekistan, the USA, Nigeria, Pakistan and Poland.

on ~400 ha of land and produced ~200 MT annually in the 2016–2020 period (cf. Table 7.1). The area allocated to lettuce is 0.0095 ha (or 95 m^2) per household and the per capita production was estimated at 5 kg/household/yr. The research system has released 12 cultivars between 2012 and 2017.

By contrast, the FAOSTAT reported ~400 ha and <300 MT, with the average yield of ~1.04 MT/ha. Figure 7.11 presents lettuce indicator trends in Ethiopia. Here, the area increased by 2.04% while the yield and production declined at an annual rate of 12.96% and 11.18%, respectively. According to this source, Ethiopia ranked 57th out 106 countries for which data were available in the 2016–2020 period (FAO, 2022a).

The area, yield and production for lettuce in Ethiopia were estimated at 2.04%, –12.96% and –11.18%, respectively. Lettuce area declined in

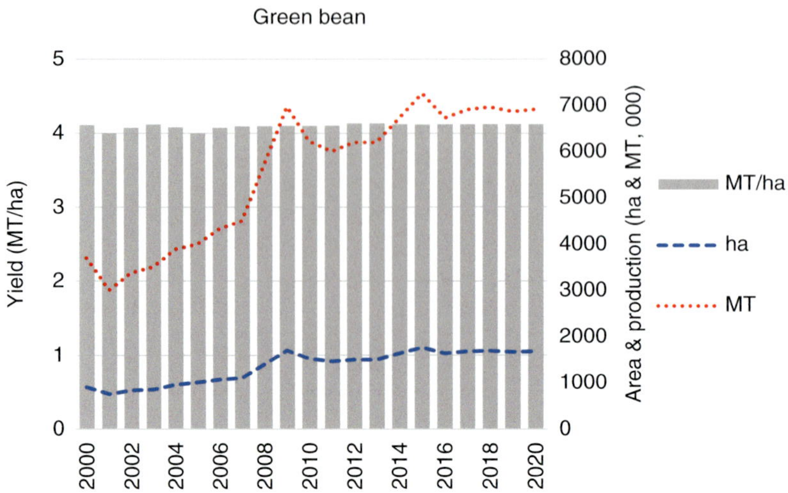

Fig. 7.10. Indicator trends for green bean in Ethiopia. Source: constructed by the author from FAO (2022b).

Table 7.10. Yield gain (b-value, kg/ha/yr) and annual rate of growth (ROG) for green bean in Ethiopia compared to the top ten cultivating countries of the world, 2000–2020. Source: constructed by the author from FAO (2022a).

	Yield gain			ROG (%)		
Country	b-value	P > |t|	R^2	Area	Yield	Production
China	628	<0.01	0.83	2.50	2.93	5.50
Türkiye	269	<0.01	0.98	–1.50	2.44	0.91
France	219	0.01	0.40	3.90	2.62	6.62
Indonesia	152	<0.01	0.90	–1.26	2.53	1.24
Bangladesh	119	<0.01	0.92	4.26	2.19	6.55
Thailand	25	<0.05	0.51	–2.57	1.62	–0.99
Ethiopia	4	<0.05	0.39	4.32	0.09	4.42
India	4	<0.01	0.92	2.22	0.13	2.36
Italy	–9	0.58	0.02	–1.72	–0.10	–1.81
Egypt	–14	0.50	0.02	0.90	–0.12	0.78
Serbia	–32	0.05	0.27	3.27	1.91	5.25

[a]The world's top ten cultivating countries (in descending order) are China, India, Thailand, Indonesia, Türkiye, Egypt, Bangladesh, France, Italy and Serbia.

five of the top ten cultivating countries of the world and production in four. In a similar fashion, production and yield declined in three and two of the top ten countries, respectively. Overall, Ethiopia's national average yield of 1.04 MT/ha was the lowest among 106 countries in the world that cultivated lettuce. Contrary to this, the yield among the top ten cultivating countries ranged from 30.12 MT/ha for Iran to 6.35 MT/ha for India (2016–2020 avg.). The world average was 18.66 MT/ha.

Lettuce yield gain for Ethiopia was estimated at –443 kg/ha/yr ($P<0.01$, $R^2 = 0.72$), as presented in Table 7.11. By contrast, yield gains were positive and highly significant for Türkiye, Italy, Mexico, Japan and China.

Other vegetables

Other vegetables cultivated in Ethiopia, but for which detailed discussions are not provided in this chapter, occupied an estimated total area of

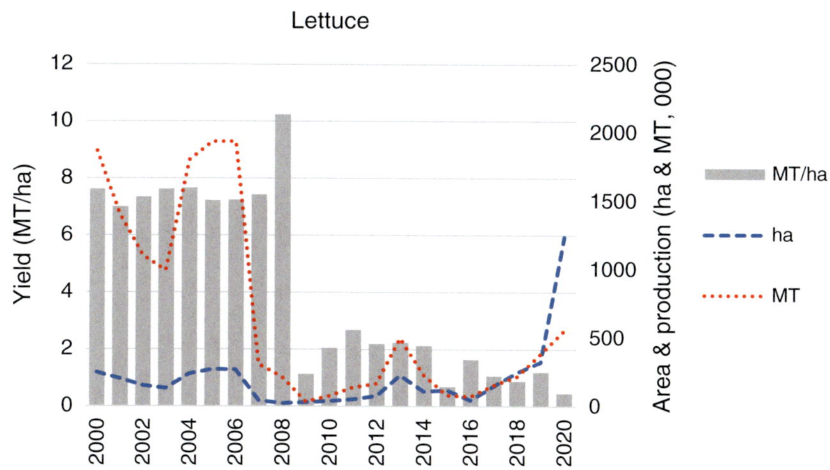

Fig. 7.11. Indicator trends for lettuce in Ethiopia. Source: constructed by the author from FAO (2022b).

Table 7.11. Yield gain (b-value, kg/ha/yr) and annual rate of growth (ROG) for lettuce in Ethiopia compared to the top ten cultivating countries of the world, 2000–2020. Source: constructed by the author from FAO (2022a).

	Yield gain			ROG (%)		
Country[a]	b-value	P > |t|	R^2	Area	Yield	Production
Türkiye	271	<0.01	0.92	0.50	–0.99	1.93
Italy	191	<0.01	0.79	–2.77	0.94	–1.85
Mexico	140	0.01	0.43	4.39	0.62	5.04
Japan	139	<0.01	0.67	–0.10	0.54	0.44
Iran	116	0.09	0.15	3.60	0.42	4.03
China	100	0.01	0.52	2.63	0.44	3.08
France	85	0.42	0.03	–2.10	0.23	–1.88
Spain	4	0.93	0.00	–0.46	0.02	–0.44
India	–9	<0.01	0.84	2.27	–0.14	2.12
USA	–54	0.47	0.03	–0.40	–0.31	–0.57
Ethiopia	–443	<0.01	0.72	2.04	–12.96	–11.18

[a]The world's top ten cultivating countries (in descending order) are: China, India, the USA, Spain, Italy, Türkiye, Japan, Mexico, France and Iran.

5700 ha with roughly 56,000 MT of production in the 2016–2020 period. These included green onion/shallot (2965 ha); other leguminous vegetables, including snap peas, snow peas and others (1575 ha); leek (564 ha); green pea (522 ha); and cucumber (75 ha) (FAO, 2022a). These vegetables are grown by the private commercial sector rather than smallholders.

Trade

Ethiopia has a multimillion-dollar vegetable industry. The country is a net exporter of vegetables at present. Ethiopia exported >58,000 MT of vegetables for an estimated value of nearly US$54 million annually in 2016–2020 (Table 7.12). In a similar fashion, vegetable products imports were >77,000 MT for an estimated value of more than US$25 million.

The main drivers of export value were fresh and dehydrated vegetables, tomato, preserved vegetables and dry onion. By contrast, dry onion accounted for nearly 78% of the vegetable import bill; it was followed by vegetable-origin oil and tomato paste.

Vegetable export in Ethiopia had its ups and downs. It rose sharply starting in 2008, with its

Table 7.12. Export and import value and quantity of vegetables in Ethiopia (2016–2020 average). Source: constructed by the author from FAO (2022d).

Export			Import		
Item	Value (US$ 000)	Quantity (MT)	Item	Value (US$ 000)	Quantity (MT)
Vegetables, fresh nes[a]	21,321	7,281	Onions, dry	14,062	67,636
Vegetables, dehydrated	14,709	1,240	Oil, vegetable origin nes	5,883	4,484
Tomato	3,875	15,246	Tomato, paste	1,361	1,459
Vegetables, preserved nes	2,964	1,503	Vegetables in vinegar	1,175	1,546
Onions, dry	2,667	15,526	Margarine, short	1,067	864
Lettuce and chicory	2,171	11,017	Vegetables, preserved nes	527	372
Beans, green	1,993	1,131	Margarine, liquid	222	155
Leeks, other alliaceous	1,554	910	Mushrooms, canned	143	134
Oil, vegetable origin nes	508	88	Vegetables, dehydrated	133	54
Peas, green	444	187	Garlic	109	112
Sub-total	52,207	54,128	Sub-total	24,682	76,816
Others	1,544	4,137	Others	525	375
Grand total	53,750	58,264	Grand total	25,207	77,191

[a]nes = not elsewhere specified.

peak at US$650 million in 2014 (Fig. 7.12). Fast declines were observed starting in 2016. In other words, annual average export earnings were US$4 million in 2000–2004, compared to US$485 million in 2011–2015 versus US$54 million in 2016–2020.

Possible factors for the decline include: (i) increased local demand owing to population growth; (ii) declines in productivity and production; and (iii) fluctuations in the international market.

Discussion and Conclusion

For a long time in the past, Ethiopia relied on cabbages (especially the Ethiopian cabbage) as its main source of vegetable requirement. The formation of the horticulture department by EIAR increased the number of crops and introduced appropriate production technologies both for smallholder and commercial production systems (Desalegne *et al.*, 1994). The MOA promoted vegetable production among smallholders (K-Yesus, 1994), while the MSFD dealt with the commercial sector (Agonafer, 1994). The private sector developed during the Ethiopian Peoples' Revolutionary Front government tenure. At present, the commercial sector produces ~34% of vegetables in Ethiopia, which accounts for the bulk of export market.

Ethiopia has made good progress in diversifying its vegetable production. Until the early 1990s there were only a few traditional vegetables – including miscellaneous fresh vegetables, onions, tomato and cabbages – mostly cultivated by smallholders. At present, there are more than 20 species of cultivated vegetables that are grown on ~312,000 ha with total production of nearly 1.6 million MT each year. Several reports (Emana and Gebremedhin, 2007; Demissie *et al.*, 2009; Wiersinga and de Jaeger, 2009; Alemayehu *et al.*, 2010; Emana *et al.*, 2015; Dube *et al.*, 2018) suggest that Ethiopia has a wide range of comparative and competitive advantages to expand vegetable production for improved nutrition and increased income in smallholder horticulture.

It appeared that increased involvement of the private sector has helped the growth of vegetable diversification in Ethiopia. For example, 18 (largely foreign) companies registered a total of

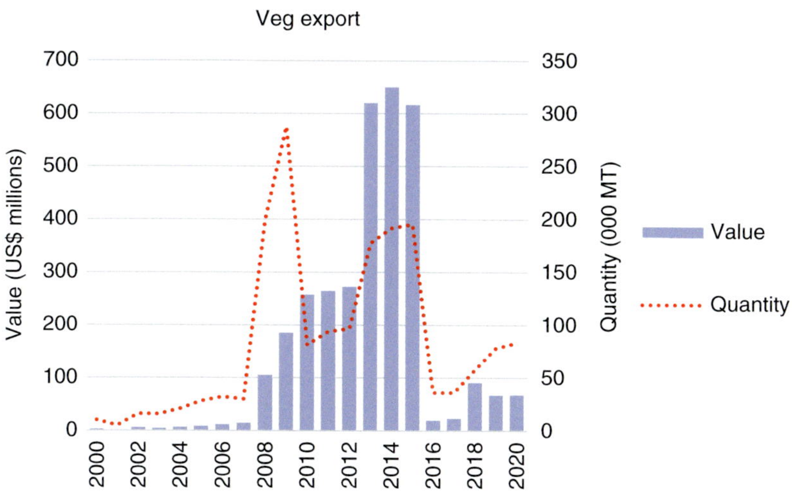

Fig. 7.12. Vegetables trade in Ethiopia. Source: constructed by the author from FAO (2022d).

90 varieties of 13 vegetables (including tomato, onion, sweet pepper, cabbage, lettuce, snap pea, cauliflower, courgette [zucchini], broccoli, carrot, beetroot, okra and sweet corn) between 2009 and 2017.

In spite of the expansion of diversity, vegetable productivity and production did not increase as much as expected. Performance varied from crop to crop, but the aggregate area, yield and production increased at an annual rate of 1.37%, 1.72% and 3.11%, respectively, between 2000 and 2020. Ethiopia's estimated aggregate yield average of 5.03–6.32 MT/ha compares unfavourably with the world average of 16.87 MT/ha.

Many of the crops showed declines in the area, yield or production. For example, green pepper area declined by 8.54 annually; yields of lettuce, carrot, tomato, garlic and dry onion declined between 12.96% and 1.15%; production of lettuce, green pepper and tomato declined between 11.16% and 0.26%. Yield gains for lettuce, garlic, tomato, dry onion and carrot declined by 443 kg/ha/yr, 190 kg/ha/yr, 178 kg/ha/yr kg/ha/yr, 117 kg/ha/yr and 54 kg/ha/yr, respectively.

One of the key challenges for vegetable production, especially under smallholders, is the small size of the area each household cultivates (and therefore the per capita production) in their garden. For example, the estimated average size of vegetable farms for onion, tomato, carrot, other fresh vegetables, garlic, lettuce and green pepper is estimated at 428 m^2, 239 m^2, 187 m^2, 111 m^2, 96 m^2, 95 m^2 and 78 m^2, respectively. In a similar fashion, the per capita production for onion, tomato, cabbages, garlic, carrot, green pepper and lettuce is estimated at 381 kg, 132 kg, 109 kg, 83 kg, 78 kg, 45 kg and 5 kg per household per season, respectively.

The low production, coupled with a lack of wider expansion among smallholder farmers across the country, appeared to be the major cause for the extremely low per capita consumption of vegetables in Ethiopia. The number of smallholder farmers cultivating vegetables in this country is perhaps less than a half of the total.

The per capita vegetable consumption in Ethiopia changed little over the last 40 years. It was 11 kg/person/yr in the 1975–1979 period. In the 1995–1999 period, it was 12 kg/person/yr and 14 kg/person/yr in the 2015–2019 period. The country ranked 176th out of 180 countries and territories in the world that cultivated vegetables in the 2015–2019 period. It was ahead of only the Democratic Republic of the Congo (DRC), Chad, the Comoros and the Solomon Islands (FAO, 2022e).

By contrast, the top five countries – China, Armenia, Albania, Uzbekistan and Guyana – had 374 kg/person/yr, 300 kg/person/yr, 296 kg/person/yr, 291 kg/person/yr and 280 kg/person/yr in in the 2015-2019 period. There were 5.50-fold, 2.64-fold, 2.34-fold, 2.14-fold and 1.91-fold increases for Guyana, Uzbekistan, Armenia, China and Albania, respectively, between 1995–1999 and 2015–2019. Lao (11th in the world) reported a nearly tenfold increase.

Compared to other African countries, Ethiopia ranked 44th out of 47 countries for which data were available in 2015–2019. It was ahead only of the DRC, Chad and the Comoros (see the discussion about nutrition in Chapter 24).

Malawi more than tripled its per capita vegetable consumption between 1995–1999 and 2015–2019, whereas Niger, Djibouti, Algeria, Senega and Cabo Verde more than doubled. Mali, Botswana, Tanzania, Sudan and Tunisia increased between 60% and 93%; Nigeria, Ethiopia, Benin, Morocco and Sierra Leone remained unchanged whereas Guinea, Mauritius, Egypt and Kenya showed varying levels of decline between 1995–1999 and 2015–2019.

Ethiopia has made good progress with its vegetable export earnings from the late 2000s onwards, peaking to US$650 in 2014, but taking a substantial downturn after that. Increased local demand owing to population growth, declines in productivity and production, and fluctuations in the international market were cited as some of the possible explanations for the decline of the country's export earnings.

This country has great potential to expand vegetable production to improve its nutrition security and foreign currency earnings from export. To achieve those, it needs to promote a wider adoption of the crops among smallholders, increase small-scale irrigation, and improve management practices against biotic and abiotic stresses.

References

Abate, T., Ayalew, G. (1994) Progress in vegetable pest management research, 1985–1992. In: Herath, E., Desalegne, L. (eds) *Horticulture Research and Development in Ethiopia*. IAR/FAO, Addis Ababa, pp. 187–193.

Abate, T., Ayalew, G. (1997) Sources of resistance in tomato against fruitworms. *Pest Management Journal of Ethiopia* 1(1&2), 1–8.

Abate, T., Shiferaw, B., Menkir, A., Wegary, D., Kebede, Y. *et al*. (2015) Factors that transformed maize productivity in Ethiopia. *Food Security* 7, 965–981. DOI: 10.1007/s12571-015-0488-z.

Agonafer, Y. (1989) Economics of horticultural production in Ethiopia. In: de Jaeger, A., Verhaegh, A. (eds) First International Symposium on Horticultural Economics in Developing Countries. 16–23 July 1989. Alemaya, Ethiopia. *Acta Horticulturae* pp. 15–19. International Society for Horticultural Science, Leuven, Belgium.

Agonafer, Y. (1994) Contribution of horticulture to the national economy. In: Herath, E., Desalegne, L. (eds) Horticulture Research and Development in Ethiopia. *Proceedings of the Second National Horticultural Workshop of Ethiopia,* 1–3 December 1992. IAR, Addis Ababa, Ethiopia.

Alemayehu, N., Hoekstra, D., Berhe, K., Jaleta, M. (2010) *Irrigated Vegetable Promotion and Expansion: The Case of Ada'a District, Oromia Region, Ethiopia*. IPMS, ILRI, Addis Ababa, Ethiopia.

Ayalew, G., Abate, T. (1993) Management of fruit worms in tomato. *First Crop Protection Conference*. Crop Protection Society of Ethiopia, Addis Ababa, Ethiopia.

CSA (Central Statistical Agency) (2004) Agricultural Sample Surveys 2003/2004 (1996 E.C.), Report on area and production for major crops (Private Peasant Holdings, *Meher* Season Volume I, Statistical Bulletin. CSA, Addis Ababa, Ethiopia.

CSA (2005) Agricultural Sample Survey 2004/2005 (1997 E.C.), Report on Area and Production of Crops (Private Peasant Holdings, *Meher* Season), Volume I, Statistical Bulletin 331. CSA, Addis Ababa, Ethiopia.

CSA (2006) Agricultural Sample Survey 2005/2006 (1998 E.C.), Report on Area and Production of Crops (Private Peasant Holdings, *Meher* Season), Volume I, Statistical Bulletin 361. CSA, Addis Ababa, Ethiopia.

CSA (2007) Agricultural Sample Survey 2006/2007 (1999 E.C.), Report on Area and Production of Crops (Private Peasant Holdings, *Meher* Season), Volume I, Statistical Bulletin.388. CSA, Addis Ababa, Ethiopia.

CSA (2008) Agricultural Sample Survey 2007/2008 (2000 E.C.), Report on Area and Production of Crops (Private Peasant Holdings, *Meher* Season), Volume I, Statistical Bulletin 417. CSA, Addis Ababa, Ethiopia.

CSA (2009) Agricultural Sample Survey 2008/2009 (2001 E.C.), Report on Area and Production of Crops (Private Peasant Holdings, *Meher* Season), Volume I, Statistical Bulletin 446. CSA, Addis Ababa, Ethiopia.

CSA (2010) Agricultural Sample Survey 2009/2010 (2002 E.C.), Report on Area and Production of Major Crops (Private Peasant Holdings, *Meher* Season), Volume I, Statistical Bulletin 446. CSA, Addis Ababa, Ethiopia.

CSA (2011) Agricultural Sample Survey 2010/2011 (2003 E.C.), Report on Area and Production of Major Crops (Private Peasant Holdings, *Meher* Season), Volume I, Statistical Bulletin. CSA, Addis Ababa, Ethiopia.

CSA (2012) Agricultural Sample Survey 2011/2012 (2004 E.C.), Report on Area and Production of Major Crops (Private Peasant Holdings, *Meher* Season), Volume I, Statistical Bulletin. CSA, Addis Ababa, Ethiopia.

CSA (2013) Agricultural Sample Survey 2012/2013 (2005 E.C.), Report on Area and Production of Major Crops (Private Peasant Holdings, *Meher* Season), Volume I, Statistical Bulletin 532. CSA, Addis Ababa, Ethiopia.

CSA (2014) Agricultural Sample Survey 2013/2014 (2006 E.C.), Report on Area and Production of Major Crops (Private Peasant Holdings, *Meher* Season), Volume I, Statistical Bulletin. CSA, Addis Ababa, Ethiopia.

CSA (2015) Agricultural Sample Survey 2014/2015 (2007 E.C.), Report on Area and Production of Major Crops (Private Peasant Holdings, *Meher* Season), Volume I, Statistical Bulletin 578. CSA, Addis Ababa, Ethiopia.

CSA (2016) Agricultural Sample Survey 2015/2016 (2008 E.C.), Report on Area and Production of Major Crops (Private Peasant Holdings, *Meher* Season), Volume I, Statistical Bulletin. CSA, Addis Ababa, Ethiopia.

CSA (2017) Agricultural Sample Survey 2016/17 (2009 E.C.), Report on Area and Production of Major Crops (Private Peasant Holdings, *Meher* Season), Volume I, Statistical Bulletin. CSA, Addis Ababa, Ethiopia.

CSA (2018) Agricultural Sample Survey 2017/18 (2010 E.C.), Report on Area and Production of Major Crops (Private Peasant Holdings, *Meher* Season), Volume I, Statistical Bulletin 586 CSA, Addis Ababa, Ethiopia.

CSA (2019) Agricultural Sample Survey 2018/19 (2011 E.C.), Report on Area and Production of Major Crops (Private Peasant Holdings, *Meher* Season), Volume I, Statistical Bulletin 589. CSA, Addis Ababa, Ethiopia.

CSA (2020) Agricultural Sample Survey 2019/20 (2012 E.C.), Report on Area and Production of Major Crops (Private Peasant Holdings, *Meher* Season), Volume I, Statistical Bulletin 587. CSA, Addis Ababa, Ethiopia.

CSA (2021) Agricultural Sample Survey 2020/21 (2013 E.C.), Report on Area and Production of Major Crops (Private Peasant Holdings, *Meher* Season), Volume I, Statistical Bulletin. CSA, Addis Ababa, Ethiopia.

Demissie, T., Ali, A., Zerfu, D. (2009) Availability and consumption of fruits and vegetables in nine regions of Ethiopia with special emphasis to vitamin A deficiency. *The Ethiopian Journal of Health Development* 23(3), 216–222.

Desalegne, L., Herath, E., Belehu, T., Lemaga, B., Gebre-Mariam, S. (1994) Horticultural research – past, present and future trends. In: Herath, E., Desalegne, L. (eds) *Horticulture Research and Development in Ethiopia. Proceedings of the Second National Horticultural Workshop of Ethiopia,* 1–3 December 1992, pp. 19–28. IAR, Addis Ababa, Ethiopia.

Dube, A.K., Ozkan, B., Govindasamy, R. (2018) Analyzing the export performance of the horticultural sub-sector in Ethiopia: ARDL bound test cointegration analysis. *Horticulturae* 4, 34. doi: 10.3390/horticulturae4040034.

Emana, B., Gebremedhin, H. (2007) *Constraints and Opportunities of Horticulture Production and Marketing in Eastern Ethiopia*. DCG (Drylands Coordination Group) Report No. 46, Addis Ababa.

Emana, E., Afari-Sefa, V., Dinssa, F., Ayana, A., Balemi, T. *et al.* (2015) Characterization and assessment of vegetable production and marketing systems in the humid tropics of Ethiopia. *Quarterly Journal of International Agriculture* 54, 163–187.

ERCA (Ethiopian Revenue and Customs Authority) (2013–2017) Available at: http://www.erca.gov.et/index.php/component/hscode/?view=importexport&iccaldate=2014-5-1 (accessed 29 Dec 2018).

FAO (2022a) Statistical databases and data-sets of the Food and Agriculture Organization of the United Nations. Available at: http://www.fao.org/faostat/en/#data (accessed 18 March 2022).

FAO (2022b) Statistical databases and data-sets of the Food and Agriculture Organization of the United Nations. Available at: http://www.fao.org/faostat/en/#data (accessed on 11 April 2022).

FAO (2022c) Statistical databases and data-sets of the Food and Agriculture Organization of the United Nations. Available at: http://www.fao.org/faostat/en/#data (accessed 6 May 2022).

FAO (2022d) Statistical databases and data-sets of the Food and Agriculture Organization of the United Nations. Available at: http://www.fao.org/faostat/en/#data (accessed 8 July 2022).

Herath, E., Desalegne, L. (eds) (1994) Horticulture research and development in Ethiopia. *Proceedings of the Second National Horticultural Workshop of Ethiopia,* 1–3 December 1992, Addis Ababa. IAR, Addis Ababa, Ethiopia.

K-Yesus, S. (1994) Horticultural development in peasant agriculture. In: Herath, E., Desalegne, L. (eds) *Proceedings of the Second National Horticultural Workshop of Ethiopia,* 1–3 December 1992, pp. 29–36. IAR, Addis Ababa, Ethiopia.

MOALR (Ministry of Agricultural and Livestock Resources) (2017) Crop Variety Register Issue No. 20. Plant Variety Release, Protection and Seed Quality Control Directorate, MOALR, Addis Ababa, Ethiopia.

Wiersinga, R.C., de Jaeger, A. (2009) *Business Opportunities in the Ethiopian Fruit and Vegetable Sector*. Report 2008 075 January 2009 Project number 40657 LEI, Wageningen UR, The Hague.

8 The Place of Fruit Crops in Improved Nutrition and Trade

Tsedeke Abate[1]* and Asmare Dagnew[2]
[1]*Homegrown Vision, Addis Ababa, Ethiopia;* [2]*Ethiopian Institute of Agricultural Research, Melkassa Research Center, Adama, Ethiopia*

Abstract
This chapter presents a critical review on the results of Ethiopia's research and development efforts in fruit crops to improve nutrition security, diversify export and improve import substitution. Smallholders accounted for 82% of the total fruit crop area (~148,000 ha in 2016–2020), while the remainder was covered by large-scale commercial growers. The aggregate fruit crops area and production increased at 6.39% and 6.11%, respectively, and declined in yield by -0.26%, indicating that increases in production resulted from area expansion rather than productivity per unit area. Yield gain was positive for banana, papaya, miscellaneous tropical fruits and orange, whereas pineapple, miscellaneous fresh fruits, grape, avocado and mango showed yield declines; there was little change for lemon/lime. Ethiopia has remained a net importer of fruit crop products between 2000 and 2020, with export earnings of ~US$7 million compared to ~US$24 million import bills. Insect pests, diseases and poor agronomic practices are the major constraints.

Introduction

From a nutritional point of view, fruit crops are sought after for their vitamin and mineral supplies. Traditional diets in Ethiopia included little or no fruit crops. People, especially children, obtained their fruit requirements from wild and cultivated trees and shrubs such as cactus, *dokma* or water berry (*Syzygium guineense*), *gishta* or wild custard apple (*Annona senegalensis*), *enjori* (*Morus mesozygia*), *koshim* (*Dovyalis abyssinca*), *shola* (*Ficus* spp.), and a large number of other sources (Bekele-Tesemma, 1993; Edwards *et al.*, 2019; Ruelle *et al.*, 2019).

Banana has long been grown and consumed as an important source of vitamins and minerals, in addition to its contribution to starch both in the rural and urban settings. Lemon/lime, peach and citron (all primarily grown in the backyards and small gardens) are perhaps among the earliest modern fruit crops known to Ethiopians. A small number of commercial citrus orchards and vineyards were established in the 1960s but were nationalized and became part of the state farms during the military-socialist government rule starting in the second half of the 1970s.

Coordinated research on fruit and other horticultural crops started in 1969 with the establishment of the then Nazareth Research Center at Melkassa under the Ethiopian Institute of Agricultural Research. The Horticultural Development Corporation (part of the state farms) that was established in the late 1970s played a key role in the commercialization of those crops.

* Corresponding author: tsedeke@homegrownvision.org

DOI: 10.1079/9781800626386.0008

They experimented with more than a dozen species of fruit crops, including citrus (orange, mandarin, lemon/lime, grapefruit), avocado, banana, grape, guava, jackfruit, mango, papaya, pineapple, and temperate fruits like apple, peach and plum in many parts of the country.

Gebre-Mariam (2003) gives the most detailed accounts of fruit crops research and development in Ethiopia. He states that 'The bulk of the fruit production in Ethiopia is in gardens and small-scale commercial farms.' Starting in the 1990s, reviews of progress have been published in successive proceedings (Godfrey-Sam-Aggrey and Tuku, 1987; Herath and Desalegne, 1994; Dagnew *et al.*, 2020; Etissa *et al.*, 2020).

This current report provides a critical review of the performance of fruit crops as a group and individually, and their impact on the national economy and nutrition security.

Approach

Sources of data

Two major sources of data on production were used in this chapter: the Central Statistical Agency (CSA) of the Government of Ethiopia (http://www.csa.gov.et) and the Food and Agriculture Organization (FAO) of the United Nations (http://www.fao.org/faostat/en/#data).

Comprehensive and complete data for fruit crops in Ethiopia by the CSA start in 2003. We also noted that both sources have similar data for most of the crops starting from this period. We gave priority to CSA data where they are available and complete, because the FAO data does not provide information on the number of households growing the crops. For long-term production data, we relied on FAOSTAT. Our information for long-term agricultural trade data is also from the FAO. Information on variety releases was obtained from the Ministry of Agriculture and Livestock Resources (MOALR), now the Ministry of Agriculture [MOA]).

Measuring progress

In order to see the effect of technological change and yield growth on production, we used economic methodology to isolate the contribution of area and yield growth on national production for each of the crops included in our assessment, as employed by Abate *et al.* (2015) and described in Chapter 3.

Results

Fruit crops in general

Different sources suggest that more than a dozen species of fruit crops are cultivated in Ethiopia. The CSA sample surveys include only eight of them (consisting of a total of 27 cultivars released by the research system), as presented in Table 8.1, and their records start from the 2003/04 *meher* season. The records show that more than 5 million smallholder households cultivated a total area of >121,000 ha and produced ~933,000 MT fruit annually in 2016–2020. The average holding was 0.0234 ha (or 234 m^2), with an estimated per capita production of about 62 kg/ha/yr. The average yield here was estimated at roughly 7 MT/ha.

The MOALR National Variety Registry includes five additional species: date palm, fig, grape, Indian jujube (*Ziziphus jujuba*) and peach, with a total of 17 cultivars released. There is at least one private company that produces unspecified amounts of strawberry, largely for export. Ethiopia also grows a small amount of passion fruit (*Passiflora edulis*) and there is low-level research being conducted at Melkassa Research Center (Dagnew *et al.*, 2020). It is not uncommon to see white sapote, also known as *casimiroa* or Mexican apple (*Casimiroa edulis*) in many gardens across the Central Rift Valley system in Ethiopia.

The FAO (2022a) data list more than a dozen individual or group of fruit crops. According to this source, fruit crops in Ethiopia occupied >148,000 ha with an average yield of 6.45 MT/ha, while the total annual production was well over 1 million MT (Table 8.2).

Performance varied according to crop type, but the aggregate area, yield and production changed at an annual rate of 6.39%, -0.26% and 6.11%, respectively, between 2000 and 2020. This means that the overall yield gain was 41 kg/ha/yr and was non-significant (not shown). The aggregate area (and production) showed

Table 8.1. Indicators for major fruit crops smallholders cultivated in Ethiopia (2016–2020 average). Source: number of holders, area, yield and production are constructed by the author from CSA sample surveys (www.csa.gov.et); area per holder and per capita production are calculated by the author from CSA data; number of holders is average for 2016–2019; number of varieties are from MOALR, 2017.

Crop	Holders (000)	Area (000 ha)	Yield (MT/ha)	Production (000 MT)	Area per holder (ha)	Production per holder (kg)	Number of varieties released
Banana	3239	70.39	8.36	594.25	0.0217	183	12
Avocado	2046	21.42	5.09	116.21	0.0105	57	6
Mango	1717	17.49	6.86	120.01	0.0102	70	4
Papaya	812	3.96	14.59	57.58	0.0049	71	3
Orange	635	4.09	8.09	32.41	0.0064	51	NA
Guava	382	2.50	1.49	3.59	0.0065	9	NA
Lemon	267	1.52	5.07	7.86	0.0057	29	NA
Pineapple	52	0.54	1.45	1.27	0.0103	24	2
Total	5209	121.90	6.96	933.18	0.0234	62	27

NA = not available.

Table 8.2. Yield gain (b-value, expressed in kg/ha/yr) and annual rate of growth (ROG) for selected fruit crops grown in Ethiopia between 2000 and 2020. Source: calculated and constructed by the authors from FAO (2022a).

	Yield gain			ROG (%)		
Fruit	b-value	P > |t|	R^2	Area	Yield	Production
Papaya	334	0.10	0.14	2.11	4.14	6.33
Banana	146	<0.01	0.69	6.92	2.01	9.07
Misc. tropical fruits	22	0.01	0.55	3.84	0.31	4.16
Orange	11	0.94	0.00	2.97	1.01	4.01
Lemon/lime	0	0.03	0.48	4.48	4.48	9.16
Mango	–28	0.63	0.01	10.24	–0.28	9.93
Avocado	–115	0.08	0.15	11.53	–2.37	8.88
Grape	–129	<0.01	0.91	2.34	–4.42	–2.18
Misc. fresh fruits	–420	<0.01	0.93	3.27	–3.61	–0.46
Pineapple	–1927	0.66	0.02	NA	NA	NA

NA = not available.

continued growth while the yield changed little (Fig. 8.1A). Even though highly significant, the yield gain for Ethiopia was one of the lowest among the top ten producing countries such as Brazil, Türkiye, China and Mexico (Fig. 8.1B).

Performance of individual crops

Banana

Banana is a unique fruit in the sense that it provides both energy as well as vitamins and minerals. It is widely cultivated throughout Ethiopia both under smallholder and commercial agriculture. Smallholders in Ethiopia grow banana for domestic consumption and cash from local sales. It is an important source of vitamin B6 and vitamin C, as well as important minerals like potassium, manganese and magnesium. Overall, nearly 98% of the banana produced in Ethiopia is consumed locally. Banana from Ethiopia is exclusively exported to the neighbouring countries of Somalia and Djibouti.

According to the CSA, ~3 million smallholder households cultivated ~70,000 ha of land and

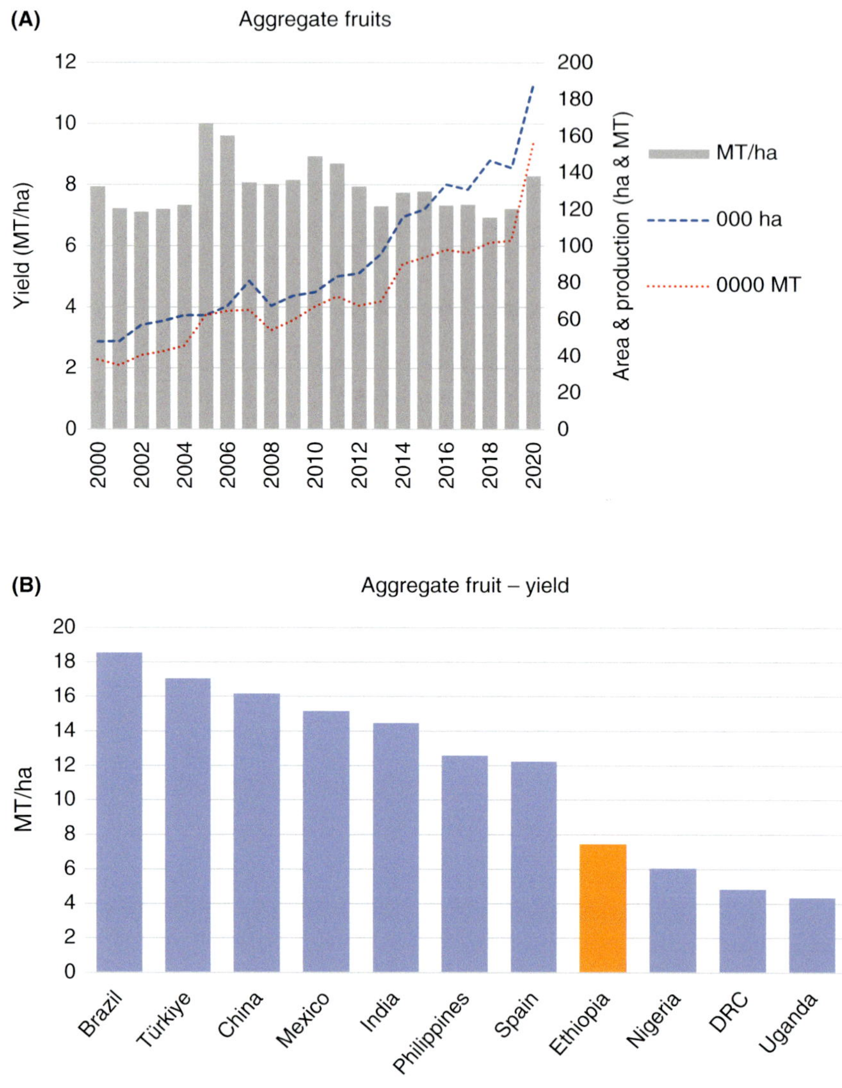

Fig. 8.1. (A) Indicator trends for aggregate fruit crops in Ethiopia. **(B)** Aggregate fruit crops yield (2016–2020 average) in Ethiopia compared to the top ten cultivating countries in the world. The countries with largest area are China, India, Brazil, Nigeria, Spain, Mexico, Democratic Republic of the Congo (DRC), Türkiye, Philippines and Uganda, respectively; Ethiopia is 72nd out of 195. Source: constructed by the lead author from FAO (2022b) and FAO (2022c).

produced ~594,000 MT of banana each year in the 2016–2020 *meher* season. The area allocated to banana under smallholder condition is ~0.0217 (or 217 m^2) ha/household and the per capita production ~183 kg/household/yr (cf. Table 8.1). The research system in Ethiopia has registered 12 banana cultivars as of 2017.

All indicators showed positive trends (Fig. 8.2A). However, the area and yield accounted for about 76% and 22%, respectively, of increases in production while the interaction factor covered the balance. Regression analysis showed that the yield gain for banana between 2000 and 2020 was 146 kg/ha/yr ($P<0.01$, $R^2 = 0.69$). Even though the yield gain was highly significant, the average yield for banana in Ethiopia is one of the lowest among the top ten countries in the world, including Ecuador, India, China and the Philippines (Fig. 8.2B).

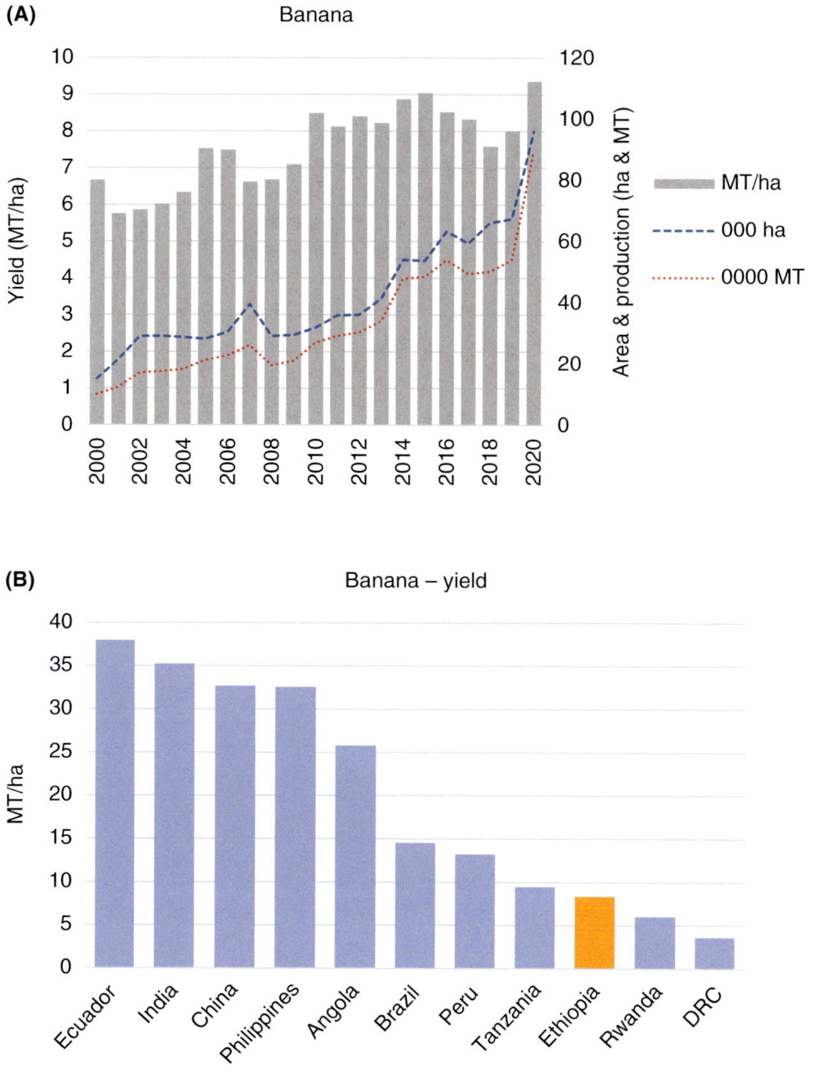

Fig. 8.2. (A) Indicator trends for banana in Ethiopia. **(B)** Banana yield (2016–2020) in Ethiopia compared to top ten cultivating countries of the world. Note: Countries with largest area are India, Brazil, Tanzania, China, Rwanda, the Democratic Republic of the Congo, Philippines, Ecuador, Peru and Angola; Ethiopia is 21st out of 120. Source: constructed by the lead author from FAO (2022a) and FAO (2022c).

Ethiopia ranked 23rd in area and 31st in production out of 120 countries in the world in the 2016–2020 period (FAO, 2022a). India, Brazil, Tanzania, China and Rwanda are on the top of the list for area harvested, while India, China, Indonesia, Brazil and Ecuador are the largest producers.

Bacterial wilt, fusarium wilt and black sigatoka have been reported as important diseases for banana in Ethiopia, but declining soil fertility and other agronomic management practices appear to be the major limiting factors for increased productivity and production.

Avocado

Avocado is a relatively newcomer to Ethiopian farming systems. Teferi and Ayanao (2016) reported that this crop was first introduced in 1938 to Hirna (eastern Ethiopia) and Wondo

Genet (southern Ethiopia) by private owners. Data recordings for this crop in Ethiopia by the FAO and CSA started in 2000 and 2003, respectively. According to the latter, an estimated ~2 million smallholders grew avocado on >21,000 ha and produced ~116,000 MT fruit annually in 2016–2020 (cf. Table 8.1). Average holdings are estimated at 0.0105 ha (or 105 m^2) and the per capita production is about 57 kg. There are six registered cultivars of avocado in Ethiopia.

Figure 8.3A presents the avocado trends in Ethiopia. With an annual rate of growth (ROG) of 11.53%, this is the fastest expanding fruit crop in the country. In other words, this crop covered an area of 9754 ha in 2000, compared to 30,588 ha in 2020, a 3.14-fold increase.

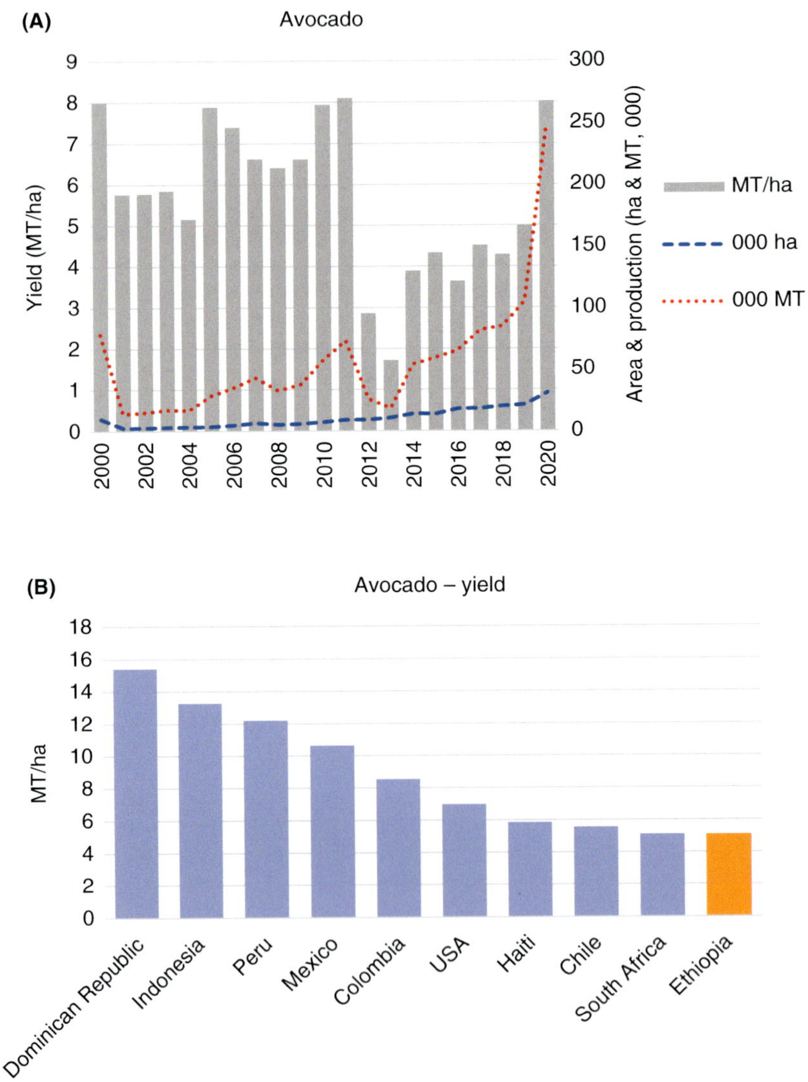

Fig. 8.3. (A) Indicator trends for avocado in Ethiopia. **(B)** Avocado yield (2016–2020) in Ethiopia compared to the top ten cultivating countries of the world. Note: Countries with largest area are Mexico, Colombia, Peru, the Dominican Republic, Haiti, Indonesia, Chile, the USA, Ethiopia and South Africa. Source: constructed by the lead author from FAO (2022a) and FAO (2022b).

The Green Legacy Initiative of the Government of Ethiopia gives high priority for this crop, and it is very likely to be expanding even faster in the coming years. However, its yield fluctuated heavily over the 21 years with a yield decline of –2.37% per annum between 2000 and 2020.

Ethiopia ranked 9th in area and 15th in production out of 65 countries in the world that cultivated avocado during the 2016–2020 period. The largest cultivating countries included Mexico, Colombia, Peru, the Dominican Republic and Haiti, while Mexico, the Dominican Republic, Peru, Colombia and Indonesia were examples of the largest producers (FAO, 2022a). Until recent years, virtually all the avocado produced in Ethiopia was for domestic consumption, but that is changing now. The national average yield of avocado for Ethiopia (~5 MT/ha) is the lowest within the top ten cultivating countries of the world, including the Dominican Republic, Indonesia, Peru and Mexico (Fig. 8.3B).

Avocado yield gain in Ethiopia declined by 115 kg/ha/yr between 2000 and 2020 (FAO, 2022a) but was non-significant. By contrast, Peru obtained a significant gain of 203 kg/ha/yr, while there were significant yield losses for Colombia and Dominican Republic (not shown). Poor crop management and limited access to roads appears to be the major limiting factors for increased expansion of avocado in Ethiopia.

Mango

The FAO records of mango for Ethiopia started in 1998 (FAO, 2022a), whereas CSA started reporting it in 2003. According to the latter, an estimated 1.7 million smallholders cultivated mango and produced ~120,000 MT of fruit annually in the 2016–2020 period, with an average yield of ~7 MT/ha (cf. Table 8.1). The average holding was 0.0102 ha (or 102 m^2) per holder, with per capita production of 70 kg. There are four registered mango cultivars in Ethiopia. Estimates from the FAO suggest the area at nearly 20,000 ha, yield <6 MT/ha and production nearly 114,000 MT (including guava). Ethiopia ranked 29th and 36th out of 97 countries in the world in area and production, respectively, in the 2016–2020 period (FAO, 2022a). India, Thailand, Indonesia, Pakistan and Mexico were the top five growers and/or producers in the world during the above period.

Mango became a popular fruit crop in this country in the 1990s following the introduction by the research system of apple mango cultivars (which are preferred by consumers because of their fibreless nature), resulting in the proliferation of juice-making businesses across the country. The CSA data show that the number of smallholder households that cultivate mango in Ethiopia increased at an annual rate of 10.60% between 2003 and 2019. In other words, there were roughly 350,000 smallholders who grew mango in 2003, compared to 1,666,000 in 2019 (a 4.76-fold increase).

Figure 8.4A depicts mango indicator trends in Ethiopia. The area showed steady growth; production also continued to grow through 2018 but declined after that. The yield took a downward trend starting in 2013. Overall, the area, yield and production changed at a rate of 10.24%, -0.28% and 9.93%, respectively.

Ethiopia's yield gain was -28 kg/ha/yr but non-significant. By contrast, yield loss for the Philippines was 134 kg/ha/yr and was highly significant ($P<0.01$). The national average yield of <6 MT/ha for Ethiopia is very low compared to many of the top ten countries (Fig. 8.4B). It was ahead only of the Cote d'Ivoire and the Philippines. The world average is ~13 MT/ha.

There is very high demand for mango within the country. Practically all of the crop produced in Ethiopia is consumed domestically.

At present, the major threats for mango production in Ethiopia are the recently introduced leaf anthracnose (caused by the fungus *Colletotrichum gloesporioides*) disease and the white mango scale (*Aulacaspis tubercularis*). Adults of the mango chafer (*Pachnoda peregrina*) and the yellow-headed chafer (*P. massajae*) devour and destroy the inflorescence sporadically in the Ethiopian Central Rift Valley. Declining soil fertility and generally poor management practices also contribute to low productivity of this crop in Ethiopia.

Papaya

Papaya has been grown in Ethiopia for a long period of time, but the CSA records only started in the 2003/04 *meher* season and the FAO provides data from 1998. There is not much difference in the figures from the two sources. An estimated 812,000 smallholder households

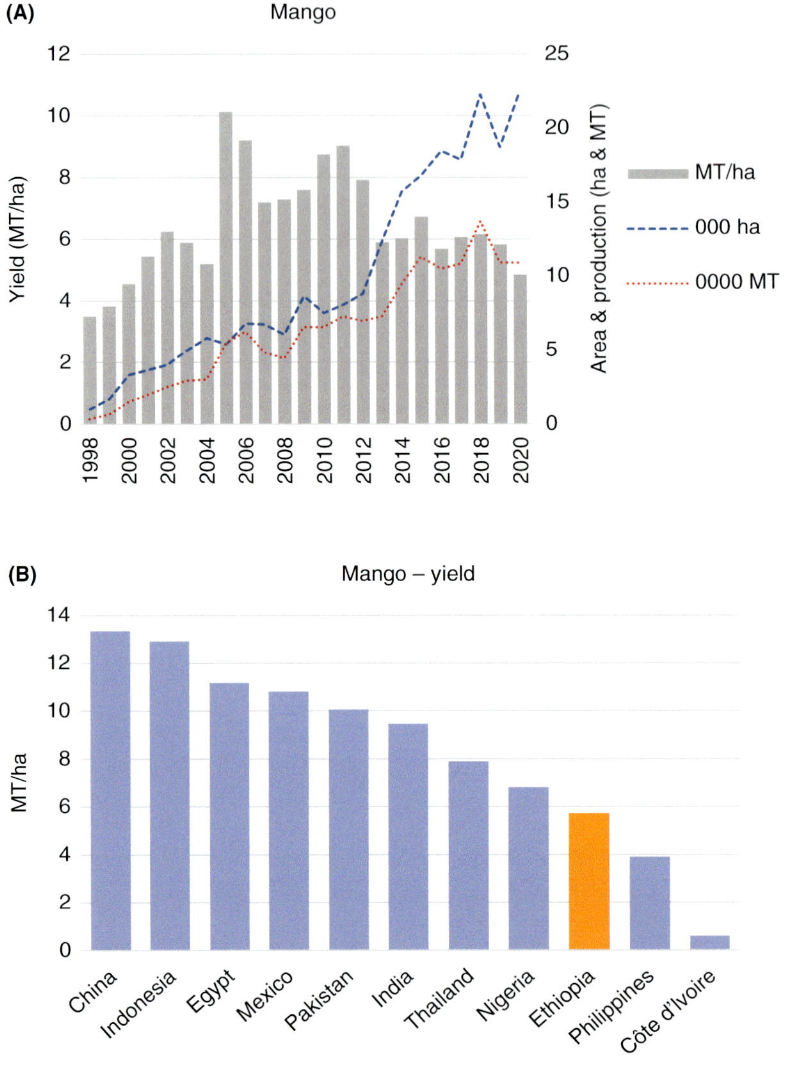

Fig. 8.4. **(A)** Indicator trends for mango in Ethiopia. **(B)** Mango yield in Ethiopia compared to the top ten cultivating countries of the world. Note: Countries with the largest area are India, Thailand, Indonesia, Pakistan, Mexico, the Philippines, China, the Cote d'Ivoire, Nigeria and Egypt; Ethiopia is 29th out of 97. Source: constructed by the lead author from FAO (2022a) and FAO (2022b).

harvested nearly 4000 ha and produced close to 58,000 MT of fruit annually in 2016–2020 (cf. Table 8.1). The average papaya area was 0.0049 ha (or 49 m^2) per holder, with an average per capita production of 71 kg/ha/holder. There are three registered cultivars under production. The number of smallholder farmers cultivating papaya nearly doubled between 2003 and 2019 – from less than 377,000 to nearly 720,000.

Figure 8.5A presents papaya trends in Ethiopia between 2000 and 2020. The area and production are showing accelerated growth especially in the most recent years. The yield jumped from about 5 MT/ha to 25 MT/ha in 2005 and 2006 but hovered around 15 MT/ha starting in the early 2010s to 2020. Papaya yields in this country more than doubled within 20 years.

Overall, the area, yield and production increased by 2.11%, 4.14% and 6.33%, respectively, between 2000 and 2020. This means that increases in yield accounted for ~65% of the increases in production while the contribution of area expansion was 33%; the interaction factor covered the balance. Ethiopia ranked 20th in area and 25th in production out of 66 countries in the world that cultivated papaya in 2016–2020. It registered a yield gain of 334 kg/ha/yr. Other countries that had higher yield gains among the top ten included Indonesia, Mexico and India.

Figure 8.5B compares national average yields of papaya in the top ten producing countries of the world. Ethiopia's average yield is ahead of almost all of its African peers (including Kenya, Malawi, Mozambique, Nigeria and the Cote d'Ivoire) but far behind the high yielding countries of the Dominican Republic, Indonesia, Mexico, India and Brazil, among others.

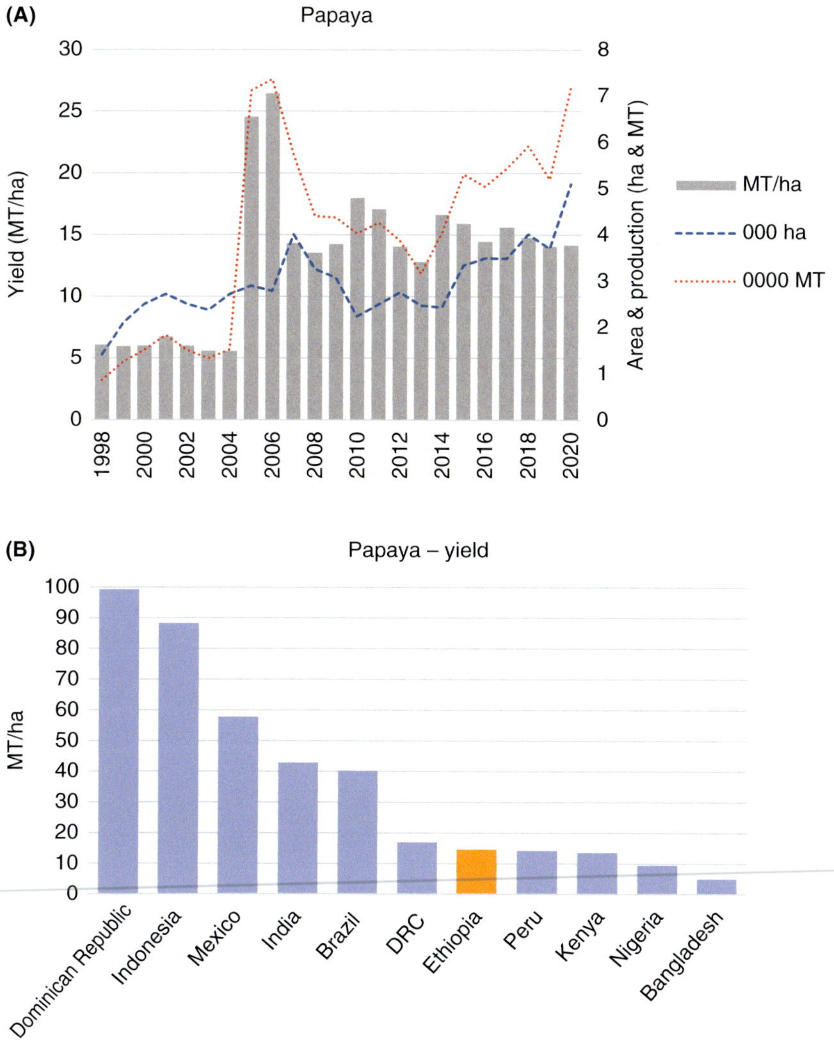

Fig. 8.5. (A) Indicator trends for papaya in Ethiopia. **(B)** Papaya yield (2016–2020) in Ethiopia compared to the top ten countries in the world. Note: countries with the largest area are India, Nigeria, Brazil, Bangladesh, Mexico, the Democratic Republic of the Congo (DRC), Peru, Indonesia, the Dominican Republic and Kenya; Ethiopia is 20th out of 66. Source: constructed by the lead author from FAO (2022a) and FAO (2022b).

Orange

Orange is among the oldest modern fruit crops grown in Ethiopia. From the literature, it appears that it has been part of the farming systems more than 100 years (Abate, 1981). The CSA sample surveys started records of orange in the 2003/04 *meher* season, whereas the FAO reports begin in 1961. According to the former, ~639,000 smallholders harvested ~4000 ha and produced ~32,000 MT in the 2016–2020 period (cf. Table 8.1). The average area allotted to orange in Ethiopia is ~0.0064 (or 64 m^2) ha per smallholder household and the per capita production is ~51 kg/household/yr.

Even though eight scions (including Parent Washington Navel, Hamlin Orange, Pineapple Orange, Campbell Valencia, Olinda Valencia, Ruby Blood Orange, Jaffa Orange, Temple Orange [Tangor]) are commercially grown on rootstocks such as Sour Orange, Troyer Citrange, Rough Lemon, Rangpur Lime and Estes Rough Lemon (Herath *et al.*, 1994), they are not officially registered by the MOA as of 2017.

It appeared that both the CSA and FAO data reports underestimate the total area (and therefore production) of orange in Ethiopia. For example, Dagnew *et al.*, (2014) reported >10,000 ha for the Bebeka Estate Farm in western Ethiopia alone. Ethiopia is ranked 64th both in area and production out of 117 countries that harvested orange in the 2016–2020 period (FAO, 2022a). Figure 8.6A presents orange indicator trends in Ethiopia between 2000 and 2020. The area has shown upward trends, especially in recent years. The yield reached 18 MT/ha in 2005 and 2006 but has declined since 2012. In general, the area, yield and production increased at 2.97%, 1.01% and 4.01%, respectively, during the above period.

Ethiopia obtained a yield gain of 22 kg/ha/yr but this was non-significant. By comparison, China registered highly significant gains of 777 kg/ha/yr (FAO, 2022a). The USA and Iraq, on the other hand, showed highly significant yield losses of 725 kg/ha/yr and 443 kg/ha/yr, respectively. The national average yield for orange in Ethiopia is lower than the world average of about 14MT/ha. It is ahead of only Iraq among the highest yielding countries including Brazil, Egypt, Spain and the USA (Figure 8.6B). South Africa, Indonesia, Türkiye, Iran and Brazil are among the top yielding countries among the top ten.

The pseudocercospora leaf and fruit spot (caused by the fungus *Pseudocercospora angolensis*) that was accidentally introduced and first reported in southwestern Ethiopia in 1990 (Derso, 1999) is one of the major constraints to orange and other citrus fruits production in this country (Yesuf, 2013; Dagnew *et al.*, 2014). Perhaps even more important are management practices, including agronomic practices such as weeding, soil fertility and crop sanitation, along with minimizing postharvest losses (Dessalegn *et al.*, 2016).

Miscellaneous tropical fresh fruits

The CSA does not have records of this group of fruits. The FAO data start from 1995. Accordingly, Ethiopia harvested ~6000 ha and produced ~46,000 MT tropical fresh fruits annually in the 2016–2020 period (c.f. Table 8.2). The area and production showed sustained growth through the mid-2010s and somewhat levelled off thereafter, while the yield showed little change over the years. In other words, the ROG for area, yield and production was 3.84%, 0.31% and 4.16%, respectively. This suggests that ~92% of the growth in production was due to increases in area expansion while increases in yield contributed ~7%, and the interaction factor accounted for the balance.

Ethiopia registered a yield gain of 22 kg/ha/yr ($P<0.01$, $R^2 = 0.55$) for tropical fruits between 2000 and 2020 (FAO, 2022a). The yield gain for India, Iran, Indonesia and Brazil ranged between 216 kg/ha/yr and 49 kg/ha/yr. By contrast, countries like Madagascar, Pakistan and the Philippines reported highly significant yield losses. The national average yield of tropical fruits for Ethiopia is close to the world average of about 8 MT/ha.

Lemon and lime

Lemon and lime (along with banana and citron) are perhaps the oldest modern fruit crops in Ethiopia. Lemon/lime, orange and citron prominently feature in the Ethiopian romantic folklore. The CSA sample survey records for lemon started from the 2003/04 *meher* season, whereas FAO records started from 1961. According to the former, ~267,000 smallholders harvested ~1500 ha and produced nearly 8000 MT annually in the 2016–2020 *meher* seasons (c.f. Table 8.1).

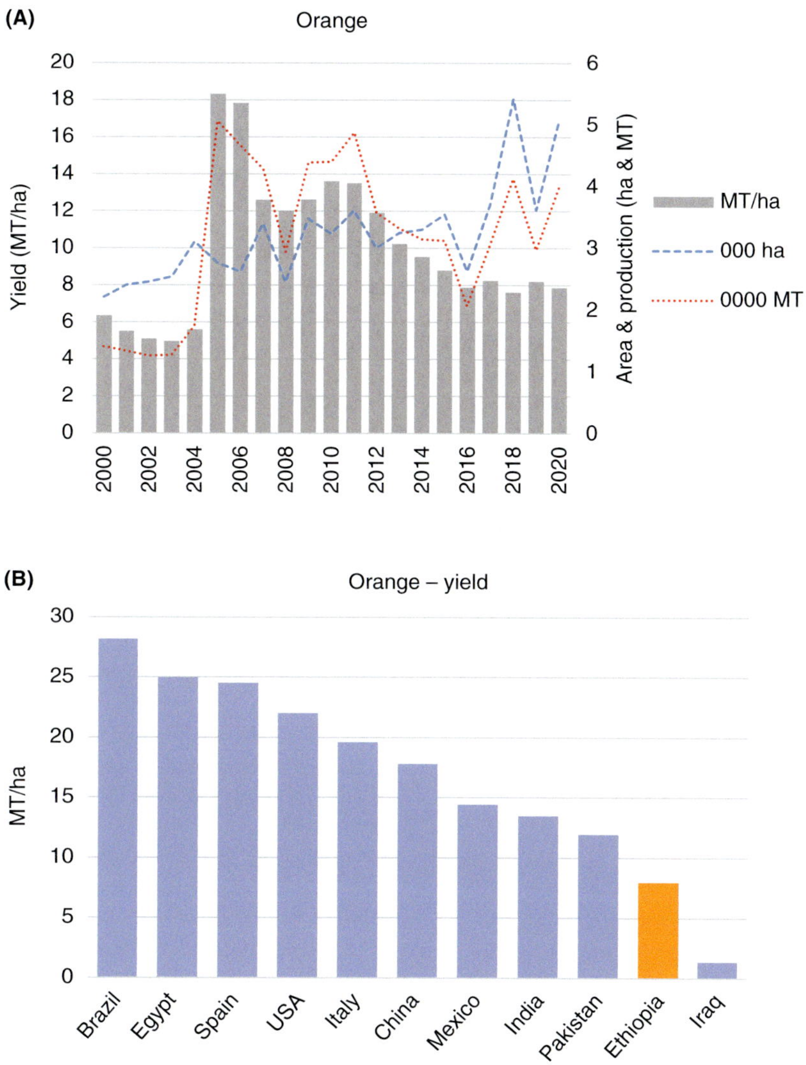

Fig. 8.6. **(A)** Indicator trends for orange in Ethiopia. **(B)** Orange yield in Ethiopia compared to the top ten largest cultivating countries. Note: Countries with the largest area are India, Brazil, China, Mexico, the USA, Spain, Pakistan, Egypt, Italy and Iraq; Ethiopia is 64th out of 117. Source: constructed by the lead author from FAO (2022a) and FAO (2022b).

Figures from the FAO (c.f. Table 8.2) are basically the same as those from CSA.

Ethiopia ranked 52nd in area and 59th in production out of 105 countries in the world that grew lemon/lime. India, Mexico, China, Argentina and Brazil are the largest producers (FAO, 2022a). The bulk of lemon/lime produced in Ethiopia is for domestic consumption and import substitution. Gebre-Mariam (2003) reports that lime/lemon juice has replaced citric acid in the manufacturing of orange marmalade at the Nura-Era Processing Plant, saving the much-needed foreign currency for imports.

The area showed rapid growth in the early 2010s but declined in 2019 and 2020. Production also followed a similar trend. The yield showed a fast growth rate starting from early 2000s but declined from 2013 onwards. Overall, the area, yield and production increased at 4.48%, 4.48% and 9.16%, respectively (c.f. Table 8.2). Regression analysis showed that Ethiopia registered a yield gain of 99 kg/ha/yr of lemon/lime but this was non-significant. South Africa, the Dominican Republic, India, Türkiye and Sudan fared slightly better than Ethiopia, whereas Brazil, Spain, Mexico, Thailand and Egypt had negative yield gains (FAO, 2022b).

Ethiopia's average yield is below the majority of the largest producing countries and the world average of >12 MT/ha. Gebre-Mariam (2003) reported 30 MT/ha at the Awara Melka state farm in the Upper Awash, indicating that this country has the potential to significantly raise its yield for lemon/lime. This is considered a minor crop in Ethiopia although it has a high demand, especially among urban consumers. Limited expansion among smallholders and generally poor crop management are among the major constraints to lemon/lime production in Ethiopia.

Grape

The history of grapes and wine making in Ethiopia goes back to at least the fourth century CE during the Axumite Kingdom (Pankhurst, 2006). Modern vineyards and wineries have been in existence since the beginning of the 20th century. The CSA sample surveys do not include grapes, whereas the FAO records are available from 1969 onwards. Accordingly, Ethiopia harvested ~2400 ha and produced ~4900 MT in the 2016–2020 period (c.f. Table 8.2). The average yield was 2.02 MT/ha. The world's top cultivating countries included Spain, France, China, Italy and Türkiye. Examples among the top producers were China, Italy, the USA, Spain and France. Ethiopia ranked 73rd in area and 79th in production out of 95 countries in the world that cultivated grape in the 2016–2020 period.

All indicators remained almost constant through the late 1980s. The ROGs for area, yield and production were 2.34%, -4.42% and -2.18%, respectively, between 2000 and 2020. Ethiopia has suffered a highly significant yield loss of 129 kg/ha/yr ($P<0.01$, $R^2 = 0.91$) between 2000 and 2020. Examples of the largest producing countries with yield losses include India, Argentina, Germany, France and Australia. By contrast, Uzbekistan, China, Afghanistan, South Africa and Türkiye obtained highly significant yield gains ($P<0.01$). With 2.03 MT/ha, the yield in Ethiopia was among the lowest in the world and way below the world average of close to 10 MT/ha. India, Brazil, China, South Africa and the USA all had upwards of 15 MT/ha. Ethiopia has great potential for expanding grape production for the local wine industry, considering its good climate and good soil. Weaver birds and poor crop management practices appeared to be the major constraint to grape production in this country.

Miscellaneous fresh fruits

The CSA does not keep records on this group of fruits in Ethiopia. The FAO data show that Ethiopia harvested ~14,000 ha and produced ~122,000 MT of these fruits annually in the 2016–2020 seasons (c.f. Table 8.2). The average yield was estimated at <9 MT/ha. The area continued to expand through the mid-2010s and levelled off thereafter. The yield continued to grow through the mid-1970s and plateaued till the mid-2000s, after which time it showed fast decline. Production depended on the area and yield. The ROGs for area, yield and production were 3.27%, -3.61% and -0.46%, respectively, between 2000 and 2020.

Miscellaneous fresh fruits in Ethiopia suffered a very highly significant yield loss of 420 kg/ha/yr ($P<0.01$, $R^2=0.93$). Iran, South Sudan, Pakistan, Bangladesh and Thailand registered significant yield gains ranging between 472 and 125 kg/ha/yr (FAO, 2022a). Ethiopia ranked 9th out of the 20 but has the potential to be more competitive with the likes of Thailand, Indonesia and Pakistan.

Tangerine

The CSA does not record tangerine separately from orange and other citrus fruits. According to the FAO (2022a), Ethiopia harvests ~2900 ha and produces ~9100 MT of tangerine each year (c.f. Table 8.2). The country ranked 38th in area and 30th in production out of 71 countries in the world that cultivated tangerine in the 2016–2020 period (FAO, 2022a). China, Spain, Morocco, Türkiye and Brazil are the top producing countries.

The area showed sustained growth, but yield declined from the early 2000s. Overall, the ROG for area, yield and production for Ethiopia was 3.49%, -2.85% and 0.53% respectively, between 2000 and 2020. This means a highly significant yield loss of 112 kg/ha/yr ($P<0.01$, $R^2=0.92$).

The national average yield for Ethiopia is much below the world average of 10.24 MT/ha and all of the top ten producing countries. Please note that Iran, the USA, Republic of Korea, Türkiye and Peru registered average yields exceeding 25 MT/ha, compared to 3.10 MT/ha for Ethiopia.

Miscellaneous citrus

The CSA sample surveys do not include these crops, but the FAO data start from 1993. Accordingly, Ethiopia harvested ~440 ha and produced ~3200 MT of miscellaneous citrus fruit annually in the 2016–2020 period (c.f. Table 8.2). This country ranked 41st in area and 43rd in production out of 67 countries during the above period. The top producing countries included China, Nigeria, India, Mexico and Guinea.

Miscellaneous citrus area and yield remained nearly constant from the early to late 1990s but showed upward trends through the early 2010s, and somewhat levelled off after that. The yield was also constant from the early 1990s to late 2000s and showed some growth through the early 2010s, after which time it did not change much. In general, the area, yield and production increased at an annual rate of 1.25%, 1.25% and 2.52%, respectively, between 2000 and 2020. Ethiopia's national average yield (>7 MT/ha) is low compared to the world average of close to 11 MT/ha or many of the top cultivating countries. China, Syria, Iran, Peru, China-Taiwan and Colombia had yields ranging between 32 and 15 MT/ha.

Pineapple

Commercial farms widely produced pineapple in southwestern Ethiopia in the 1960s and 1970s where there was an ideal climate for the crop. Examples included Gojeb Farm and the state farm at Bebeka, southwestern Ethiopia. Production declined in the late 1970s and early 1980s, primarily because of the pineapple mealybug (*Dysmicoccus brevipes*) devastation and poor management. The pineapple cultivar 'Smooth Cayenne' was the only variety that was grown at the time, and it was susceptible to the insect. The variety 'Tafache' was released in 2017 but the extent of its production is too early to predict. Farmers abandoned this crop and switched to coffee (Gebre-Mariam, 2003) because the latter was more feasible.

According to the CSA, 52,000 smallholders planted 540 ha of land and produced nearly 1300 MT of fruit, with an average yield of 1.45 MT/ha, in the 2016–2020 period. The average holding size was 0.0103 ha (or 103 m^2) per holder and the per capita production 24 kg. According to the FAO (2022a), the average area, yield and production for 2016–2020 were estimated at 633 ha, 2.89 MT/ha and 1855 MT, respectively. However, long-term data from either source were incomplete and unreliable to conduct any meaningful further analysis.

Peach

Peach has been traditionally grown in higher altitudes for many years in Ethiopia. The exact time of its introduction is not definitive, but some authors suggest it might have been introduced by the Portuguese around the 16th century. The CSA sample surveys do not include peach, but FAO (2022b) data are available starting from 1995. Accordingly, Ethiopia harvested ~27 ha and produced ~155 MT of peaches annually in the 2016–2020 period. Ethiopia ranked 80th in area and 82nd in production out of 82 countries in the world, with China, Spain, Italy, Türkiye, Greece and Tajikistan among the top producing countries.

The area and production showed steady growth through the mid-2010s and levelled off after that, whereas the yield increased marginally. For example, the ROGs for area, yield and production were estimated at 4.37%, 0.87% and 5.30%, respectively. Ethiopia's national average yield of <6 MT/ha is well below the world average of >10 MT/ha. Several of the world's top cultivating countries obtain more than 15 MT/ha (FAO, 2022a). The yield gain for Ethiopia between 2000 and 2020 was roughly 48 kg/ha/yr and was significant ($P<0.01$, $R^2=0.68$). The peach aphid, *Myzus persicae*, is the only major insect pest on peach in Ethiopia.

The current level of peach production is the lowest among all recorded fruit crops, but recent studies suggest that Ethiopia has good potential for expansion of peach production and marketing (Firde *et al.*, 2019; Mosie *et al.*, 2019). This is particularly true in the light of declining production in some of the major producing countries such as the USA, Italy, Argentina, Greece, Mexico and Brazil, all of which are among the top producing countries of the world.

Per capita consumption

The FAO (2022c) provides the per capita consumption data for 180 countries and territories in the world in 2015–2019. Ethiopia ranked 175th. Its per capita was nearly 9 kg, compared to 387 kg, 317 kg, 258 kg, 216 kg and 191 kg for the Dominican Republic, Dominica, Papua New Guinea, Sao Tome and Principe, and Ghana, respectively. This country ranked 45th out of 50 countries in Africa; it was only ahead of Togo, Zambia, Chad, Gambia and Burkina Faso. See Chapter 23 (Nutrition) for further details.

International trade

Ethiopia exports and imports large numbers of fruits and fruit products. In 2016–2020, the country exported approximately 12,000 MT of fruits/fruit products for an estimated value of US$7 million, while its imports included nearly 28,000 MT, valued at nearly US$24 million (Table 8.3).

The number of both export and import items/products increased over the years. For example, the number of items exported (imported) in 1976–1980 was 8 (11), while it rose to 11 (23) and 20 (43) in 1996–2000 and 2016–2020, respectively. Strawberries accounted for roughly 47% of export earnings in 2016–2020; this was followed by oranges and bananas, respectively. Processed products (mainly juices and prepared fruit) constituted about 16% of the total export earnings.

Different kinds of fruit juices accounted for about 59% of the import bill while the share of dates was about 13% (Table 8.3). Overall, about 62% of the import bill was paid for processed/semi-processed products (mainly juices and canned pineapple). Export earnings from fruits increased from 2000 to 2015, declined after 2016, and levelled off after that. Overall, Ethiopia remained a net importer of fruit and fruit products throughout the last two decades (Fig. 8.7). The country's fruit import expenditure for 2016–2020 was 3.35-fold that of its earnings from exports.

Discussion and Conclusion

Ethiopia started introducing modern horticulture (fruit crops) in the 1980s with the primary aim of import substitution and improved export earnings, in addition to supplying raw material for the

Table 8.3. Export and import value and quantity of fruit crops and their products in Ethiopia (2016–2020). Source: constructed by the lead author from FAO (2022a).

	Export			Import	
Item	US$ (000)	MT	Item	US$ (000)	MT
Strawberry	3,361	1,037	Misc. fruit juice	6,892	8,618
Orange	869	2,006	Dates	3,081	4,835
Banana	802	3,729	Misc. prepared fruit	2,860	3,056
Citrus juice (single strength)	614	360	Concentrated orange juice	2,134	2,810
Mango	394	1,781	Apples	1,911	1,681
Misc. prepared fruit	343	190	Concentrated citrus juice	1,543	1,937
Papaya	217	1,431	Raisins	847	562
Misc. juice	212	140	Citrus juice (single strength)	685	864
Lemons and limes	182	1,175	Canned pineapple	629	669
Misc. fresh fruit	30	15	Mango	560	668
Others	67	345	Others	2,587	2,049
Grand total	7,089	12,209	Grand total	23,729	27,749

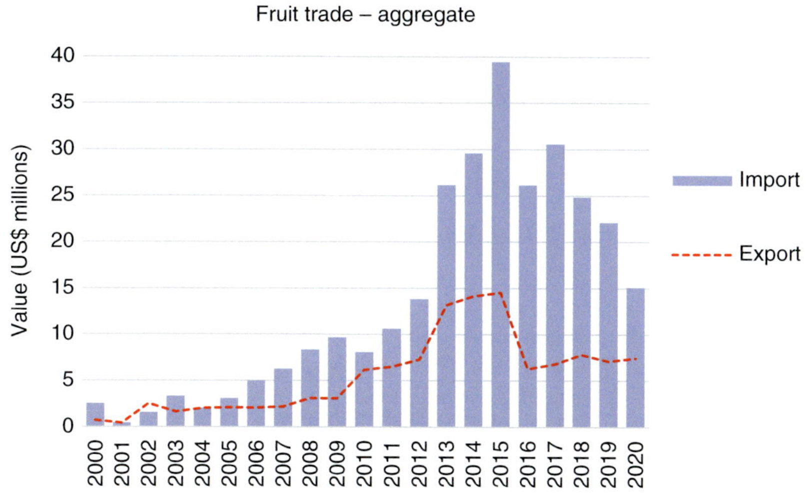

Fig. 8.7. Export and import value trends for fruits in Ethiopia. Source: constructed by the lead author from FAO (2022a).

local industry and improving nutrition (as important sources of vitamins and minerals) for the people. Fruit crops are also expected to serve as sources of income both for smallholders and commercial farmers. Because of their perennial nature and wide canopies, they also contribute to environmental protection by minimizing runoff and soil erosion. This study presents in-depth analyses of the performance of fruit crops in Ethiopia vis-à-vis the perceived objectives of the sector.

Our study shows that this country grows well over a dozen species of fruit crops. The total area expanded from approximately 48,000 ha in 2000 to 188,000 ha in 2020 – an ROG of 6.39% per annum – while the yield showed a slight decline of ~0.26%. However, results were variable according to specific crops. With an ROG of 11.53% and 10.24% per annum, respectively, avocado and mango showed the fastest growth in area; banana followed with 6.92%.

Papaya was the only fruit crop for which increases in production were more the result of increases in yield (65%) rather than increases in area (33%). Its yield gain was estimated at 334 kg/ha/yr but was non-significant. Banana registered a highly significant yield gain of 146 kg/ha/yr ($P<0.01$, $R^2=0.69$). The ROG was negative for grape, fresh fruit, tangerine, avocado and mango. This means yield losses of 420 kg/ha/yr, 129 kg/ha/yr, 115 kg/ha/yr, 112 kg/ha/yr and 28 kg/ha/yr for fresh fruits, grape, avocado, tangerine and mango, respectively; however, the values were highly significant only for fresh fruit, grape and tangerine while those for avocado and mango were non-significant.

Ethiopia has been a net importer of fruit products since the late 1990s, but its costs showed fast growth since the mid-2000s, reaching a peak of US$39 million in 2015. The country's per capita consumption of fruit is among the lowest in Africa. These situations are attributed to low productivity and production, coupled with the increasing human population.

There are biotic and abiotic factors that limit fruit production in Ethiopia, which include diseases such as leaf anthracnose on mango and pseudocercospora leaf/fruit spot on orange and other citrus species; insects such as the white mango scale on mango and pineapple mealybug on pineapple; and general poor crop management.

References

Abate, T. (1981) Preliminary notes on the citrus insect and mite pest situation in Ethiopia. *Ethiopian Journal of Agricultural Sciences* 3, 121–128.

Abate, T., Shiferaw, B., Menkir, A., Wegary, D., Kebede, Y. *et al.* (2015) Factors that transformed maize productivity in Ethiopia. *Food Security.* DOI: 10.1007/s12571-015-0488-z.

Bekele-Tesemma, A., Birnie, A. and Tengnäs, B. (1993) *Useful Trees and Shrubs for Ethiopia. Identification, Propagation and Management for Agricultural and Pastoral Communities*. Regional Soil Conservation Unit (RSU), Swedish International Development Authority.

Dagnew, A., Belew, D., Admassu, B. and Yesuf, M. (2014) Citrus production, constraints and management practices in Ethiopia: The case of pseudocercospora leaf and fruit spot disease. *Science, Technology and Arts Research Journal* 3(2), 4–18.

Dagnew, A., Assefa, W., Kebede, G., Etissa, E., Ayele, L. *et al.* (2020) Research achievements, gaps and future prospects of tropical fruit crops at Melkassa Agricultural Research Center (MARC). In: Ayalew, G. (ed.) *Fifty Years of Service for Dryland and Irrigated Agriculture*, EIAR, Addis Ababa, Ethiopia, pp. 33–48.

Derso, E. (1999) Occurrence, prevalence and control methods of *Phaeoramularia* leaf and fruit spot disease of citrus in Ethiopia. *Fruits* 54, 225–232.

Dessalegn, Y., Assefa, H., Derso, T. and Haileslassie, A. (2016) Assessment of fruit postharvest handling practices and losses in Bahir Dar, Ethiopia. *African Journal of Agricultural Research* 11(52), 5209–5214. DOI: 10.5897/AJAR2016.11731.

Edwards, S.B., Gebre-Egziabher, T., Tewolde-Berhan, S., Asfaw, Z., Ruelle, M., *et al.* (2019) *A Guide to Edible Legumes Found in Ethiopia: For Extension Officers and Researchers*. Published by Mekelle University Press, Mekelle, Ethiopia, 109 p. ISBN 978-99944-74-56-1.

Etissa, E., Dagnew, A., Assefu, W., Ayele, L., Damtew, M. *et al.* (2020) Achievements, gaps and prospects of sub-tropical fruit crops research at Melkassa Agricultural Research Center. In: Ayalew, G. (ed.) *Fifty Years of Service for Dryland and Irrigated Agriculture*. EIAR, Addis Ababa, Ethiopia, pp. 49–60.

FAO (Food and Agriculture Organization) (2022a) Statistical Databases and Data-Sets of the Food and Agriculture Organization of the United Nations. Available at: http://www.fao.org/faostat/en/#data (accessed 21 March 2022).

FAO (2022b) Statistical Databases and Data-Sets of the Food and Agriculture Organization of the United Nations. Available at: http://www.fao.org/faostat/en/#data (accessed 12 April 2022).

FAO (2022c) Statistical Databases and Data-Sets of the Food and Agriculture Organization of the United Nations. Available at: http://www.fao.org/faostat/en/#data (accessed 4 August 2022).

Firde, K., Setu, H., Eshete, T., Mosie, T., Sileshi, G. and Negash, E. (2019) Growth, yield, and fruit quality performance of peach varieties. *Ethiopian Journal of Agricultural Sciences* 29(2), 45–58.

Gebre-Mariam, S. (2003) *Status of Commercial Fruit Production in Ethiopia*. IAR, Addis Ababa, Ethiopia.

Godfrey-Sam-Aggrey, W. and Tuku, B.T. (1987) Review of deciduous fruits research in Ethiopia and proposal for future research and development direction. In: *First Ethiopian Horticultural Workshop, 20–22 February 1985.* IAR, Addis Ababa, Ethiopia. pp. 39–51.

Herath, E. and Desalegne, L. (eds) (1994) Horticulture research and development in Ethiopia. In: *Proceedings of the Second National Horticultural Workshop of Ethiopia, 1–3 December 1992*. IAR, Addis Ababa, Ethiopia.

Herath, E., Bullo, A., Gebre, E. and Gebre-Mariam, S. (1994) Fruit crops improvement research. In: Herath, E. and Desalegne, L. (eds) *Horticulture Research and Development in Ethiopia. Proceedings of the Second National Horticultural Workshop of Ethiopia, 1–3 December 1992.* IAR, Addis Ababa, Ethiopia.

MOALR (Ministry of Agricultural and Livestock Resources) (2017) Crop Variety Register Issue No. 20. Plant Variety Release, Protection and Seed Quality Control Directorate. MOALR, Addis Ababa, Ethiopia.

Mosie, T., Woldetsadik, K. and Azene, A. (2019) Quality of peach (*Prunus persica* L.) genotypes packed in LDPE plastic packaging under different storage conditions. *Agriculture and Food Sciences Research* 6(2), 182–197.

Pankhurst, R. (2006) The history of grapes, vineyards and wine in Ethiopia, prior to the Italian invasion. *Journal of Ethiopian Studies* 39(1/2), 35–54.

Ruelle, M.L., Asfaw, Z., Dejen, A., Tewolde-Berhan, S., Nebiyu, A., Tana, T. and Power, A.G. (2019) Inter- and intraspecific diversity of food legumes among households and communities in Ethiopia. *PLoS ONE* 14(12), e0227074. DOI: 10.1371/ journal.pone.0227074.

Teferi, D. and Ayano, A. (2016) Significance of *Phytophthora* disease with special emphasis to avocado and pineapple in Southwestern Ethiopia: A review. *Journal of Natural Sciences Research* 6(7), 76–80.

Yesuf, M. (2013) Pseudocercospora leaf and fruit spot disease of citrus: Achievements and challenges in the citrus industry: A review. *Agricultural Sciences* 4(7), 324–328. DOI: 10.4236/as.2013.47046.

9 Spices for Export Diversification

Tsedeke Abate*
Homegrown Vision, Addis Ababa, Ethiopia

Abstract

Ethiopia embarked on diversification of its agricultural export away from its heavy reliance on coffee in the early 1970s and tried to capitalize on its rich spice diversity as a major contribution to achieving that goal. This chapter provides an assessment of the progress made in hitting that target. There are nearly 50 species of spices and herbs cultivated in Ethiopia, but data are available for miscellaneous spices, dry chilli, fenugreek, peppers, coriander, ginger and cardamoms, with the total cultivated area estimated at 511,000 ha (2016–2020). Yields for all spices were low compared to their potential and many of the major producing countries. Ethiopia exported nearly 10,000 MT valued at ~US$16.4 million annually in the 2016–2020 period. Export earnings continually declined starting from 2015, perhaps because of increasing domestic demand. Poor agronomic practices and a lack of widespread use of improved varieties are some of the major constraints.

Introduction

Spices have a long history and huge significance in Ethiopian culture and its economy. Anyone familiar with the country's cuisine would understand how important these commodities are. Ethiopia is well known for its diversity of spices (Mengesha and Alemaw, 2010; Awas *et al.*, 2016; Tesfa *et al.*, 2017). Peethambaran *et al.* (2016) suggest that Ethiopia grows 50 of the 109 International Standards Organization-listed species and herbs in the world. They further noted that nine of those, including the Ethiopian cardamom, chilli, coriander, fenugreek, black cumin, basil, long pepper, rue and bishop weed have their centre of origin and/or diversity in this country. Other mentions can be found in Jansen (1981), Agize and van der Zouwen (2016), and Hailemichael *et al.* (2016).

Table 9.1 presents a list of the more commonly grown spice species in Ethiopia, including their scientific, Ethiopian and English names. Spices in Ethiopia are grown for local consumption, export earnings (mainly as part of coffee diversification) and import substitution.

Spice research and development in Ethiopia started in the early 1970s with the main aim of diversification of exports dominated by coffee. The Ethiopian Institute of Agricultural Research established Tepi, southwestern Ethiopia, as its testing site both for spices and coffee in 1973, and it has subsequently been upgraded to serve

* Corresponding author: tsedeke@homegrownvision.org

 Untold Stories of African Agriculture: Lessons from Ethiopia (T. Abate ed.)
DOI: 10.1079/9781800626386.0009

Table 9.1. Spices and herbs cultivated in Ethiopia.

Name			
Scientific	አማርኛ	English	Source(s)
Aframomum corrorima	ኮረሪማ	Eth. cardamom	Jansen (1981), Eyob *et al.* (2008)
Anethum foeniculum	እንስላል	Fennel	Jansen (1981)
Anethum graveolens	እንስላል	Dill	Jansen (1981)
Brassica nigra	ሰናፍጭ	Mustard	Tesfa *et al.* (2017)
Capsicum frutescens	በርበሬ/ሚጥሚጣ	Chilli	Jansen (1981)
Cinnamomum verum	ቀረፋ	Cinnamon	FAO (2021)
Coriandrum sativum	ድምብላል	Coriander	Jansen (1981)
Cuminum cyminum	ከሙን	Cumin	MOALR (2017)
Curcuma domestica	ዕርድ	Turmeric	MOALR (2017)
Cymbopogon citratus	ጠጅ ሳር	Lemon grass	Tesfa *et al.* (2017)
Elettaria cardamom	የፈረንጅ ኮረሪማ	Cardamom	MOALR (2017)
Illicium verum	አኒቼ	Badian, star anise	FAO (2021)
Lepidium sativum	ፌጦ	Garden cress	Tesfa *et al.* (2017)
Lippia abyssinica	ኮሰረት	Eth. marjoram	Tesfa *et al.* (2017)
Mentha arvensis	ናና	Japanese mint	MOALR (2017)
Mentha piperita	ናና	Peppermint	MOALR (2017)
Mentha spicata	ናና	Spearmint	MOALR (2017)
Myristica fragrans	ገውዝ	Nutmeg (long)	FAO (2021)
Nigella sativa	ጥቁር አዝሙድ	Black cumin	MOALR (2017)
Ocinum basilicum	በሶብላ	Basil	Jansen (1981)
Origanum vulgaris	ኦሬጋኖ	Oregano	MOALR (2017)
Pimpinella anisum	አኒቼ	Anise	FAO (2021)
Piper capense	ጥምዝ	Long pepper	Jansen (1981), FAO (2021)
Piper guineense	አሻንቲ ጥምዝ	Ashanti pepper	Avlon (2008)
Piper nigrum	ቁንዶ በርበሬ	Black pepper	Jansen (1981), FAO (2021)
Piper umbellatum	ሰኮና ቅጠል	Cow-foot leaf	Avlon (2008)
Rosemarina officinalis	ሮዝማሪና	Rosemary	MOALR (2017)
Ruta chalepensis	ጤናዳም	Rue	Jansen (1981)
Salvia officinalis	ከሴ	Common sage	MOALR (2017)
Syzygium aromaticum	ቁርንፉድ	Clove	FAO (2021)
Thymus vulgaris	ጦስኝ	Thyme	Tesfa *et al.* (2017)
Trachyspermum ammi[a]	ነጭ አዝሙድ	Cumin	MOALR (2017)
Trigonella foenum-graecum	አብሽ	Fenugreek	MOALR (2017)
Xylopia aethiopica	ቁንዶ በርበሬ ዛፍ	Eth. pepper	Zeven and de Wet (1982)
Zingiber officinale	ዝንጅብል	Ginger	Jansen (1981), MOALR (2017)

[a]Mistakenly reported as *Nigella sativa*
Eth., Ethiopian.

as the national coordinating centre for spice research. At about the same time, the then Alemaya College of Agriculture introduced a course on coffee and diversification to educate students in its plant sciences department (Haile-Michael Kidane-Mariam, pers. comm.), with spices as an important commodity group as most of them have similar habitats to coffee.

This chapter assesses the progress made on spices research and development to date in Ethiopia, with the objective of helping priority setting for the ongoing government initiative to transform Ethiopian agriculture.

Approach

Data sources

Like in many of the chapters in this book, the primary sources of production data for spices in Ethiopia are the sample surveys of the Central

Statistical Agency (CSA) of the Government of Ethiopia (http://www.csa.gov.et; CSA, 2004–2021) and the statistical database (FAOSTAT) of the Food and Agriculture Organization (FAO) of the United Nations (http://www.fao.org/faostat/en/#data). Trade data were accessed from FAO as well as the Ethiopian Revenue and Customs Authority (ERCA; www.erca.gov.et, 2018).

Spices are not well represented in CSA sample surveys. Data on production and number of holders are available only for fenugreek (placed under legumes) and garlic (under roots and tubers). Production data for fenugreek start from the 1994/95 *meher* (main rain) season, whereas for the rest of the commodities they start from the 2003/04 *meher* season. Data on the number of households are available for the period 2003/04 to 2019/20.

The FAOSTAT provides production data for seven of the spices (or groups of them) starting from different years. These include dry chilli and peppers from 1961, coriander and its allies (including anise, fennel and dill) from 1968, miscellaneous spices from 1972, cardamoms from 1979, cinnamon – only for 1979 and 1980 – and ginger from 1993.

Measuring productivity gains in spices

In order to see the effect of technological change and yield growth on production, economic methodology was used to isolate the contribution of area and yield growth on national production for each of the crops included in the assessment, as employed by Abate *et al.* (2011, 2015). See Chapter 3 for further details on the methods.

Results

Spices in general

Our study has revealed that there are about three dozen species of spices and herbs grown in Ethiopia, as shown in Table 9.1 above. Table 9.2 presents the status of seven spices (or groups of them) for which data are available either from the CSA, FAOSTAT, or both. These are (in descending order of area harvested) miscellaneous spices (designated by FAO as 'Spices, nes', meaning, spices not elsewhere specified), dry chilli, fenugreek, coriander, pepper, ginger and cardamoms. In total, these spice commodities occupied nearly 511,000 ha and produced more than 624,000 MT annually in the 2016–2020 period.

The area and production showed upward trends for the spices (various); flattened starting in the early 2010s for chilli, coriander, ginger and cardamoms, and fluctuated heavily for fenugreek. Yields for miscellaneous spices, coriander, peppers and cardamoms showed upward trends from the early 2000s but plateaued starting from the early 2010s; dry chilli and fenugreek yields changed little whereas ginger yield declined. Overall, increases in production resulted more from area expansion rather than from increases in productivity (Table 9.2). Further details for each crop are presented below.

Performance of individual crops

Miscellaneous spices

These refer to all spices other than those specifically mentioned in Table 9.1. The FAO data for these spices starts from 1972; they are not included in the CSA sample surveys. According to the former, miscellaneous spices in Ethiopia occupied ~292,000 ha, with total production of nearly 261,000 MT per annum in the 2016–2020 period. Their yields are estimated at ~900 kg/ha. Ethiopia ranked second (after India) both in area harvested and production, out of 67 countries that were reported during the 2016–2020 period. Other major producers included Bangladesh, Türkiye, China, Pakistan, Myanmar, Indonesia, Syria and Bolivia. Ethiopia ranked 23rd out of 65 producing countries of the world.

Figure 9.1A presents the miscellaneous spices indicator trends in Ethiopia. All indicators showed upward trends from the early 2000s, but little has changed since the early 2010s. Overall, however, the area, yield and production increased at an annual rate of 21.43%, 2.47% and 24.43%, respectively, between 2000 and 2020. This means that the area and yield contributed ~88 and ~10%, respectively, to increases in production, while the interaction factor accounted for the remaining ~2%.

Table 9.2. Indicators and annual rate of growth (ROG) for selected spices cultivated in Ethiopia. Source: calculated and constructed by the author from FAO (2022a), except fenugreek which was from CSA data.

	Indicators (2016–2020)			% ROG (2000–2020)		
Plant name	Area (ha)	Yield (kg/ha)	Production (MT)	Area	Yield	Production
Miscellaneous	292,048	893	260,760	21.43	2.47	24.43
Dry chilli	169,681	1779	302,216	6.89	0.39	7.30
Fenugreek	31,895	1301	41,274	3.77	4.19	8.12
Peppers	7,128	624	4,447	9.01	3.06	12.34
Coriander[a]	6,083	800	4,867	11.62	1.76	13.59
Ginger	3,562	2928	10,432	6.48	–0.84	5.58
Cardamoms[b]	399	386	154	2.09	1.03	3.14
Total	510,797	NA	624,150	NA	NA	NA

[a]Includes coriander, fennel, anise; [b]includes nutmeg.
NA = not applicable (or not available).

Ethiopia has registered a yield gain of ~18 kg/ha/yr ($P<0.05$, $R^2=0.46$) between 2000 and 2020. However, the average yield (roughly 900 kg/ha) is low, compared to what some of the major producing countries like Türkiye, Pakistan and Bangladesh obtain (Fig. 9.1B).

Dry chilli

Dry chilli in Ethiopia is produced in the mid-altitudes (1500–2000 m above sea level) in south-central (mainly in the Central Rift Valley) and southwestern parts of the country. It is one of the most widely used spices in Ethiopian diets. According to the CSA, an estimated 2.3 million households grew dry chilli annually on ~170,000 ha and produced ~302,000 MT annually during 2016–2020 *meher* seasons. Ethiopia ranked second in area (after India) and fourth in production (after India, Thailand and China) out of 64 countries in the world that grew dry chilli in the 2016–2020 period. Other countries among the top ten included Myanmar, Bangladesh, Thailand, Viet Nam, Pakistan, China, Nigeria and Mexico.

Almost all of the chilli produced in Ethiopia is used for domestic consumption – mainly as part of the daily diet and partly to feed the domestic spice extraction industry. The latter operates at a much less capacity because there is no adequate local supply of chilli.

Figure 9.2A depicts the indicators for chilli in Ethiopia. The FAO data are available from 1961 but those through to 2002 are unreliable; therefore, we have only used data from 1993 to the latest available – 2020. The area expanded at an annual rate of 6.89% whereas the yield increased at a rate of 0.39%, with the production growing at 7.30% in the 21 years between 2000 and 2020. This means that the contribution of area expansion to increases in production was ~94% and increases in yield accounted for ~5%, and the remaining ~1% was from the interaction factors.

Ethiopia has registered a non-significant yield gain of ~7 kg/ha/yr, compared to 106 kg/ha/yr, 64 kg/ha/yr, 35 kg/ha/yr, 34 kg/ha/yr, 30 kg/ha/yr and 21 kg/ha/yr for Thailand, India, Bangladesh, Myanmar, China and Mexico, respectively. Those were all highly significant ($P<0.01$). By contrast, Viet Nam showed a loss of 5 kg/ha/yr ($P<0.01$, $R^2=0.84$).

Figure 9.2B presents the national average yields of dry chilli in the top ten cultivating countries of the world. Ethiopia's yield was comparable with some of its peers but was low in comparison to China and Thailand, in particular.

One of the major constraints for chilli production in Ethiopia appears to be the lack of maintaining good quality seed. The risk of deterioration in varietal seed purity is very high for an open pollinated crop like chilli. At present, there are six chilli varieties that are reported to be under production. Melka Zala, Oda Haro, Melka Awaze and Melka Shote were released between 2004 and 2006, whereas Melka Dera and Melka Oli were both released in 2016. However, it is not clear how many of the improved varieties are adopted by smallholders so far.

(A)

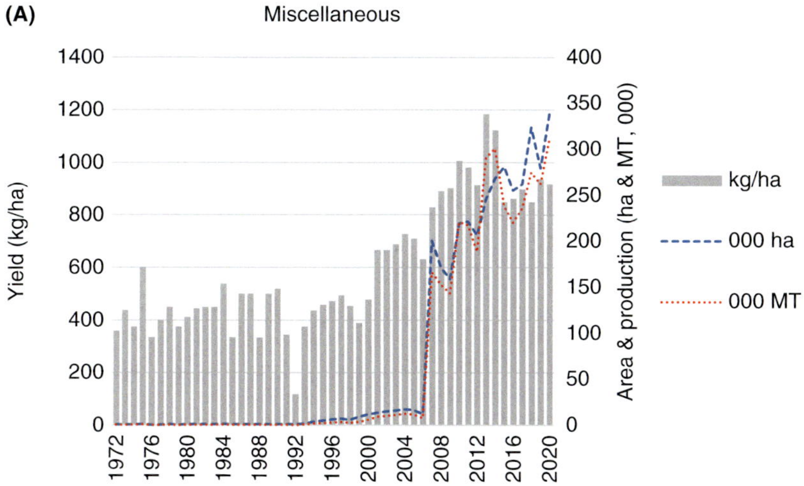

(B)

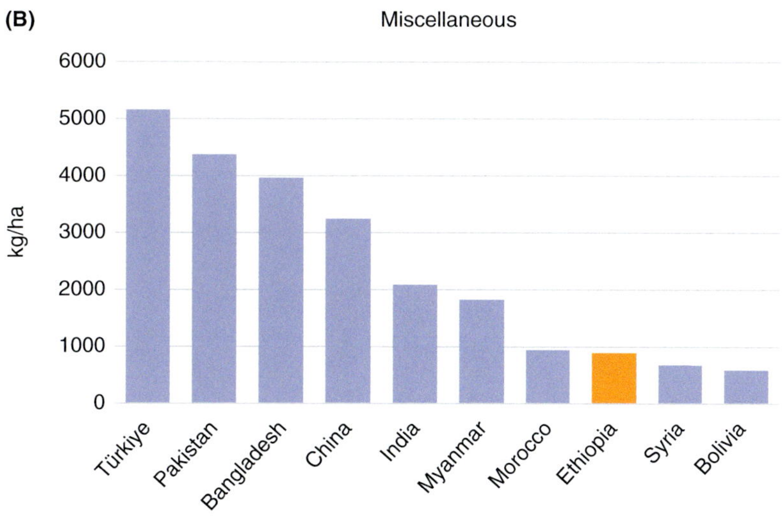

Fig. 9.1. (A) Indicator trends for miscellaneous spices in Ethiopia. **(B)** Miscellaneous spices yield in the top ten cultivating countries. Countries with the largest area are India, Ethiopia, Bangladesh, Türkiye, China, Pakistan, Myanmar, Indonesia, Syria and Bolivia. Indonesia is not included because yield data were questionable; it was replaced by Morocco. Source: constructed by the author from FAO (2022a).

Fenugreek

The CSA sample surveys report fenugreek under grain legumes, starting in the 1994/95 *meher* season. An estimated 450,000 smallholders grew fenugreek on ~32,000 ha and produced ~41,000 MT of grain annually during the 2016–2020 *meher* crop seasons; the average yield was ~1300 kg/ha. The average holding is calculated to be roughly 0.0708 ha (or about 700 m^2) per holder, and the per capita production was estimated at ~92 kg/holder/yr. Fenugreek area increased at an annual rate of 1.51% between 2003 and 2019; however, there was a sudden drop from more than 521,000 holders in 2018 to 381,000 (a 27% decline) in 2019. Whether this is a one season occurrence, or a continuing trend is to be seen in the subsequent seasons.

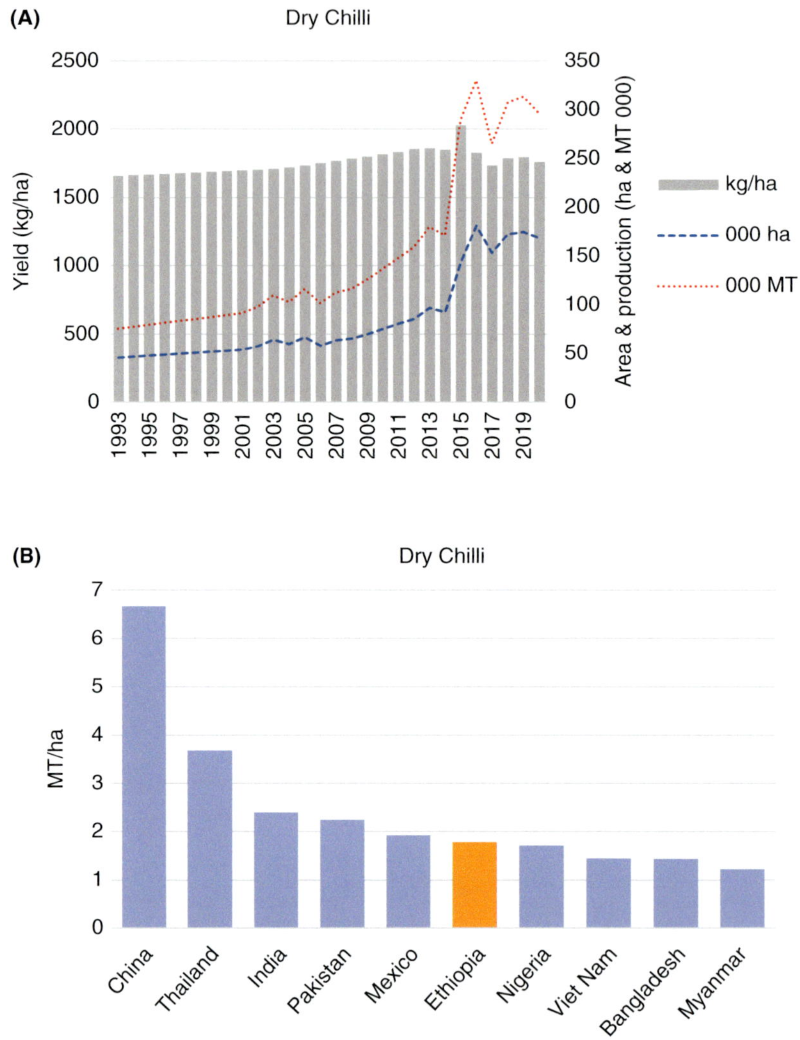

Fig. 9.2. **(A)** Dry chilli indicator trends in Ethiopia. **(B)** Dry chilli yield (2016–2020) in the top ten cultivating countries. Countries with largest area are India, Ethiopia, Myanmar, Bangladesh, Thailand, Viet Nam, Pakistan, China, Nigeria and Mexico, out of 64 countries. Source: constructed by the author from FAO (2022a).

According to the CSA, the Amhara, Oromia and Tigray regional states accounted for about 51%, 47% and 1%, respectively, of the total cultivated area of fenugreek in the 2018/19 *meher* season. This crop is cultivated in the altitude range of 1600–2300 m above sea level.

Fenugreek production in Ethiopia is characterized by heavy seasonal fluctuations (Fig. 9.3), especially of area and production, perhaps depending on the weather conditions and other constraints such as disease incidences. On average, the area, yield and production increased at an annual rate of 3.77%, 4.19% and 8.12%, respectively. In other words, increases in yield, area expansion and the interaction factor accounted for 52%, 46% and 2%, respectively, of the change in production.

The fungal disease powdery mildew (*Erysiphe polygoni* [*Oidium* sp.]) is the major limiting constraint for fenugreek production in Ethiopia (Tadesse *et al.*, 2008). Developing and promoting an integrated disease management based on

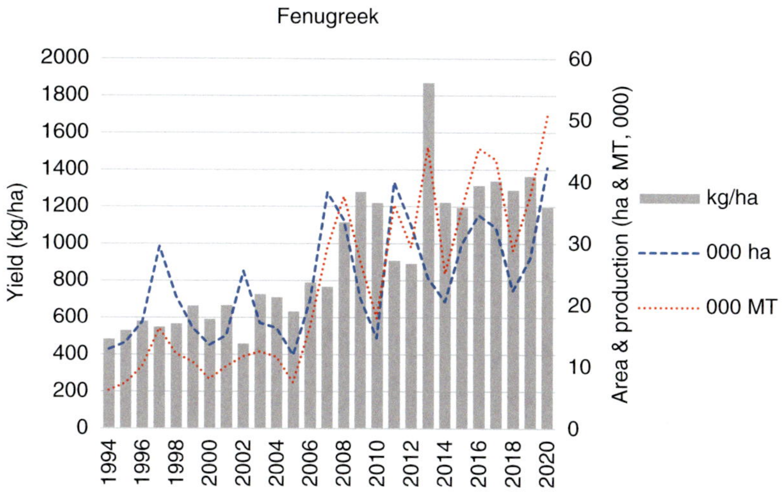

Fig. 9.3. Fenugreek indicator trends in Ethiopia. Source: constructed by the author from CSA (2004–2021).

high-yielding cultivars with high levels of resistance is a recommended approach.

Pepper

There are four species of peppers known to grow in Ethiopia (Avlon, 2008): *Piper capaense*, *P. nigrum*, *P. guineense* and *P. umbellatum*. Jansen (1981) recorded data from only the first two species. Peppers are traditional crops in Ethiopia and are essential ingredients of the country's cuisine. The CSA sample surveys do not mention peppers. The FAO provides production data on peppers from 1961 onwards, but reliable data start from 1994. The country grew peppers on >7100 ha and produced well over 4400 MT annually in 2016–2020 period. Ethiopia ranked 9th in area and 11th in production out of 41 countries that produced peppers during the above period. The top ten countries with the largest area are Indonesia, India, Viet Nam, Sri Lanka, Brazil, China, Malaysia, Madagascar, Ethiopia and Ghana.

Both smallholders and commercial growers cultivate peppers in southern and southwestern parts of the country, with the largest production coming from Bench Maji, Dawro, Keffa and Sheka zones. These are the largest foreign currency earning spices in Ethiopia (as can be seen from the section on trade later in the chapter).

Figure 9.4A presents indicator trends for peppers in Ethiopia. The area and production showed rapid growth from the early 1990s; the yield started upward trends from the early 2000s but slowed down thereafter. The annual rate of growth for area, yield and production was 9.01%, 3.06% and 12.30%, respectively. This means that 73% of the increases in production came from area expansion whereas increases in productivity accounted for 23%, with the interaction factor covering the balance.

Ethiopia registered a modest but highly significant ($P<0.01$, $R^2=0.82$) yield gain of 15 kg/ha/yr in the two decades between 2000 and 2020. With a yield gain of 24 kg/ha/yr, China was the second country out of the top ten that has registered highly significant gains. Viet Nam and Indonesia incurred yield losses of 38 kg/ha/yr and 10 kg/ha/yr, respectively, during the same period.

Ethiopia's average yield of ~600 kg/ha is lower than the world average of ~1800 kg/ha. It was also among the lowest compared to the top ten producing countries (Fig. 9.4B). Malaysia, Brazil, Viet Nam and China are among the top yielding countries.

Diminishing forest habitats and poor agronomic management appear to be the major constraints to pepper production in Ethiopia. In addition, only two modern varieties of black pepper are reported to be under production (MOALR, 2017). These are Tato and Gacheb, both released in 2007. It is not clear how much promotion has been carried out to popularize these modern cultivars and how many farmers are now growing

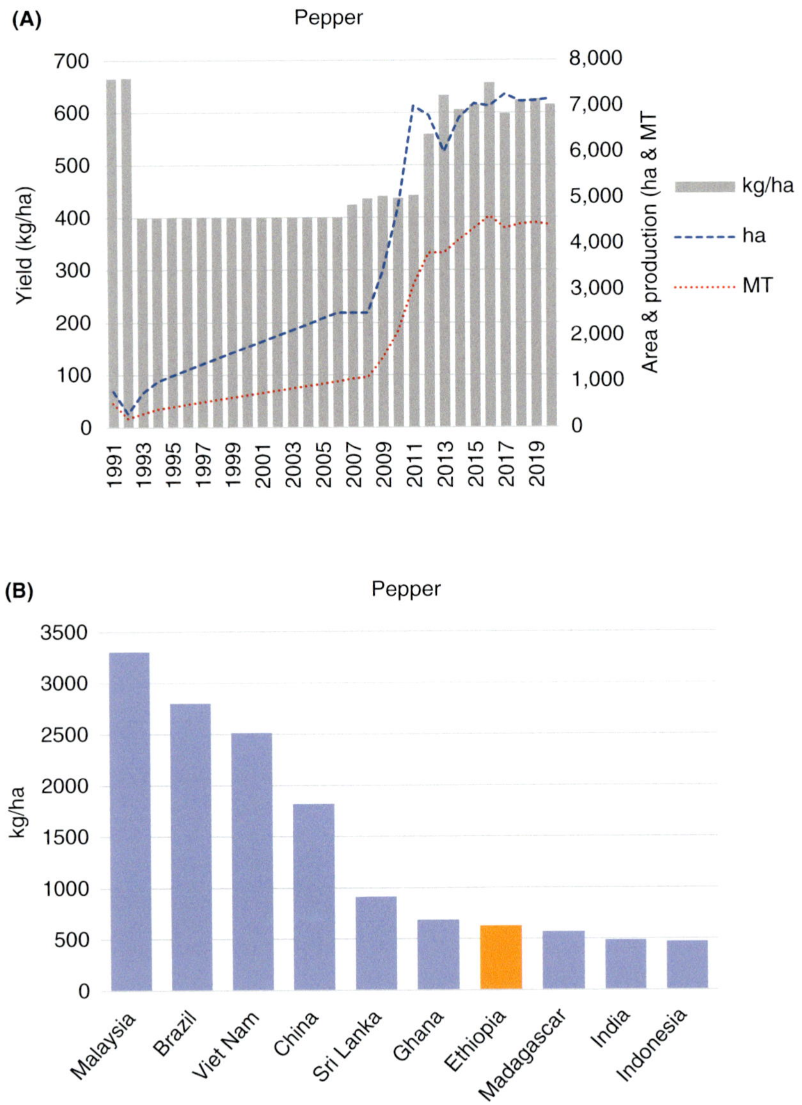

Fig. 9.4. **(A)** Pepper indicator trends in Ethiopia. **(B)** Peppers yield in the top ten cultivating countries. Countries with largest area are Indonesia, India, Viet Nam, Sri Lanka, Brazil, China, Malaysia, Madagascar, Ethiopia and Ghana. Source: constructed by the author from FAO (2022a).

them. It would not be surprising if most of the farmers are still growing their own traditional varieties.

Coriander

The figures for coriander are apparently inclusive of herbs such as fennel, dill, anise and badian (also known as star anise). The CSA sample surveys do not include these herbs. The FAO data are available from 1968. This group is considered minor and there has been little research conducted on their improvement (Mengesha and Alemaw, 2010; Awas *et al.*, 2016; Kassu *et al.*, 2018) even though they have great potential for local consumption and export earnings. Smallholder farmers' innovations, rather than any government effort, have kept the tradition of these crops alive in Ethiopia.

Available data show that Ethiopia grew ~6000 ha of coriander (and its allies) and produced nearly 5000 MT in the 2016–2020 period. The average yield was estimated at ~800 kg/ha. According to the FAO (2021), Ethiopia ranked 16th in area and 15th in production out of 39 countries in the world that produced coriander during 2016–2020, with India, Syria, Türkiye, the Russian Federation, China, Egypt, Afghanistan, Morocco, Tunisia and Canada being the top ten.

Figure 9.5A depicts the coriander indicator trends in Ethiopia. The area and production showed upward trends from the late 1990s but levelled off from the early 2010s. The yield rose from about 600 kg/ha in the late 1990s to 800 kg/ha in the early 2010s, then little change after that. The dips in 1991 and 1992 are perhaps due to the violent takeover by the Ethiopian Peoples' Revolutionary Democratic Front from the socialist/military government.

Overall, the area, yield and production increased at an annual rate of 11.62%, 1.76% and 13.59%, respectively, in the two decades between 2000 and 2020. Ethiopia's annual rates of growth in area and production are the fastest in the world. However, it should be noted that area expansion contributed to 86% of the increase in production compared to 13% from increases in yield, with the interaction factor covering the balance. Ethiopia has registered a yield gain of about 12 kg/ha/yr between 2000 and 2020, which was highly significant ($P<0.01$, $R^2=0.82$). Out of the top ten cultivating countries in the world, India and China showed highly significant yield gains ($P<0.01$) of about 27 and 18 kg/ha/yr, respectively. Ethiopia's average yield is very low (Fig. 9.5B), compared to the world average of close to 1500 kg.

Ginger

The CSA has no record for ginger in Ethiopia; FAOSTAT records start from 1993. This spice is grown in the humid, forest, mid-altitude areas of southern and southwestern Ethiopia, including Dawro, Konta, Keffa, Sheko and Maji. It occupied nearly 3600 ha with production of >10,000 MT in 2016–2020 (cf. Table 9.2). Ethiopia ranked 10th in area and 18th in production out of 38 countries that produced ginger during the above period. The top ten cultivating countries are: India, Nigeria, China, Nepal, Indonesia, Thailand, Bangladesh, Cameroon, Philippines and Ethiopia.

Figure 9.6A presents the indicator trends for ginger in Ethiopia. Overall, the annual rate of growth for area, yield and production were 6.48%, –0.84% and 5.58%, respectively, between 2000 and 2020. The yield decline was particularly fast starting from the early 2010s. Both area expansion and production somewhat levelled off during this period.

Ethiopia's yield loss for the above period was estimated at 26 kg/ha/yr ($P<0.01$, $R^2=0.65$). By contrast, Nigeria, India, Cameroon and Bangladesh obtained highly significant yield gains of 468 kg/ha/yr, 343 kg/ha/yr, 273 kg/ha/yr and 151 kg/ha/yr, respectively. This country's average yield of <3 MT/ha is the lowest among the top ten producing countries of the world (Fig. 9.6B). By contrast, Indonesia, Thailand, Nepal and China all get >10 MT/ha. The bulk of ginger produced in Ethiopia (~93%) is used for domestic consumption. This current study shows that Ethiopia is a net importer of ginger, even though the amount is very minimal (cf. Table 9.3).

There are two varieties of ginger that have been released in Ethiopia as of 2017. These are Yali and Boziab, both released in 2007. There is no information on their adoption. It is possible that almost all of smallholder farmers in Ethiopia produce their own landraces. There is an urgent need for the country to modernize its varieties, preharvest and postharvest management, and the value chain to reap full benefits of its potential for ginger.

Cardamoms

Ethiopia grows two species of cardamoms: the Ethiopian cardamom (*Aframomum corrorima*) and the common cardamom (*Elettaria cardamomum*). Data on their relative importance in Ethiopia is not available but it appears that the former is more widely grown. The research system has released only one improved cultivar, Gene, of the latter in 2007, but the extent of its production is unknown. The CSA sample surveys do not include cardamoms. The FAO (2021) records for cardamoms and allies (including nutmeg and mace) in Ethiopia start from 1979.

The country grew ~400 ha and produced ~154 MT annually in the 2016–2020 *meher*

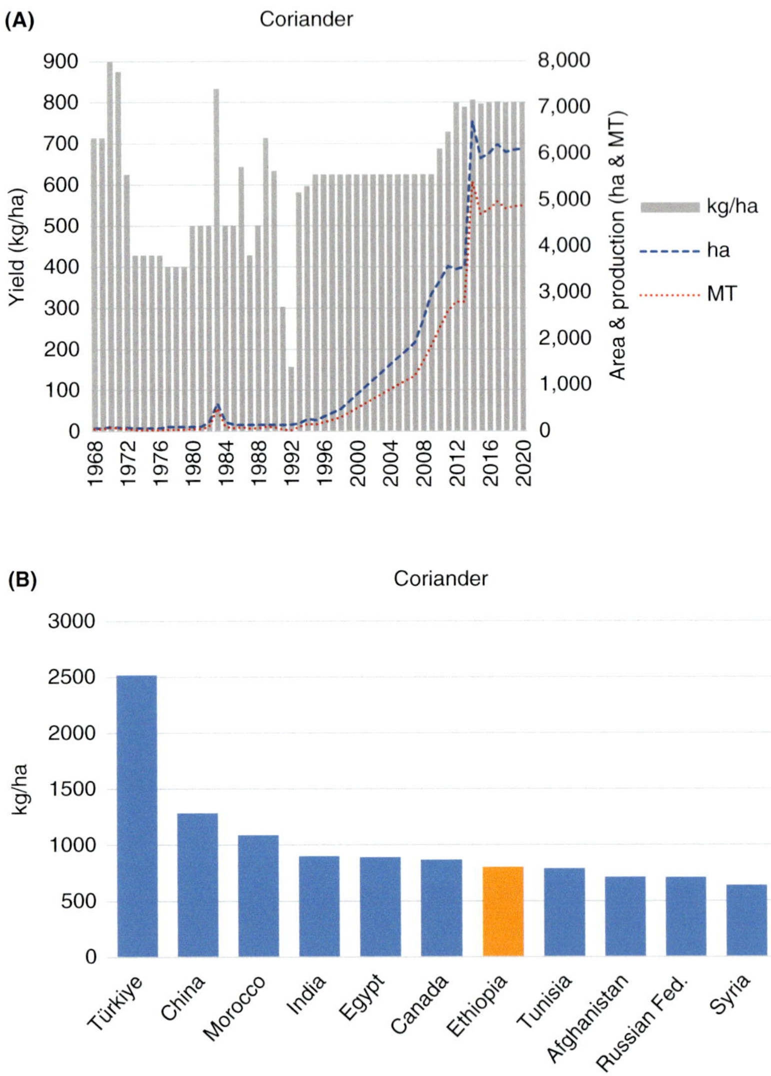

Fig. 9.5. (A) Indicator trends for coriander and allies in Ethiopia. **(B)** Coriander yield in Ethiopia compared to the top ten cultivating countries. Note: countries with the largest area are India, Syria, Türkiye, Russian Federation, China, Egypt, Afghanistan, Morocco, Tunisia and Canada; Ethiopia is 16th out of 39. Source: constructed by the author from FAO (2022a).

seasons (c.f. Table 9.2). Ethiopia ranked 11th in area and 13th in production out of 21 countries in the world that produced cardamoms and their allies during the above period. Indonesia, India, Guatemala, Lao, Bhutan, Sri Lanka, Tanzania and Grenada are the top ten producers of cardamoms and their allies in the world.

Figure 9.7A presents the indicator trends for cardamoms in Ethiopia. All indicators showed upward trends starting in the early 2000s, and they did not change much thereafter. In general, the area, yield and production increased at an annual rate of 2.09%, 1.03% and 3.14%, respectively, between 2000 and 2020. That is to say, the area expansion and increases in productivity accounted for ~67% and ~33%, respectively, of the changes in production, with the interaction factor covering the remaining <1%.

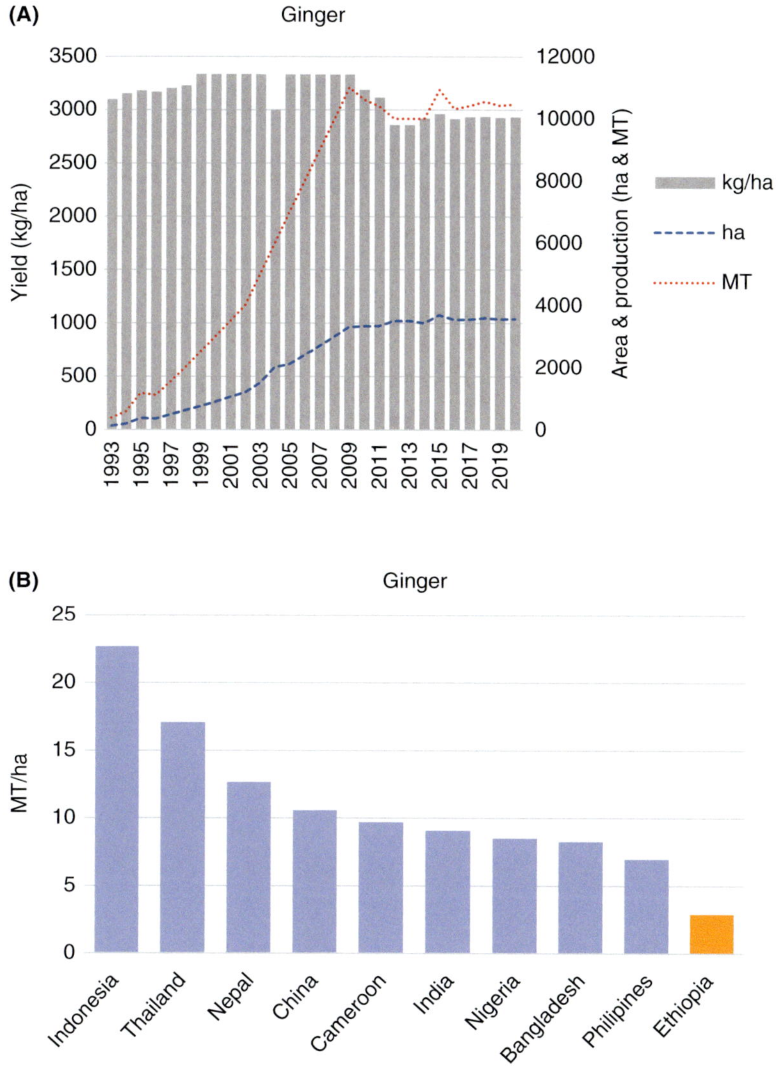

Fig. 9.6. **(A)** Ginger indicator trends in Ethiopia. **(B)** Ginger yield (2016–2020) in the top ten cultivating countries. Countries with the largest area are India, Nigeria, China, Nepal, Indonesia, Thailand, Bangladesh, Cameroon, Philippines and Ethiopia. Source: constructed by the author from FAO (2022a).

Ethiopia's yield gain was 4 kg/ha/yr ($P<0.01$, $R^2=0.79$), compared to Sri Lanka, India and Guatemala with 37 kg/ha/yr, 12 kg/ha/yr and 10 kg/ha/yr, respectively, among the top ten cultivating countries. The average yield of ~400 kg/ha for Ethiopia is far below the world average of ~736 kg/ha and among the lowest of the top ten producing countries (Fig. 9.7B). With ~1000 kg/ha, Sri Lanka has the highest yield of cardamoms among the top ten producing countries. Habitat destruction, lack of improved cultivars and poor agronomic management practices are the major constraints to cardamom production in Ethiopia.

International trade

Ethiopia has a long tradition of spices export and import. According to the FAO data for 2016–2020, this country was a net exporter of spices

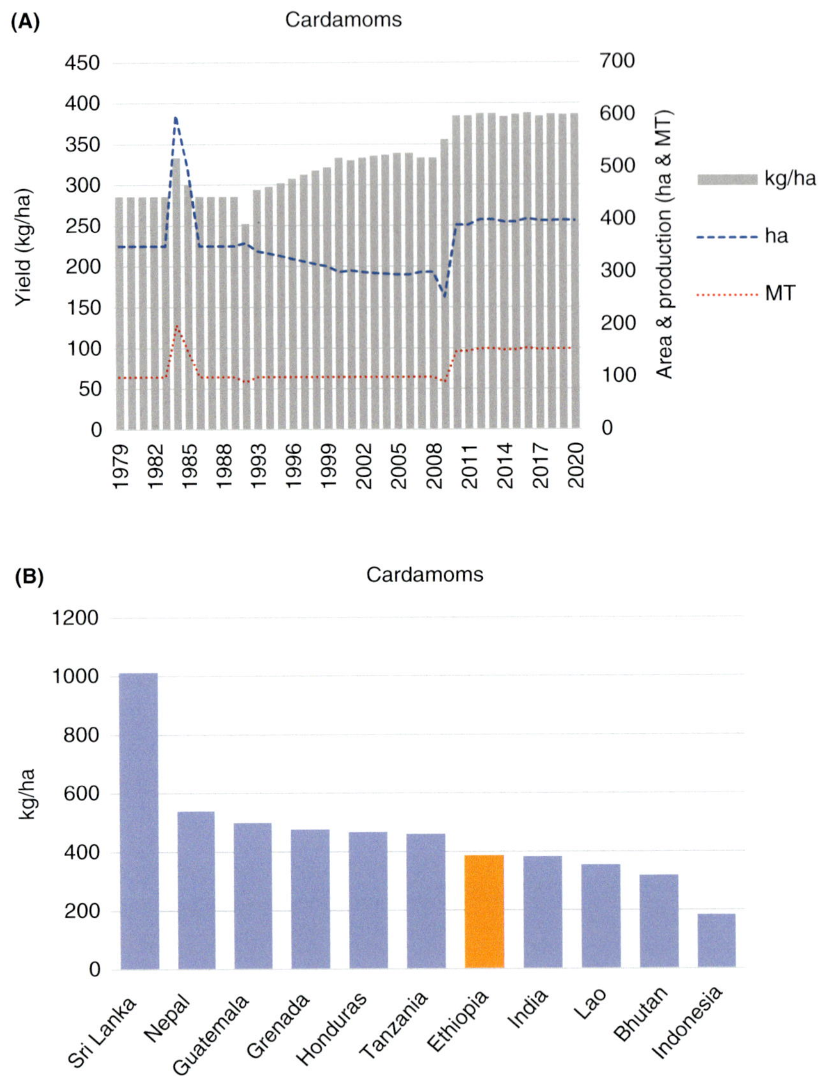

Fig. 9.7. **(A)** Cardamoms indicator trends in Ethiopia. **(B)** Cardamoms yield in Ethiopia compared to the top ten cultivating countries. Note: countries with the largest area are Indonesia, India, Guatemala, Lao, Bhutan, Sri Lanka, Tanzania and Grenada; Ethiopia is 11th out of 21. Source: constructed by the author from FAO (2022a).

in general. For example, it exported nearly 10,000 MT of spices and herbs at an estimated value of about US$16.4 million, while the import quantity was close to 3200 MT for an estimated value of nearly US$5.9 million in the 2016–2020 period (Table 9.3).

However, the status varied from crop to crop. The country was a net exporter for peppers, miscellaneous spices, coriander and chilli. On the other hand, it was a net importer for cinnamon, clove, cardamoms and ginger (Table 9.3).

Pepper, miscellaneous spices, coriander and dry chilli constituted roughly 46%, 33%, 17% and just over 1%, respectively, of the total export earnings. In contrast, cinnamon, cloves, cardamoms and ginger accounted for about 35%, 14%, 12% and 7%, respectively, of the total import costs.

Table 9.3. Value and quantity of spices trade in Ethiopia (2016–2020 average). Source: constructed by the author from FAO (2022b).

	Export		Import	
Item	Value (US$ 000)	Quantity (MT)	Value (US$ 000)	Quantity (MT)
Pepper (*Piper* spp.)	7,525	2,714	568	220
Miscellaneous	5,419	5,296	525	329
Coriander	2,851	1,617	723	420
Cardamom	244	36	688	175
Dry chilli	236	52	136	75
Cinnamon (canella)	60	19	2,038	1,273
Ginger	47	28	397	261
Clove	1	0	801	423
Total	16,383	9,762	5,877	3,176

ERCA reported annual export earnings of close to US$11 million from pepper in the 2013–2017 period. Approximately 87% of Ethiopia's pepper production was destined for export markets in the 2014–2018 period. There are more than 40 countries around the world that buy pepper from Ethiopia. With 71% and 12%, respectively, of the total quantity shipped in 2013–2017, Sudan and Djibouti were the largest buyers. Other important destinations include Saudi Arabia, the USA, South Africa, Israel, the United Arab Emirates and Germany.

Ethiopia is a net exporter of spices when one considers the latest 5 years' average. However, Fig. 9.8 tells us that export earnings grew starting from the early 2000s but took a downturn starting from 2015. The downturn trend reached its lowest ebb in 2020, where the import costs outstripped export earnings for the first time.

Increased domestic consumption (due to rapid population growth), decline or slow growth of production, and fluctuations in the international market situation could be some of the causes for the decline in export earnings and increase in imports.

Discussion and Conclusions

Spice research and development in Ethiopia started in the early 1970s as part of the government drive to diversify coffee export. The major spices considered initially included cardamoms, peppers (*Piper* spp.), turmeric and ginger, all of which are adapted to coffee-growing agroecologies in southern and southwestern parts of the country. Current literature shows that Ethiopia grows close to 50 species that are classified as spices and herbs; several of them have medicinal uses. This study has identified a total of 35 of those plants (cf. Table 9.1).

Production data are available only for seven of those listed, even though there is some level of cultivar development-cum-agronomic research going on for the vast majority of them (as judged from the National Variety Register of the Ministry of Agriculture). In the majority of the cases, increases in production were accrued from area expansion rather than from increases in productivity. Productivity actually declined for some of the spices over the past two decades.

The major limiting factors for spice production in Ethiopia are the declines in forest cover and biodiversity (as most of the spice crops are adapted to forest conditions), coupled with declining soil fertility mainly due to removal of organic matter, poor agronomic practices and lack of widespread use of improved varieties.

Basically, there are three broad categories of constraints to spice production in Ethiopia, depending on the crops or group of crops (Tesfa *et al.*, 2017; Zakir, 2018; Abera *et al.*, 2019; Hassen *et al.*, 2019). First, the major challenge to those spices adapted to forest conditions – i.e. cardamoms, peppers, turmeric and ginger – is habitat deterioration and decline in soil fertility caused by the encroachment and destruction of forests from fast growing population. In addition, there is a lack of widespread use of improved varieties and agronomic practices. For example, there was only one improved variety of

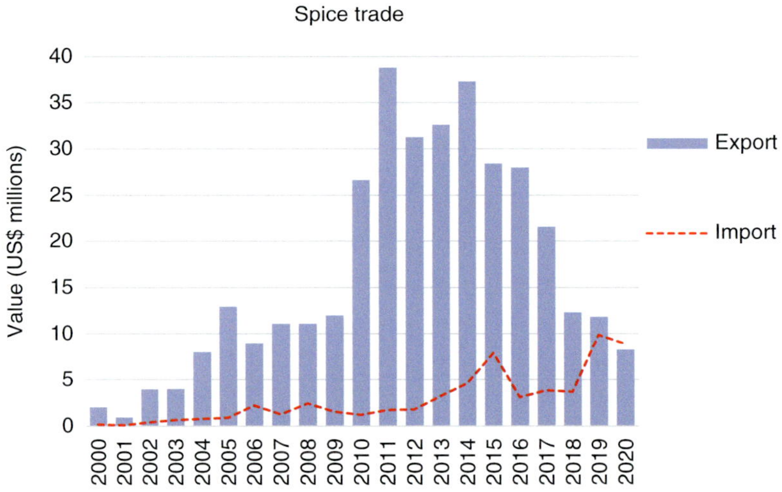

Fig. 9.8. Spice export and import values in Ethiopia. Source: constructed by the author from FAO (2022b).

cardamom, Gene (released in 2007); one turmeric, Tepi-1 (released in 2017); two black peppers, Gacheb and Tato (both released in 2007); and two ginger, Boziab and Yali (both released in 2007). It is not even obvious how much promotion work was undertaken to popularize these varieties and how many of them are actually being grown by farmers. Furthermore, there is no incentive for skilled high calibre researchers to stay for a long period in a remote research centre like Tepi and mentor the younger generation of scientists.

Second, for spices like fenugreek the major constraints are fungal diseases – powdery mildew caused by *Erysiphe polygoni* (*Oidium* spp.). These require an integrated disease management approach that heavily relies on developing and promoting high-yielding modern varieties that are resistant to the diseases and with highly desirable commercial traits. Currently, there are seven varieties released as of 2017, but it is not clear how many have been promoted and adopted by farmers.

Third, dry chilli is affected by growers in the Central Rift Valley switching to more profitable cereals – mainly *teff* and perhaps maize – away from chilli. This crop has an additional burden of keeping the purity of varieties because of its high level of cross pollination.

Currently, Ethiopia is a net exporter of overall spices but import values are growing faster and overtaking exports. Considering its great potential to expand production, and therefore maximize export, the present level of low export and the growing number of imports should not be acceptable for this country. Future priorities should focus on creating incentives for spices in Ethiopia across the value chain to overcome the constraints listed above. Some specific suggestions are given below.

First, the current level of research staffing is inadequate both in terms of experience and numbers and needs to be upgraded to accelerate spice research and development. This can be achieved by creating incentives for the more skilled staff to continue on the job, recruiting more staff and upgrading their skills through both short- and long-term training opportunities.

Second, efforts to disseminate improved technologies need to take a more active and comprehensive approach – similar to what has been achieved with grain legumes (Abate *et al.*, 2011) and maize (Abate *et al.*, 2015). That is, researchers should work *ejj-le-ejj* (hand-in-hand) with all stakeholders across the value chain.

Third, incentives should be in place for spice producers to specialize in the production of their crops or groups of crops by guaranteeing market access and price assurance. The recent government initiative to establish development such as the Koisha project in the southwestern corridor of the country is expected to boost road infrastructure and improve market access for spices and other agricultural products.

Fourth, possibilities for expanding spice production both within and outside the current geographical areas with suitable agroecologies should be explored. This can be tied with the prime minister's initiative to plant abundant trees across the country. What needs to be done is identification of appropriate tree species compatible with spice production.

Fifth, spices are marginally covered in the CSA sample surveys. There is a strong case to improve the national database on these and other crops that have not received the attention they deserve.

References

Abate, T., Shiferaw, B., Gebeyehu, S., Amsalu, B., Negash, K. *et al.* (2011) A systems and partnership approach to agricultural research for development: Lessons from Ethiopia. *Outlook on Agriculture* 40(3), 213–220.

Abate, T., Shiferaw, B., Menkir, A., Wegary, D., Kebede, Y. *et al.* (2015) Factors that transformed maize productivity in Ethiopia. *Food Security* 7, 965–981. DOI 10.1007/s12571-015-0488-z.

Abera, S., Gezahegn, B. and Hirko, B. (2019) Nutritional composition of cardamom (*Elettaria cardamomum*) cultivated in Ethiopia. *International Journal of Modern Pharmaceutical Research* 3(5), 50–53.

Agize, M. and van der Zouwen, L. (2016) Spice and medicinal plants production and value chain analysis from South-West Ethiopia. *Journal of Pharmacy and Alternative Medicine* 10, 126–144.

Avlon, M. (2008) *A Study Case on Timiz (Piper capense).* SupAgro, Institut des Regions Chaudes, Montpellier, France, 62 p.

Awas, G., Mekbib, F. and Ayana, A. (2016) Genetic diversity analysis of Ethiopian coriander (*Coriandrum sativum* L.) genotypes for seed yield and oil content. *Journal of Experimental Agriculture International* 14(6), 1–8.

CSA (Central Statistical Agency) (2004) Agricultural Sample Surveys 2003/2004 (1996 E.C.), Report on area and production for major crops (Private Peasant Holdings, *Meher* Season Volume I, Statistical Bulletin. CSA, Addis Ababa, Ethiopia.

CSA (2005) Agricultural Sample Survey 2004/2005 (1997 E.C.), Report on Area and Production of Crops (Private Peasant Holdings, *Meher* Season), Volume I, Statistical Bulletin 331. CSA, Addis Ababa, Ethiopia.

CSA (2006) Agricultural Sample Survey 2005/2006 (1998 E.C.), Report on Area and Production of Crops (Private Peasant Holdings, *Meher* Season), Volume I, Statistical Bulletin 361. CSA, Addis Ababa, Ethiopia.

CSA (2007) Agricultural Sample Survey 2006/2007 (1999 E.C.), Report on Area and Production of Crops (Private Peasant Holdings, *Meher* Season), Volume I, Statistical Bulletin.388. CSA, Addis Ababa, Ethiopia.

CSA (2008) Agricultural Sample Survey 2007/2008 (2000 E.C.), Report on Area and Production of Crops (Private Peasant Holdings, *Meher* Season), Volume I, Statistical Bulletin 417. CSA, Addis Ababa, Ethiopia.

CSA (2009) Agricultural Sample Survey 2008/2009 (2001 E.C.), Report on Area and Production of Crops (Private Peasant Holdings, *Meher* Season), Volume I, Statistical Bulletin 446. CSA, Addis Ababa, Ethiopia.

CSA (2010) Agricultural Sample Survey 2009/2010 (2002 E.C.), Report on Area and Production of Major Crops (Private Peasant Holdings, *Meher* Season), Volume I, Statistical Bulletin 446. CSA, Addis Ababa, Ethiopia.

CSA (2011) Agricultural Sample Survey 2010/2011 (2003 E.C.), Report on Area and Production of Major Crops (Private Peasant Holdings, *Meher* Season), Volume I, Statistical Bulletin. CSA, Addis Ababa, Ethiopia.

CSA (2012) Agricultural Sample Survey 2011/2012 (2004 E.C.), Report on Area and Production of Major Crops (Private Peasant Holdings, *Meher* Season), Volume I, Statistical Bulletin. CSA, Addis Ababa, Ethiopia.

CSA (2013) Agricultural Sample Survey 2012/2013 (2005 E.C.), Report on Area and Production of Major Crops (Private Peasant Holdings, *Meher* Season), Volume I, Statistical Bulletin 532. CSA, Addis Ababa, Ethiopia.

CSA (2014) Agricultural Sample Survey 2013/2014 (2006 E.C.), Report on Area and Production of Major Crops (Private Peasant Holdings, *Meher* Season), Volume I, Statistical Bulletin. CSA, Addis Ababa, Ethiopia.

CSA (2015) Agricultural Sample Survey 2014/2015 (2007 E.C.), Report on Area and Production of Major Crops (Private Peasant Holdings, *Meher* Season), Volume I, Statistical Bulletin 578. CSA, Addis Ababa, Ethiopia.

CSA (2016) Agricultural Sample Survey 2015/2016 (2008 E.C.), Report on Area and Production of Major Crops (Private Peasant Holdings, *Meher* Season), Volume I, Statistical Bulletin. CSA, Addis Ababa, Ethiopia.

CSA (2017) Agricultural Sample Survey 2016/17 (2009 E.C.), Report on Area and Production of Major Crops (Private Peasant Holdings, *Meher* Season), Volume I, Statistical Bulletin. CSA, Addis Ababa, Ethiopia.

CSA (2018) Agricultural Sample Survey 2017/18 (2010 E.C.), Report on Area and Production of Major Crops (Private Peasant Holdings, *Meher* Season), Volume I, Statistical Bulletin 586 CSA, Addis Ababa, Ethiopia.

CSA (2019) Agricultural Sample Survey 2018/19 (2011 E.C.), Report on Area and Production of Major Crops (Private Peasant Holdings, *Meher* Season), Volume I, Statistical Bulletin 589. CSA, Addis Ababa, Ethiopia.

CSA (2020) Agricultural Sample Survey 2019/20 (2012 E.C.), Report on Area and Production of Major Crops (Private Peasant Holdings, *Meher* Season), Volume I, Statistical Bulletin 587. CSA, Addis Ababa, Ethiopia.

CSA (2021) Agricultural Sample Survey 2020/21 (2013 E.C.), Report on Area and Production of Major Crops (Private Peasant Holdings, *Meher* Season), Volume I, Statistical Bulletin. CSA, Addis Ababa, Ethiopia.

ERCA (Ethiopian Revenue and Customs Authority) (2018) Available at: http://www.erca.gov.et/index.php/component/hscode/?view=importexport&iccaldate=2014-5-1 (accessed 29 December 2018).

Eyob, S., Appelgren, M., Tsegaye, A. and Messele, G. (2008). Traditional medicinal uses and essential oil composition of leaves and rhizomes of korarima (*Aframomum corrorima* (Braun) P.C.M. Jansen) from southern Ethiopia. *South African Journal of Botany* 74(2), 181–185.

FAO (Food and Agricultural Organization of the United Nations) (2021) Statistical Databases and Data-Sets of the Food and Agriculture Organization of the United Nations. Available at: http://www.fao.org/faostat/en/#data (accessed 25 February 2021).

FAO (2022a) Statistical Databases and Data-Sets of the Food and Agriculture Organization of the United Nations. Available at: http://www.fao.org/faostat/en/#data (accessed 30 August 2022).

FAO (2022b) Statistical Databases and Data-Sets of the Food and Agriculture Organization of the United Nations. Available at: http://www.fao.org/faostat/en/#data (accessed 4 September 2022).

Hailemichael, G., Kifelew, H. and Mitiku, H. (2016) Spices research achievements, challenges and future prospects in Ethiopia. *Academic Research Journal of Agricultural Science and Research* 4(1), 9–17.

Hassen, S., Tsegaye, M., Getachew, S., Abebe, A. and Biruk, H. (2019) Enhancement of seed germination and seedling growth of cardamom (*Elettaria cardamomum*) at Tepi south-western part of Ethiopia. *Academic Research Journal of Agricultural Science and Research* 7(6), 303–306.

Jansen, P.C.M. (1981) *Spices, Condiments and Medicinal Plants in Ethiopia, Their Taxonomy and Agricultural Significance*. Centre for Agricultural Publishing and Documentation, Wageningen, the Netherlands.

Kassu, K.T., Dawit, H.H., Wubengeda, A.Y., Almas, A.T. and Asrat, M.T. (2018) Yield and yield components of coriander under different sowing dates and seed rates in tropical environment. *Advances in Horticultural Science* 32(2), 193–203.

Mengesha, B. and Alemaw, G. (2010) Variability in Ethiopian coriander accessions for agronomic and quality traits. *African Crop Science Journal* 18(2), 43–49.

MOALR (Ministry of Agriculture and Livestock Resources) (2017) Crop Variety Register Issue No. 20. Plant Variety Release, Protection and Seed Quality Control Directorate. MOALR, Addis Ababa, Ethiopia.

Peethambaran, C.K., Hailemichael, G., Mitiku, H. and Kifelew, H. (2016) In search of spices in Ethiopia. *Spice India* 29(7), 4–10.

Tadesse, N., Ahmed, S., Gorfu, G., Beshir, T., Fininsa, C. et al. (2008) Review of research on diseases of food legumes. In: Tadesse, A. (ed.) *Increasing Crop Production through Improved Plant Protection—Volume* I. *Plant Protection Society of Ethiopia (PPSE),* 19–22 December 2006, Addis Ababa, Ethiopia. PPSE and EIAR, Addis Ababa, Ethiopia, 598 p.

Tesfa, T., Bayu, W., Gashaw, A. and Beshir, H. (2017) Spice production, marketing, and utilization in South Wollo, Ethiopia. *East African Journal of Sciences* 11(1), 27–36.

Zakir, M. (2018) Review on korarima (*Aframomum corrorima* (Braun) P.C.M. Jansen) research achievements, challenges and future prospective in Ethiopia. *International Journal of Research Studies in Agricultural Sciences* 4(1), 29–36.

Zeven, A.C. and de Wet, J.M.J. (1982) Dictionary of cultivated plants and their regions of diversity. In: *Excluding Most Ornamentals, Forest Trees and Lower Plants*. Centre for Agricultural Publishing and Documentation, Wageningen.

10 Coffee and Other Plantation and Industrial Crops: Opportunities for Enhancing Productivity and Unlocking their Full Potential

Tsedeke Abate,[1]* Ashenafi Ayano[2] and Mikael Fesseha Yohannes[1]
[1]Homegrown Vision, Addis Ababa, Ethiopia; [2]Ethiopian Institute of Agricultural Research, Jimma Research Center, Jimma, Ethiopia

Abstract

This chapter gives a critical assessment of major plantation/industrial crops (including coffee, *ch'at* [*Catha edulis*], cotton, *gesho* or buckthorn [*Rhamnus prinoides*], tea and castor) research and development in Ethiopia. Yields of coffee, *ch'at* and *gesho* changed little between 2000 and 2020, while cotton, tea and castor showed significant gains, even though they were generally low compared to the top ten cultivating countries across the globe. Sugarcane yield has been falling (from the highest in the world of 177 MT/ha in 1976–1980 to 45 MT/ha in 2016–2020), ever since the nationalization by the military-socialist government, following which, personnel with both managerial and field level skills were lost and never replaced. Soil organic matter decline and high turnover of qualified researchers are among the major constraints. There are opportunities to develop and enter into the lucrative international market for naturally decaffeinated coffee that is projected to reach US$28.86 billion by 2030.

Introduction

Ethiopia grows close to a dozen species of plantation/industrial crops. These include coffee, *ch'at*, cotton, *gesho* (commonly known as shiny-leaf buckthorn [erroneously referred to as hops]), sugarcane, tea, castor, unspecified fibre crops, tobacco, bast fibres, sisal, kenaf (*Hibiscus cannabinus*), vernonia (*Vernonia galamensis*), among others. This chapter discuss the first seven as they are considered to be the major crops of the country. Depending on the crop, plantation/industrial crops have great value as major export, import substitution, as raw material for the local industry, sources of income for smallholders, and huge sociocultural benefits for the country.

The Central Statistical Agency (CSA) of the Government of Ethiopia records (CSA, 2004–2021) only have data for coffee, *ch'at* (also known as *khat* or *meera* outside of Ethiopia; scientific name *Catha edulis*), sugarcane and *gesho*. The FAOSTAT provides data for more extended periods, but it does not include *ch'at*. Alemu *et al.* (2016) published a review of some of these crops through the mid-2010s.

This chapter presents the results of our analyses on the performance of the plantation and individual crops, with particular emphasis

* Corresponding author: tsedeke@homegrownvision.org

 Untold Stories of African Agriculture: Lessons from Ethiopia (T. Abate ed.)
DOI: 10.1079/9781800626386.0010

on the period covering 2000 to 2020. We used scientific methods to measure progress accruing from research and development efforts over the years, identify the major constraints limiting progress and suggest the path forward. It is our hope that the information contained here would feed the priority setting and policy decision processes for agricultural transformation in Ethiopia.

Approach

We used two major sources of data on production: the CSA (http://www.csa.gov.et) and the Food and Agriculture Organization of the United Nations (FAO) (http://www.fao.org/faostat/en/#-data). The latter and ERCA (Ethiopian Revenues and Customs Authority) were our main source for data on trade. Other details on the methods are as presented in Chapter 3.

Results

Plantation crops in general

Table 10.1 presents the indicators and their annual rate of growth (ROG) for the selected crops. These crops occupied more than 1.2 million ha of land during the 2016–2020 period. Performance varied from crop to crop.

All crops showed increases in area. The fastest area growth was observed for coffee, followed by tea, *ch'at* and *gesho*. Cotton and sugarcane had modest growth for area. The latter suffered the highest decline in yield; there was a slight decline in coffee yield. Sugarcane was the only crop that suffered declines in production. Further details for individual crops are presented below.

Performance of individual crops

Coffee

Coffee is affectionately known by Ethiopians as 'Our gift to the world', as Ethiopia is Vavilov's primary gene centre (or centre of origin and diversity) for this crop (Vavilov, 1951; Westphal, 1975). According to the CSA, >5.7 million smallholders grew coffee on >761,000 ha of land and produced >496,000 MT of beans annually during 2016–2020. The average yield was estimated at 651 kg/ha. The calculated coffee area per smallholder is ~0.1328 ha and the per capita production ~87 kg/holder/yr.

This crop is grown mostly in the mid-altitudes of southern, southwestern and western parts of the country. Coffee in Ethiopia is produced under four systems: forest coffee, semi-managed forest coffee, garden coffee and plantation coffee (Amamo, 2014; Wolde *et al.*, 2017). Forest coffee used to be the dominant system until the past several decades, but currently it accounts for ~10% of the total area, whereas garden, semi-managed forest and plantation coffee cover ~50%, ~30% and ~10%, respectively.

Oromiya and the Southern Nations, Nationalities and Peoples Region together account for ~97% of the total area occupied by coffee in Ethiopia, with Amhara, Gambella, Dire Dawa,

Table 10.1. Indicators and annual rate of growth (ROG) for plantation crops in Ethiopia. Source: constructed by the lead author from FAO (2022c), except *ch'at*, which was from CSA (www.csa.gov.et), for the period from 2003 to 2020.

	Indicators (2016–2020)			ROG (2000–2020)		
Crop	Area (ha)	Yield (kg/ha)	Production (MT)	Area	Yield	Production
Coffee	761,283	651	496,049	7.45	–0.36	7.07
Ch'at	292,053	891	260,850	3.83	0.34	6.83
Cotton	70,400	2,115	148,400	1.91	4.17	6.16
Gesho	33,707	1,294	43,606	2.89	0.70	3.61
Sugarcane	29,627	45,046	1,336,469	1.70	–5.08	–3.47
Tea	9,900	1,042	10,317	5.45	0.84	6.34
Castor[a]	5,962	1741	12,600	1.67	3.83	5.56

[a]Data accessed on 30 June 2023.

Tigray and Harari (in descending order) regions covering the balance. In 2016–2020, Ethiopia ranked fourth (after Brazil, Indonesia and Colombia) in area and fifth (after Brazil, Viet Nam, Colombia and Indonesia) in production out of 78 countries that grew/produced coffee in the world. Further details on coffee production in Ethiopia can be found in Kufa *et al.* (2016).

Figure 10.1A presents the coffee indicator trends for Ethiopia. The area showed ups and downs through the years, with sharp declines starting in the mid-1960s that continued through the early 2000s, and sharp upward trends since then. It increased at an annual rate of 7.45% between 2000 and 2020. The sharp decline in area in the 1970s is believed to be because of the farmers replacing coffee with *ch'at*, which is supposed to be more profitable.

There were continued upward trends in the yield through the 2000s. This growth was

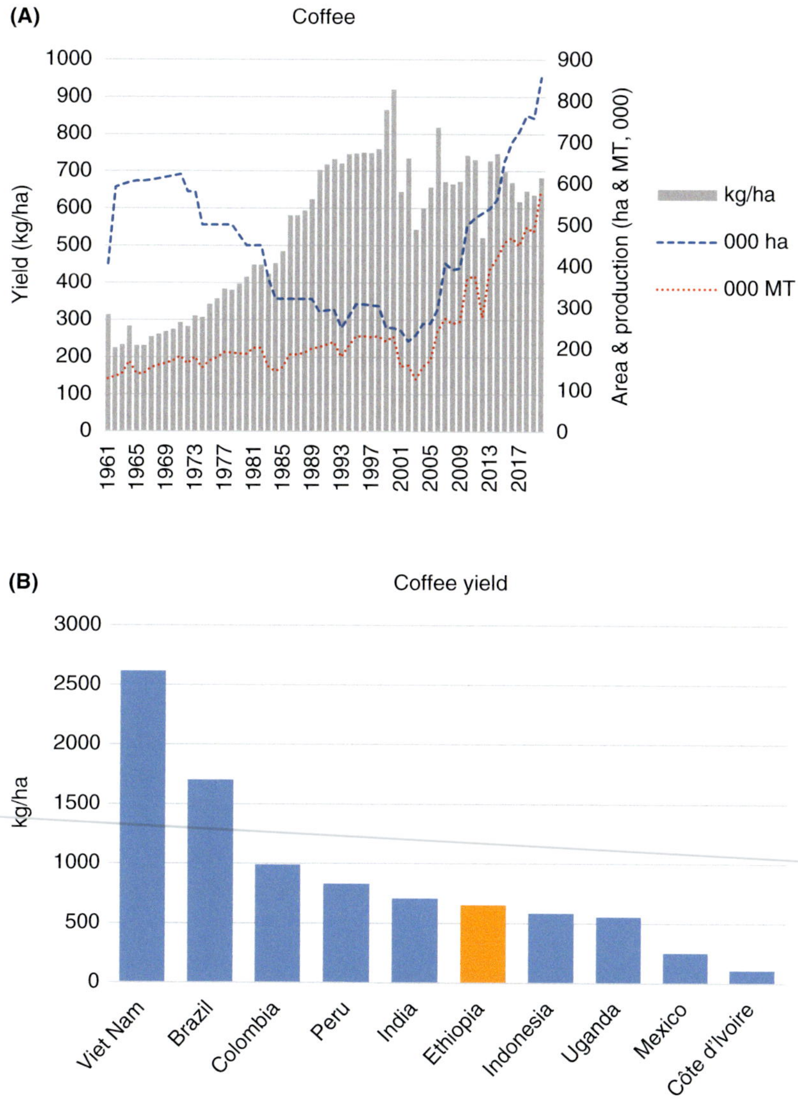

Fig. 10.1. **(A)** Coffee indicator trends in Ethiopia. **(B)** Coffee yield in the top ten cultivating countries of the world. Source: constructed by the lead author from FAO (2022c).

spurred by the identification and introduction of coffee lines resistant to the notorious coffee berry disease that ravaged the industry in the 1970s (van der Graaff, 1981) in various parts of the country. The socialist government also ensured that coffee was the backbone of the Ethiopian economy and accorded high priority to its research and extension by establishing a very strong research team and popularizing coffee management practices – particularly stumping to renew old mother trees.

The annual ROG for yield was -0.36% for the period 2000 to 2020; production increased at a rate of 7.07% per annum. Obviously, the growth in production was brought about largely by area expansion. On the world scene, coffee area showed little change whereas the yield (and therefore production) increased steadily over the years (FAO, 2022c).

The top ten coffee-cultivating countries were Brazil, Indonesia, Colombia, Ethiopia, Mexico, Viet Nam, Cote d'Ivoire, Uganda, India and Peru. Highly significant yield gains were obtained in Viet Nam (~56 kg/ha/yr, $P<0.01$, $R^2=0.82$) and Brazil (~49 kg/ha/yr, $P<0.01$, $R^2=0.87$). Yield gain was also significant for Indonesia but was low (<7 kg/ha/yr). By contrast, highly significant yield losses were incurred in Cote d'Ivoire (~17 kg/ha/yr, $P<0.01$, $R^2=0.85$), India (~11 kg/ha/yr, $P<0.01$, $R^2=0.70$) and Mexico (~10 kg/ha/yr, $P<0.01$, $R^2=0.89$).

Figure 10.1B presents the national average yields of the top ten countries. Ethiopia's average yield of 651 kg/ha is less than the world average of 761 kg/ha. Viet Nam and Brazil have the highest yields among the top ten countries. However, it should be noted that all of Viet Nam's and 41% of Brazil's coffee is robusta type, which naturally yields higher than arabica type.

Ethiopia produces exclusively arabica-type coffee – the preferred type worldwide because of its superior cup quality. By contrast, many of the major producing countries have substantial proportions of robusta type coffee. For example, the latter accounts for 100% of the total production for Viet Nam, Uganda and Cote d'Ivoire; 81% and 70%, respectively, of coffee in Indonesia and India is robusta type.

Coffee in Ethiopia is produced both for domestic consumption and the export market. Through the mid-1990s, domestic consumption outstripped export. However, this situation was reversed from the early 2000s through to 2015. The 2016–2020 data show that Ethiopia exports about 38% of its total production.

Coffee is Ethiopia's largest agricultural export earner. Export volume and value showed fast growth starting from the early 2000s (Fig. 10.2). The country exported more than 190,000 MT coffee valued at ~US$658 million annually during the 2016–2020 period, making it the world's 13th largest exporter by quantity and 9th by

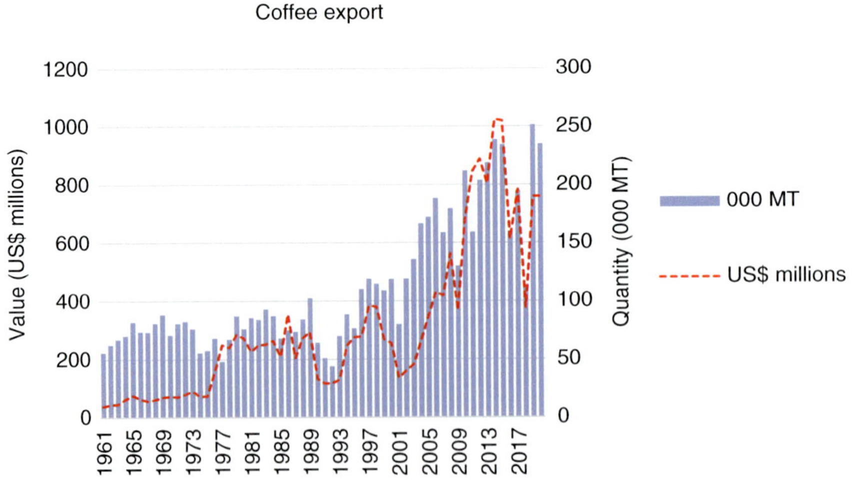

Fig. 10.2. Coffee export value and quantity in Ethiopia. Source: constructed by the lead author from FAO (2022c).

value, after Brazil, Viet Nam, Colombia, Germany, Indonesia, Honduras, India, Belgium, Uganda, Italy, Peru and Guatemala (FAO, 2022c).

Ethiopia's coffee export included 98.15% green coffee, 1.32% substitutes containing coffee, 0.52% roasted coffee and 0.01% coffee extracts in 2016–2020. Compare this to Germany's export of ~52% green and ~48% processed coffee (consisting of 34% roasted and the remaining 14% comprising extracts, husk and substitutes) in 2016–2018. Germany, Saudi Arabia, the USA, Japan and Belgium were the major buyers of Ethiopian coffee, according to ERCA data for 2013–2017. Further details on coffee can be found in Teshome *et al.* (2018).

Ch'at

According to the CSA, ~3.6 million smallholders harvested ~292,000,000 ha and produced ~261,000 MT of *ch'at* annually in the 2016–2020 period.

Figure 10.3 shows *ch'at* indicator trends in Ethiopia. *Ch'at* area and yield increased by 3.83% and 0.34%, respectively, between 2003 and 2020. In other words, the area expansion explained ~56% of the change in production whereas the contribution of increases in yield was ~5%, with the interaction factor accounting for the remaining ~39%. The lack of progress in productivity is perhaps attributed to the lack of government support for research and development of the crop. Further information on *ch'at* can be found in Kandari *et al.* (2014), Binalfew (2017) and Gebre *et al.* (2018).

Because of its narcotic nature, *ch'at* is a controversial crop in Ethiopia. On the other hand, it is an important source of foreign currency from its export to many countries. Recent data on its trade are not available. The report by Belwal and Teshome (2011) suggest that *ch'at* export accounted for ~11% of the total agricultural export earnings between 1997 and 2003. The ERCA data show that Ethiopia's annual export revenues from *ch'at* were >US$4.4 million between 2008 and 2011.

During the above period, Ethiopia exported *ch'at* to a total of 35 countries and territories in different parts of the globe. The top ten of those were: China, Kenya, the UK, Australia, the USA, Thailand, the Netherlands, Malaysia, Djibouti and Hong Kong (according to ERCA).

Cotton

Cotton is the most important fibre crop in Ethiopia. The CSA does not keep records of this crop, perhaps because it is mainly produced by the commercial sector. According to the FAO, Ethiopia cultivated ~70,000 ha of cotton and produced ~148,000 MT of seed cotton, with an average yield of ~1300 kg/ha in the 2016–2020 period (cf. Table 10.1). This country ranked 34th in area and 21st in production out of 86 countries

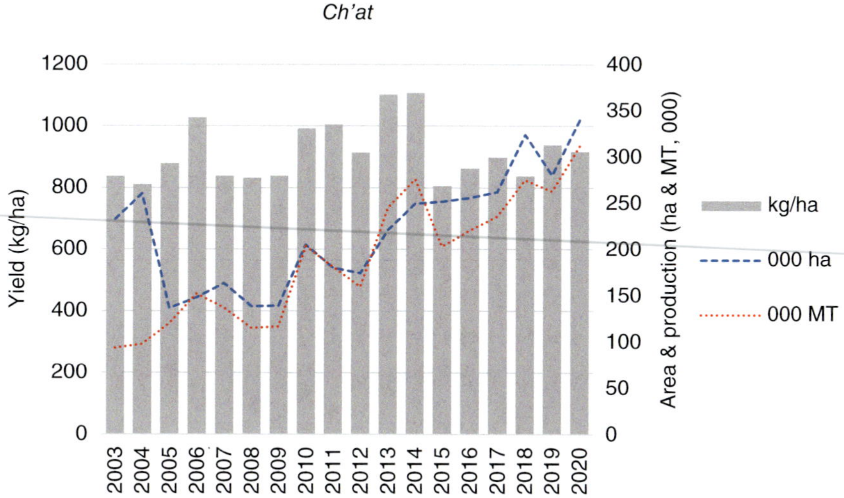

Fig. 10.3. Indicator trends for *ch'at* in Ethiopia. Source: constructed by the lead author from CSA, (www.csa.gov.et).

and territories that grew cotton in 2016–2020. India, the USA, China, Pakistan, Brazil, Uzbekistan, Burkina Faso, Mali, Benin and Turkmenistan are the top ten cultivating countries (FAO, 2022b).

The USDA (2019) report shows the current and projected government plan to expand cotton area, yield and production along with the quality and commerce (Annex Table 10.1). The 2020 plan shows the total area of 250,000 ha, yield of 2008 kg/ha and production of 502,000 MT. Available data show that this goal has not been achieved. Further details are available in Gudeta *et al.* (2016).

Figure 10.4A presents seed cotton trends for Ethiopia. The area and production showed upward trends from the early 1960s through to the early 1980s and either declined or levelled off through 2000. The area declined fast from

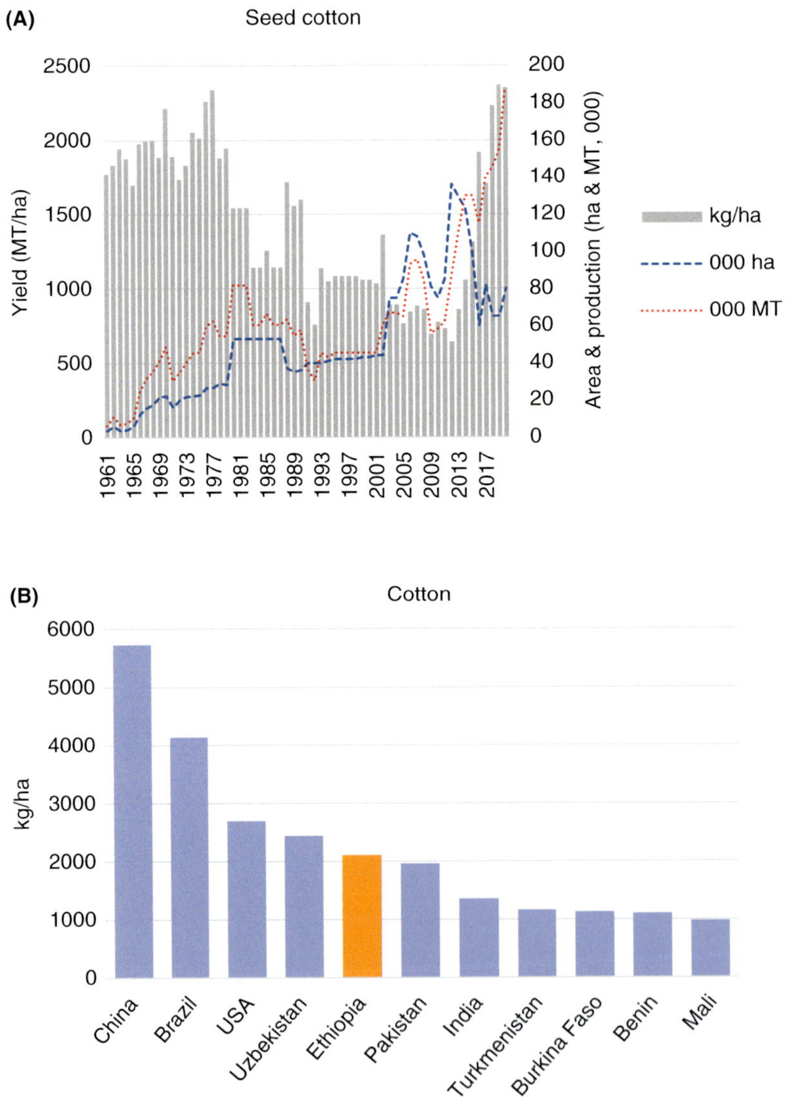

Fig. 10.4. (A) Indicator trends for cotton in Ethiopia. Source: constructed by the lead author from FAO (2022a). **(B)** Cotton yield in Ethiopia compared to the top ten cultivating countries in the world. Source: constructed by the lead author from FAO (2022d).

2014 onwards. By contrast, the yield continued to decline until early 2010 but showed an upward trend from 2018 onwards. Overall, the area, yield and production increased at annual rates of 1.91%, 4.17% and 6.16%, respectively. What this means is that increases in productivity contributed nearly 68% and area expansion about 31% to increases in production; the interaction factor covered the remaining 1% or so.

Simple regression analysis of the data for 2000 to 2020 showed that Ethiopia registered a yield gain of about 61 kg/ha/yr. China made by far the highest gain of 169 kg/ha/yr ($P<0.01$, $R^2=0.56$), while Brazil registered 74 kg/ha/yr ($P=0.01$, $R^2=0.84$) and the USA ~43 kg/ha/yr ($P<0.01$, $R^2=80$) among the top ten countries. Three of the top ten countries – Turkmenistan, Benin and Mali – showed non-significant yield losses of about 34 kg/ha/yr, 8 kg/ha/yr and 5 kg/ha/yr, respectively.

With >5700 kg/ha, China showed the highest national yield average, followed by Brazil with >4100 kg/ha (Figure 10.4B). This was followed by the USA and Uzbekistan. Ethiopia's average yield was >2100 kg/ha.

Ethiopia is a net exporter of cotton. For example, it exported 5063 MT valued at US$5.6 million annually in 2016–2020, compared to its import of about 282 MT for an estimated value of US$1.2 million. This country's cotton export fluctuated heavily over the years, especially from 2000 onwards (Fig. 10.5). The values peaked at about US$13 million for the first time in 2003, and then again in 2007 and 2008.

Gesho

Gesho is exclusively a garden crop that smallholders in Ethiopia grow for use in making local beer (*tella*) and honey wine (*tej*). Both the CSA and FAO erroneously report *gesho* as hops. Even though the purpose is the same, *gesho* is a native Ethiopian crop that belongs in the buckthorns family of Rhamnaceae, whereas the common hop (*Humulus lupulus*) belongs in the hemp family of Cannabaceae.

More than 2.6 million smallholders produced ~44,000 MT of *gesho* on ~34,000 ha of land annually in the 2016–2020 period; the estimated yield was ~1300 kg/ha (cf. Table 10.1). This crop is grown in gardens as a source of income for smallholder farmers from sales of leaves, branches and stems that are used for brewing the local beverages mentioned above. The average area per household is estimated at roughly 150 m², with a per capita production of ~19 kg/household/yr.

The area planted to *gesho*, its yield and production increased at an annual rate of 2.89%, 0.70% and 3.61%, respectively (cf. Table 10.1). This shows that the increases in production were obtained more from increases in area rather than from increases in productivity, as presented

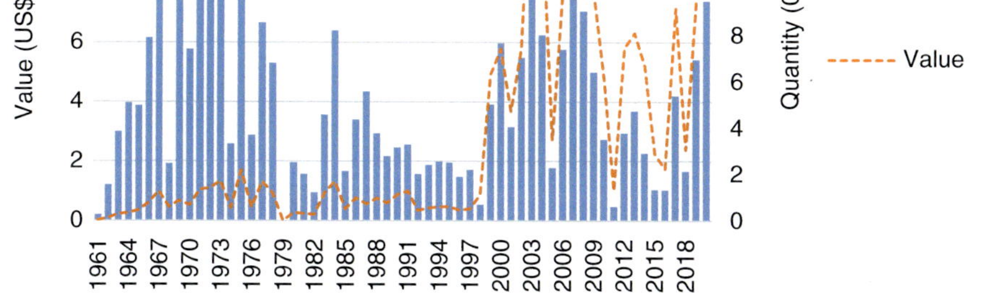

Fig. 10.5. Cotton export patterns in Ethiopia. Source: constructed by the lead author from FAO (2022d).

in Fig. 10.6A. More specifically, about 80% of the increase in production was attributed to area expansion while increases in yield accounted for 19%; the interaction factor covered the remaining 1% or so.

The demand for *gesho* has increased in recent times (Fig. 10.6B) because of its export to the Ethiopian diaspora. According to the FAO (2022c), there was fast growth in export starting from 2009. The country exported about 906 MT of *gesho* for an estimated value of >US$1.18 million each year between 2012 and 2017. However, data from ERCA show that Ethiopia earned >US$1.80 million annually in 2016–2020 period; the earnings in 2020 alone were US$2.66 million (Fig. 10.6B).

There are 15 export destinations for *gesho* from Ethiopia – the USA, Israel, Sudan, Canada, the UK, Germany, the Netherlands, Australia, Norway, Sweden, Switzerland, New Zealand,

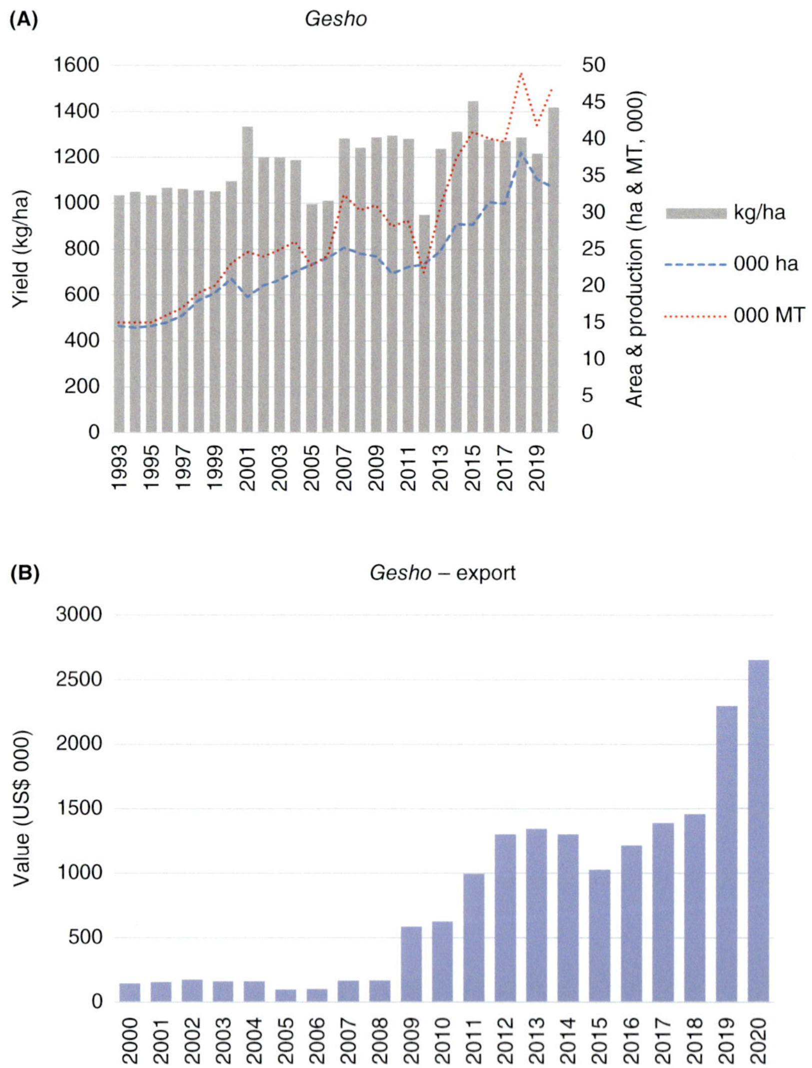

Fig. 10.6. (A) Indicator trends for *gesho* in Ethiopia. Source: constructed by the lead author from FAO (2022a). **(B)** Revenue from *gesho* export in Ethiopia. Source: constructed by the lead author from ERCA, www.erca.gov.et.

France, South Africa and Finland. The first three countries accounted for 84% of the total export revenues annually.

Currently, there is little or no research on this crop and it is unlikely that its status will change dramatically any time soon. However, there is a possibility for expansion of this crop if there should be large-scale, industrial production of *tella* and *tej* in the near future.

Sugarcane

Sugarcane in Ethiopia is grown both at the smallholder and commercial level. Smallholder sugarcane production in this country is believed to have started in the 16th century (EIA, 2012). Commercial production of sugar in Ethiopia began with a Dutch company, named HVA, that grew its first crop in 1951 and started sugar production in 1954 at Wonji Sugar Estate. At present, there are a total of nine sugar estates and development projects, including Arjo-Didessa, Finchaa, Kuraz, Kessem, Metehara, Tana-Beles, Tendaho, Wolkait and Wonji (Kebede *et al.*, 2011). For further details please refer to EIA (2012), Degefa and Saito (2017) and Gashaw *et al.* (2018).

According to the FAO, sugarcane occupied approximately 30,000 ha with production of more than 1.3 million MT. The yield was estimated at 45 MT/ha. According to the CSA, there were nearly 1.2 million smallholders cultivating sugarcane annually in Ethiopia in the 2016–2020 period. Their numbers increased at an annual rate of 3.05% between 2003 and 2020. The calculated area allocated to sugarcane is ~0.0241 ha (or about 241 m^2) per household. Ethiopia ranked 49th and 54th in area and production, respectively, out of 95 countries that cultivated sugarcane in 2016–2020. The top ten cultivating countries were Brazil, India, Thailand, China, Pakistan, Mexico, Argentina, Colombia, Indonesia and Australia.

Sugarcane in Ethiopia is produced as raw material for the sugar industry that manufactures sugar, ethanol, molasses and confectioneries; it is also produced for local consumption, i.e. chewing the cane. The commercial growers have their own estates and also have agreements with contract growers (Wendimu *et al.*, 2015). They have their own research centres, whereas there has been little or no research by the public sector to help smallholder producers (Tena *et al.*, 2016). Further details on sugar production in Ethiopia can be found in Semie *et al.* (2019).

Figure 10.7A presents sugarcane indicator trends in Ethiopia. Even though the area continued to rise for most of the time period, there has been progressive decline in yield, especially starting in the 1980s, thus causing declines in production. Overall, the ROGs for area, yield and production were 1.70%, –5.08% and –3.47%, respectively, during the 2000 to 2020 period (cf. Table 10.1).

The current national average yield of sugarcane in Ethiopia is 45 MT/ha, lower than the world average of 56 MT/ha. Australia, China, India, Brazil and Mexico all obtain >70 MT/ha (Fig. 10.7B). Ethiopia's yield during the 1976–1980 period was 177 MT/ha, the highest in the world, followed by Indonesia (122 kg/ha), Colombia, Australia, Mexico, Brazil, India, Argentina, Thailand and Pakistan; China did not have records until 1991.

Ethiopia suffered highly significant ($P<0.01$, $R^2=0.54$) yield loss of nearly 3.6 MT/ha/yr between 2000 and 2020. However, yield losses in sugarcane were not unique to Ethiopia. For example, five of the top ten countries – Argentina, Colombia, Indonesia, Australia and Mexico – also suffered yield losses during the same period; the loss for Argentina (1905 kg/ha/yr, $P<0.01$, $R^2= 0.62$) was highly significant. By contrast, Pakistan (1009 kg/ha/yr, $P<0.01$, $R^2=0.90$) and China (846 kg/ha/yr, $P<0.01$, $R^2=0.83$) obtained highly significant yield gains. The world average yield declined by 0.65% annually between 2000 and 2020, while the area increased by 1.87%.

Ethiopia's current sugarcane production could not satisfy the growing demand by the local industry and the country imports huge amounts of sugar each year, whereas it has good potential to expand production (Amrouk *et al.*, 2013). The country has experienced growing import quantity and value especially over the last two decades (Fig. 10.8). For example, in 1996–2000 it imported nearly 21,000 MT of sugar at an estimated cost of <US$10 million vs ~441,000 MT at ~US$225 million in 2016–2020.

The imports consisted of roughly 73.4%, 25.2%, 1.3% and 0.5% (by quantity), respectively,

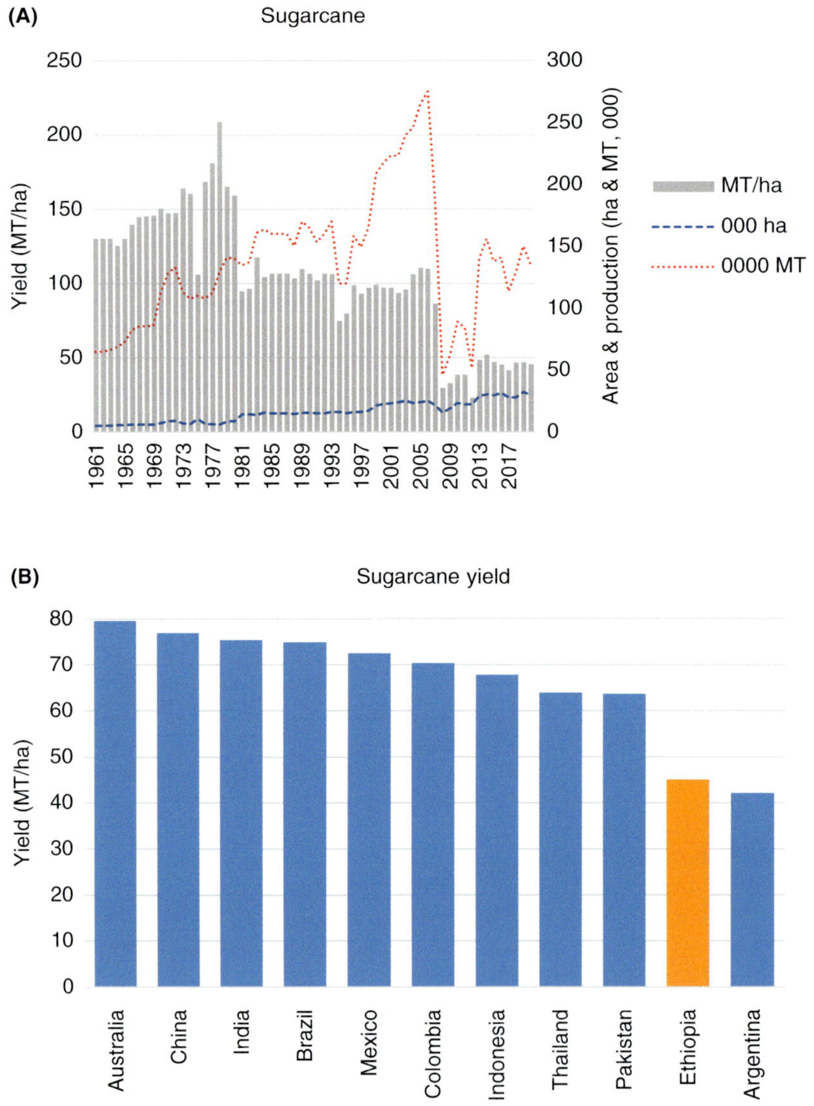

Fig. 10.7. (A) Sugarcane indicator trends in Ethiopia. Source: constructed by the lead author from FAO (2022a). **(B)** Sugarcane yield (2016–2020) in Ethiopia compared to the top ten cultivating countries in the world. Source: constructed by the lead author from FAO (2022d).

of refined sugar, raw sugar, confectioneries and miscellaneous. Their shares by value were 64.1%, 32.3%, 3.4% and 0.3%, respectively. The experience with exports has been somewhat sporadic. In 1984–1989, the country exported ~24,000 MT of sugar valued at ~US$5 million, while in 2000–2004 it was about 51,000 MT, valued at US$13 million. There has been little or no exports since 2009.

Tea

Tea is a relatively new plantation crop in Ethiopia. Seeds were known to be first introduced by a Canadian Catholic Missionary by the name George Holland in 1927, and planted at Bonga (Coultas, 1968). Subsequent introductions were made from Sri Lanka in 1928 through the British Consul General, Captain Erskine, who resided in

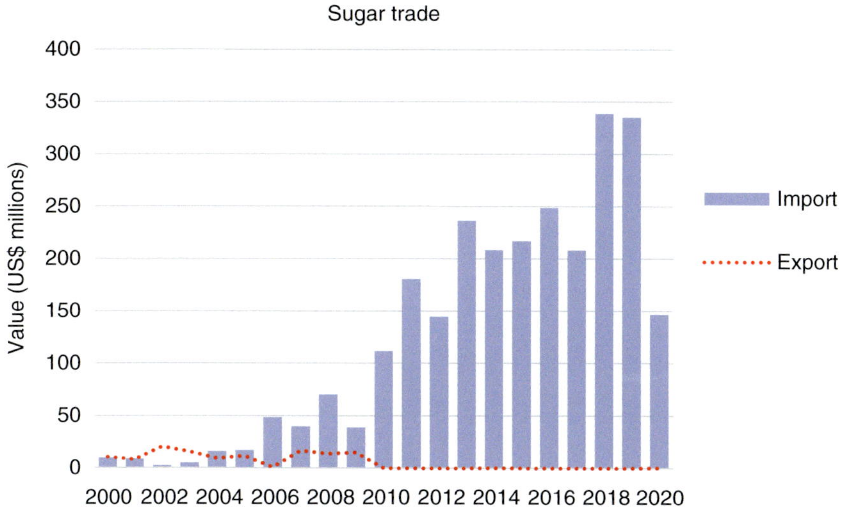

Fig. 10.8. Sugar export and import value trends in Ethiopia. Source: constructed by the lead author from FAO (2022d).

Gore at that time, and spread to the Gumaro, the Gore, the Wushwush, Jimma, the Gojeb valley and the Abekoran areas, all in southwestern Ethiopia.

The socialist government undertook a major tea expansion initiative in 1981 to boost earnings from export and for import substitution. Currently, the two major tea estates are under the ownership of Ethio Agriceft, a company owned by an Ethiopian-born Saudi billionaire Sheikh Mohammed Al Amoudi.

The CSA does not include tea in its annual sample surveys, perhaps because it is grown mainly by commercial farmers. The FAO data for tea in Ethiopia starts from 1976. The country grew ~10,000 ha and produced a little over 10,000 MT of tea each year in the 2016–2020 period. Ethiopia ranked 22nd both in area and production out of 47 countries in the world that produced tea during the above period. China, India, Kenya, Sri Lanka, Viet Nam, Indonesia, Myanmar, Türkiye, Bangladesh and Japan are the world's leading tea-growing countries. Ethiopia's tea production is the seventh largest in Africa, after Kenya, Uganda, Malawi, Mozambique, Rwanda and Burundi.

The government plan under GTP-II was to expand tea plantation to 1.4 million ha and generate US$1.5 billion from export earnings by 2020. However, available data show that this has not materialized.

Figure 10.9A presents the tea indicator trends for Ethiopia. Yields started upward trends, albeit slowly, starting from the early 1990s. Area expansion showed fast growth through the early 2010s and appeared to be levelling off after that. Overall, the area, yield and production increased at an annual rate of 5.45%, 0.84% and 6.34%, respectively, between 2000 and 2020 (cf. Table 10.1). That is, the area and yield contributed ~86% and ~13%, respectively, to changes in production while the interaction factor covered the remaining ~1%. The tea trends for Ethiopia followed somewhat similar trends to the world total.

Ethiopia's average yield of a little over 1.0 MT/ha is lower than the world average of about 1.9 MT/ha and is among the lowest of the top ten countries (Fig. 10.9B). Türkiye, India, Viet Nam and Kenya all have more than 2.0 MT/ha yield averages. Ethiopia's yield is in the same category as that of China. The FAO records show that Argentina, Iran and Thailand obtain more than 9.9 MT/ha, 5.9 MT/ha and 5.4 MT/ha, respectively.

Ethiopia obtained a highly significant ($P<0.01$, $R^2=0.82$) yield gain of over 8 kg/ha/yr between 2000 and 2020. Türkiye, Viet Nam and India also reported highly significant yield gains. By contrast, Mozambique showed highly significant yield loss of 63 kg/ha/yr ($P<0.01$, $R^2=0.74$). Kenya, Sri Lanka and Indonesia also suffered non-significant yield losses.

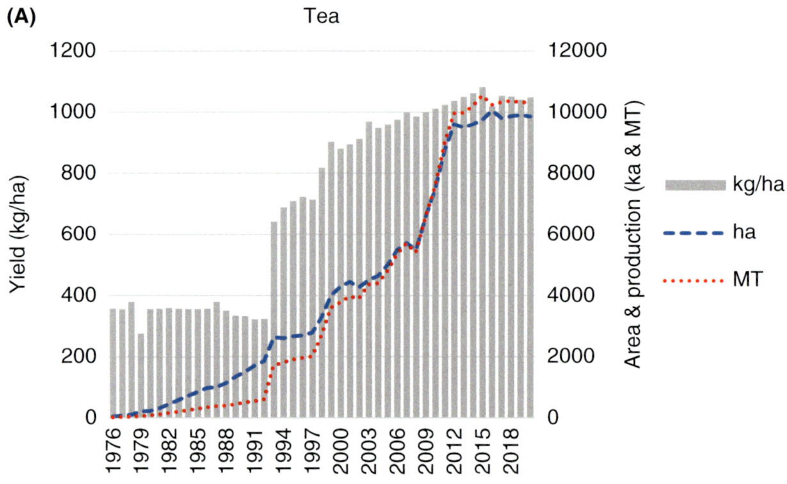

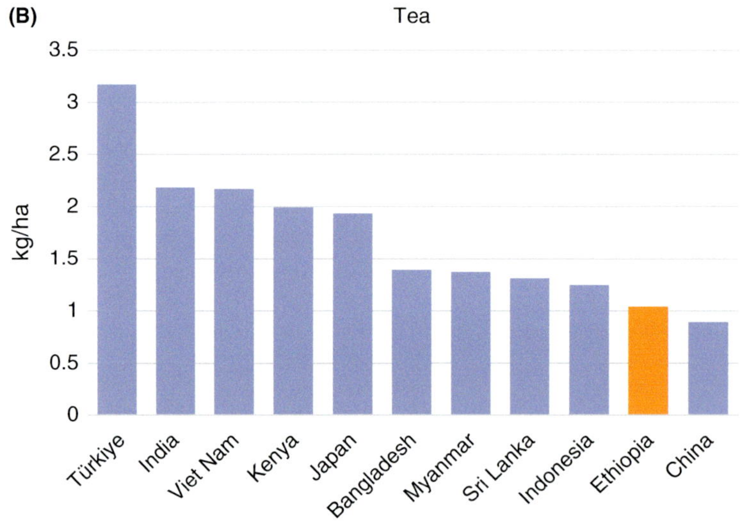

Fig. 10.9. (A) Tea indicator trends in Ethiopia. Source: constructed by the lead author from FAO (2022a). **(B)** Tea yield in Ethiopia, compared to the top ten cultivating countries in the world. Source: constructed by the lead author from FAO (2022d).

Ethiopia has achieved a significant milestone with its tea industry. It has transitioned from net importer to net exporter status (Fig. 10.10). Between 1981 and 2000, the country imported 835 MT of tea valued at ~US$1.5 million annually, compared to an export quantity of 110 MT valued at US$187,000. By contrast, each year between 2001 and 2020 it exported 955 MT of tea valued at ~US$1.5 million in comparison to an import quantity of 55 MT for an estimated value of US$232,000.

This country exported about 12% of its total tea production in 2016–2020, while the remaining portion was used for domestic consumption.

Castor

Smallholders in Ethiopia widely grow castor in small gardens in many parts of the country. Ethiopia is a secondary gene centre for this crop (Vavilov, 1951) and there is wide genetic diversity.

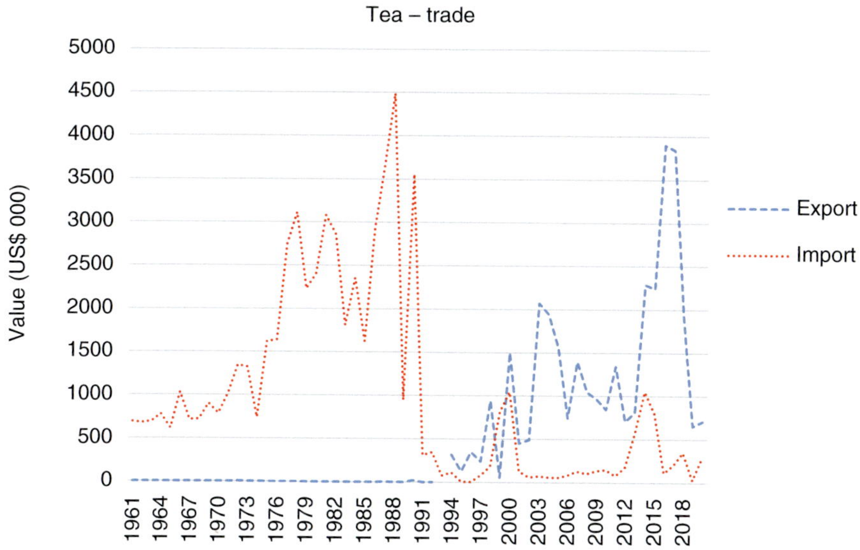

Fig. 10.10. Tea export and import value in Ethiopia. Source: constructed by the lead author from FAO (2022e).

The CSA sample surveys do not include castor. The FAO (2022f) data indicate that this crop occupies a little over 5500 ha with production estimates of >10,000 MT (cf. Table 10.1). The national research system has released only two cultivars as of 2017, indicating the low level of emphasis given to castor research in the country (Alemaw *et al.*, 2014; Alemaw, 2016).

Ethiopia ranked 13th in area, 6th in production and 3rd in yield out of 31 countries that produced castor in 2016–2020. India, Mozambique, Brazil, China, Angola, Myanmar, Kenya, Thailand, South Africa and Viet Nam have the top ten largest areas. However, the area declined in three of the top ten countries: China, Brazil and Thailand; the yield declined in Brazil, Kenya and Angola, while the production showed downward trends in China, Brazil and Thailand.

Figure 10.11A presents the castor bean indicator trends for Ethiopia, which has remained a subsistence crop for a long time. The area and production have declined for the most part, while the yield started to improve starting in the early 2010s but has slowly declined in recent years. This country's national average yield was a little over 1700 kg/ha in 2016–2020, the highest among top ten and the third highest in the world, after Mexico and Syria, with about 2600 kg/ha and 2300 kg/ha, respectively. It is in the same category as China and India (Fig. 10.11B). Ethiopia's yield gain of 52 kg/ha/yr ($P<0.01$, $R^2=0.71$) was the highest among the top ten countries.

According to the FAO, Ethiopia has only been in the world trade market of castor bean since 2018. With just over 3000 MT at an estimated value of more than US$2.3 million (2018–2020 average), it has become the highest exporter of castor bean in the world, ahead of Myanmar, India, Pakistan, Paraguay, Kenya, Tanzania, Israel, the USA and Lao People's Republic. Ethiopia's castor bean oil import during the same period was about 7 MT for an estimated value of US$21,000.

Data from ERCA also show that sustained trade of castor bean in Ethiopia started in 2018, although there were sporadic reports of small amounts between 2001 and 2007. This country exports only beans, with little or no processed castor oil. The ERCA data cover from 2018 to 2022. Accordingly, Ethiopia's annual revenue from castor bean export was estimated at nearly US$4.33 million during this period. Out of ten countries that imported castor bean from this country, Viet Nam accounted for about 68%, with the United Arab Emirates and China reporting 15% and 14%, respectively.

There is fast-growing international trade in castor bean seed and its oil. The total castor bean oil export in 2021 was estimated at ~US$1.23

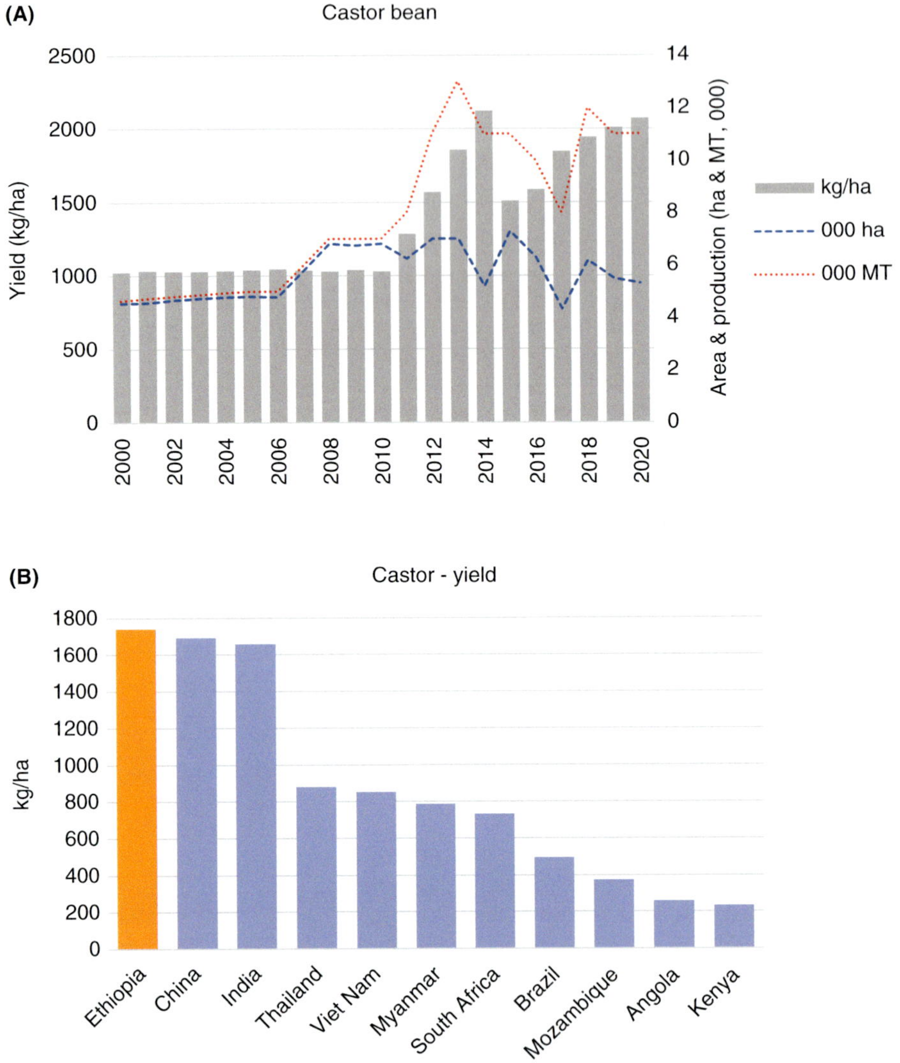

Fig. 10.11. **(A)** Castor bean indicator trends in Ethiopia. Source: constructed by the author from FAO (2022f). **(B)** Castor yield in Ethiopia compared to the top ten cultivating countries in the world. Source: constructed by the lead author from FAO (2022b).

billion whereas the import was ~US$1.26 billion; in a similar fashion, castor seed export was ~US$18.33 million, compared to ~US$20.63 million for import. India supplied 87% of the world's castor bean oil market whereas China imported 42% of the world's total castor bean oil, with France, Germany, the USA and the Netherlands accounting for about 10%, 10%, 9% and 8%, respectively.

Ethiopia, Myanmar, India and Pakistan accounted for about 26%, 24%, 22% and 20%, respectively, of the world's total caster beans export. On the other hand, China was responsible for 84% of the total import, with Brazil, Viet Nam and Thailand taking a distant second, third and fourth place (FAO, 2022e). The combined share of 39 African countries that imported castor bean oil was 0.75% of the world total.

Vernonia

Vernonia (common/English name: ironweed; scientific name: *Vernonia galamensis*) is famed for its oil, which is rich in vernolic acid, that is used for industrial purposes such as manufacturing plastic, rubbery coatings and drying agents. Neither the CSA nor FAO keep records of this crop in Ethiopia. There was so much enthusiasm among the research community to help commercialize vernonia beginning in the early 1990s, but there is no evidence it has gone very far. Only one cultivar (Boke Kuni, released in 2005) is available as of 2017 (MOALR, 2027), but there is no information on how much of it or the landraces that are found in Ethiopia are commercially produced at present. As a source of genetic diversity for vernonia (Perdue *et al.*, 1986; Baye, 2002; Baye and Becker, 2005), Ethiopia has good opportunities to develop and commercialize this crop for the international market.

Trade

Plantation/industrial crops are important sources of agricultural commodities trade in Ethiopia, mainly because of coffee, the country's biggest export earner. Available data show that the country earned more than US$666 million annually in the 2016–2020 period from export of the above crops (Table 10.2). Nearly 98% of the revenue was generated from coffee. By contrast, sugar accounted for 99% of the total imports among these group of crops.

Data from various sources suggest that *ch'at* also makes a major contribution to Ethiopia's agricultural commodity export earnings, which was ~US$ 400 million in the previous decade. The revenue for 2022 was estimated at US$248 million.

Discussion and Conclusions

This study has revealed several critical issues regarding plantation/industrial crops in Ethiopia. With the exception of sugarcane, production showed varying levels of positive growth for all the crops, even though area expansion rather than increases in productivity was the main driver. Particularly worrisome are the declines in cotton and sugarcane yields starting in the early 1980s and coffee yield from the early 2000s. It appeared that yield declines for cotton and sugarcane followed nationalization of commercial production by the military-socialist government of the time, which means that the long-term expertise and experience along with institutional memory were lost. In a similar fashion, there has been a heavy attrition rate of senior researchers on coffee over the last two decades. There are no seasoned researchers to provide leadership and mentor young scientists.

Loss of long-established crop management practices is particularly critical for these crops. For example, maintaining a 'closed season' – i.e. not growing any crop at least for 1 month during the off season – was the major component of the integrated pest management strategy of pink bollworm in cotton in the Awash valley, but that system does not exist anymore because of government policy to introduce irrigated wheat production.

Traditionally, coffee in Ethiopia was grown 'wild' in the forest where there is plenty of organic matter, adequate moisture and abundant biodiversity that kept insect pests and diseases in check (Yadessa *et al.*, 2020); at present, forest coffee accounts for only 10% of the total area.

Table 10.2. Selected plantation/industrial crops trade in Ethiopia (2016–2020). Source: constructed by the lead author from FAO (2022c).

	Export		Import	
Crop	Value (US$ 000)	Quantity (MT)	Value (US$ 000)	Quantity (MT)
Coffee	658,034	190,320	985	195
Cotton	5,581	5,063	1,177	282
Tea	2,205	1,254	198	31
Gesho	616	438	0	0
Sugar	0	0	255,480	441,667
Castor (oil)	0	0	12	3
Total	666,436	197,074	257,851	442,178

It should also be pointed out that Ethiopia achieved a major milestone with its tea. The yield gain for this crop was 8 kg/ha/yr and highly significant (R^2=0.82, P<0.01) and the country transitioned from a net importer to a net exporter starting from the early 1990s.

Besides improving the productivity of its plantation/industrial crops, Ethiopia has to strive to unlock the full potential of the crops for which it is the primary (or secondary) centre of genetic diversity – coffee, castor bean and *vernonia*. With its naturally decaffeinated coffee, this country has ample opportunities to develop and enter into the very lucrative and growing world market of decaffeinated coffee that was estimated to be US$18.67 billion in 2021 and projected to reach US$28.86 billion by 2030. This market is currently dominated by European and North American countries (Switzerland, Italy, Germany, France, the Netherlands, the USA, Canada, Poland, the UK and Belgium are the top ten countries and account for ~87 % of the world total volume; Ethiopia is 52nd) where they use artificial methods to extract the caffeine. Ethiopia also has a unique advantage to develop and market castor bean and *vernonia* in the world market.

This country has faced unrealistic goals for some of its plantation/industrial crops. For example, the National Cotton Development Strategy aimed to harvest 250,000 ha and produce 502,000 MT of cotton by 2020 (Annex Table 10.1). However, the FAO data show that the area harvested and production for that year were 80,000 ha and 188,000 MT, respectively. In a similar fashion, it was planned to harvest 1.4 million ha of tea and obtain export earnings of US$1.5 billion. However, import earnings for 2020 were US$709,000.

Many countries (including China, Brazil and Thailand) that were once major producers of castor bean appeared to have been phasing it out over the last many years. Ethiopia's rich genetic diversity, coupled with its high yield potential, may open up opportunities for it to develop this crop and be competitive in the international market.

References

Alemaw, G. (2016) Diversity of castor (*Ricinus communis* L.) in Ethiopia. *Ethiopian Journal of Agricultural Sciences* 26(2), 57–67.

Alemaw, G., Kassahun, B.M., Taye, G. and Endalemaw, C. (2014) Phenotypic variability in Ethiopian castor (*Ricinus communis* L.) accessions. *International Journal of Advanced Biological and Biomedical Research* 2(12), 2909–2914.

Amamo, A.A. (2014) Coffee production and marketing in Ethiopia. *European Journal of Business and Management* 6(37), 109–121.

Amrouk, E.M., Rakotoarisoa, M.A. and Kaison Chang, K. (2013) Structural Changes in the Sugar Market and Implications for Sugarcane Smallholders in Developing Countries. Country Case Studies for Ethiopia and the United Republic of Tanzania. FAO Commodity and Trade Policy Working Paper No. 37, FAO, Rome.

Baye, T. (2002) Genotypic and phenotypic variability in *Vernonia galamensis* germplasm collected from eastern Ethiopia. *Journal of Agricultural Science* 139, 161–168. DOI: 10.1017/S0021859602002459.

Baye, T. and Becker, H.C. (2005) Exploration of *Vernonia galamensis* in Ethiopia, and variation in fatty acid composition of seed oil. *Genetic Resources and Crop Evolution* 52, 805–811. DOI: 10.1007/s10722-003-6086-5.

Baye, T., Kebede, H. and Belete, K. (2001) Agronomic evaluation of *Vernonia galamensis* germplasm collected from Eastern Ethiopia. *Industrial Crops and Products* 14(3), 179–190. DOI: 10.1016/S0926-6690(01)00082-6

Belwal, R. and Teshome, H. (2011) Ch'at exports and the Ethiopian economy: Opportunities, dilemmas and constraints. *African Journal of Business Management* 5(9), 3635–3648. DOI: 10.5897/AJBM09.342

Binalfew, T. (2017) The expansion of production, marketing and consumption of ch'at in Ethiopia. *Global Journal of Science Frontier Research: C Biological Science* 17(2), 16–22.

Coultas, W.H.W. (1968) *A Report on the Investigation and Possibility of Establishing Tea Industry in Ethiopia*. Tea Adviser, Ministry of Overseas Development, London (available at EIAR library: IAR 633.7206093, A Cou., 1968. Ac210).

CSA (Central Statistical Agency) (2004) Agricultural Sample Surveys 2003/2004 (1996 E.C.), Report on area and production for major crops (Private Peasant Holdings, *Meher* Season Volume I, Statistical Bulletin. CSA, Addis Ababa, Ethiopia.

CSA (2005) Agricultural Sample Survey 2004/2005 (1997 E.C.), Report on Area and Production of Crops (Private Peasant Holdings, *Meher* Season), Volume I, Statistical Bulletin 331. CSA, Addis Ababa, Ethiopia.

CSA (2006) Agricultural Sample Survey 2005/2006 (1998 E.C.), Report on Area and Production of Crops (Private Peasant Holdings, *Meher* Season), Volume I, Statistical Bulletin 361. CSA, Addis Ababa, Ethiopia.

CSA (2007) Agricultural Sample Survey 2006/2007 (1999 E.C.), Report on Area and Production of Crops (Private Peasant Holdings, *Meher* Season), Volume I, Statistical Bulletin.388. CSA, Addis Ababa, Ethiopia.

CSA (2008) Agricultural Sample Survey 2007/2008 (2000 E.C.), Report on Area and Production of Crops (Private Peasant Holdings, *Meher* Season), Volume I, Statistical Bulletin 417. CSA, Addis Ababa, Ethiopia.

CSA (2009) Agricultural Sample Survey 2008/2009 (2001 E.C.), Report on Area and Production of Crops (Private Peasant Holdings, *Meher* Season), Volume I, Statistical Bulletin 446. CSA, Addis Ababa, Ethiopia.

CSA (2010) Agricultural Sample Survey 2009/2010 (2002 E.C.), Report on Area and Production of Major Crops (Private Peasant Holdings, *Meher* Season), Volume I, Statistical Bulletin 446. CSA, Addis Ababa, Ethiopia.

CSA (2011) Agricultural Sample Survey 2010/2011 (2003 E.C.), Report on Area and Production of Major Crops (Private Peasant Holdings, *Meher* Season), Volume I, Statistical Bulletin. CSA, Addis Ababa, Ethiopia.

CSA (2012 Agricultural Sample Survey 2011/2012 (2004 E.C.), Report on Area and Production of Major Crops (Private Peasant Holdings, *Meher* Season), Volume I, Statistical Bulletin. CSA, Addis Ababa, Ethiopia.

CSA (2013) Agricultural Sample Survey 2012/2013 (2005 E.C.), Report on Area and Production of Major Crops (Private Peasant Holdings, *Meher* Season), Volume I, Statistical Bulletin 532. CSA, Addis Ababa, Ethiopia.

CSA (2014) Agricultural Sample Survey 2013/2014 (2006 E.C.), Report on Area and Production of Major Crops (Private Peasant Holdings, *Meher* Season), Volume I, Statistical Bulletin. CSA, Addis Ababa, Ethiopia.

CSA (2015) Agricultural Sample Survey 2014/2015 (2007 E.C.), Report on Area and Production of Major Crops (Private Peasant Holdings, *Meher* Season), Volume I, Statistical Bulletin 578. CSA, Addis Ababa, Ethiopia.

CSA (2016) Agricultural Sample Survey 2015/2016 (2008 E.C.), Report on Area and Production of Major Crops (Private Peasant Holdings, *Meher* Season), Volume I, Statistical Bulletin. CSA, Addis Ababa, Ethiopia.

CSA (2017) Agricultural Sample Survey 2016/17 (2009 E.C.), Report on Area and Production of Major Crops (Private Peasant Holdings, *Meher* Season), Volume I, Statistical Bulletin. CSA, Addis Ababa, Ethiopia.

CSA (2018) Agricultural Sample Survey 2017/18 (2010 E.C.), Report on Area and Production of Major Crops (Private Peasant Holdings, *Meher* Season), Volume I, Statistical Bulletin 586 CSA, Addis Ababa, Ethiopia.

CSA (2019) Agricultural Sample Survey 2018/19 (2011 E.C.), Report on Area and Production of Major Crops (Private Peasant Holdings, *Meher* Season), Volume I, Statistical Bulletin 589. CSA, Addis Ababa, Ethiopia.

CSA (2020) Agricultural Sample Survey 2019/20 (2012 E.C.), Report on Area and Production of Major Crops (Private Peasant Holdings, *Meher* Season), Volume I, Statistical Bulletin 587. CSA, Addis Ababa, Ethiopia.

CSA (2021) Agricultural Sample Survey 2020/21 (2013 E.C.), Report on Area and Production of Major Crops (Private Peasant Holdings, *Meher* Season), Volume I, Statistical Bulletin. CSA, Addis Ababa, Ethiopia.

Degefa, S. and Saito, O. (2017) Assessing the impacts of large-scale agro-industrial sugarcane production on biodiversity: A case study of Wonji Shoa Sugar Estate, Ethiopia. *Agriculture* 7, 99. DOI: 10.3390/agriculture7120099.

EIA (Ethiopia Investment Agency) (2012) Investment Opportunity Profile for Sugar Cane Plantation and Processing in Ethiopia. EIA, Addis Ababa.

ERCA (Ethiopian Revenue and Customs Authority) (2013–2017). Available at: http://www.erca.gov.et/index.php/component/hscode/?view=importexport&iccaldate=2014-5-1 (accessed 29 December 2018).

FAO (Food and Agriculture Organization of the United Nations) (2022a) Statistical Databases and Data-Sets of the Food and Agriculture Organization of the United Nations. Available at: http://www.fao.org/faostat/en/#data (accessed 13 March 2022).

FAO (2022b) Statistical Databases and Data-Sets of the Food and Agriculture Organization of the United Nations. Available at: http://www.fao.org/faostat/en/#data (accessed 30 June 2022).

FAO (2022c) Statistical Databases and Data-Sets of the Food and Agriculture Organization of the United Nations. Available at: http://www.fao.org/faostat/en/#data (accessed 15 August 2022).

FAO (2022d) Statistical Databases and Data-Sets of the Food and Agriculture Organization of the United Nations. Available at: http://www.fao.org/faostat/en/#data (accessed 18 August 2022).

FAO (2022e) Statistical Databases and Data-Sets of the Food and Agriculture Organization of the United Nations. Available at: http://www.fao.org/faostat/en/#data (accessed 20 August 2022).

FAO (2022f) Statistical Databases and Data-Sets of the Food and Agriculture Organization of the United Nations. Available at: http://www.fao.org/faostat/en/#data (accessed 25 August 2022).

FAO (2023a) Statistical Databases and Data-Sets of the Food and Agriculture Organization of the United Nations. Available at: http://www.fao.org/faostat/en/#data (accessed 23 March 2023).

FAO (2023b) Statistical Databases and Data-Sets of the Food and Agriculture Organization of the United Nations. Available at: http://www.fao.org/faostat/en/#data (accessed 30 June 2023).

Gashaw, E.T., Mekbib, F. and Ayana, A. (2018) Sugarcane landraces of Ethiopia: Germplasm collection and analysis of regional diversity and distribution. *Advances in Agriculture* 2018, 18 p. Article ID 7920724. DOI: 10.1155/2018/7920724.

Gebre, A., Alebel, A., Zegeye, A. and Tesfaye, B. (2018) Prevalence and predictors of khat chewing among Ethiopian university students: A systematic review and meta-analysis. *PLoS One* 13(4), e0195718. DOI: 10.1371/journal. pone.0195718.

Gudeta, B., Workie, A., Shonga, E., G/Egziabher, A., Gebre, D. and Girma, B. (2016) Cotton research in Ethiopia: Achievements, challenges, opportunities and prospects. In: Alemu *et al.* (eds) *Proceedings of the National Conference on Agricultural Research for Ethiopian Renaissance,* 26–27 January 2016, Addis Ababa, Ethiopia.

Kandari, L.S., Yadav, H.R., Ashok, K., Thakur, A.K. and Kandari, T. (2014) Ch'at (*Catha edulis*): A socio economic crop in Harar Region, Eastern Ethiopia. *Springer Plus* 3, 579.

Kebede, S., Damtie, A. and Yirefu, F. (2011) Trends of sugar industry development in Ethiopia: Challenges and prospects. *Presented at Ethiopian Academy of Sciences Conference, 28-30 November 2011*. Theme: The State of Science and Technology in Agriculture. Ethiopian Academy of Sciences, Addis Ababa, Ethiopia.

Kufa, T., Addisu, M., Teferi, D. and Ayano, A. (2016) Review of coffee and tea research achievements and prospects in Ethiopia. In: Alemu, D., Derso, E., Assefa, G. and Kirub, A. (eds) *Proceedings of the National Conference on Agricultural Research for Ethiopian Renaissance.* 26–27 January 2016, Addis Ababa, Ethiopia.

Perdue, R.E., Carlson, K.D. and Gilbert, M.G. (1986) *Vernonia galamensis*, potential new crop source of epoxy acid. *Economic Botany* 40, 54–68. DOI: 10.1007/BF02858947

Semie, T.K., Silalertruksa, T., Shabbir, T.S. and Gheewala, H. (2019) The impact of sugarcane production on biodiversity related to land use change in Ethiopia. *Global Ecology and Conservation* 18, e00650.

Tena, E., Mekbib, F., Shimelis, H. and Mwadzingeni, L. (2016) Sugarcane production under smallholder farming systems: Farmers preferred traits, constraints and genetic resources. *Cogent Food & Agriculture* 2, 1191323. DOI: 10.1080/23311932.2016.1191323

Teshome, K., Girma, Z. and Eshetu, B. (2018) Assessment of pre- and post-harvest management practices on coffee (*Coffea arabica* L.) quality determining factors in Gedeo Zone, Southern Ethiopia. *African Journal of Agricultural Research* 14(28), 1216–1228. DOI: 10.5897/AJAR2019.14116.

USDA (United States Development Agency) (2019) *Ethiopia Cotton Production Annual*. GAIN Report Number: ET1906. FAS Issued on 29 May 2019. Addis Ababa, Ethiopia.

van der Graaff, N.A. (1981) *Selection of Arabica Coffee Types for Resistance to Coffee Berry Disease in Ethiopia*. Med. Landbouwhogeschool, Wageningen, The Netherlands.

Vavilov, N.I. (1951) The origin, variation, immunity and breeding of cultivated plants. *Chronica Botanica* 13(1–6), 1–364.

Wendimu, M.A., Henningsen, A. and Gibbon, P. (2015) *Sugarcane Out Growers in Ethiopia: 'Forced' to Remain Poor?* IFRO Working Paper 2015/06, Department of Food and Resource Economics (IFRO), University of Copenhagen, Frederiksberg, Denmark.

Westphal, E. (1975) *Agricultural Systems in Ethiopia*. Joint Publication of the College of Agriculture, Haile Selassie I University, Ethiopia, and the Agricultural University, Wageningen, the Netherlands.

Wolde, Z., Tefera, A., Yared, S., Gezahagn, T. and Tadesse, T. (2017) A review on coffee farming, production potential and constraints in Gedeo Zone, Southern Ethiopia. *Journal of Natural Sciences Research* 7(23), 1–8.

Yadessa, A., Burkhardt, J., Bekele, E., Hundera, K. and Goldbach, H. (2020) The major factors influencing coffee quality in Ethiopia: The case of wild arabica coffee (*Coffea arabica* L.) from its natural habitat of southwest and southeast afromontane rainforests. *African Journal of Plant Science* 14(6), 213–230. DOI: 10.5897/AJPS2020.1976.

Annex Table 10.1. Targets of Ethiopia's National Cotton Development Strategy (2017–2032). Source: adapted from USDA (2019), GAIN Report Number: ET1906.

Element	2017	2020	2025	2032
Area				
Smallholder, rainfed	26,000	48,000	260,000	290,000
Smallholder, irrigated	1,000	2,000	5,000	10,000
Large farm, rainfed	33,000	100,000	200,000	350,000
Large farm, irrigated	20,000	100,000	200,000	350,000
Total	**80,000**	**250,000**	**665,000**	**1,000,000**
Productivity (kg/ha)				
Smallholder, rainfed	1,500	1,600	2,000	2,200
Smallholder, irrigated	2,400	2,600	3,000	3,300
Large farm, rainfed	1,500	1,600	2,000	2,200
Large farm, irrigated	2,400	2,600	3,000	3,300
Average	**1,736**	**2,008**	**2,308**	**2,596**
Production (MT seed cotton)				
Smallholder, rainfed	39,000	76,800	520,000	638,000
Smallholder, irrigated	2,400	5,200	15,000	33,000
Large farm, rainfed	49,500	160,000	400,000	770,000
Large farm, irrigated	48,000	260,000	600,000	1,155,000
Total	**138,900**	**502,000**	**1,535,000**	**2,596,000**
Ginning outturn (%)	37	39	42	43
Lint production (MT)	51,393	195,780	644,700	1,116,280
Lint yield (kg/ha)	642	783	969	1,116
Lint classed by machine (%)	5	20	50	100
Share of higher grades (%)	10	25	50	75
Identity/sustainable cotton (%)	5	25	50	100
Domestic mill use (MT lint)	40,000	100,000	350,000	600,000
Export (MT lint)	11,393	95,780	294,700	516,280

11 Towards Realizing the Potential of Apiculture for Honey Production and Crop Pollination

Tsedeke Abate[1]* and Workneh Ayalew[2]

[1]*Homegrown Vision, Addis Ababa, Ethiopia;* [2]*International Centre for Insect Physiology and Ecology, Addis Ababa, Ethiopia*

Abstract

This chapter explores the status of apiculture research and development in Ethiopia. With more than six million hives, the country ranks first in Africa and fifth in the world. About 95% of beehives in smallholder apiaries are traditional. The national average honey yield (2015–2019) is 9 kg/colony compared to the world average of 23 kg/colony. Even though the external parasitic mite *Varroa destructor* that is devastating colonies of the European honeybee in North America and Europe has been recorded in parts of Ethiopia, it is unlikely to be the major factor for the poor performance here where a different subspecies that is tolerant to the mite is raised. The most likely major factors could include loss of abundance and diversity of flowers. The importance of honeybees in crop pollination has been overlooked so far in Ethiopia and there is a strong need to quantify it and create awareness.

Introduction

Ethiopia has a long tradition of beekeeping (also known as apiculture) and production of honey and beeswax. Abebe (2011) suggests that Ethiopian farmers have been practicing apiculture since about 3500 to 3000 BCE. Honey is primarily used for making *tej*, the traditional honey wine (the equivalent of mead), while the main use of beeswax is for making wicks for lighting homes in the rural areas and in Ethiopian Orthodox Church services. Beeswax has long been a source of foreign currency earnings from its export. In recent decades, honey also has become an important grocery item in urban areas as well as foreign currency earner, albeit in small amounts.

The bulk of honey in this country is produced by smallholder households. All regions of the country have beekeeping facilities even though the potential and level of production vary. There are three types of honey production systems recognized in Ethiopia: (i) traditional; (ii) transitional; and (iii) modern beekeeping. According to the Central Statistical Agency (CSA, 2017), there were over 6 million colonies and nearly 48,000 MT of honey produced in the 2016/17 season. These estimates showed that traditional, modern and transitional systems accounted for >95%, >3% and >1%, respectively, of the total number of colonies, and roughly 90%, 6% and 4%, respectively, of the total production.

* Corresponding author: tsedeke@homegrownvision.org

DOI: 10.1079/9781800626386.0011

In most instances, people think of only honey and wax when talking about the importance of beekeeping. The more significant importance of crop pollination (Bareke and Addi 2019; Sagwe *et al.*, 2021; Fekadie *et al.*, 2023) and other potential areas such as commercialization of propolis and bee sting venom are often forgotten.

Beekeeping in Ethiopia faces several constraints (Tesfaye *et al.*, 2017; Sahle *et al.*, 2018; Sebho and Baraki, 2018; Teferi, 2018; Fikadu, 2020; Wakgari and Yigezu, 2021). Chief among these is the low level of modern colonies, habitat degradation and decline of floral diversity, and diseases, pests and predators.

Comprehensive and coherent reviews on honeybee in Ethiopia are lacking at present. This report attempts to close those gaps.

Approach

We used the latest available Food and Agriculture Organization of the United Nations (FAO) data to make graphical presentations of indicator trends for honey and beeswax production. To calculate the annual rate of growth (ROG) for honey and beeswax production, per capita consumption and dietary energy supply of honey, we used methods employed by Abate *et al.*, (2015) for maize in Ethiopia. Further details are given in Chapter 3.

Results

This study has revealed that Ethiopia had a little over 6 million colonies, the fifth highest number in the world after India, China, Türkiye and Iran (Annex Table 11.1) out of 115 countries that produced natural honey (average for 2015–2019). It was followed by the Russian Federation, Argentina, Tanzania, the USA and Spain. The average production was a little over 52,000 MT, tenth out of 132 countries (after China, Türkiye, Canada, Iran, Argentina, the USA, the Russian Federation, India and Mexico). The average yield for Ethiopia was 9 kg/colony (96th out of 108 countries; the Ukraine, Latvia, Belarus, Fiji and Rwanda were not included in the comparison because their reported yields did not make sense – i.e. reported yields were roughly 25, 20, 13, 12 and 1 MT/colony), compared to 23 kg/colony for the world average.

Figure 11.1 presents indicator trends for honey in Ethiopia. The number of hives and production showed slow increases until the early 1990s; they increased a little faster thereafter, but both yield and production fluctuated in the years that followed and showed declining trends in recent years.

The number of colonies and production increased at a rate 3.04% and 1.91%, respectively, annually between 2002 and 2019, whereas the yield declined by 0.56%. By contrast, the world number of colonies, yield and production increased by 1.43%, 1.16% and 2.71%, respectively (Annex Table 11.1).

Available data show that Ethiopia produces approximately 5600 MT of beeswax annually, second highest in the world after India, followed by Argentina, Türkiye, Republic of Korea, Kenya, Angola, Tanzania, Brazil and Mexico (Annex Table 11.2). Its average yield is 0.90 kg/colony, compared to 2.26 kg/colony for the world average. Ethiopia is ranked 30th out of 53 countries for which data were available in 2015–2019. Other countries that obtain high yield include Uganda, Guinea-Bissau, Rwanda, Dominican Republic, Venezuela, Jamaica, Madagascar, Paraguay, Haiti and Costa Rica (Annex Table 11.2).

The per capita consumption of natural honey in Ethiopia showed declining trends over the years. It was 0.52 kg/person/yr in the 1976–1980 period, compared to 0.45 and 0.33 kg/person/yr in 1996–2000 and 2016–2020 periods, respectively. It ranked 64th out of 166 countries and territories that produced honey in the last period. The Central African Republic has the highest per capita consumption of natural honey in the world, with 3.03 kg/person/yr. It is followed by Greece, Switzerland, the Ukraine, Austria, Türkiye, Angola, New Zealand, Slovenia and Samoa.

We present long-term trends of beeswax indicators in Fig. 11.2. Colony numbers and production showed relatively fast growth starting from the early 2000s. The yield showed upward trends starting from the early 1990s but declined since the mid-2010s. The number of colonies and production increased at an annual rate of 3.04% and 2.47%, respectively, whereas the yield declined at 0.67% between 2002 and

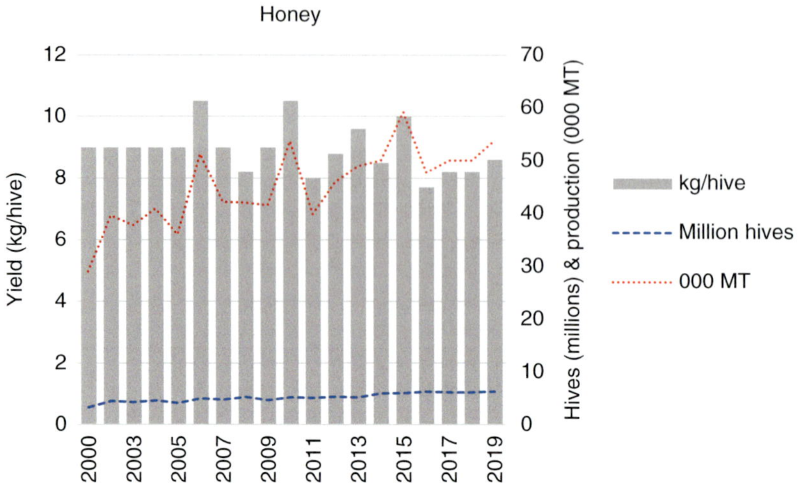

Fig. 11.1. Indicators for natural honey in Ethiopia. Note: 2001 data were incomplete and are not included. Source: constructed by the lead author from FAO (2021).

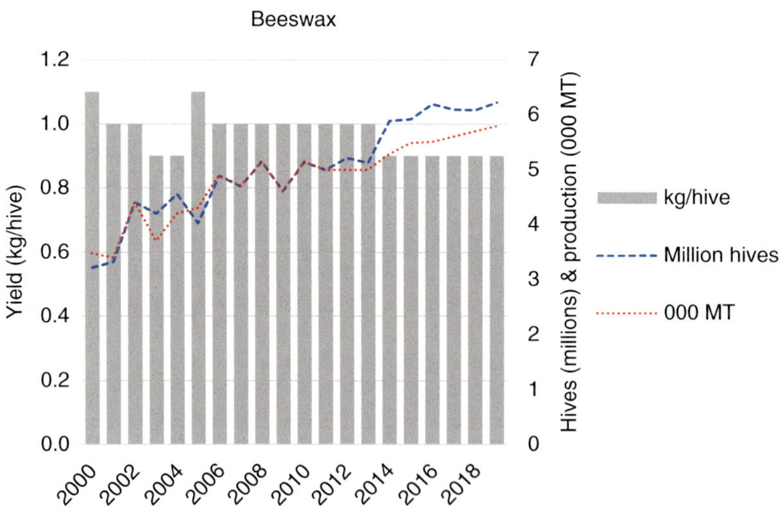

Fig. 11.2. Indicators for beeswax in Ethiopia. Source: constructed by the lead author from FAO (2021).

2019 (Annex Table 11.3). The annual ROG for the number of colonies and production in the world between 2000 and 2019 was 1.43% and 0.85%, respectively, whereas the yield declined to 1.08%.

Ethiopia exported little or no honey until the mid-2000s; the export quantity and value showed fast growth through to 2015 and has been declining since (Fig. 11.3A). By contrast, there has been a fair amount of beeswax export, even though it has fluctuated from time to time (Fig. 11.3B).

The bulk of honey and beeswax produced in Ethiopia is used domestically. For example, approximately 5.4% and 0.7% of the total production of beeswax and honey, respectively,

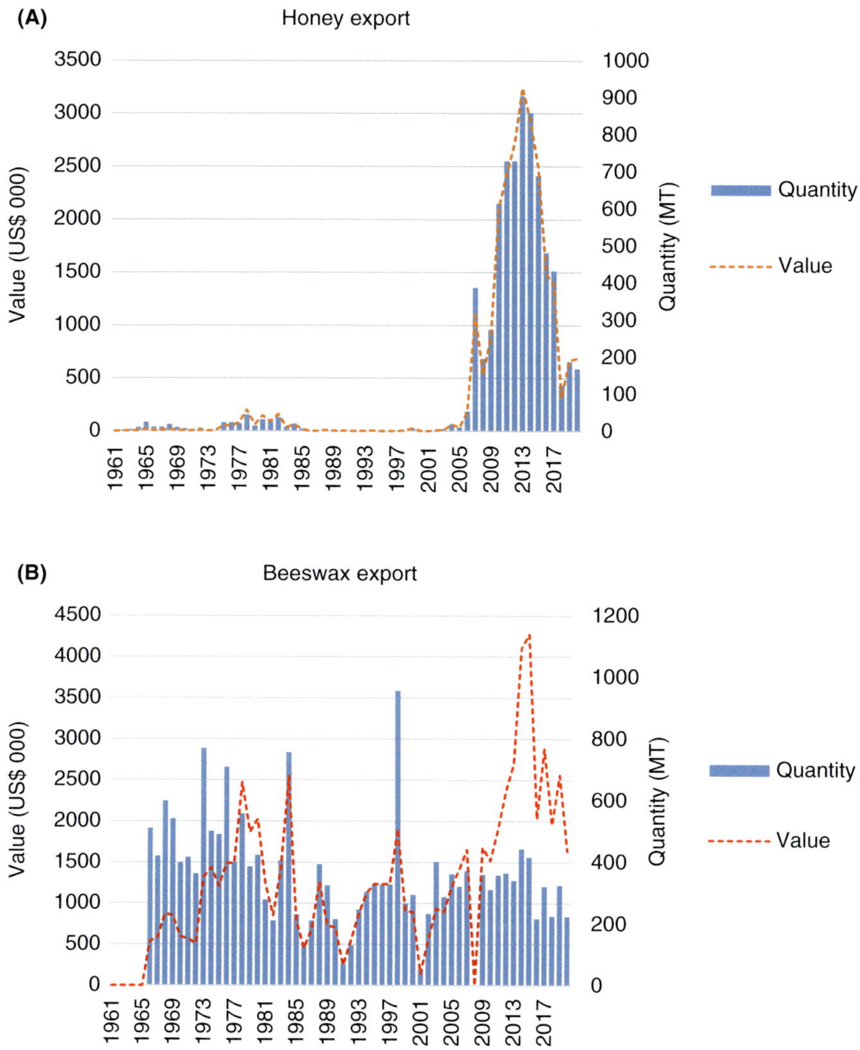

Fig. 11.3. Value and quantity of **(A)** honey and **(B)** beeswax export in Ethiopia 1961–2000. Source: constructed by the lead author from FAO (2022).

was exported annually during the 2015–2019 period. Overall, the country earned a combined total of US$4 million each year from beeswax and honey export, according to FAO (2021). This means export of approximately 300 MT and 380 MT valued at US$2.7 million and US$1.3 million, respectively, of beeswax and honey. In other words, beeswax and honey accounted for 68% and 32%, respectively, of the total export of honeybee products.

According to the Ethiopian Revenues and Customs Authority, Ethiopia exported about 1 million MT of beeswax and natural honey combined, valued at US$5.1 million annually between 2013 and 2017. During the peak year of 2014, the country exported 1242 MT of both, valued at US$6.4 million.

Wax was exported to Germany, the USA, Japan, UK, Italy, the Netherlands and Canada, whereas the honey export included 28 countries across the world. Norway and Sudan accounted for approximately 35% and 31%, respectively, of the total quantity; these were followed by Germany, the UK and Somalia, with about 7% each.

Discussion and Conclusions

It is obvious from this study that Ethiopia is one of the major natural honey- and beeswax-producing countries in the world. However, its yield for both products was low and showed negative growth rates, albeit slow, in recent times. The local literature on honeybees in Ethiopia has largely been reviews focusing on problem identification in different parts of the country, such as challenges and opportunities (Fikru, 2015; Lijalem *et al.*, 2017; Sahle *et al.*, 2018; Sebho and Baraki, 2018; Teferi, 2018; Wakgari and Yigezu, 2021); indigenous knowledge (Bogale, 2009; Abebe, 2011; Abejew and Zeleke, 2017); occurrence of pests and enemies of honeybee (Tesfaye, *et al.*, 2017; Gebremedhin *et al.*, 2019); and pesticide use (Fikadu, 2020). Studies dealing with original research on various other aspects are unavailable.

Even though Begna (2015) reported the first record of the external parasitic mite *Varroa destructor* and the 'most likely occurrence of the attendant disease' Deformed Wing Virus (DWV) from parts of Tigray in 2010, data relating the mite damage with yield are not available. His report showed an average infestation of 82% (ranging from 40% in Shire Endaselassie to 100% in Alamata, Kilte Awlalo and Ganta Afshum) out of 200 colonies sampled, with an average number of about 7 (4 in Shire Endaselassie to 27 in Adwa) varroa per colony.

It is widely believed that varroa is the biggest threat to honeybee health in the world (Locke, 2015; Traynor *et al.*, 2020), but this refers mostly to the European honeybee, *Apis mellifera scutellata*, the species exclusively used in North America and Europe. By contrast, the subspecies used in Ethiopia, *Apis mellifera simensis*, and the Africanized bees in Brazil and elsewhere round the world are known to be tolerant or resistant to varroa because of their defence mechanism. There are strains of even the European subspecies that are known to be resistant to the mite and raised without acaricidal treatment (Locke, 2015; Le Conte *et al.*, 2020).

The decline of honeybee populations resulting in a phenomenon termed as 'colony collapse disorder' is attributed to not only varroa, but to a number of complex and interacting factors that include loss of abundance and diversity of flowers, other pests, pathogens, stock diversity of bees, environmental change, exposure to pesticides (including neonicotinoids, especially in Europe and North America), management practices and nutrition (Cepero *et al.*, 2014; Stavely *et al.*, 2014; DeGrandi-Hoffman *et al.*, 2016b; van Engelsdorp *et al.*, 2017; Rhodes, 2018; Lopez-Uribe and Simone-Finstrom, 2019; Beaurepaire *et al.*, 2020; Gregorc, 2020).

Considering the fact that the bee subspecies raised in Ethiopia is resistant to varroa, the low density of varroa in the samples tested by Begna (2015) and the dominance of traditional hives where the mite is less common than in frame hives (Gebremedhin *et al.*, 2019), it is highly unlikely that this pest is a major factor in the decline of honey and beeswax yields in this country. Other factors such as a decline in the abundance and diversity of flowers, management practices and other unknown factors may be more important. Even the highest mite prevalence of 27 per hive reported by Begna (2015) in Adwa is too low to be the cause for colony death or decline in honey/beeswax yield. Peak mite numbers can reach as high as nearly 850 per colony in parts of the USA (DeGrandi-Hoffman *et al.*, 2016a).

For Ethiopia, well-designed scientific studies need to be initiated and carried out to generate information that would help to isolate major factors affecting the honeybee population and provide appropriate recommendations. It is important to emphasize here that future studies should include the more significant importance of crop pollination (Rhodes, 2018; Bareke and Addi, 2019; Fekadie *et al.*, 2023) and other potential areas such as commercialization of propolis and bee sting venom that are often forgotten.

References

Abate, T., Shiferaw, B., Menkir, A., Wegary, D., Kebede, Y. *et al.* (2015) Factors that transformed maize productivity in Ethiopia. *Food Security* 7, 965–981. DOI: 10.1007/s12571-015-0488-z.

Abebe, W. (2011) Identification and documentation of indigenous knowledge of beekeeping practices in selected districts of Ethiopia. *Journal of Agricultural Extension and Rural Development* 3(5), 82–87.

Abejew, T.A. and Zeleke, Z.M. (2017) Study on the beekeeping situation, the level of beekeepers knowledge concerning local honeybee subspecies, their productive characteristics, and behavior in eastern Amhara region, Ethiopia. *Advances in Agriculture* 2017, 6354250. DOI: 10.1155/2017/6354250.

Bareke, T. and Addi, A. (2019) Effect of honeybee pollination on seed and fruit yield of agricultural crops in Ethiopia. *MOJ Ecology & Environmental Sciences* 4(5), 205–209.

Beaurepaire, A., Piot, N., Doublet, V., Antunez, K., Campbell, E. *et al.* (2020) Diversity and global distribution of viruses of the western honeybee, *Apis mellifera. Insects* 11, 239. DOI: 10.3390/insects11040239.

Begna, D. (2015) Occurrences and distributions of honeybee (*Apis mellifera Jemenetica*) varroa mite (*Varroa destructor*) in Tigray region, Ethiopia. *Journal of Fisheries & Livestock Production* 3, 1. DOI: 10.4172/2332-2608.1000126.

Cepero, A., Ravoet, J., Gomez-Moracho, T., Bernal, J.L., Del Nozal, M.J. *et al.* (2014) Holistic screening of collapsing honey bee colonies in Spain: A case study. *BMC Research Notes* 7, 649.

DeGrandi-Hoffman, G., Ahumada, F., Zazueta, V., Chambers, M., Hidalgo, G. and de Jong, E.W. (2016a) Population growth of *Varroa destructor* (Acari: Varroidae) in honey bee colonies is affected by the number of foragers with mites. *Experimental and Applied Acarology* 69, 21–34.

DeGrandi-Hoffman, G., Chen, Y., Rivera, R., Carroll, M., Chambers, M., Hidalgo, G. and de Jong, E.W. (2016b) Honey bee colonies provided with natural forage have lower pathogen loads and higher over-winter survival than those fed protein supplements. *Apidologie* 47, 186–196. DOI: 10.1007/s13592-015-0386-6.

Bogale, S. (2009) Indigenous knowledge and its relevance for sustainable beekeeping development: A case study in the highlands of southeast Ethiopia. *Livestock Research for Rural Development* 21, 184.

CSA (Central Statistical Agency) (2017) Agricultural Sample Survey 2016/17 Report on Livestock and Livestock Characteristics. Addis Ababa, Ethiopia.

FAO (Food and Agriculture Organization) (2021) Statistical Databases and Data-Sets of the Food and Agriculture Organization of the United Nations. Available at: http://www.fao.org/faostat/en/#data (accessed 27 April 2021).

FAO (2022) Statistical Databases and Data-Sets of the Food and Agriculture Organization of the United Nations. Available at: http://www.fao.org/faostat/en/#data (accessed 14 August 2022).

Fekadie, B., Getachew, A., Ayalew, W. and Jenberie, A. (2023) Evaluating the effect of honeybee pollination on production of watermelon (*Citrullus lanatus*), in northern Ethiopia. *International Journal of Tropical Insect Science* 43,1431–1449. DOI: 10.1007/s42690-023-01042-2.

Fikadu, Z. (2020) Pesticides use, practice and its effect on honeybee in Ethiopia: A review. *International Journal of Tropical Insect Science* 40, 473–481. DOI: 10.1007/s42690-020-00114-x.

Fikru, S. (2015) Review of honey bee and honey production in Ethiopia. *Journal of Animal Science Advances* 5(10), 1413–1421. DOI: 10.5455/jasa.20151019083635.

Gebremedhin, H., Amssalu, B., De Smet, L. and de Graaf, D.C. (2019) Factors restraining the population growth of *Varroa destructor* in Ethiopian honey bees (*Apis mellifera simensis*). *PLOS One* 14(9), e0223236. DOI: 10.1371/journal.pone.0223236.

Gregorc, A. (2020) Monitoring of honey bee colony losses: A special issue (editorial). *Diversity* 12, 403.

Le Conte, Y., Meixner, M.D., Brandt, A., Carreck, N.L., Costa, C., Mondet, F. and Büchler, R. (2020) Geographical distribution and selection of European honey bees resistant to *Varroa destructor. Insects* 11(12), 873. DOI: 10.3390/insects11120873.

Lijalem, T., Zereu, G. and Tebeje, M. (2017) Opportunities and constraints of beekeeping in Wolaita and Dawro zones, southern Ethiopia. *African Journal of Agricultural Research* 122(18), 1587–1592.

Locke, B. (2015) Natural Varroa mite-surviving *Apis mellifera* honeybee populations. *Apidologie* 47, 467–482. DOI: 10.1007/s13592-015-0412-8.

Lopez-Uribe, M.M. and Simone-Finstrom, M. (2019) Honey bee research in the US: Current state and solutions to beekeeping problems. *Insects* 10, 22. DOI:10.3390/insects10010022.

Rhodes, C.J. (2018) Pollinator decline—An ecological calamity in the making. *Science Progress* 101(2), 121–160. DOI: 10.3184/003685018X15202512854527.

Sagwe, R.N., Peters, M.K., Dubois, T., Steffan-Dewenter, I. and Lattorff, H.M.G. (2021) Pollinator supplementation mitigates pollination deficits in smallholder avocado (*Persea americana* Mill.) production systems in Kenya. *Basic and Applied Ecology* 56, 392–400.

Sahle, H., Enbiyale, G., Negash, A. and Neges, T. (2018). Assessment of honey production system, constraints and opportunities in Ethiopia. *Pharmacy & Pharmacology International Journal* 6(1), 42–47.

Sebho, H.K. and Baraki, Y. (2018) Challenges and prospects of honey bee production in Ethiopia: A review. *Journal of Biology, Agriculture and Healthcare* 8(13), 72–77.

Staveley, J.P., Law, S.A., Fairbrother, A. and Menzie, C.A. (2014) A causal analysis of observed declines in managed honey bees (*Apis mellifera*). *Human and Ecological Risk Assessment: An International Journal* 20(2), 566–591. DOI: 10.1080/10807039.2013.83126.
Teferi, K. (2018) Status of beekeeping in Ethiopia—A review. *Journal of Dairy and Veterinary Sciences* 8(4), 555743. DOI: 10.19080/JDVS.2018.08.555743.
Tesfaye, B., Dadi, G. and Gelgelu, T. (2017) Assessment of honeybee enemies (pests and predators) in Bale zone, southeastern Ethiopia. *Journal of Agricultural Extension and Rural Development* 9(4), 53–61. DOI: 10.5897/JAERD2016.0844.
Traynor, K., Mondet, F.R., de Miranda, J., Techer, M., Kowallik, V. *et al.* (2020) *Varroa destructor*: A complex parasite, crippling honey bees worldwide. *Preprints* 2020, 2020020374. DOI: 10.20944/preprints 202002.0374.v2.
van Engelsdorp, D., Traynor, K., Andree, M., Lichtenberg, E.M., Chen, Y., Saegerman, C. and Cox-Foster, D.L. (2017) Colony collapse disorder (CCD) and bee age impact honey bee pathophysiology. *PLoS One* 12(7), e0179535. DOI: 10.1371/journal. pone.0179535.
Wakgari, M. and Yigezu, G. (2021) Honeybee keeping constraints and future prospects. *Cogent Food & Agriculture* 7(1), 1872192. DOI: 10.1080/23311932.2021.1872192.

Annex Table 11.1. Natural honey indicators and their percentage annual rate of growth (ROG) in the top ten producing countries of the world. Source: constructed by the lead author from FAO (2021).

	Indicators (2015–2019)			ROG (2000–2019)		
Country	000 colonies	kg/colony	000 MT	Number of colonies	kg/colony	Production
India	12,080	5	66	1.50	0.17	1.66
China	9,180	55	500	1.10	2.95	4.09
Türkiye	7,975	14	109	4.29	–1.07	3.16
Iran	7,289	10	72	7.67	–0.66	6.95
Ethiopia	6,100	9	52	3.04	–0.56	1.91
Russian Federation	3,299	20	66	–0.22	2.03	1.80
Argentina	2,976	24	71	0.27	–1.55	–1.28
Tanzania	2,956	10	30	0.67	0.23	0.90
USA	2,752	26	71	0.55	–1.64	–1.10
Spain	2,287	9	26	–[a]	–	–
World total/average	88,847	23	1,869	1.47	1.08	2.40

[a]Dashes indicate data not available.

Annex Table 11.2. Beeswax indicators and their percentage annual rate of growth (ROG) in the top ten producing countries of the world. Source: constructed by the lead author from FAO (2021).

	Indicators (2015–2019)			ROG (2000–2019)		
Country	000 colonies	kg/colony	000 MT	Number of colonies	kg/colony	Production
India	12,080	2.08	25	1.50	0.24	1.75
China	9,180	–[a]	–	1.10	–	–
Türkiye	7,975	0.60	1	4.29	–2.39	0.91
Iran	7,289	–	–	7.67	–	–
Ethiopia	6,100	0.90	6	3.04	–0.67	2.47
Russian Fed.	3,299	–	–	–0.22	–	–
Argentina	2,976	1.64	5	0.27	–0.03	0.26
Tanzania	2,956	0.60	2	0.67	–0.98	0.29
USA	2,752	0.60	2	0.55	–0.50	–0.17
Spain	2,287	0.34	1	–	–	–
World total/avg.	88,847	2.26	67	1.47	–1.08	0.85

[a]Dashes indicate data not available.

Annex Table 11.3. Per capita consumption and dietary energy supply of natural honey and their respective annual rate of growth (ROG) in Ethiopia compared to selected countries in the world. Source: constructed by the lead author from FAO (2021).

	Per capita consumption		Dietary energy	
Country	kg/person/yr	ROG (%)	kcal/person/day	ROG (%)
CAR	3.49	0.43	30	0.44
Greece	1.55	0.41	13	0.49
Switzerland	1.34	0.33	11	0.40
Ukraine	1.32	0.93	11	0.90
New Zealand	1.16	–2.68	10	–2.60
Slovenia	1.14	1.52	9	1.42
Samoa	1.10	–3.69	9	–3.57
Germany	1.01	–0.74	8	–0.46
Albania	0.93	14.65	8	–
Bosnia and Herzegovina	0.92	11.33	8	12.03
Montenegro	0.92	–	8	–
Slovakia	0.83	4.52	7	4.47
Uruguay	0.82	10.02	7	–
Georgia	0.80	12.51	7	–
Czechia	0.79	2.60	7	2.86
Canada	0.77	–1.24	6	–1.08
Sweden	0.77	2.63	6	2.72
Ethiopia	0.50	0.72	4	0.29
World average	0.35	0.78	4	0.73

Dashes indicate data not available.
CAR = Central African Republic.

12 What Modern Science Can Offer to Accelerate Sustained Agricultural Productivity: The Role of Biotechnology

Tsedeke Abate[1]* and Endashaw Bekele[2]
[1]*Homegrown Vision, Addis Ababa, Ethiopia;* [2]*Addis Ababa University, Addis Ababa, Ethiopia*

Abstract

This chapter provides an assessment of progress made in the application of various aspects of biotechnology in Ethiopia. The country has made substantial progress in building institutions' research capacity that now includes close to 300 researchers. Applications of tissue culture, disease cleaning, molecular markers and biofertilizer development have been successfully used in one or more of 32 plant species tested in the country. Attempts are now being made to introduce genetically modified crops, but there is no scientific evidence that these have significant yield advantage and other desirable traits over the conventionally bred available crop varieties/hybrids. Institutions in Ethiopia and elsewhere in Africa need to adopt the emerging biotechnology tools, such as gene editing using CRISPR-Cas9 or TALENs that may not involve transfer of foreign genes, to help break the low-yield deadlock of important traditional crops.

Introduction

Ethiopia's concerted attempt to modernize its agriculture started with the establishment of the Imperial Ethiopian College of Agriculture and Mechanical Arts (IECAMA) in 1952 (with the Ambo and Jimma Agricultural Technical schools established much earlier). IECAMA's initial mandates included teaching, research and extension. The mandate changed with the establishment of the Ethiopian Institute of Agricultural Research (EIAR) in 1966 and transfer of extension to the Ministry of Agriculture in the mid-1960s.

The Ethiopian agricultural research and development system has undergone substantial structural changes over the years. The EIAR's initial efforts concentrated on institution building, i.e. infrastructure and human capacity. Through the 1990s EIAR followed conventional breeding techniques and was able to release a large number of crop varieties and management practices that are still impacting on productivity today.

The serious attempts and efforts made in developing agricultural biotechnology by the Ethiopian government and the Ethiopian people with their diverse culture was presented as evolutionary development of biotechnology in Ethiopia with their respected institutional memories (Mekbib, 2006; Bekele, 2012).

The role of biotechnology in improving crop and livestock development processes was realized in the late 1990s and the government made investment in infrastructure and human

* Corresponding author: tsedeke@homegrownvision.org

DOI: 10.1079/9781800626386.0012

capacity building, especially starting from the early 2000s. Abebaw *et al.* (2021) provide a detailed review of tissue culture research and development capacity in Ethiopia. Birmeta *et al.* (2022) give a detailed review of problems associated with repeatability and standardization of research results reported and offer recommendations to overcome the observed shortcomings.

The research system has made progress in developing and introducing biotechnology tools for micropropagation (tissue culture) of several crop species; use of molecular markers for hastening agricultural technology development and release; and use of this noble technique for disease diagnosis and cleaning to produce disease-free planting materials.

While there are no issues regarding the use of biotechnology tools for the above purposes, there has been a raging debate surrounding the use of genetically modified (GM) crops in Ethiopia in recent years. The aim of this chapter is to provide an overview of the status of biotechnology advances in this country and share experiences of GM crops use from selected countries to create public awareness and inform priority setting and investment in biotechnology in this country.

Results

Current capacity

Ethiopia has made commendable progress in building its agricultural biotech institutions, as reviewed by Abebaw *et al.*, 2021. The centrepiece of its infrastructure capacity building is the National Biotechnology Research Center at Holetta. Bekele (2015) lists 15 public universities with varying levels of research and development capacity and facilities: Adama Science and Technology University, Addis Ababa University, Addis Ababa Science and Technology University, Arba-Minch University, Ambo University, Axum University, Bahir Dar University, Dilla University, Haramaya University, Hawassa University, University of Gondar, Jimma University, Mekelle University, Wolkite University and Wollega University. At present, there are two private agricultural biotech companies in Ethiopia. These are the Animal Production and Productivity Improvement Services Plc and the Nares Biotech and Industry.

Table 12.1 presents the human capacity for biotech research and development in Ethiopia. There are close to 300 personnel working in biotechnology and related areas of research and development. Of these, 46%, 39% and 15%, respectively, are PhD, MSc and BSc degree holders.

Addis Ababa University has the largest share of personnel; its work covers agriculture, health and other biotechnology research. Similarly, EIAR has the largest number of researchers in agricultural biotech (Table 12.1). EIAR and the regional agricultural research institutes (RARIs) conduct research on agricultural biotechnology for both crops and livestock.

Target plants

We present the list of plants targeted for biotechnology research in Ethiopia in Table 12.2. Out of 37 plant species included in biotech research in Ethiopia so far, 27 (73%) are food crops and 10 (27%)

Table 12.1. Agricultural biotechnology research capacity in Ethiopia. Source: constructed by the authors from various sources.

	Qualification			
Organization	BSc	MSc	PhD	Grand total
EIAR centres	33	25	4	62
Holetta	19	11	3	33
Ambo	6	7	1	14
Jimma	3	3	0	6
Melkassa	3	3	0	6
Debre-Zeit	2	1	0	3
RARIs	12	4	3	19
SARI	4	2	1	7
TARI	3	1	1	5
ORARI	2	1	1	4
ARARI	3	0	0	3
Universities	1	87	129	217
Addis Ababa	0	70	116	186
Haramaya	0	8	8	16
Ambo	0	4	2	6
Wolkite	0	4	1	5
Jimma	1	1	2	4
Grand total	46	116	136	298

ARARI = Amhara Regional Agricultural Research Institute; EIAR = Ethiopian Institute of Agricultural Research; ORARI = Oromiya Regional Agricultural Research Institute; RARI = regional agricultural research institute; SARI = Southern Agricultural Research Institute; TARI = Tigray Agricultural Research Institute.

Table 12.2. Plant species targeted for implementing biotech knowledge in Ethiopia.

Plant name			
Common/local	Scientific	Main target	Source
Black pepper	*Piper nigrum*	TC	Abraham (2009)
Cardamom	*Elettaria cardamom*	TC	Abraham (2009)
Ethiopian cardamom	*Aframomum corrorima*	TC	Abraham (2009)
Geranium	*Geranium* sp.	TC	Abraham (2009)
Ginger	*Zingiber officinale*	TC	Abraham (2009)
Grapevine	*Vitis vinifera*	TC	Abraham (2009)
Pineapple	*Ananaus comosus*	TC	Abraham (2009)
Ethiopian potato	*Plectranthus edulis*	TC	Hanumanthaia and Alemu (2021)
Sweet potato	*Ipomoea batatas*	TC	Abraham (2009)
Banana	*Musa* spp.	TC; VC	Abraham (2009)
Brassicas	*Brassica* spp.	TC; VC	Abraham (2009)
Cassava	*Manihot esculenta*	TC; VC	Abraham (2009)
Citrus	*Citrus* spp.	TC; VC	Abraham (2009)
Garlic	*Allium sativum*	TC; VC	Abraham (2009)
Noug	*Guizotia abyssinica*	TC; VC	Abraham (2009); Hanumanthaia and Alemu (2021)
Potato	*Solanum tuberosum*	TC; VC	Abraham (2009)
Teff	*Eragrostis tef*	TC; VC	Abraham (2009)
Coffee	*Coffea arabica*	TC; disease cleaning; molecular markers	Abraham (2009); Hanumanthaia and Alemu (2021)
Aloe	*Aloe vera*	Molecular markers	Hanumanthaia and Alemu (2021)
Artemisia	*Artemisia annua*	Molecular markers	Hanumanthaia and Alemu (2021)
Broom rape	*Orobanche crenata*	Molecular markers	Hanumanthaia and Alemu (2021)
Chilli pepper	*Capsicum* spp.	Molecular markers	Hanumanthaia and Alemu (2021)
Ere meraat	*Aloe percarsa*	Molecular markers	Hanumanthaia and Alemu (2021)
Jatropha	*Jatropha curcas*	Molecular markers	Hanumanthaia and Alemu (2021)
Lupin	*Lupinus* spp.	Molecular markers	Hanumanthaia and Alemu (2021)
Napier grass	*Cenchrus purpureus*	Molecular markers	Hanumanthaia and Alemu (2021)
Rice	*Oryza sativa*	Molecular markers	Hanumanthaia and Alemu (2021)
Securidaca	*Securidaca longipedunculata*	Molecular markers	Hanumanthaia and Alemu (2021)
Sesame	*Sesamum indicum*	Molecular markers	Hanumanthaia and Alemu (2021)
Sorghum	*Sorghum bicolor*	Molecular markers	Hanumanthaia and Alemu (2021)
Sugarcane	*Saccharum officinarum*	Molecular markers	Hanumanthaia and Alemu (2021)
Wheat	*Triticum* spp.	Molecular markers	Hanumanthaia and Alemu (2021)
Yeheb	*Cordeauxia edulis*	Molecular markers	Hanumanthaia and Alemu (2021)
Soybean	*Glycine max*	Microbial fertilizer	EIAR (2017)
Enset	*Ensete ventricosum*	TC; disease cleaning; GMO (vs bacterial wilt)	Abraham (2009); Hanumanthaia and Alemu (2021)
Cotton	*Gossypium hirsutum*	GMO, insect resistance	Hanumanthaia and Alemu (2021)
Finger millet	*Eleusine coracana*	GMO, insect resistance	Hanumanthaia and Alemu (2021)
Maize	*Zea mays*	GMO, insect resistance	Hanumanthaia and Alemu (2021)

GMO = genetically modified organism; TC = tissue culture; VC = virus cleaning.

are non-food crop plant species. The latter group includes medicinal plants (e.g. *Securidaca longipedunculata*, *Artemisia annua*), ornamentals (*Geranium* sp.), parasitic weed (*Orobanche crenata*), forage (*Cenchrus purpureus*) and biodiesel (*Jatropha curcas*).

The majority of biotechnology work in Ethiopia has targeted tissue culture for a rapid multiplication of planting materials, virus and other diseases cleaning, use of molecular markers for speeding up the crop variety development process, and development of biofertilizer. In general, there is no controversy regarding the non-transgenic application aspect of biotechnology (Dwivedi *et al.*, 2018).

Four crops (i.e. cotton, *enset*, maize and finger millet) have been targeted for GM crops in Ethiopia (Table 12.2). Each of these crops has been dealt with in detail in earlier chapters under their respective crop groups. This chapter provides a quick summary of the current state of knowledge on the management of each crop, in order to put the GM crop work agenda into perspective.

Cotton

Cotton in Ethiopia is grown mostly under irrigated commercial agriculture, even though there are pockets of rain-fed production by smallholders in some parts of the country. Historically, insect pests, particularly the bollworms complex – including African bollworm (*Helicoverpa armigera*) and the pink bollworm (*Pectinophora gossypiella*) – were the key pests limiting cotton production in this country. An integrated pest management (IPM) programme based on strict adherence to a closed (or dead) season (achieved by removing crop residue and avoiding growth of any crop or wild host plant species at least for 1 month after harvest), coupled with the destruction of larvae on a trap crop (cotton strips surrounding the main fields, planted in mid-April), timely sowing and judicious use of effective pesticides (only when necessary) kept the pests below damaging level (Abate, 1982, 1996). By using this approach, Ethiopia has avoided the 'pesticide treadmill' phenomenon that occurred in many of the major cotton-growing countries in the 1950s and 1960s. There has never been excessive use of pesticides in Ethiopia.

The current attempts by EIAR are aimed at introducing the *Bacillus thuringiensis* (Bt) cotton from India where the partner company intends to import seed developed there every year. There is no evidence that there have been background studies to justify the use of Bt cotton in Ethiopia in the first place.

Enset

Enset (Ensete ventricosum) is a banana-like tree crop (the pulverized and fermented mixture of the trunk and rhizome is eaten), with its centre of origin in Ethiopia. It is grown in south-central, southern and southwestern Ethiopia and feeds perhaps up to 40 million people across the country. It has one of the few African indigenous sustainable cropping systems. In spite of its huge importance and potential expansion to other parts of the country and beyond, *enset* has not received government investment commensurate with its stature. The production system is not well understood even though its ethnobotany has been well covered so far. There are no decent statistics on this crop.

The bacterial wilt disease (caused by *Xanthomonas musacearum*) is an important constraint to *enset* production, especially in areas outside the optimum adaptation zone of the crop. Serious outbreaks and epidemics rarely occur in the cooler altitudes 1800 m above sea level where the crop is widely grown.

The *enset* GM crop work planned by EIAR aims to develop clones resistant to the above disease. However, critics pose several questions regarding this plan. These shall be elaborated in the discussion section later in this chapter.

Maize

Maize is the most important food security crop grown by nearly 10 million smallholders, the highest of all crops in Ethiopia. The national research system has developed and released high-yielding and widely adapted varieties such as BH 611, BH 546 and BH 547 over the last decade. These are extensively adopted and grown across the country.

Maize productivity in Ethiopia has doubled in less than 10 years (Abate *et al.*, 2015), with the 2020 national average yield of 4.24 MT/ha (FAO, 2022), only second highest to South Africa on the continent, whereas the African average was 2.10 MT/ha.

From a historic perspective, stalk borers, including *Busseola fusca*, *Chilo partellus* and *Sesamia calamistis*, were the major pests that constrained maize (and sorghum) production in Ethiopia, but they are well managed by an IPM programme established over the last three decades (Gebre-Amlak, 1988; Abate, 1996; Azerefegne *et al.*, 2001; Abate, 2006, 2011).

Finger millet

Finger millet has its centre of origin and centre of diversity in Ethiopia (Zeven and De Wet,

1982). According to the Central Statistical Agency (CSA), an estimated 1.7 million smallholder households grew finger millet on close to half a million ha of land and produced more than 1 million MT of grain annually in 2016–2020. The average area per smallholder household for this crop is about 0.27 ha, and each household produces an average of close to 600 kg/yr. The research system has released 21 improved varieties between 1998 and 2017 (MOALR, 2017). Records show that finger millet in Ethiopia is grown mainly for food security mostly in Amhara, Oromiya, and Tigray regional states. It is highly valued for its *injera*–making quality (Anchala *et al.* 2006) as a substitute for or mixed with *teff*.

The average yield for 2016–2020 was estimated at 1200 kg/ha (FAO, 2022), even though the CSA data report a much higher figure. According to the FAO (2022), the yield gain for finger millet in Ethiopia was 88 kg/ha/yr and highly significant ($P<0.01$, $R^2=0.95$).

Hanumanthaia and Alemu (2021) reported the use of six primers for genetic diversity investigation of 80 finger millet accessions, where it was observed that there was high genetic diversity that can be used for improving the productivity and conservation. It was concluded that research on genetic modification of finger millet in Ethiopia is not a priority.

The controversy

There have been controversies over the use of GM crops since they were first commercialized in the USA in 1996. There continues to be very vocal proponents and opponents of this technology worldwide. NAS (2016) published a comprehensive review of potential negative effects and benefits of GM crops.

In general, the committee avoided any sweeping statements either way. Here are some of the salient points of their findings: (i) GM soybean, cotton and maize had generally favourable economic outcomes, but outcomes were variable depending on pest abundance, farming practices and agricultural infrastructure; (ii) Bt crops generally decreased yield losses and the use of insecticides; (iii) damaging levels of resistance evolved where resistance management strategies were not followed; (iv) herbicide spraying with glyphosate resulted in 'small increases in yield'; and (v) there was no substantiated evidence on impacts on livestock and human health. Detailed reviews of GM crops in Africa can be found in Falc-Zepeda *et al.* (2013).

Proponents claim so many advantages and benefits of GM crops compared to conventionally bred varieties and hybrids. Examples include reduced impacts of agriculture on biodiversity (Adenle, 2011; Carpenter, 2011); increased crop productivity, conservation of crop biodiversity, providing better environment, reducing CO_2 emissions and helping poverty alleviation (ISAAA, 2018); and higher insect biodiversity and a powerful tool for IPM (Romeis *et al.*, 2018). Paarlberg (2013) wrote that GM crops are Africa's choice. Raman (2017) claimed that genetic modification started 10,000 years ago in southwest Asia where farmers first bred plants through artificial selection, the implication being that GM crops are harmless. This is a misleading claim.

On the opponents' side, Kuyek (2002) argued that GM crops are only suited to the green revolution technologies – i.e. irrigated agriculture with access to chemical inputs (such as mineral fertilizer and pesticides) – and they do not benefit smallholder farmers. FEI (2004) pointed out that the promises of GM crops have not materialized, and many questions are being asked about its relevance. Makanya (2004) cited a dozen reasons to reject GM crops in Africa, including contamination of non-GM crops, fostering dependence on a corporate seed supply, favouring industrial agriculture systems and the destruction of biodiversity.

Fagan *et al.* (2014), Kuper (2014) and Gaffney et al. (2019) present well-researched and comprehensive scientific reviews of GM crops.

Experiences with GM crops in other countries

Dowd-Uribe and Schnurr (2016) reviewed the status of *Bacillus thuringiensis* (Bt) cotton in South Africa and Burkina Faso. The first commercially grown GM crop, Bt cotton, on the African continent was introduced to South Africa in the rural district of Makhathini Flats, near the border with Mozambique, in 1997. Initial successes with increased yield and profits were hailed as an example of GM crop success in smallholder agriculture. However, contrary to the report by Areal *et al.* (2012), this was not sustainable because the company that provided the farmers with credit to purchase the more expensive Monsanto seed went bankrupt. The number of smallholder farmers growing Bt cotton in the 2014/15 season was below 5% of what it was during the peak seasons when it was introduced. Most growers abandoned the technology within 10 years.

The authors suggested three lessons from the South African experience for other countries who intend to introduce GM crops in Africa. First, there needs to be enabling institutions; second, evaluations of GM crops performance should be based on long-term results as short-term results might be misleading; and third, there is a need for an independent body (rather than the company itself) to assess the performance of GM crops.

The government of Burkina Faso signed an agreement with Monsanto in 2003 to test Bt cotton in experimental field trials. It was realized that the Monsanto cotton quality was inferior to the native variety, so the company inserted the Bt gene into the local variety. The new crop covered 70% of Burkinabe cotton area by 2013, with a yield advantage of 22% over the conventional variety. However, the Bt cotton showed a 40% decline in staple length between 2005/6 and 2013/14 seasons, which means that the prices went down because of the poor staple length and low ginning ratio. Cotton growers in Burkina Faso sued Monsanto for US$280 million for their loss of revenue.

Bt cotton technology adoption surged rapidly following its approval for central and southern India in 2002 and northern India in 2005. Ninety-three per cent of India's total area was covered by the hybrid Bt cotton by 2011, replacing the conventional seed (historically, India used to grow its indigenous 'desi' cotton *Gossypium arboreum* until it shifted to the new species, *G. hirsutum*, in the 1990s). The GM cotton was credited, unfairly, for India's tripling of production and doubling of the global market share between 2002 and 2014.

However, the increase in yield was actually the result of the government policy to increase the use of fertilizers rather than the Bt cotton. The latter controlled pink bollworm (*Pectinophora gossypiella* Saunders) and resulted in reduction of pesticide use initially. In the long-term, however, resistance developed in another pest, African bollworm (*Helicoverpa armigera* [Hübner]) and there was a resurgence of non-target pests such as the cotton mealybug (*Phenacoccus solenopsis* Tinsley). As a result, cotton farmers in India spent more on pesticides in 2020 than before the introduction of Bt.

Discussion and Conclusions

When Ethiopia identified agricultural biotechnology as one of the vehicles for national socio-economic development following its Science, Technology and Industry policy, it was with the understanding of the requirements of a high level of skilled manpower, infrastructure and level of technology, as well as the required institutional structures and arrangements as detailed in the National Biotechnology Roadmap. However, as of 2023, Ethiopia does not have a holistic and coherent policy for biotechnology as anticipated in the roadmap, and is left only with patchy, specific regulations and unclear institutional relationships at international, regional and national levels. Thus, a shared vision by various institutions for a collaborative approach and effective mechanisms for implementation of responsibilities and control is desperately required.

The recently developed research and development (R&D) programmes based on the roadmap developed earlier for agricultural biotechnology consists of improvement in crop yield to control biotic and abiotic stresses, as well as increases in nutritional qualities. These programmes underline the need for any research to be contextualized, taking into account the various systems of agroecological diversity of the country and the need of the people in these ecologies for sustainable agricultural production and productivity. The programmes are also expected to be linked with the new advanced bioeconomy that require a critical mass of trained manpower, infrastructure for the initiation of competitive agricultural research programmes interfaced with other emerging technologies in order to occupy a global niche in the growing market of agricultural biotechnology products. The developed R&D programme recommends periodic assessments of situations at the global, continental and country level for sustainable utilization of the agricultural biotechnology in Ethiopia. The bioeconomy suggested includes the use of plants for biorefinery (PRI, 2009) manufacturing models and the '4f' (food, feed, fuel, fibre) for the new agricultural development view. Thus, value is added to primarily produced food, feed, fuel and fibre through various research-based agricultural biotechnology processes, particularly when the technological base is widened and developed.

Knowledge of important technologies like gene editing and pyramiding and the landscape of genomics and proteomics data in relation to the structure of plants, animals and microbes are also suggested to be carefully utilized since these impact agricultural biotechnology research results of economic importance with short, medium- and long-term effects. As these technologies are interconnected with systems biology, mathematical modelling, epigenetic imaging, regenerative biology, synthetic biology, metabolomics pathways engineering and nano-agricultural sciences, considering them through avoiding institutional fragmentations is recommended for the development of the next-generation lead platforms for agricultural biotechnology.

In addition to what has been stated above, since the agricultural system is defined by its structure, function, equilibrium and change, and interaction with the natural ecosystem (Bekele, 2007), it is suggested to carefully consider the place of agriculture and agricultural technology within the current Ethiopian agroecology landscape before addressing agricultural biotechnology R&D. Such consideration minimizes hard trade-offs among farm productivity, stability and sustainability, embracing a widening span of disciplines across the natural and social sciences including the existing rich indigenous knowledge (Mekbib, 2006). Agricultural biotechnology thus has to consider agricultural system complexity for its sustainable production. The Ethiopian environment is very heterogeneous and consists of numerous cultures and a wealth of genetic resources. In all these environments, food and nutrition security is associated with several spectrums of issues. Such an approach begs for a change in the existing Ethiopian science and technology landscape with appropriate human, institutional and infrastructural critical capacities.

Agricultural biotechnology research programmes are expected to focus on the country's biological resources, including resources in agricultural systems. They should incorporate the tried and tested application of knowledge to develop solutions to some of the pressing challenges. Ethiopian agriculture benefits from some modern biosciences, and such benefits integrated with the collective indigenous knowledge and past research achievements are not only cardinal to the survival of the nation but can also provide models for reversing some of the environmentally harmful effects of modern commercial farming practices (Bekele, 2012).

It is argued that GM *enset* is not a priority at present. First, even though the ethnobotany is relatively better understood, we do not yet have total understanding on this extremely complex, sustainable indigenous system; there are not even decent production statistics on it. Second, Ethiopia is the only country where *enset* is domesticated as food crop and there is no experience elsewhere to learn from others; the perceived risks are too high. For example, ISAAA (2021) lists 31 GM crops commercialized worldwide but there is not a single one even remotely related to *enset* (such as banana) on that list (Annex Table 12.1). Fourth, it is questionable whether there is currently adequate capacity and skills to undertake such work in Ethiopia.

With regards to GM cotton, the initiative was started without sufficient background knowledge on the existing insect pest management system. An IPM programme based on strict adherence to closed (or dead) season, coupled with the destruction of larvae on a trap crop (cotton strips surrounding the main fields, planted one month before the main season[1]), timely sowing and judicious use of effective insecticides (only when necessary) kept the bollworm populations below damaging levels (Abate, 1982, 1996). By using this approach, Ethiopia has avoided the 'pesticide treadmill' phenomenon that plagued many of the major cotton-growing countries in South and North America in the 1950s and 1960s.

Even if there were enough justification for the launch of GM cotton, one would not expect much change in the productivity of the Bt cotton, as has been witnessed in South Africa, Burkina Faso (Dowd-Uribe and Schnurr, 2016) and India (Kranthi and Stone, 2020). Furthermore, the corporate-controlled Bt cotton seed would be too expensive to use. The primary issue for cotton production in Ethiopia should be good crop management rather than GM crops.

Just as in the case of cotton, the GM maize work in Ethiopia was also launched without a thorough background check and enough justification. First, the cereal stalk borers (the target of this initiative) were well managed by sound IPM programmes (Gebre-Amlak, 1988; Abate, 1996; Azerefegne *et al.*, 2001; Abate, 2006, 2011). There have not been cases where the pest populations have reached damaging levels similar to the 1990s and earlier. Second, there are adequate locally developed maize hybrids in Ethiopia that outperform the Bt maize. Third, the corporate-controlled seed would be too expensive for smallholder maize growers to afford. Fourth, similar attempts to introduce this technology in Kenya and Tanzania have failed.

Finally, it should be noted that there is not a single GM crop that has been adopted and sustained in smallholder agriculture on the whole of the African continent. The so-called 'big four' GM crops (maize, soybean, cotton and canola) worldwide occupied 99% of the 191.7 million ha of the area in 2018, and 91% of the total area was occupied by the USA, Brazil, Argentina, Canada and India (ISAAA, 2018). This is indicative of the fact that GM crops are more suited to commercial agriculture than to smallholder situations. It appears that most of the European Union countries have remained free of GM crops since the beginning of their commercialization (Burke, 2004). Overall, the available literature (Rock and Schurman, 2020) suggests that agricultural biotechnology focusing on GM crops has yet to see success on the African continent.

Note

[1] Recommended planting dates for upper and lower Awash are mid-April to mid-May and mid-May to mid-June, respectively.

References

Abate, T. (1982) Cotton pests and their control in Ethiopia. In: Abebe, M. (ed.) *Cotton Production under Irrigation in Ethiopia*. IAR, Addis Ababa, Ethiopia, pp. 111–128.

Abate, T. (1996) Pest management research in Ethiopia: An overview. In: Gebrekidan, B., Amirault, J-P. and Abate, T. (eds) *Proceedings of the IPM Networking in Sub-Saharan Africa Workshop, 14-16 October 1996*. Addis Ababa/Nazareth, Ethiopia.

Abate, T. (2006) IPM in Ethiopia: The current status. In: Bekele, E., Azerefegne, E. and Abate, T. (ed.) *Facilitating the Implementation and Adoption of Integrated Pest Management (IPM) in Ethiopia*. Drylands Coordination Group (DCG), Oslo, Norway.

Abate, T. (2011) Maize stalk borers of Ethiopia: Quantitative data on ecology and management. In: Regassa, W.M., Twumasi-Afriye, S., Wolde, L., Tadesse, B., Demissie, G., *et al.* (eds) *Meeting the Challenges of Global Climate Change through Innovative Maize Research*. EIAR/CIMMYT, Addis Ababa, Ethiopia, pp. 174–184.

Abate, T., Shiferaw, B., Menkir, A., Wegary, D., Kebede, Y. *et al.* (2015) Factors that transformed maize productivity in Ethiopia. *Food Security* 7, 965–981. DOI: 10.1007/s12571-015-0488-z.

Abebaw, Y.M., Tobiaw, D.C., Birhan Addisie Abate, B.A., Eshete, B.K., Seymour, S.K. and Tesfaye, K. (2021) Plant tissue culture research and development in Ethiopia: A case study on current status, opportunities, and challenges. *Advances in Agriculture* 2021, 9979549. DOI: 10.1155/2021/9979549.

Abraham, A. (2009) Agricultural biotechnology research and development in Ethiopia. *African Journal of Biotechnology* 8(25), 7196–7204.

Adenle, A.A. (2011) Response to issues on GM agriculture in Africa: Are transgenic crops safe? *BMC Research Notes* 4, 388.

Areal, F.J., Riesgo, L. and Rodriguez, E. (2012) Economic and agronomic impact of commercialized GM crops: A meta-analysis. *Journal of Agricultural Science* 151, 7–33. DOI: 10.1017/S0021859612000111.

Anchala, C., Kidane, H.S., Mulatu, T. (2006) The impact of improved finger millet technology promotion in the central Rift Valley of Ethiopia. In: Abate, T. (ed.) *Successes with Value Chain. Proceedings of Scaling up and Scaling out Agricultural Technologies in Ethiopia: An International Conference*, 9–11 May 2006, Addis Ababa. EIAR, Addis Ababa, Ethiopia, 275 pp.

Azerefegne, F., Kitaw, D. and Amare, B. (2001) Major insect pests of maize and their management. In: Mandefro, N., Tanner, D. and Twumasi-Afriyie, S. (eds) *Enhancing the Contribution of Maize to Food Security in Ethiopia: Proceedings of the Second National Maize Workshop of Ethiopia, 12–16 November 2001*. EARO and International Maize and Wheat Improvement Center (CIMMYT), Addis Ababa, Ethiopia, pp. 89–96.

Bekele, E. (2007) *Enset and Musa Culture Indigenous Knowledge International Symposium*. Kyoto University, Kyoto, Japan.

Bekele, E. (2012) African crop diversity for global food security and safe livelihood. In: Inter-Graduate School Program for Sustainable Development and Survivable Societies. First International Advisors Conference, Kyoto University, Kyoto, Japan.

Bekele, E. (2015) Ethiopian Biotechnology Education, Training and Human Resources Development Roadmap. Policy Research Institute, Addis Ababa, Ethiopia.

Bekele, E., Azerefegne, E. and Abate, T. (eds) (2006) *Facilitating the Implementation and Adoption of Integrated Pest Management (IPM) in Ethiopia*. Drylands Coordination Group (DCG), Oslo, Norway.

Bhardwaj, A. and Nain, V. (2021) TALENs—An indispensable tool in the era of CRISPR: A mini review. *Journal of Genetic Engineering and Biotechnology* 19, 125. DOI: 10.1186/s43141-021-00225-z.

Birmeta, G., Safawo, T., Dida, M.G. and Bekele, E. (2022) Critical review on plant micropropagation of Ethiopian plants reported so far: Existing gaps, required standardization, and future research direction. *Hindawi Advances in Agriculture* 2022, 5874899. DOI: 10.1155/2022/5874899.

Burke, B. (2004) GM food and cops: What went wrong in the UK? *EMBO Reports* 5(5), 432–436.

Carpenter, J.E. (2011) Impact of GM crops on biodiversity. *GM Crops* 2(1), 7–23.

Dowd-Uribe, B. and Schnurr, M. (2016) Briefing: Burkina Faso's reversal on genetically modified cotton and the implications for Africa. *African Affairs* 115(458), 161–172. DOI: 10.1093/afraf/adv063.

Dwivedi, S.L., Siddique, K.H.M., Muhammad Farooq, M., Thornton, P.K. and Ortiz, R. (2018) Using biotechnology-led approaches to uplift cereal and food legume yields in dryland environments. *Frontiers in Plant Science* 9. DOI: 10.3389/fpls.2018.01249.

EIAR (Ethiopian Institute of Agricultural Research) (2017) National Plant Biotechnology Research: Strategic Plan (2016–2030). EIAR, Addis Ababa, Ethiopia.

Fagan, J., Antoniou, M. and Robinson, R. (2014) *GMO Myths and Truths: An Evidence-Based Examination of the Claims Made for the Safety and Efficacy of Genetically Modified Crops and Foods*, 2nd edn, Version 1.0. Earth Open Source, London.

Falck-Zepeda, J., Gruère, G. and Sithole-Niang, I. (eds) (2013) *Genetically Modified Crops in Africa. Economic and Policy Lessons from Countries South of the Sahara*. IFPRI, Washington, DC, USA.

FAO (Food and Agriculture Organization of the United Nations) (2022) Statistical Databases and Data-Sets of the Food and Agriculture Organization of the United Nations. Available at: http://www.fao.org/faostat/en/#data (accessed 5 November 2022).

FEI (Friends of the Earth International) (2004) Genetically Modified Crops: A Decade of Failure [1994–2004]. February 2004, Issue 105, Penang, Malaysia.

Gaffney, J., Bing, J., Byrne, P.F., Cassman, K.G., Ciampetti, I. *et al.* (2019) Science-based intensive agriculture: Sustainability, food security, and the role of technology. *Global Food Security* 23, 236–244.

Gebre-Amlak, A. (1988) Ecology and management of maize stalk borer, *Busseola fusca* (Fuller) (Lepidoptera: Noctuidae) in southern Ethiopia. Dissertation, Department of Plant and Forest Protection, Uppsala, Sweden.

Hanumanthaia, P. and Alemu, C. (2021) Plant biotechnology in Ethiopia: Current status, opportunities and challenges. *Plant Cell Biotechnology and Molecular Biology* 22(13 & 14), 136–156.

ISAAA (International Service for the Acquisition of Agri-biotech Applications) (2018) *Global Status of Commercialized Biotech/GM Crops in 2018: Biotech Crops Continue to Help Meet the Challenges of Increased Population and Climate Change*. ISAAA Brief No. 54. ISAAA, Ithaca, NY, USA.

ISAAA (2021) *Breaking Barriers with Breeding: A Primer on New Breeding Innovations for Food Security*. ISAAA Brief No. 56. ISAA, Ithaca, NY, USA.

Kranthi, K.R. and Stone, G.D. (2020) Long-term impacts of Bt cotton in India. *Nature Plants* 6, 188–196.

Kuper, R. (2014) The politics of GMOs. Write-up and reworking of a talk given to the Red Green Study Group on 18 January 2014 (unpublished PDF).

Kuyek, D. (2002) *Genetically Modified Crops in Africa: Implications for Small Farmers*. Genetic Resources Action International, Barcelona, Spain.

Makanya, Z. (2004) 12 reasons for Africa to reject GM crops. *Seedling* 18–22.

Mekbib, F. (2006) Farmers and formal breeding of sorghum (*Sorghum bicolor* (L.) Moench) and the implications for integrated plant breeding. *Euphytica* 152, 163–176.

MOALR (Ministry of Agriculture and Livestock Resources) (2017) Crop Variety Register Issue No. 20. Plant Variety Release, Protection and Seed Quality Control Directorate, MOALR, Addis Ababa, Ethiopia.

NAS (National Academy of Sciences) (2016) *Genetically Engineered Crops. Experiences and Prospects*. The National Academies Press, Washington, DC, USA.

Paarlberg, R. (2013) Genetically modified foods and crops: Africa's choice. In: Falck-Zepeda, J., Gruère, G. and Sithole-Niang, I. (eds) *Genetically Modified Crops in Africa. Economic and Policy Lessons from Countries South of the Sahara*. IFPRI, Washington, DC, USA.

PRI (Policy Research Institute) (2009) Biorefinery Roadmap. The Byproduct of Sugar Cane Industry. Policy Research Institute, Addis Ababa, Ethiopia.

Raman, R. (2017) The impact of genetically modified (GM) crops in modern agriculture: A review. *GM Crops & Food* 8(4), 195–2008. DOI: 10.1080/21645698.2017.1413522.

Rock, J. and Schurman, R. (2020) The complex choreography of agricultural biotechnology in Africa. *African Affairs* 119/477, 499–525. DOI: 10.1093/afraf/adaa021

Romeis, J., Naranjo, S.E., Meissle, M. and Shelton, A.M. (2018) Genetically engineered crops help support conservation biological control. *Biological Control* 130, 136–154. DOI:10.1016/j.biocontrol.2018.10.001.

Zeven, A.C. and de Wet, J.M.J. (1982) *Dictionary of cultivated plants and their regions of diversity. Excluding most Ornamentals, Forest Trees and Lower Plants*. Centre for Agricultural Publishing and Documentation, Wageningen, The Netherlands.

Annex Table 12.1. Genetically modified crops worldwide. Source: adapted by the authors from ISAAA (2021).

Name		
Common/English	Scientific	Group/use
Maize	*Zea mays*	Cereal
Rice	*Oryza sativa*	Cereal
Wheat	*Triticum aestivum*	Cereal
Cotton	*Gossypium* spp.	Fibre
Alfalfa	*Medicago sativa*	Fodder
Chicory	*Chicorium intybus*	Fodder
Apple	*Malus domestica*	Fruit
Melon	*Cucumis melo*	Fruit
Papaya	*Carica papaya*	Fruit
Pineapple	*Ananas comosus*	Fruit
Plum	*Prunus domestica*	Fruit
Creeping bentgrass	*Agrostis stolonifera*	Lawn (sports fields)
Common bean	*Phaseolus vulgaris*	Grain legume
Cowpea	*Vigna unguiculata*	Grain legume
Sugarcane	*Saccharum officinarum*	Industrial
Tobacco	*Nicotiana tabacum*	Industrial
Argentine canola	*Brassica napus*	Oilseed
Flax	*Linum usitatissimum*	Oilseed
Polish canola	*Brassica rapa*	Oilseed
Safflower	*Carthamus tinctorius*	Oilseed
Soybean	*Glycine max*	Oilseed
Carnation	*Dianthus caryophyllus*	Ornamental
Petunia	*Petunia hybrida*	Ornamental
Rose	*Rosa hybrida*	Ornamental
Potato	*Solanum tuberosum*	Root crop
Eucalyptus	*Eucalyptus* spp.	Tree
Poplar	*Populus* sp.	Tree
Eggplant	*Solanum melongena*	Vegetable
Squash	*Cucurbita pepo*	Vegetable
Sweet pepper	*Capsicum annuum*	Vegetable
Tomato	*Solanum lycopersicum*	Vegetable

13 Protecting the Ethiopian Flora from the Ravages of Arthropod Pests

Gashawbeza Ayalew,[1] Tsedeke Abate,[2*] Ferdu Azerefegne,[3] Melaku Wale[4] and Muluken Goftishu[5]

[1]Ethiopian Institute of Agricultural Research, Melkassa Research Center, Adama, Ethiopia; [2]Homegrown Vision, Addis Ababa, Ethiopia; [3]Hawassa University, Hawassa, Ethiopia; [4]Bahir Dar University, Bahir Dar, Ethiopia; [5]Haramaya University, Haramaya, Ethiopia

Abstract

This chapter reviews the progress in entomological research on major cultivated crops in Ethiopia and offers suggestions for future priorities. More than 740 arthropod pest species that attack a variety of cultivated crops in Ethiopia have been recorded, of which about 20% are regarded as major. The focus in the 1980s and 1990s was on the development and implementation of integrated pest management approaches, with a major emphasis on the use of cultural practices and host plant resistance. The efforts on biological control are limited mainly to the determination of associated indigenous natural enemies. The need for prioritizing pest problems based on the country's development agenda and available resources is stressed by listing pest species that need to be given special emphasis by research in the immediate future. The research system needs to gear itself towards tackling pest problems associated with expanding commercial farming using existing and emerging scientific advances.

Introduction

Since the establishment of the Ethiopian Institute of Agriculture Research (EIAR) in the late 1960s, crop protection research has been in the forefront of fighting crop losses due to arthropod, disease and weed pests. Disease and weed pests are covered in Chapters 14 and 15, respectively, in this book. Here, we present the current state of knowledge and the way forward for arthropod (insect and mite) pest research and development in Ethiopia.

Initial work on arthropod pests focused on their identity and bionomics through the 1980s (cf. various papers in Tadesse, 2008, 2009), followed by loss assessment studies to determine the economic importance. Figure 13.1 shows estimated crop losses due to arthropod pests on selected groups of crops in Ethiopia. Chemical control was the first line of defence against these pests during this initial period, as has been customary across the continent and elsewhere. Considering that Ethiopian agriculture is largely smallholder, reliance on pesticides was not deemed tenable and major efforts were undertaken to introduce an integrated pest management (IPM) system starting from the early 1980s, not only for smallholders but also for commercial agriculture.

IPM in the Ethiopian context means the use of cultural practices such as timely sowing, crop

* Corresponding author: tsedeke@homegrownvision.org

DOI: 10.1079/9781800626386.0013

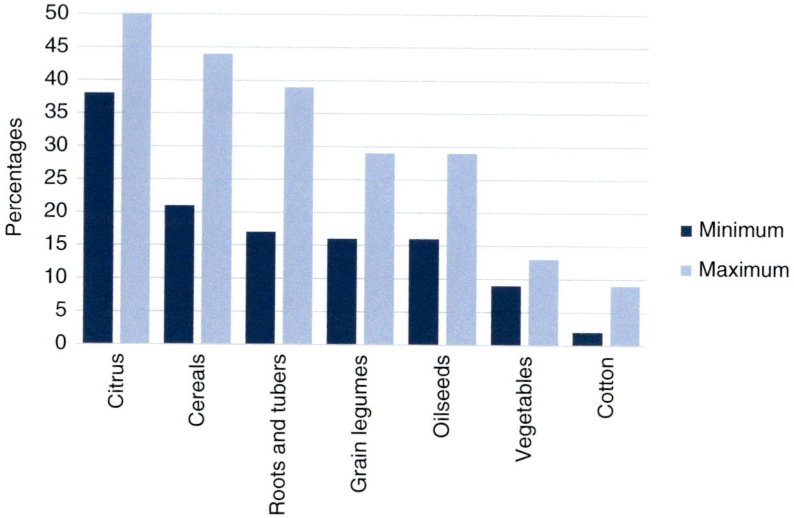

Fig. 13.1. Estimated minimum and maximum crop losses due to arthropod pests on selected group of crops in Ethiopia. Source: constructed by the authors from Getaneh *et al.* (2016).

diversity (to encourage natural biological control), host plant resistance and the use of pesticides only when needed (Abate *et al.*, 2000). Experimentation with the use of botanicals was initiated in the early 1990s with marginal successes. Trap crops were used as a means of crop diversity and in cultural practices in the overall IPM system.

The department of crop protection has been very active and dynamic in organizing comprehensive reviews of research and development studies since its early years of establishment. The first ever comprehensive review in the whole of EIAR was organized in February 1985 and its proceedings published in 1986 (Abate, 1986). A similar conference was organized in December 2006 and its proceedings were published (Tadesse, 2008, 2009). Most recently, Getaneh *et al.* (2016) reviewed crop protection research efforts and achievements at the 50th anniversary of the EIAR.

In this chapter, we make a critical assessment of the status of entomological research and development in Ethiopia and forward suggestions to inform government policy decision and priority setting.

Arthropod Pests of Economic Importance

The number of arthropod pests recorded in Ethiopia varied from a few to many, depending on the crop. In a similar manner, the proportion of pest status (major, minor, sporadic, unknown) also depended on the crop, as presented in Figure 13.2. For convenience, we divided arthropod pests into regular and emerging and/or invasive species. The latter are inadvertently introduced into the country as a result of human movement with agricultural products, among others.

Regular pests

These are pests that are known to be associated with cultivated crops for a long period of time and occur frequently and regularly. They can inflict heavy damage depending on the environment and control measures practiced. For example, stalk borers of various species regularly inflict damage on maize and sorghum at various levels in different parts of Ethiopia. Additional examples can be found in Tadesse (2008, 2009), Getaneh *et al.* (2016) and Ayalew *et al.* (2019).

There is ample literature dealing with the various aspects of management of important crop pests on different groups of crops or individual crops in Ethiopia. Providing detailed reviews of those pests is beyond the scope of this chapter. We therefore present only highlights of major milestones for arthropod pest management in Ethiopia, followed by brief

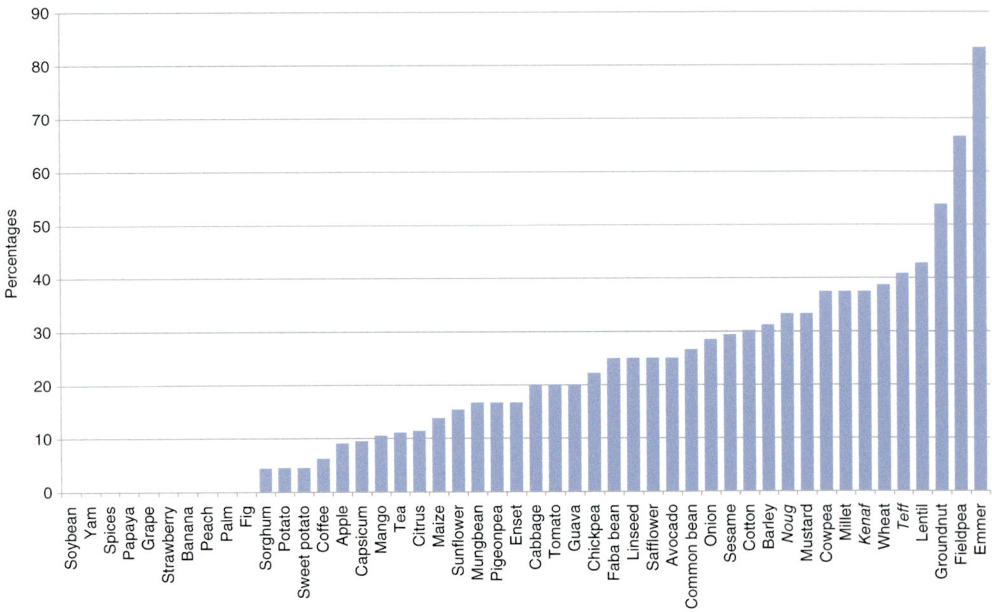

Fig. 13.2. Proportions of arthropod pests that are designated as major on selected crops cultivated in Ethiopia. Source: data from various sources.

descriptions of emerging and/or invasive pests in the following two sections.

Milestones in insect pest management research and development in Ethiopia

As pointed out earlier, there is a lot of information on several aspects of major pests of cultivated crops in Ethiopia. This information and technologies are believed to have significantly contributed to minimizing losses due to the pests and hence in the productivity of crops. A few examples of major milestones in managing crop pests in this country are highlighted below.

INTEGRATED MANAGEMENT OF RED SCALE ON CITRUS. Among fruit insect pests, the red scale *(Aonidiella aurantii)* was the most important in the late 1970s and early 1980s. Organophosphate insecticides such as methidathion were routinely applied to control this and other insects in orchards of state farms during the late 1970s without much success. IPM of these insects was developed through understanding of their population dynamics and selection of insecticides that are compatible with the IPM approach. It had been recommended that a maximum of three sprays (one or two in October–November, one in March–April) of white oil be applied annually (Abate, 1994). These recommendations have been followed and the pest has remained less important to the citrus industry of Ethiopia since then.

RESPONDING TO THE INVASION OF THE TOMATO LEAF MINER. There was a huge public outcry as almost all insecticides available on the market for controlling other vegetable insect pests did not provide any level of control against the pest threatening the continuity of tomato production in the Central Rift Valley (CRV) and other tomato-producing regions of Ethiopia following its detection in CRV in 2013 (Addis Fortune, 2013; Reporter, 2013). Supplying effective insecticides to growers to ensure continuity of the tomato industry was an urgent task. Insecticides reported effective in countries where the pest is established were obtained through different ways and evaluated in the same year, 2013. This effort led to the identification of insecticides that can give effective control (Ayalew, 2015), and national recognition was given for this achievement by the Ministry of Science and Technology (MOST, 2015). Soon after the identification of effective insecticides, studies towards IPM of the pest were initiated. Sequential application of insecticides from different classes in rotational schemes, use of pheromone lure and safe disposal of infested fruits helped to reduce the pest damage and losses on quality and quantity of produce both on research and farmers' fields (Ayalew, 2019).

CLASSICAL BIOCONTROL OF DIAMONDBACK MOTH ON CABBAGE. The diamondback moth (DBM) (*Plutella xylostella*) is a key insect pest of head cabbage with yield loss of 36–91% (Ayalew, 2006). Chemical control is the most common method of its control wherever the crop is produced. However, biological control forms an integral component of the pest's management in several countries of the world. Studies conducted on the natural enemies of the pest between 2000 and 2003 in Ethiopia identified the indigenous parasitoids associated with the pest and determined the need for introducing parasitoids with a proven track record of success in a classical biological control programme (Ayalew and Ogol, 2006). Hence, the larval parasitoid *Diadegma semiclausum* (Hymenoptera: Ichneumonidae) was imported and released in head cabbage fields of Kofele highland in June 2008. Before release, DBM numbers fluctuated between 4.2 and 11.2 per plant and parasitism ranged between 6.5% and 24.7%. DBM density declined to 2.8, 0.9 and 0.7 per plant while parasitism levels increased successively to 21%, 39% and 38% in 2008, 2009 and 2010, respectively (Ayalew and Hopkins, 2013). This decline of DBM density following the release of the introduced parasitoid and its establishment ensured production of *Brassica* spp. without pesticide use against DBM in the affected area. A national recognition was given for this achievement by the Ministry of Science and Technology (MOST, 2015).

BIOLOGICAL CONTROL OF SPIDER MITES IN COMMERCIAL FARMS. The two-spotted spider mite (*Tetranychus urticae* Koch) is a major arthropod pest of cut roses in flower farms of Ethiopia. Because of restrictions on the level of pesticide use, there was a need to develop a biological control programme. The predatory mite *Phytoseiulus persimilis* Athias-Henriot was obtained from Koppert, the Netherlands, and evaluated against the pest. Spider mite levels were lowered to nil due to the predatory effect of the introduced bioagent (Belder *et al.*, 2009). The technology was adopted by several farms in the CRV and contributed to the strengthening of biological control of greenhouse pests in Ethiopia (Getaneh *et al.*, 2016).

MANAGEMENT OF MAIZE AND SORGHUM STALK BORERS. Maize and sorghum stalk borers have been the focus of research since the start of agricultural research in Ethiopia. Commendable results have been generated on the biology, ecology and management of stalk borers and our knowledge has improved through time. The in-depth studies made on the bionomics of the insect stalk borer helped to better understand the phenology of the insect, including its population ecology, diapause and survival during the dry period (Gebre-Amlak, 1988; Yitaferu and Gebre-Amlak, 1994). The insect completes two to three generations per year, among which the second generation is the most abundant and more harmful. As a result, maize planted early in the growing season usually escapes the attack of the destructive second generation. For example, maize planted before the end of April at Hawassa does not need any insecticide treatment, while complete losses are not uncommon on late planted crops (Gebre-Amlak, 1989). It was reported that early planted maize and sorghum suffered less than the late planted in many areas of the country including Haramaya (Yitaferu and Gebre-Amlak, 1994). Moreover, a study on the oviposition site of *Busseola fusca* enabled researchers to focus the sampling on the leaf sheath of the lower half of the plant at any growth period of maize and sorghum (Azerefegne and Gebre-Amlak, 1994). On the other hand, horizontal placement of the stalks for 4–6 weeks was found to effectively destroy the diapausing larvae and reduce the level of infestation in the upcoming growing season (Gebre-Amlak, 1988). The push and pull strategy is a new addition integrated in the management of stalk borers. It encompasses planting Napier grass or *Brachiaria* around the field to attract ovipositing female stem borer moths (pull) and larvae fail to develop. *Desmodium* planted in between the rows of maize serve as the push factor. This method has been especially attractive to smallholders with crop–livestock.

Emerging and/or invasive pests

The occurrence of a new invasive species is generally noticed in countries with weak quarantine facilities after the damage is inflicted by the introduced pest. The absence of a periodically updated pest list has contributed to this problem. For example, the tomato leaf miner (*Tuta absoluta*) was believed to be introduced into Ethiopia in 2012, although its presence was noted after its outbreak in tomato fields of the Ethiopian CRV in early 2013 (Ayalew, 2015). The presence of white mango scale (*Aulacaspis tubercularis*) was also noticed after significant damage occurred on mango trees in the Wollega region of western Ethiopia in 2010 (Dawd *et al.*, 2012). The Center for Agriculture and Bioscience International (CABI) has recently launched a tool called invasive species Horizon Scanning Tool, which is a decision support aid to help users identify potential invasive species threats to a country, state or province. This helps to identify dangerous pest species to a country anywhere in the world. Quarantine pest lists can be developed by using this along with the European Plant Protection Organization global database, CABI distribution maps and literature searches on published arthropod pests lists of the target country and crop species (Ayalew and Azerefegne, 2019). Table 13.1 provides examples of recently introduced invasive insect pests of crops with significant impact on crop production and productivity in Ethiopia. This is followed by a brief account on the history of their introduction to Ethiopia and efforts made to generate information on their ecology and control methods.

Woolly whitefly

The woolly whitefly (WWF), *Aleurothrixus floccosus* (Maskell) (Homoptera: Aleyrodidae), was first described from Jamaica in 1986 and is currently found in several countries in the Americas, Asia, Europe and Africa. Its occurrence in Ethiopia was detected from leaf samples collected from orange trees planted in home gardens around the towns of Wonji and Adama in December 2000 (Ayalew and Tilahun, 2012). Initial work on WWF in Ethiopia concentrated on understanding the ecology of the pest (such as seasonal abundance and geographic distribution studies) and development of chemical control measures (Belay *et al.*, 2011; Ayalew and Tilahun, 2012). In 2014, the WWF parasitoid *Cales noacki* (Hymenoptera: Aphelinidae) was introduced from Israel and released in selected citrus orchards around Adama in the CRV region of Ethiopia. Although the parasitoid was not recovered from the initial release, it was recovered from releases made in the ensuing year. The level of

Table 13.1. Examples of invasive alien insect pests accidentally or intentionally introduced to Ethiopia between 2000 and 2019.

Scientific name	Common name	Main host	Origin/nativity	Year	Reference
Aleurothrixus floccosus	Citrus woolly aphid	Citrus	South America	2000	Ayalew and Tilahun (2012)
Dactylopius coccus	Cactus cochineal	Cactus	Mexico	2001	Belay (2011)
Leptocybe invasa	Blue gum chalcid	Eucalyptus	Australia	2002	Giliomee (2011)
Cinara cupressivora	Cypress aphid	Conifer tree	C & S America	2003	Dejen (2020)
Bactrocera invadens	Mango fruit fly	Mango	Asia	2004	Azerefegne *et al.* (2009)
Phenacoccus solenopsis	Cotton mealybug	Cotton	C & S America	2010	Abate and Mulatu (2020)
Aulacaspis tubercularis	White mango scale	Mango	Asia	2010	Dawd *et al.* (2012)
Prostephanus truncatus	Larger grain borer	Maize	America	2011	Tadesse *et al.* (2012)
Tuta absoluta	Tomato leaf miner	Tomato	America	2012	Ayalew (2014)
Spodoptera frugiperda	Fall armyworm	Maize	America	2017	Sisay *et al.* (2018)
Ophelimus maskelli	Eucalyptus gall wasp	Eucalyptus	Australia	2018	Asayehegne (2018)
Glycaspis brimblecomei	Red gum lerp psyllid	Eucalyptus	Australia	2019	Yirgu and Anjulo (2019)

pest infestation showed a progressive decline in areas where the parasitoid was released, although the level of parasitism appeared to be inconsistent. The pest does not currently seem to be important in citrus orchards of the CRV areas (Ayalew *et al.*, 2019).

Cactus cochineal

The cactus cochineal *Dactylopius coccus* Costa (Homoptera: Dactylopidae) was intentionally introduced to the Tigray region of Ethiopia in 2011 for the production of carminic acid (Belay, 2011). According to this author, a Chilean company called 'Food Safe' was granted 300 ha of cactus pear infested plains in southern parts of the region but was later forcibly closed because of the invasion of the insect and total damage of cactus pear in other parts of Tigray where the wild plant is regarded economically important. The insect infested and destroyed cactus pear, an edible wild fruit widely consumed in the region, in a short period because of favourable climatic conditions, abundance of the host cactus pear and the absence of natural enemies of the pest, among others.

Cypress aphid

The cypress aphid, *Cinara cupressivora* (Buckton) (Homoptera: Aphididae), attacks conifer trees causing major damage to plantations and hedges (FAO, n.d.). The insect was accidentally introduced to Africa from Europe in the 1980s. The presence of cypress aphids in Ethiopia was first detected in Addis Ababa in 2003 following severe infestation on cypress. A braconid parasitoid *Pauesia juniperorum* that was successfully introduced for the control of the pest in Malawi from Western Europe in 1994, was introduced to Ethiopia from Kenya for biological control of the pest. The release was made by Arsi Forest Enterprise to forests and hedges of infested areas in November 2008. The parasitoid was recovered in 2 to 3 months and effective control was observed in June 2009 (FAO, n.d.)

Cotton mealybug

The cotton mealybug, *Phenacoccus solenopsis* Tinsley (Homoptera: Pseudococcidae), originally described in America in 1897, spread to several Latin American, Asian and African countries in the first decade of the 2000s (see the references in Getahun *et al.*, 2019). Its occurrence in Ethiopia was reported from the Melka Sadi area in the Awash valley in July 2010 (Abate and Mulatu, 2020). These authors reported 42 host plants of *P. solenopsis* from cotton-growing areas of the Middle Awash. Their report included 28 weed species, four tree species, four vegetables, three field crops and one medicinal plant. The pest inflicts damage on several horticultural and field crops. Gebregergis (2018) reported heavy incidence and damage on sesame from the pest attack. Information on the insect biology has been generated locally (Getahun *et al.*, 2019).

Larger grain borer

The larger grain borer, *Prostephanus truncatus* (Horn) (Coleoptera: Bostrichidae), was accidentally introduced from Central America into Tanzania in the late 1970s (Infonet, n.d.). Its occurrence in Ethiopia was not known until 2008 (Tadesse *et al.*, 2012). It was confirmed in 2009 through trap catches deployed at the Moyale Ethio-Kenyan border by entomologists of Melkassa Agricultural Research Center using sex pheromones (Ayalew, 2014). However, according to Tadesse *et al.* (2012), the pest was not found in any of the grain samples obtained from different stores and market places in suspected areas in the districts of Moyale, Bulo Hora and Arsi Negele, suggesting that detecting the presence of the pest at low populations is difficult (Ayalew and Azerefegne, 2019).

White mango scale

The white mango scale (WMS), *Aulacaspis tubercularis* Newstead (Homoptera: Diaspididae) is of Asian origin. It occurs in several countries of West Africa and African countries south of Ethiopia, including the East African countries of Uganda, Kenya and Tanzania. The time of its introduction to Africa is unclear, but the occurrence of WMS in Ethiopia was known in 2010 (Dawd *et al.*, 2012). It had remained confined to western Ethiopia where local mango trees of old age were found until a few years back. It is increasing its geographical coverage and is currently the major pest in mango-producing areas of western, southern and northern regions of the country. Information on the ecology of the pest and control measures have been generated (Ayalew *et al.*, 2019; Habtegebriel, 2019) even though effective control of the pest is not yet in place.

Tomato leaf miner

The tomato leaf miner, *Tuta absoluta* (Meyrick) (Lepidoptera: Gelechiidae) is native to South America and was first described in 1917 by Meyrick from Peru. It appeared in Spain in 2006 and spread to all the Mediterranean countries, Northern Europe and the Middle East between 2006 and 2010. The first record in northeastern Africa was from Sudan in 2010. Its presence in Ethiopia was confirmed in February 2013 (Ayalew, 2014). Commonly available insecticides used for the control of a large number of vegetable arthropod pests failed to provide control of this insect during the period of its outbreak in early 2013. Later, a pesticide screening programme led to the identification of effective insecticides from the diamide and spinosyn classes (Ayalew, 2015). Efforts are being made to develop an IPM programme involving the use of pheromones, biopesticides and rational use of insecticides (Ayalew, 2019).

Fall armyworm

The fall armyworm (FAW), *Spodoptera frugiperda* (JE Smith) (Lepidoptera: Noctuidae), is native to tropical and subtropical regions of the Americas and is the key insect pest of maize in tropical regions. The occurrence of FAW was reported in Africa for the first time in late 2016 in West Africa (Goergen *et al.*, 2016; Abrahams *et al.*, 2017). Subsequently, FAW has spread rapidly throughout Africa south of the Sahara; recently, its occurrence has been confirmed in 44 African countries (Rwomushana *et al.*, 2018). FAW is a highly polyphagous insect pest that attacks more than 350 plant species, including maize, sorghum, millet, sugarcane and vegetables (Montezano *et al.*, 2018). Ecological studies, including the pest and associated natural enemies and chemical and botanical control studies have been conducted in Ethiopia and useful information has been generated (Sisay *et al.*, 2018; Demissie *et al.*, 2019).

Asian fruit fly

The invasive fruit fly *Bactrocera invadens* Drew, Tsuruta & White (Diptera: Tephritidae) is an Asian species first detected in eastern Africa in 2003 (Lux *et al.*, 2003). Its presence in Ethiopia was first mentioned in Azerefegne *et al.* (2009). It displaced the earlier dominant fruit fly species such as the Mediterranean fruit fly (*Ceratitis capitata*) and became the dominant fruit fly species in several horticultural crops, particularly on mango (Massebo and Tefera, 2015). Studies on this and other fruit fly species have mainly been on their distribution and species composition. The use of the para-pheromone methyl eugenol in mass annihilation technique (MAT) is widely practiced at the Upper Awash Agro-Industry enterprise in Ethiopia, along with field sanitation through regular collection and destruction of infested fruits. Although limited efforts are underway to promote the IPM package involving the use of the fruit fly parasitoids, spot application of food bait, use of biopesticides, MAT and field sanitation (Ekesi *et al.*, 2015), the management method in smallholdings in Ethiopia is so far limited to field sanitation as the pheromones and other methods are not readily available.

Blue gum chalcid, red gum lerp psyllid, eucalyptus gall wasp

Planting of exotic tree species in Ethiopia started with the introduction of eucalyptus, approximately 110 years ago. This country grows eucalyptus trees extensively in plantations and woodlots to provide fuel wood and construction timber, and it is an important source of income for farmers. Eucalyptus is the preferred tree due to its fast growth, adaptability to a wide range of environments and its versatile uses. Over time, insect pests on eucalyptus are accumulating. Since 2000 three new pests have been recorded, all of them with origins in Australia. The blue gum chalcid (*Leptocybe invasa* Fisher and La Salle) has been reported in Ethiopia since 2002 (Giliomee, 2011). It was observed on seedlings in the field and nurseries, coppiced young trees and new shoots of old *Eucalyptus camaldulensis*. It is found in north, south and central Ethiopia. Insecticides have been screened for the control of this pest on eucalyptus seedlings and effective products have been recommended. However, management of such introduced pests in forest settings is more appropriately done using classical biological control methods rather than synthetic insecticides (Yirgu and Anjulo, 2019).

The eucalyptus gall wasp (EGW), also known as blade gall wasp *(Ophelimus maskelli)* (Ashmead), is another gall-making insect pest newly introduced to Ethiopia. It was first detected in February 2017 on 1-year and above aged coppiced *E. camaldulens* at Kombolcha, South Welo, Amhara National Regional State (Asayehegne, 2018). EGW develops in a single-cell-type gall. The galls are pimple-like, nearly round and occur only on the leaf blade. Under epidemic conditions the entire leaf surfaces are usually densely covered with galls. Heavy galling by this insect results in premature shedding of the leaves, soon after the emergence of the wasp.

The red gum lerp psyllid (*Glycaspis brimblecombei)* was identified for the first time from infested leaves of *E. camaldulensis* from infested leaves collected in October 2018 in Ginchi (located at 90 01′ N latitude and 380 10′ E longitude, 2200 m above sea level) and Ambo (80 59′ N latitude and 370 51′ E longitude, 2101 m above sea level), West Shewa (Yirgu and Anjulo, 2019). No infestation of *G. brimblecombei* has occurred in other eucalyptus species.

Other examples of successful integrated pest management programmes in Ethiopia

In addition to the milestones mentioned above, there were several examples of successful IPM development and demonstration efforts undertaken and technologies adopted in Ethiopia. We have listed a few examples below:

- IPM of bean stem maggots (*Ophiomyia* spp.) relied heavily on the use of timely planting, host plant resistance and increased seeding rate (Abate, 1990). These were demonstrated at Jimma and Shalla in 2014 and 2015 and are now practiced across bean-growing areas of the country (Wondimu *et al.*, 2017).
- IPM programmes were developed and implemented against major vegetable pests, including fruit worms (*Phthorimaea operculella* and *Heliothis armigera*) on tomato, onion thrips (*Thrips tabaci*) on onion, and DBM (*Plutella xylostella*) on cabbage. They were demonstrated on-farm by establishing farmers' research groups at Wonji, near the Melkassa Research Center, between 1999 and 2001, and the technology has been adopted and is in use (Yesuf *et al.*, 2006).
- The use of fruit worms-resistant tomato variety 'Serio' along with application of recommended insecticides at critical growth stages (once each at early flowering and early fruiting) resulted in comparable performance with plots treated with frequent insecticide applications. This helped farmers realize the benefits of need-based pesticide application that also reduced control costs (Yesuf *et al.*, 2006).
- Reducing thrips infestation on onion and DBM on cabbage with the use of neem seed extracts was demonstrated. Farmers were made aware of the availability of non-chemical options of pest control and the benefit that can be obtained from the options demonstrated. It was learned that on-farm demonstrations of available pest management technologies would improve farmers' perception of pest problems, their knowledge on insect pest recognition and rational use of insecticides (Ayalew *et al.*, 2009).
- The IPM components of tomato leaf miner (*Tuta absoluta*) included the use of pheromone traps for mass trapping of adult moths, rational use of insecticides through rotational applications of registered insecticides from different classes, and safe disposal of infested fruits by keeping them in a polythene bag under the sun for at least a week. The fruits were used as poultry feed or spread over the soil surface to enhance soil organic matter and improve fertility (Ayalew, 2019).

Research Priorities in Agricultural Entomology

As stated earlier in this chapter, there are large numbers of arthropod pests with varying degrees of economic importance recorded in Ethiopia. Research has focused on the major pests and a considerable amount of knowledge has been gained and management practices (with emphasis on cultural practices and host plant resistance) are in place for the majority of them. Efforts to manage these pests will continue as long as they are of major economic importance for Ethiopian agriculture.

Meanwhile, it is important to set priorities for those arthropod pests that are known to be an immediate threat to crops. Accordingly, we have identified the top ten that need particular attention for the coming years. These are presented in Table 13.2.

We used the following criteria to come up with the list of priority arthropod pests: (i) government priority for the crop; (ii) current and potential importance of the crop (in terms of food security, export earning, import substitution and raw material supply for the local industry); (iii) importance of the pest; and (iv) availability or the lack of control measures in place at present.

Examples for our justification of the choices follow:

- Even though fieldpea is a traditional crop in Ethiopia, its productivity and production is mainly constrained by the pea aphid (*Acyrthosiphon pisum*). According to the Food and Agriculture Organization of the United Nations (FAO, 2022), the country imported upwards of 69,000 MT of dry peas, valued at >US$50 million each year during 2016–2020.
- The WMS (*Aulacaspis tubercularis*) on mango, the fruit fly (*Bacterocera invadens*) on avocado and mango, the fall armyworm (*Spodoptera frugiperda*) on maize and sorghum, and the tomato leaf miner (*Tuta absoluta*) on tomato are pests of recent introduction in Ethiopia and their management practices are being refined. Policy makers here attach very high priority on avocado as an important source of local industry and export market; maize and sorghum are accorded high priorities for food security in Ethiopia.
- The Russian aphid (*Diuraphis noxius*) and wheat aphid (*Schizaphis graminum*) are the major threats for wheat where the government gives high priority for food security and import substitution. *S. graminum* is particularly important in irrigated wheat.
- The sesame seed bug (*Elasmolomus sordidus*) continues to be the major pest of sesame, Ethiopia's largest oilseed export.
- The African bollworm (*Heliothis armigera*) and pink bollworm (*Pectinophora gossypiella*) are the key pests on cotton; the latter has been under a strict IPM regime that is based on 'closed season' or 'dead season' – i.e. avoidance of growing crops in the months following harvest and removal of crop residue to deny carryover of the pest. This system is now under threat because of government policy to grow wheat in these areas under irrigation.

Discussions and Conclusions

The crop protection department of EIAR (that included entomology, plant pathology, weed science and vertebrate pest sections) pioneered comprehensive research reviews in 1985 and set examples for all other research departments. This tradition has continued, albeit with less frequency and regularity. It took 21 years until the

Table 13.2. Top ten research priority arthropod pests in Ethiopia.

Scientific name	Common name	Host(s)
Acyrthosiphon pisum	Pea aphid	Fieldpea
Aulacaspis tubercularis	White mango scale	Mango
Bactrocera invadens	Fruit fly	Avocado, mango
Diuraphis noxius	Russian wheat aphid	Wheat
Elasmolomus sordidus	Sesame seed bug	Sesame
Helicoverpa armigera	African bollworm	Cotton, tomato
Pectinophora gossypiella	Pink bollworm	Cotton
Schizaphis graminum	Green bug	Wheat
Spodoptera frugiperda	Fall armyworm	Maize, sorghum
Tuta absoluta	Tomato leaf miner	Tomato

next such event, and 23 years until the proceedings were finally published. There is a need for more frequent comprehensive reviews and publication of proceedings. Considering the rapid policy and environmental changes taking place in recent times, there is a compelling need to plan a 10-year calendar for such events. This should be an important forum for reflecting the past and bringing in new thinking and direction for the future.

The emphasis for pest management so far has focused on IPM approaches that relied heavily on cultural practices and host plant resistance; the biocontrol aspect has not received the due attention it deserved. That might have been excusable considering the smallholder agriculture dominating the crop production system in Ethiopia so far, but big changes are anticipated because of the government policy to encourage agricultural intensification and commercial farming. This should open up opportunities for the public sector to engage in biocontrol research and development, and the private sector to commercialize the technology.

Intensification is bound to bring a resurgence of existing pests and increased use of synthetic pesticides. There are already indications that pesticide resistance is developing in some pests in Ethiopia (Negash *et al.*, 2018, 2020), and this is likely to continue following the modernization efforts of agriculture by the government. Researchers should start looking for this phenomenon and addressing it properly.

On the international scene, more and more pesticides are banned or put on the watch list. Twenty-two of the 184 active ingredients registered in Ethiopia are banned elsewhere; a further 20 are on the watch list because of international conventions and their acute and chronic toxicity to humans and the environment, including the honeybee (cf. Chapter 18 in this book). The research system should engage in screening more environmentally friendly and IPM-compatible pesticides as a matter of urgency.

There is a general feeling that researcher skills and the quality of research are declining despite the increase in the number of universities offering postgraduate degrees in agriculture. Stronger emphasis should be given to upgrading the quality of research through provision of needs-based training and improving research facilities.

On-farm validation and demonstration of research results will help to facilitate adoption of the developed technologies. An inventory of research results and the re-evaluation and promotion of the promising treatments are needed to make use of the outputs. Surveys should be needs based and undertaken periodically to detect changes in pest status and develop management options in a timely manner. The establishment of a national reference biosystematics centre and using the necessary expertise are critical to strengthen entomological research.

Systematic documentation of research results on arthropod pests of crops in Ethiopia is lacking, as is the case with many other fields. Information is gathered from published documents, including journals and conference proceedings. Electronic media-based information exchange platforms ensure proper documentation of research results. This, in turn, is useful to evaluate and validate generated information and technologies for demonstration and extension of technologies to the end users.

References

Abate, T. (ed.) (1986) *A Review of Crop Protection Research in Ethiopia, 4–7 February 1985.* Crop Protection Department, IAR, Addis Ababa, Ethiopia, 685 pp.

Abate, T. (1990) Studies on genetic, cultural and insecticidal controls against the bean fly, *Ophiomyia phaseoli* (Tryon) (Diptera: Agromyzidae), in Ethiopia. PhD dissertation, Simon Fraser University, Burnaby, British Columbia, Canada, 177 pp.

Abate, T. (1994) Entomological studies on fruit crops. In: Herath, E. and Desalegne, L. (eds) *Horticultural Research and Development in Ethiopia. Proceedings of the Second International Workshop of Ethiopia*, 1–3 December 1992, IAR, Addis Ababa, Ethiopia. pp. 177–186.

Abate, S. and Mulatu, B. (2020) Survey of host plants of cotton mealybug *Phenacoccus solenopsis* Tinsley (Homoptera: Pseudococcidae) on cotton crops in middle Awash, Ethiopia. *Greener Journal of Agricultural Science* 10(3), 145–151.

Abate, T., van Huis, A. and Ampofo, J.K.O. (2000) Pest management strategies in traditional agriculture: An African perspective. *Annual Review of Entomology* 45, 631–659.

Abrahams, P., Bateman, M., Beale, T., Clottey, V., Cock, M. *et al.* (2017). *Fall Armyworm: Impacts and Implications for Africa*. Evidence Note (2). CABI, Oxfordshire, UK.

Addis Fortune (2013) *Tomato in Danger. Addis Fortune*, June 16, 13(685).

Asayehegne, B. (2018) First record of Eucalyptus Gall Wasp (EGW) (*Ophelimus maskelli*) (Ashmead) on *Eucalyptus camaldulensis* in Ethiopia. *International Journal of Agriculture and Forestry* 8(2), 112–117. DOI: 10.5923/j.ijaf.20180802.10.

Ayalew, G. (2006) Comparison of yield loss on cabbage from diamondback moth, *Plutella xylostella* L. (Lepidoptera: Plutellidae) using two insecticides. *Crop Protection* 25, 915–919.

Ayalew, G. (2014) Alien invasive and emerging resident pests of crops in Ethiopia. In: Shonga, E. (ed.) *Proceedings of the 21st Annual Conference of the Plant Protection Society of Ethiopia,* 25 to 26 December. EIAR, Addis Ababa, Ethiopia, pp. 37–73 (ISBN 978-99944-952-5-2).

Ayalew, G. (2015) Efficacy of selected insecticides against the South American Tomato Moth, *Tuta absoluta* (Meyrick) (Lepidoptera: Gelechiidae) on tomato in the Central Rift valley of Ethiopia. *African Entomology* 23(2), 410–417.

Ayalew, G. (2019) *On the Support of Horti-LIFE on the Project Activities to the Integrated Pest Management (IPM) on Farm Testing'*. SNV Ethiopia, Addis Ababa, Ethiopia, p. 26.

Ayalew, G. and Azerefegne, F. (2019) Trends in research on arthropod pests of vegetables in Ethiopia. *Pest Management Journal of Ethiopia* 22, 111–123.

Ayalew, G. and Hopkins, R. (2013) Selecting the right parasitoid for the environment in classical biological control programmes: The case of *Diadegma semiclausum* (Hymenoptera: Ichneumonidae) and *Plutella xylostella* (Lepidoptera: Plutellidae) in the Kofele highland of Ethiopia. *Biocontrol Science and Technology* 23, 1284–1295.

Ayalew, G. and Ogol, C. (2006) Occurrence of the Diamondback moth, *Plutella xylostella* L. and its parasitoids in Ethiopia: Influence of geographical region and agronomic traits. *Journal of Applied Entomology* 130, 343–348.

Ayalew, G. and Tilahun, A. (2012) Research progress on invasive woolly whitefly (Homoptera: Aleyrodidae). In: Mulatu, B. (ed.) *Invasive Plant Pests Threatening Ethiopian Agriculture. Proceedings of the 17th Annual Conference of the Plant Protection Society of Ethiopia,* 26–27 November 2010. PPSE, Addis Ababa, Ethiopia, pp. 67–75 (ISBN 978-99944-987-3-4).

Ayalew, G., Mulatu, B., Negeri, M., Merene, Y., Sitotaw, L., Ibrahim, A. and Tefera, T. (2009) Progress in vegetable arthropod insect and mite pests management research between 1985 and 2005. In: Tadesse, A. (ed.) *Increasing Crop Production through Improved Plant Protection-Volume II. Plant Protection Society of Ethiopia (PPSE)*. PPSE and EIAR, Addis Ababa, Ethiopia, pp. 47–67.

Ayalew, G., Wondimu, M., Sisay, B. and Fekadu, A. (2019) Review of entomological research on lowland horticultural and field crops at Melkassa Agricultural Research Center: Achievements and Prospects. In: Ayalew, G. (ed.) *Fifty Years of Service for Dryland and Irrigated Agriculture in Ethiopia. Proceedings of the 50th Anniversary of Melkassa Agricultural Research Center, 27–31 August 2019*. Melkassa Agricultural Research Center, Melkassa-Adama, Ethiopia, pp.167–199.

Azerefegne, F. (1999) Biology and economic importance of the sweet potato butterfly in Ethiopia. PhD thesis, Swedish University of Agricultural Sciences, Uppsala, Sweden.

Azerefegne, F. and Gebre-Amlak, A. (1994) Oviposition pattern and preference by *Busseola fusca* (Fuller) moths to different stages of maize. *Insect Science and Its Application* 15, 269–273.

Azerefegne, F., Dawd, M., Belay, D. and Mekonen, B. (2009) Review of entomological research on fruit crops in Ethiopia. In: Tadesse, A. (ed.) *Increasing Crop Production through Improved Plant Protection-Volume II*. PPSE and EIAR, Addis Ababa, Ethiopia, pp. 69–92.

Belay, T. (2011) Commercialization of carmine cochineal in northern Ethiopia. Paper Presented at the Third Biennial Conference of the Ethiopian Horticultural Science Society, 4–5 February 2011, Jimma University, College of Agriculture and Veterinary Medicine, Jimma, Ethiopia.

Belder, E., Elings, A., Yilma, Y., Dawd, M. and Lemessa, F. (2009) *On-Farm Evaluation of Integrated Pest Management of Red Spider Mites in Cut Roses*. Final Report to the Ministry of Agriculture and Rural Development, WUR report 296, Wageningen, the Netherlands.

Belay, D., Zewdu, A. and Foster, J. (2011) Ecology and management of the woolly whitefly (Hemiptera: Aleyrodidae), a new invasive citrus pest in Ethiopia. *Journal of Economic Entomology* 104(4), 1329–1338.

Dawd, M., Habtegebriel, B., Ayele, L., Feleke, K., Hailemariam, S. and Burka, T. (2012) White mango scale: A new insect pest of mango in Western Ethiopia. In: Derso, E., Zelleke A., Desalegne, L., Worku, Z.,

Kidane-Mariam, H., Tabor, G., Getachew, Y. (eds) *Proceedings of the 3rd Biennial Conference of Ethiopian Horticultural Science Society,* 4–5 February 2011, Addis Ababa. Ethiopian Horticulture Science Society, Addis Ababa, Ethiopia, pp. 257–267.

Dejen, A. (2020) Status of cypress aphid on *Cupressus lusitanica* and *Juniperus procera* in protected and cultivated forests of south Wollo, Ethiopia. *Journal of Forestry Research* 31, 333–337.

Demissie, G., Salato, Z., Ayalew, G., Mengistu, F. and Sisay, B. (2019) Fall armyworm, *Spodoptera frugiperda* (J. E. Smith): A new threat to maize production in Ethiopia (status, impact, interventions, lessons learned and prospects). In: Habtegebriel, B. (ed.) *Emerging and Reemerging Plant Pests of Ethiopia: Status, Interventions and Future Prospects in a Changing Climate. Proceedings of the 24th Annual Conference and Silver Jubilee,* 16–17 March 2018, Haramaya University, Plant Protection Society of Ethiopia, Addis Ababa, Ethiopia. pp. 137–158.

Ekesi, S., De Meyer, M. and Malacrida, A. (2015) Taxonomy, bioecology, and management of native and exotic fruit fly species in Africa. *Annual Review of Entomology* 61(1), 219–238.

FAO (Food and Agriculture Organization of the United Nations) (n.d.) *Cypress Aphid Pest Control in Ethiopia* Available at: http://www.fao.org/3/av255e/av255e.pdf (accessed 19 March 2024).

FAO (2022) Statistical Databases and Data-Sets of the Food and Agriculture Organization of the United Nations. Available at: http://www.fao.org/faostat/en/#data (accessed 14 August 2022).

Gebre-Amlak, A. (1988) Survival of maize stemborer, *Busseola fusca* (Fuller) in crop residues in Ethiopia. *Crop Protection* 7, 183–185.

Gebre-Amlak, A. (1989) Termination of diapause in the maize stemborer *Busseola fusca* (Fuller) in Awassa, southern Ethiopia. *Journal of Applied Entomology* 107, 160–165.

Gebregergis, Z. (2018) Incidence of a new pest, the cotton mealybug *Phenacoccu solenopsis* Tinsley, on sesame in north Ethiopia. *International Journal of Zoology* 2018, 3531495. DOI: 10.1155/2018/3531495.

Getahun, S., Seid, N., Alemu, Z., Taye, W., Abate, S. and Hemba, M. (2019) Biology of cotton mealybug (*Phenacoccus solenopsis* (Tinsley) on cotton plants under the laboratory conditions. *Academic Research Journal of Agricultural Science and Research* 7(7), 560–568.

Getaneh, G., Ayalew, G., Derso, E., Mulatu, B., Hailu, E. and Hunduma, T. (2016) Plant Protection Research in Ethiopia: Major Achievements, Challenges and Future Directions In: Alemu, D., Derso, E., Assefa, G., Kirub, A. (eds) *Proceedings of Agricultural Research for Ethiopian Renaissance: Challenges, Opportunities and Directions*, 26–27 January 2016. EIAR, Addis Ababa, Ethiopia, pp. 193–212.

Giliomee, J.H. (2011) Recent establishment of many alien insects in south Africa—A cause for concern. *African Entomology* 19(1), 151–155.

Goergen, G., Lava, K.P., Sankung, S.B., Togola, A. and Tamò, M. (2016) First report of outbreaks of the fall armyworm, *Spodoptera frugiperda* (J E Smith) (Lepidoptera, Noctuidae): A new alien invasive pest in west and central Africa. *PLoS One* 11(10), e0165632. DOI:10.1371/journal.pone.0165632.

Habtegebriel, B. (2019) White mango scale in Ethiopia: The problem, management options and interventions. In: Habtegebriel, B. (ed.) *Emerging and Reemerging Plant Pests of Ethiopia: Status Interventions and Future Prospects in a Changing Climate. Proceedings of the 24th Annual Conference,* 16–17 March 2018, Haramaya University. Plant Protection Society of Ethiopia, Addis Ababa, Ethiopia, pp. 195–209.

Infonet.com. Large grain borer. Available at: https://infonet-biovision.org/plant_pests/larger-grain-borer-lgb (accessed 19 March 2024).

Lux, S.A., Copeland, R.S., White, I.M., Manrakhan, A. and Billah, M.K. (2003) A new invasive fruit species from the *Bactrocera dorsalis* (Hendel) group detected in east Africa. *Insect Science and Its Application* 23, 355–360.

Massebo, F. and Tefera, Z. (2015) Status of *Bactrocera invadens* (Diptera: Tephritidae) in mango-producing areas of Arba Minch, southwestern Ethiopia. *Journal of Insect Science* 15(1), 3. DOI: 10.1093/jisesa/ieu166.

Montezano, D.G., Sosa-Gomez, D.R. and Roque-Specht, V.F. (2018) Host plants of *Spodoptera frugiperda* (Lepidoptera: Noctuidae) in the Americas. *African Entomology* 26, 286–300.

MOST (Ministry of Science and Technology) (2015) *Sixth National Award on Science, Technology and Innovation* (in Amharic). MOST, Addis Ababa, Ethiopia, p. 122.

Negash, B., Azerefegne, F. and Ayalew, G. (2018) Field evaluation of commonly used synthetic insecticides against *Thrips tabaci* (Thysanoptera: Thripidae) on onion (*Allium cepa*) in the central Rift Valley of Ethiopia. *Pest Management Journal of Ethiopia* 21, 23–34.

Negash, B., Azerefegne, F. and Ayalew, G. (2020) Insecticide resistance management against thrips (Thysanoptera: Thripidae) on onion in the central Rift Valley of Ethiopia. *International Journal of Tropical Insect Sciences* 40, 759–767. DOI: 10.1007/s42690-020-00127-6.

Reporter (2013) *Tuta absoluta*, Section 1, page 11, June 23 (In Amharic).

Rwomushana, I., Bateman, M., Beale, T., Beseh, P., Cameron, K. *et al.* (2018) *Fall Armyworm: Impacts and Implications for Africa*. Evidence Note Update, CABI, Oxfordshire, UK.

Sisay, B., Simiyu, J., Malusi, P., Likhayo, P., Mendesil, E. *et al.* (2018) First report of the fall armyworm, *Spodoptera frugiperda* (Lepidoptera: Noctuidae), natural enemies from Africa. *Journal of Applied Entomology* 142(8), 800–804. DOI: 10.1111/jen.12534.

Tadesse, A. (ed.) (2008) Increasing crop production through improved plant protection. Volume I. In: *Plant Protection Society of Ethiopia (PPSE),* 19–22 December 2006, Addis Ababa, Ethiopia. PPSE and EIAR, Addis Ababa, Ethiopia, 598 pp.

Tadesse, A. (ed.) (2009) Increasing crop production through improved plant protection—Volume II. In: *Plant Protection Society of Ethiopia (PPSE)*. PPSE and EIAR, Addis Ababa, Ethiopia, 542 pp.

Tadesse, A., Lemma, H. and Negeri, M. (2012) The larger grain borer, *Prostephanus truncatus* (Horn) (Coleoptera: Bostrichidae, an invasive pest of maize and cassava in Africa: A review. In: Mulatu, B. (ed.) *Invasive Plant Pests Threatening Ethiopian Agriculture. Proceedings of the 17th Annual Conference of the Plant Protection Society of Ethiopia,* 26–27 November 2010. PPSE Addis Ababa, Ethiopia, pp.19–42.

Wondimu, M., Getu, E., Kidanu, S. and Fufa, A. (2017) Aging effect of two-decade old technologies of bean stem maggot (*Ophiomyia* spp.) management options in Ethiopia. Baseline information generation for future research direction. *Journal of Agricultural Biotechnology and Sustainable Development* 9(3), 16–21.

Yesuf, M., Dessalegne, L., Ayalew, G., Deressa, A., Bekele, A. *et al.* (2006) Farmers awareness building on integrated pest management options of major vegetable pests in the central Rift Valley region of Ethiopia. In: Bekele, E., Azerefegne, F. and Abate, T. (eds) *Proceedings of a Planning Workshop on Facilitating the Implementation and Adoption of Integrated Pest Management (IPM) in Ethiopia, 13–15 October 2003*. ASAI/EARO, Nazareth, pp 46–52.

Yirgu, A. and Anjulo, A. (2019) First record of *Glycaspis brimblecombei* Moore (Hem.: Psyllidae) on *Eucalyptus camaldulensis* in Ethiopia. *Phytoparasitica* 47, 67–70. DOI: 10.1007/s12600-018-00711-0.

Yitaferu, K. and Gebre-Amlak, A. (1994) Phenology and survival of the maize stalk borer, *Busseola fusca* in eastern Ethiopia. *International Journal of Tropical Insect Science* 15, 177–184.

14 Priorities for Plant Pathology Research and Development

Tsedeke Abate*
Homegrown Vision, Addis Ababa, Ethiopia

Abstract

More than 530 crop diseases have been recorded on cultivated crops in Ethiopia, of which 19% are considered major. The proportion of diseases designated as such on traditional crops tended to be lower than introduced ones. The research system has accumulated adequate knowledge and skills to tackle existing plant diseases based largely on host plant resistance and cultural practices. Information on control using chemical sprays is available but this has its limitations, including lack of availability, affordability, knowhow and human hazard risk. This study has identified the top 20 major diseases on more than a dozen crops to be given immediate priority. Disease dynamics are bound to change with the government's intent to expand agriculture in the coming years. There may be a need to develop, demonstrate and implement different sets of disease management practices for smallholders on the one hand and for commercial growers on the other.

Introduction

Ethiopia is a country of great agroecological, crop and cultural diversity. In 2020, the country grew 72 cultivated crops covering a total area of roughly 16 million ha (FAO, 2022a). Topping the list are *teff*, maize, sorghum, wheat, barley, coffee, faba bean, common bean, millet and sesame. Even though cereals in general and some of the grain legumes have shown increased productivity and production, many others, including yams, several vegetables and fruits have declined.

Damage by diseases plays an important role in determining the performance of cultivated crops. Substantial progress has been made in understanding the identity and economic importance of plant diseases, especially following the establishment of the Ethiopian Institute of Agricultural Research (EIAR) in 1966. Two major reviews of progress in crop protection (entomology, plant pathology, weed science, and migratory and vertebrate pests) research and development have been undertaken over the years. The results have been documented in Abate (1986) and Tadesse (2008, 2009). There is also rich literature in plant pathology (and other crop protection fields) published since the last review.

Initial efforts in plant pathology research concentrated on understanding the biology and epidemiology of the diseases. This was followed

* Corresponding author: tsedeke@homegrownvision.org

DOI: 10.1079/9781800626386.0014

by devising control measures. Short-term emphasis was on screening chemical pesticides. Disease management practices based on host plant resistance and cultural practices have attracted much attention in the medium and long terms. At present, integrated disease management practices are in place for the majority of major plant diseases in Ethiopia.

This current review presents a critical assessment of the state of plant pathology research and development in Ethiopia and identifies high priority areas of research in the immediate future.

Approach

The data presented in this chapter are based on literature reviews and my personal observations. The bulk of the information on the status of the diseases was the two volumes of proceedings of the Plant Protection Society of Ethiopia (Tadesse, 2008, 2009), and to some extent, proceedings of the First Crop Protection Symposium in Ethiopia (Abate, 1986). References were also made to the latest publications available on plant diseases in question in this country.

Results

Available data show that there were at least 538 species of plant diseases (not including forest diseases as the available records do not categorize status for forest diseases) recorded in Ethiopia as of December 2006. Of those, only 97 (or 19%) were designated as 'major'. This means that a large number of diseases are associated with each group of crops or crop species, but only a limited number are considered major, with much larger proportions designated as 'minor', 'sporadic' or 'status unknown'. This present study showed that 53% and 50%, respectively, of diseases of ornamentals and fruit crops were designated as major, in contrast to 8% for medicinal plants, 13% for brassicas, 14% for cereals, 16% for grain legumes and 17% for oilseeds. Figure 14.1 depicts the proportions of what are considered major diseases of selected crops in Ethiopia.

As mentioned above, disease status varied from crop to crop. For example, less than 10% of the diseases on safflower, sorghum, *noug*, potato and fieldpea are designated as major, whereas 40%, 38% and 38%, respectively, of the diseases on common bean, citrus and wheat were given such status. This graph suggests that well-established,

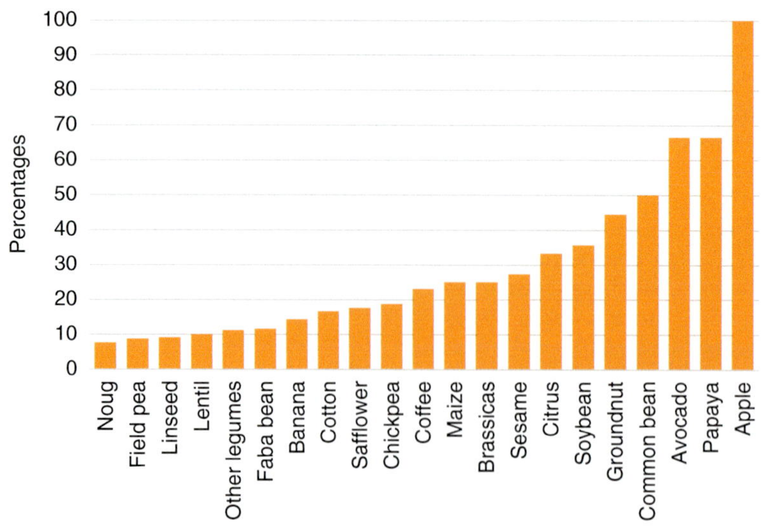

Fig. 14.1. Proportions of plant diseases that are recognized as major among selected crops cultivated in Ethiopia. Source: constructed by the author from various sources.

traditional crops might suffer from relatively lower numbers of major diseases than introduced ones. Such crops may not be highly productive, but they can be used as great source of desirable traits in breeding programmes. While researchers in Ethiopia have not adequately used these resources, their use is expected to be even more important in the future when climate change is affecting crop productivity. Their desirability stems from the fact that they have a broad genetic base and are less susceptible to diseases and therefore more sustainable.

Priorities for disease research

While research on all major diseases presented here (shown in Annex Tables 14.1 to 14.8) and those that may be inadvertently introduced in the future should continue, I believe that some of them need to be given special emphasis. I have picked what I consider the top 20 major diseases of 13 crops for this purpose (Table 14.1). My choice is based on one or more of the following criteria:

- government priority for the crop;
- current and potential importance (in terms of food security, export earning, import substitution, and raw material supply for the local industry);
- importance of the disease;
- availability/unavailability of control measures in place at present; and
- nature of the disease, e.g. how frequently does it form new races?

The next section presents the highlights for each of the diseases associated with the selected crops.

Avocado

According to the Food and Agriculture of the United Nations (FAO, 2022b), there is little avocado export from Ethiopia at present. The country exported a little over 150 MT of avocado for an estimated value of just a little over US$100,000 annually in 2016–2020. However, this crop has great potential for expansion of production and export, given the large suitable agroecological areas for its production available in this country. Avocado is one of the crops accorded the highest priority for export by the government of Ethiopia.

Root rot (or decline) caused by the fungal pathogen *Phytophthora cinnamomi* is one of the major biotic factors limiting production of

Table 14.1. Top 20 diseases on selected crops in Ethiopia that need to be given special emphasis by research in the immediate future. Source: author's own judgement based on available literature and pers. com. with Dr Mohamed Yusuf of EIAR.

Scientific name	Common name	Host(s)
Phytophthora cinnamomi	Root rot/decline	Avocado
Alternaria porri	Purple blotch	Onion/shallot
Sclerotium cepivorum	White rot/bulb rot	Onion/shallot/garlic
Puccinia alli	Garlic rust	Garlic
Fusarium spp.	Wilt complex	Pepper
Phytophthora spp.	Wilt complex	Pepper
Pseudocercospora angolensis	Leaf & fruit spot	Citrus
Colletotrichum gloeosporioides	Anthracnose	Mango
Pseudomonas phaseolicola	Halo blight	Mungbean
Ascochyta boltshauseri	Brown leaf spot	Mungbean
Ascochyta phaseolorum	Ascochyta blight	Cowpea
Botrytis fabae	Chocolate spot	Faba bean
Physoderma viciae	Faba bean gall	Faba bean
Mycoplasma-like organism	Phyllody	Sesame
Xanthomonas campestris pv. *sesami*	Bacterial blight	Sesame
Unidentified	Dubti Syndrome	Cotton
Colletotrichum kahawae	Coffee berry disease	Coffee
Hemileia vastatrix	Coffee rust	Coffee
Puccinia graminis f. sp. *tritici*	Stem/black rust	Wheat
Puccinia striiformis	Yellow/stripe rust	Wheat

avocado widely distributed in Ethiopia (Yesuf, 2007; Teferi and Ayano, 2016). The work of Teferi and Weldemichael (2015) shows that no effective control measures are available in Ethiopia on infected trees, and therefore farmers need to be advised to use preventive measures. Mitiku (2017) reported a 77% infection by powdery mildew of avocado in Metser *wereda* of Ari in southern Ethiopia. He also reported 85% and 40%, respectively, infection of avocado by anthracnose (*Colletotrichum gloeosporioides*) in Metser and Bytsemal *weredas*.

Onion/shallot

Ethiopia went from a net exporter of onion through the 1990s to a net importer in the most recent years. For example, the country imported close to 68,000 MT of dry onion with an estimated value of US$14 million annually during 2016–2020 while its export stood close to 16,000 MT for an estimated value of US$2.7 million (FAO, 2022b).

Purple blotch (caused by *Alternaria porii*) and white rot or bulb rot (caused by *Sclerotium cepivorum*) are the major biotic factors limiting the productivity and production of onion in Ethiopia (Mengesha *et al.*, 2022). Management of white rot is difficult because of persistence of the pathogen in the soil for more than 40 years (Mandefro *et al.*, 2009). These diseases are particularly severe when the crop is produced under wet conditions during the main rainy season, as has been reported in India (Dar *et al.*, 2020). Jula *et al.* (2021) conducted a year-long experiment on farmers' fields in 2017/18, and reported that its severity is influenced by season, i.e. it is more severe under wetter conditions. Getinet and Yalew (2022) demonstrated that two or three sprays of recommended fungicides resulted in lowest disease severity and highest yield.

The government policy now aims to intensify production twice a year, unlike what has been practiced so far of major production during the dry season under irrigation. It is therefore mandatory that solutions are found to manage these diseases in the immediate future.

Garlic

Ethiopia has great agroecological potential to develop and intensify production of garlic for domestic consumption and the export market. The country's current production is lagging behind the domestic demand and export is negligible. For example, Ethiopia exported about 550 MT for an estimated value of US$317,000 annually in the 2016–2020 period.

Garlic rust (caused by *Puccinia allii*) has remained perhaps the single most important biotic limiting factor affecting the production of this crop in almost all producing regions of Ethiopia (Endale *et al.*, 2021) where it causes nearly 59% yield loss (Worku and Dejene 2012). Garlic is also reported to suffer from the white rot or bulb rot disease (*S. cepivorum*) mentioned for onion/shallot above. Research needs to find a remedy to these diseases to help the country boost productivity and production for the domestic and export market.

*Hot pepper (*Capsicum *spp.)*

Hot pepper in Ethiopia is a high-value crop produced for home consumption and as raw material for the local spice extraction industry. A large number of diseases are recorded on this crop (Mandefro *et al.*, 2009). Fusarium wilt (caused by *Fusarium oxysporum* f. sp. *capsici*) stands out as one of the key diseases limiting the productivity and production of hot pepper in this country.

Fusarium wilt attacks all stages of the crop – from seedling to harvest. Gabrekiristos and Demiyo (2020) reported up to 90% disease incidence and 68–71% yield loss due to this disease. Gabrekiristos *et al.* (2020) observed that the disease incidence varied from location to location and agronomic practices.

Citrus

Ethiopia is a net importer of citrus and citrus products. According to the FAO, Ethiopia imported citrus and its products (mainly juices) valued at >US$4.87 million compared to its export earnings of approximately US$1.07 million annually during 2016–2020 (FAO, 2022b).

Diseases are among the biotic agents limiting productivity and production of citrus in Ethiopia, as they are elsewhere. Yesuf *et al.* (2009) provide a list of important diseases on citrus in this country. Three of those, viz. leaf and fruit spot (caused by *Pseudocercospora angolensis* [synonym: *Phaeoramularia angolensis*]), anthracnose (*Colletotrichum gloeosporioides*) and fruit rot (*Penicillium italicum*) are considered most important.

P. angolensis is perhaps the most devastating of them all. Yesuf (2013) reported up to nearly 97% infection on leaves and 94% on fruit. The disease was widely distributed in the wetter, more humid areas in the south, southwest, northwest and north-central parts of the country but not in the lower-lying, drier areas of the Central Rift Valley and eastern Ethiopia. It caused total crop failure under severe conditions, as is the case in other parts of Africa where the disease occurs (Pretorius, 2005). Dagnew *et al.* (2014) surveyed citrus-growing areas of Ethiopia and reported that more than 63% of the areas surveyed were infected. This disease is found only in tropical Africa and Yemen and occurs at altitudes ranging from 80–1800 m above sea level (EFSA Panel, 2017).

Mango

Mango is among the fastest expanding fruit crops in Ethiopia, where its area increased at an annual rate of more than 10% between 2000 and 2020. However, its productivity showed a slight decline during this period, mainly because of diseases (cf. Chapter 8). Anthracnose (caused by the fungus pathogen *Colletotrichum gloeosporioides*) is the limiting biotic constraint for productivity and production of mango in Ethiopia (Yesuf *et al.*, 2009). The disease also infects citrus and papaya, but it is not as serious as it is in mango.

This disease is widely distributed throughout mango-growing agroecologies in Ethiopia (Chala *et al.*, 2014; Tucho *et al.*, 2014; Abera *et al.*, 2016; Mitiku, 2017). It affects both leaves and fruits. Incidences of up to more than 95% and severity of nearly 71% have been reported, depending on the location. Chala *et al.* (2014) reported 100% inhibition of fungal growth by in vitro treatment of palmarosa oil, but the level of effort being made to further develop such treatments is unknown.

Mungbean

Mungbean was once a neglected, minor crop but it has gained significance in recent years because of its demand in the international market. Its area expanded at an annual rate of nearly 25% between 2013 and 2020 alone, the fastest of all crops (cf. Chapter 5). It fetched an estimated US$35.8 million in export earnings in the 2015/16 season (www.ecx.com) and has great potential to grow further.

Halo blight (caused by *Pseudomonas phaseolicola*) has long been considered the only major disease limiting mungbean production in Ethiopia (Assefa and Gorfu, 1986), but recent studies show that the brown leaf spot (caused by *Ascochyta phaseolorum*) is another biotic constraint (Teferie *et al.*, 2020). These diseases need to be tackled in the immediate future in order to boost mungbean productivity and production in Ethiopia.

Cowpea

Ethiopia is a secondary centre of origin for cowpea (Westphal, 1975) and possesses wide genetic diversity. Even though it is considered a minor crop at present, it has high potential to expand in the near future (Etana *et al.*, 2013; Alemu *et al.*, 2016; Beshir *et al.*, 2019; Walle *et al.*, 2019; Kebede and Bekeko, 2020; Asrat *et al.*, 2021). Neither the Ethiopian Central Statistical Agency nor the FAO have data on this crop in Ethiopia. The work of Beshir *et al.* (2019) suggests that cowpea covers about 70,000 ha (does not cover the whole country; data are only for surveyed areas), with about 56,000 MT of production in Ethiopia.

Various reports also show that Ethiopia has the potential for higher cowpea productivity than any of its peers in Africa. This crop is grown in parts of Amhara, Gambella, Southern Nations, Tigray and Oromia regional states for improved nutrition, export and fodder for animals.

Ascochyta blight (caused by the fungus pathogen *Ascochyta phaseolorum*) is a major disease limiting the production of cowpea in Ethiopia. Research needs to be intensified to overcome this and other biotic constraints (e.g. bruchids) for the country to exploit the full potential of cowpea.

Faba bean

Faba bean occupies the largest area of all grain legumes in Ethiopia and is produced across wide areas of the highlands. It is produced mainly for nutrition security, but Ethiopia exported close to 22,000 MT valued at US$9 million annually during 2016–2020 (FAO, 2022b).

Chocolate spot (caused by the fungus pathogen *Botrytis fabae*) has continued to be the major disease limiting the production of faba bean in Ethiopia (Eshetu *et al.*, 2018; Teshome and Kora, 2021). Wubshet and Chala (2021) estimated yield loss of nearly 35% in unsprayed fields. Faba bean gall (caused by *Physoderma viciae*) has appeared in recent years and is becoming the major threat for faba bean production in Ethiopia (You *et al.*, 2021). This disease can cause total crop loss. It is also known to infect other legumes such as fieldpea and *Trifolium*.

Sesame

Sesame accounted for 85% of the total oilseeds export (estimated at nearly US$333 million) or 23% of the overall export earnings (second only to coffee) annually in the 2016–2020 period (FAO, 2022b). Bacterial blight complex (caused by the pathogens *Xanthomonas campestris* pv. *sesami* and *Pseudomonas syringae* pv. *sesami*) and phyllody are the major biotic constraints to sesame productivity and production in Ethiopia.

Disease incidence and severity may vary according to locations and climatic conditions (Terefe *et al.*, 2009; Kindeya *et al.*, 2018; Endalew *et al.*, 2020; Kefale *et al.*, 2021). These diseases need particular attention across the major growing parts of the country. The first two may occur together and are usually more severe under more humid conditions.

Cotton

Ethiopia exported approximately 1400 MT of cotton, valued at US$2.1 million annually in 2016–2020 compared to nearly 3800 MT, at an estimated value of nearly US$3.6 million in the 1996–2000 period (FAO, 2022b). By contrast, its import value was more than US$3.2 million in 2016–2020 compared to US$5000 in the 1996–2000 period.

Dubti Syndrome (caused by an unidentified entity) has been a major disease of cotton (Terefe *et al.*, 2009) going back at least the last four decades. Research needs to provide a solution to this disease for the country to benefit from improved productivity and production of cotton.

Coffee

The successful identification and implementation of varietal resistance against coffee berry disease (CBD), caused by the fungal pathogen *Colletotrichum kahawae*, in the mid-1970s (van der Graaff, 1981) was the first major breakthrough for the EIAR. CBD still continues to be a major disease of coffee, with its incidence and severity varying from place to place (Getachew, 2017; Adem and Amin, 2021). Coffee leaf rust (caused by *Hemileia vastatrix*) is another major disease, especially in the more humid, wetter higher altitudes. Continued research is needed to keep these and other diseases in check for increased productivity and production of coffee in Ethiopia.

Wheat

Ethiopia is the second largest producer (after Egypt) and third largest grower (after Morocco and Algeria) of wheat in Africa (FAO, 2022c). The government of Ethiopia plans to expand and intensify wheat production by expanding irrigation systems. This would mean growing wheat twice a year, i.e. both in the main rainy and dry seasons.

Wheat rusts are the main biotic constraints to wheat productivity and production in Ethiopia (Tadesse *et al.*, 2008; Meyer *et al.*, 2021). Substantial knowledge exists on wheat rust management in Ethiopia but the ability of the pathogens to change into different races (e.g. Ug99) necessitates continuous monitoring and developing integrated management systems that are based on resistant cultivars and improved cultural practices.

Discussion and Conclusions

It is obvious that concerted research efforts on plant pathology that commenced with the establishment of the EIAR in the mid-1960s are paying dividends for Ethiopian agriculture. There is adequate knowledge accumulated and experience gained to tackle existing and potential disease problems. Please refer to Annex Tables 14.1 to 14.8 to see examples of major diseases of various groups of crops cultivated in Ethiopia.

The research system has developed disease management practices mainly based on host plant resistance and cultural measures. This is logical for a country like Ethiopia whose agriculture is predominantly smallholder based. The available data show that a large number of diseases could be controlled with chemical sprays, but this has its limitations. There are problems of availability, affordability, knowhow, logistics, human hazard risk and environmental concerns.

Disease dynamics are bound to change more quickly with the government's intent to expand and intensify agriculture in the coming years. The research community needs to be ready to provide practical solutions as quickly as possible. There may be a need to develop, demonstrate and implement disease management practices for smallholders on the one hand and for commercial agriculture on the other.

The research system gave particular emphasis to diseases associated with cereals and grain legumes, while many of the diseases on other crop groups did not receive the due attention they deserve. This needs to change. It is very likely that priorities would change as the government enters into a new phase of agricultural development, and researchers should make themselves ready to keep pace with the dynamics.

References

Abate, T. (ed.) (1986) A review of crop protection research in Ethiopia. In: *Proceedings of the First Ethiopian Crop Protection Symposium,* 4–7 February 1985. Institute of Agricultural Research, Addis Ababa, Ethiopia, 685 p.

Adem, A. and Amin, M. (2021) Incidence and severity of major diseases of coffee in highland of eastern Ethiopia. *Journal of Food, Nutrition and Agriculture* 4, 13–15. DOI: 10.21839/jfna.2021.v4.6944.

Abera, A., Lemessa, F. and Adugna, G. (2016) Prevalence and intensity of mango (*Mangifera indica* L.) anthracnose caused by *Colletotrichum* species in south-western of Ethiopia. *Ethiopian Journal of Applied Science and Technology* 8(1), 1–10.

Adugna, G., Jefuka, C., Zeru, A. and Tesfaye, A. (2009) Advances in coffee disease research in Ethiopia. In: Tadesse, A. (ed.) *Increasing Crop Production through Improved Plant Protection—Volume II*. Plant Protection Society of Ethiopia (PPSE), Addis Ababa, Ethiopia, pp. 275–304.

Ahmed, E., Gorfu, D., Tadesse, A. and Dawd, M. (2009) Pest problems and their management practices in flower farms in Ethiopia. In: Tadesse, A. (ed.) *Increasing Crop Production through Improved Plant Protection—Volume II*. Plant Protection Society of Ethiopia (PPSE), Addis Ababa, Ethiopia, pp. 440–461.

Alemu, M., Asfaw, Z., Woldu, Z., Fenta, B.A. and Medvecky, B. (2016) Cowpea (*Vigna unguiculata* (L.) Walp.) (Fabaceae) landrace diversity in northern Ethiopia. *International Journal of Biodiversity and Conservation* 8(11), 297–309. DOI: 10.5897/IJBC2016.0946.

Asrat, Z., Begna, T. and Tariku, A. (2021) Evaluation of yield and yield related performance of cowpea [Vigna unguiculata (L.) Walp] varieties at west Hararghe zone, eastern Ethiopia. *International Journal of Research Studies in Agricultural Sciences* 7(7), 42–48. DOI: http://dx.doi.org/10.20431/2454-6224.0707005.

Assefa, H. and Gorfu, D. (1986) A review of food legume diseases research in Ethiopia. In: Abate, T. (ed.) *A Review of Crop Protection Research in Ethiopia. Proceedings of the First Ethiopian Crop Protection Symposium*, 4–7 February 1985. Institute of Agricultural Research, Addis Ababa, Ethiopia, 685 p.

Badebo, A., Bekele, E., Bekele, B., Hundie, B., Degefu, M., *et al.* (2008). Review of two decades of research on diseases of small cereal crops. In: Tadesse, A. (ed.) *Increasing Crop Production through Improved Plant Protection—Volume I*. Plant Protection Society of Ethiopia (PPSE), PPSE, Addis Ababa, pp. 375–431.

Beshir, B., Amsalu, B., Teamir, M., Tsegaye, D., Yilma, B. and Ketema, S. (2019) *Cowpea: Production, Marketing and Utilization in Ethiopia*. EIAR Research Report 121, EIAR, Addis Ababa.

Chala, A., Getahun, M., Alemayehu, S. and Tadesse, M. (2014) Survey of mango anthracnose in southern Ethiopia and in-vitro screening of some essential oils against *Colletotrichum gloeosporioides*. *International Journal of Fruit Science* 14, 157–173. DOI: 10.1080/15538362.2013.817899.

Dagnew, A., Belew, D., Admassu, B. and Yesuf, M. (2014) Citrus production, constraints and management practices in Ethiopia: The case of *Pseudocercospora* leaf and fruit spot disease. *Science, Technology and Arts Research Journal* 3(2), 4–18. DOI: 10.4314/star.v3i2.2.

Dar, A.A., Sharma, S., Mahajan, R., Mushtaq, M., Salathia, A., Ahmad, S. and Sharma, J.P. (2020) Overview of purple blotch disease and understanding its management through chemical, biological and genetic approaches. *Journal of Integrative Agriculture* 19(12), 3013–3024. DOI: 10.2903/j.efsa.2017.4883

EFSA (European Food Safety Authority) Panel on Plant Health (2017) Pest categorisation of *Pseudocercospora angolensis*. *EFSA Journal* 15(7), e04883. DOI: 10.2903/j.efsa.2017.4883.

Endale, Y., Ambaw, A., Worku, M. and Gelaye, M. (2021) Assessment of major insect pests and diseases of sesame (*Sesamum orientale* L) in west Gondar zone, Ethiopia. *Abyssinia Journal of Science and Technology* 6(1). DOI: 10.20372/ajst.2021.6.1.61.

Endalew, Z., Terefe, H. and Dejene, M. (2020) Management of rust (*Puccinia allii* R.) through host resistance and alternate application of fungicides on garlic (*Allium sativum* L.) in north Wollo, Ethiopia. *East African Journal of Sciences* 14(1), 65–82.

Eshetu, G., Bimrew, Y. and Shifa, H. (2018) Association of chocolate spot and faba bean rust epidemics with climate change resilient cultural practices in Bale Highlands, Ethiopia. *Advances in Agriculture* 2018, 13 p. DOI: 10.1155/2018/6042495. Article ID 6042495.

Etana, A., Tadesse, E., Mengistu, A. and Hassen, A. (2013). Advanced evaluation of cowpea (*Vigna unguiculata*) accessions for fodder production in the central Rift Valley of Ethiopia. *Journal of Agricultural Extension and Rural Development* 5(3), 55–61.

FAO (2022a) Statistical Databases and Data-Sets of the Food and Agriculture Organization of the United Nations. Available at: http://www.fao.org/faostat/en/#data (accessed 14 January 2022).

FAO (2022b) Statistical Databases and Data-Sets of the Food and Agriculture Organization of the United Nations. Available at: http://www.fao.org/faostat/en/#data (accessed 14 August 2022).

FAO (2022c) Statistical Databases and Data-Sets of the Food and Agriculture Organization of the United Nations. Available at: http://www.fao.org/faostat/en/#data (accessed 8 October 2022).

Gabrekiristos, E. and Demiyo, T. (2020) Hot pepper fusarium wilt (*Fusarium oxysporum* f. sp. *capsici*): Epidemics, characteristic features and management options. *Journal of Agricultural Science* 12(10), 347–360. DOI: 10.5539/jas.v12n10p347.

Gabrekiristos, E., Teshome, D. and Ayana, G. (2020) Distribution and relative importance of hot pepper fusarium wilt (*Fusarium oxysporum* f. sp. *capsici*) and associated agronomic factors in the central Rift Valley of Ethiopia. *Advances in Crop Science and Technology* 8, 437.

Getachew, S. (2017) Status of forest coffee (*Coffea arabica* L.) diseases in the Afromonte rainforest areas of Ethiopia: A review. *Greener Journal of Agricultural Sciences* 7(1), 19–31. DOI: 10.15580/GJAS.2017.1.122616223.

Getinet, A. and Yalew, D. (2022) Demonstration of recommended fungicides against onion purple blotch disease in Fogera plains, Ethiopia. *Journal of Plant Biology and Crop Research* 5(1), 1046.

Jula, A., Abi, D. and Adugna, G. (2021) Purple blotch disease under different level of fungicide spray frequency and irrigation interval in Adami Tulu Jido Kombolcha district of Ethiopia. *International Journal of Novel Research in Life Sciences* 8(6), 1–6.

Kebede, E. and Bekeko, B. (2020) Expounding the production and importance of cowpea (*Vigna unguiculata* (L.) Walp.) in Ethiopia. *Cogent Food & Agriculture* 6, 1769805. DOI: 10.1080/23311932.2020.1769805.

Kefale, Y., Azanew, A., Worku, M. and Gelaye, M. (2021) Assessment of major insect pests and diseases of sesame (*Sesamum orientale* L) in west Gondar zone, Ethiopia. *Abyssinia Journal of Science and Technology* 6(1), 6–11.

Kindeya, Y.B., Golla, W.N., Kebede, A.A. and Sibhatu, F.B. (2018) Survey and identification of major sesame diseases in low land areas of western zone of Tigray, Ethiopia. *Journal of Biomaterials* 2(2), 58–64. DOI: 10.11648/j.jb.20180202.16.

Mandefro, W., Ahmed, E., Yesuf, M., Lencho, A., Hiskias, Y., *et al.* (2009) Review of vegetable diseases research in Ethiopia. In: Tadesse, A. (ed.) *Increasing Crop Production through Improved Plant Protection—Volume II*. Plant Protection Society of Ethiopia (PPSE), PPSE, Addis Ababa, pp. 203–230.

Mengesha, G.G., Yae, A.J. and Cheleko, D.C. (2022) Effect of integrated management on purple blotch (*Alternaria porri* (Ellis) Cif.) progression and bulb yield of onion at Arba Minch in southern Ethiopia. *Archives of Agriculture and Environmental Science* 7(1), 31–41. DOI: 10.26832/24566632.2022.070106.

Meyer, M., Bacha, N., Tesfaye, T., Alemayehu, Y., Abera, E. *et al.* (2021) Wheat rust epidemics damage Ethiopian wheat production: A decade of field disease surveillance reveals national-scale trends in past outbreaks. *PloS One* 16(2), e0245697. DOI: 10.1371/journal. Pone.0245697.

Mitiku, M. (2017) Assessment of important diseases of major fruit crops (mango, avocado, papaya and orange) in south Omo zone of Ethiopia. *Journal of Natural Sciences Research* 7(5), 68–71.

Pretorius, M.T. (2005) Epidemiology and control of *Pseudocercospora angolensis* fruit and leaf spot disease on citrus in Zimbabwe. Thesis presented in partial innamomic of the requirements for the degree of Master of Science in Agriculture at the University of Stellenbosch, 70 p.

Tadesse, A. (ed.) (2008) Increasing crop production through improved plant protection—Volume I. In: *Plant Protection Society of Ethiopia (PPSE), 19–22 December 2006*, Addis Ababa, Ethiopia. PPSE and EIAR, Addis Ababa, Ethiopia, 598 p.

Tadesse, A. (ed.) (2009) Increasing crop production through improved plant protection—Volume II. In: *Plant Protection Society of Ethiopia (PPSE)*. PPSE and EIAR, Addis Ababa, Ethiopia, 542 p.

Tadesse, N., Ahmed, S., Gorfu, D., Beshir, T., Fininsa, C. *et al.* (2008) Review of research on diseases of food legumes. In: Tadesse, A. (ed.) *Increasing Crop Production through Improved Plant Protection. Proceedings of Plant Protection Society of Ethiopia (PPSE)*. Volume I, PPSE, Addis Ababa, Ethiopia, pp. 85–132.

Teferi, D. and Ayano, A. (2016) Significance of phytoptora disease with special emphasis to avocado and pineapple in south western Ethiopia: A review. *Journal of Natural Sciences Research* 6(7), 76.

Teferi, D. and Weldemichael, G. (2015) Evaluation of avocado root rot (*Phytophthora cinnamomi* Rands) control options in southwest Ethiopia. *Journal of Biology, Agriculture and Healthcare* 5(11), 10–13.

Teferie, B.B., Admasu, M.A. and Damessa, G.G. (2020) Assessment and characterization of mung bean (*Vigna radiata*) bacterial brown spot in eastern Amhara, Ethiopia. *African Journal of Agricultural Research* 16(5), 606–621. DOI: 10.5897/AJAR2019.14681.

Tegegne, G., Abebe, F., Hussein, T., Tilahun, T., Belete, E., *et al.* (2008) Review of maize, sorghum and millet pathology research. In: Tadesse, A. (ed.) *Increasing Crop Production through Improved Plant Protection—Volume I*. Plant Protection Society of Ethiopia (PPSE), Addis Ababa, Ethiopia, pp. 245–302.

Terefe, G. and Tesfaye, D. (2009) Review of research on diseases of fiber crops in Ethiopia. In: Tadesse, A. (ed.) *Increasing Crop Production through Improved Plant Protection—Volume II*. Plant Protection Society of Ethiopia (PPSE), Addis Ababa, Ethiopia, pp. 305–314.

Terefe, G., Gorfu, D., Tesfaye, D. and Abebe, F. (2009) Review of research on diseases of oil crops in Ethiopia. In: Tadesse, A. (ed.) *Increasing Crop Production through Improved Plant Protection—Volume II*. Plant Protection Society of Ethiopia (PPSE), Addis Ababa, Ethiopia, pp. 253–274.

Teshome, E. and Kora, D. (2021) Integrated management of chocolate spot (*Botrytis fabae* Sard) through host resistance and fungicide application in the highlands of Bale, southeastern Ethiopia. *Journal of Agricultural Science and Food Technology* 7(1), 86–91. DOI: 10.17352/2455-815X.000093.

Tucho, A., Lemessa, F. and Berecha, G. (2014) Distribution and occurrence of mango anthracnose in humid agro-ecology of southwest Ethiopia. *Plant Pathology Journal* 13(4), 268–277.

van der Graaff, N.A. (1981) *Selection of Arabica Coffee Types for Resistance to Coffee Berry Disease in Ethiopia*. Med. Landbouwhogeschool, Wageningen, The Netherlands.

Westphal, E. (1975) *Agricultural Systems in Ethiopia*. Joint Publication of the College of Agriculture, Haile Sellassie I University, Ethiopia, and the Agricultural University, Wageningen, the Netherlands.

Worku, Y. and Dejene, M. (2012) Effects of garlic rust (*Puccinia allii*) on yield and yield components of garlic in Bale Highlands, southeastern Ethiopia. *Journal Plant Pathology & Microbiology* 3, 2. DOI: 10.4172/2157-7471.1000118.

Wubshet, M.L. and Chala, A. (2021) Management of faba bean chocolate spot (*Botrytis fabae*) through varieties and fungicide application frequencies in southern Tigray, Ethiopia. *Archives of Phytopathology and Plant Protection* 54(19–20), 2233–2246. DOI.org/10.1080/03235408.2021.1925516

Yesuf, M. (2007) Distribution and management of *Phaeoramularia* leaf and fruit spot disease of citrus in Ethiopia. *Fruits* 62, 99–106.

Yesuf, M. (2013) Pseudocercospora leaf and fruit spot disease of citrus: Achievements and challenges in the citrus industry: A review. *Agricultural Sciences* 4(7), 324–328. DOI:10.4236/as.2013.47046.

Yesuf, M., Mandefro, W., Ahmed, E., Adugna, G., Tadesse, D., Hussen, T. and Shehabu, M. (2009) Review of research on fruit crop diseases in Ethiopia. In: Tadesse, A. (ed.) *Increasing Crop Production through Improved Plant Protection—Volume II*. Plant Protection Society of Ethiopia (PPSE), Addis Ababa, Ethiopia, pp. 231–252.

You, M.P., Eshete, B.B., Kemal, S.A., van Leur, J. and Barbetti, M.J. (2021) *Physoderma*, not *Olpidium*, is the true cause of faba bean gall disease of *Vicia faba* in Ethiopia. *Plant Pathology* 70(5), 1180–1194. DOI: 10.1111/ppa.133.

Annex Table 14.1. Examples of major diseases of major cereals in Ethiopia. Source: constructed by the author from Badebo *et al.* (2008), Tadesse *et al.* (2008) and Tegegne *et al.* (2008).

Scientific name	Common name	Host(s)
Rhyncosporium secalis	Scald	Barley
Pyrenophora teres	Net blotch	Barley
Cochliobolus sativus	Spot blotch	Barley
Puccinia hordei	Rust	Barley
Cercospora zeae-maydis	Grey leaf spot	Maize
Exserohilum turcicum	Turcicum leaf blight	Maize
Maize Streak Virus	Streak virus	Maize
Colletotrichum graminicola	Anthracnose	Sorghum
Puccinia purpurea	Rust	Sorghum
Sphacelotheca sorghi	Covered kernel smut	Sorghum
Sphacelotheca cruenta	Loose kernel smut	Sorghum
Puccinia striiformis	Yellow/stripe rust	Wheat
Puccinia graminis f. sp. *tritici*	Stem/black rust	Wheat
Puccinia recondita f. sp. *triticina*	Leaf/brown rust	Wheat
Mycosphaerella graminicola	Septoria/glume blotch	Wheat

Annex Table 14.2. Examples of major diseases of major grain legumes in Ethiopia. Source: constructed by the author from Tadesse *et al.* (2008).

Scientific name	Common name	Host(s)
Fusarium oxysporum f. sp. *ciceris*	Fusarium wilt	Chickpea
Rhizoctonia bataticola	Dry root rot	Chickpea
Ascochyta rabiei	Ascochyta blight	Chickpea
Colletotrichum lindemuthianum	Anthracnose	Common bean
Uromyces appendiculatus	Rust	Common bean
Xanthomonas campestris pv. *phaseoli*	Common bacterial blight	Common bean
Rhizoctonia solani	Web blight	Common bean
Phaeoisariopsis griseola	Angular leaf spot	Common bean
Ascochyta phaseolorum	Ascochyta blight	Cowpea
Botrytis fabae	Chocolate spot	Faba bean
Fusarium solani	Black root rot	Faba bean
Uromyces viciae-fabae	Rust	Faba bean
Erysiphe polygoni	Powdery mildew	Fieldpea
Mycosphaerella pinodes	Spot/blight	Fieldpea
Fusarium oxysporum f. sp. *lentis*	Fusarium wilt	Lentil
Uromyces viciae-fabae	Rust	Lentil
Oidium spp.	Powdery mildew	Lentil, grasspea
Pseudomonas phaseolicola[a]	Halo blight	Mungbean

[a]Own observation.

Annex Table 14.3. Examples of major diseases of major oilseeds in Ethiopia. Source: constructed by the author from Tadesse *et al.* (2009) and Terefe *et al.* (2009).

Scientific name	Common name	Host(s)
Alternaria spp.	Stem & leaf blight	*Noug*
Albugo candida	White rust	Brassicas
Alternaria brassicae	Leaf and pod spot	Brassicas
Cercospora arachidicola	Early leaf spot	Groundnut
Cercospora personata	Late leaf spot	Groundnut
Puccinia arachidis	Rust	Groundnut
Fusarium oxysporum	Wilt	Linseed
Ramularia carthami	Leaf spot	Safflower
Phytophthora dreschleri/P. parasitica	Root rot	Safflower
Pseudomonas sesami	Blight	Sesame
Xanthomonas sesami	Blight	Sesame
Mycoplasma-like organism	Phyllody	Sesame
Pyrenochaeta glycines	Leaf spot	Soybean
Xanthomonas phaseoli pv. *sojense*	Bacterial pustule	Soybean
Plasmopara halstedii	Downy mildew	Sunflower
Puccinia helianthi	Rust	Sunflower
Sclerotinia sclerotiorum	Stem & head rot	Sunflower
Septoria helianthi	Leaf spot	Sunflower

Annex Table 14.4. Examples of major diseases of major vegetables in Ethiopia. Source: constructed by the author from Mandefro *et al.* (2009).

Scientific name	Common name	Host(s)
Albugo candida	White rust	Brassicas
Alternaria brassicae	Leaf & pod spot	Brassicas
Puccinia allii/P. porii	Garlic rust	Allium
Perenospora destructor	Downy mildew	Onion/shallot
Alternaria porii	Purple blotch	Onion/shallot
Fusarium oxysporum	Fusarium wilt	Pepper (capsicum)
Leveillula taurica	Powdery mildew	Pepper (capsicum)
Sclerotium rolfsii	Southern blight	Pepper (capsicum)
Botrytis cinerea	Grey mould	Snap bean
PVX, PVY, CMV	Virus diseases	Vegetables

CMV = Cucumber Mosaic Virus; PVX = Potato Virus X; PVY = Potato Virus Y.

Annex Table 14.5. Examples of major diseases of major fruit crops in Ethiopia. Source: Yusuf *et al.* (2009).

Scientific name	Common name	Host(s)
Venturia inaequalis	Scab	Apple
Agrobacterium tumefaciens	Crown gall	Apple
Podosphaera leucotricha	Powdery mildew	Apple
Phytophthora spp.	Crown and root rot	Apple
Phytophthora cinnamomi	Root rot/decline	Avocado
Xanthomonas campestris pv. *musacearum*	Banana wilt	Banana
Phaeoramularia angolensis	Leaf and fruit spot	Citrus
Colletotrichum gloeosporioides	Citrus anthracnose	Citrus
Xanthomonas campestris pv. *citri*	Citrus canker	Citrus
Penicillium italicum	Fruit rot	Citrus
Colletotrichum gloeosporioides	Anthracnose	Mango
Asperisporium caricae	Black spot	Papaya

Annex Table 14.6. Examples of major diseases of coffee, sugarcane, and fibre crops in Ethiopia. Sources: constructed by the author from Adugna *et al.* (2009), Tadesse *et al.* (2008) and Terefe and Tesfaye (2009).

Scientific name	Common name	Host
Colletotrichum kahawae	Coffee berry disease	Coffee
Gibberella xylarioides	Coffee wilt	Coffee
Hemileia vastatrix	Coffee leaf rust	Coffee
Bipolaris sacchari	Eye spot	Sugarcane
Ustilago scitaminea	Smut	Sugarcane
Leifsonia xyli subsp. *xyli*	Ratoon stunting disease	Sugarcane
Xanthomonas campestris pv. *malvacearum*	Bacterial blight	Cotton
Fusarium oxysporum f. sp. *vasinfectum*	Fusarium wilt, root rot	Cotton
Unidentified	Dubti Syndrome	Cotton
Botrytis cinerea	Stem rot	Kenaf
Colletotrichum capsici	Crown rot	Kenaf
Phoma spp.	Collar rot	Kenaf

Annex Table 14.7. Examples of major diseases of commercial flowers in Ethiopia. Source: Ahmed *et al.* (2009).

Scientific name	Common name	Host
Fusarium oxysporum	Blight	Chrysanthemum
Melampsora ricini	Rust	Castor
Oidium spp.	Powdery mildew	Jatropha
Oidium spp.	Powdery mildew	Fenugreek
Puccinia menthae	Rust	Mint
Puccinia nakanishikii	Rust	Lemongrass
Puccinia punctiformis	Rust	Vernonia

Annex Table 14.8. Examples of major diseases of aromatic, medicinal, and non-edible oil plants in Ethiopia. Source: constructed by the author from Gorfu *et al.* (2009).

Scientific name	Common name	Host
Peronospora sparsa	Downy mildew	Roses
Sphaerotheca macularis	Powdery mildew	Roses
Sphaerotheca pannosa var. *rosae*	Powdery mildew	Roses
Oidium spp.	Powdery mildew	Roses
Botrytis cinerea	Grey mould	Roses
Rhizoctonia solani	Stem rot	Carnation, geranium, gypsophylla, hypericum
Fusarium oxysporum f. sp. *dianthi*	Wilt	Carnation, hypericum
Botrytis cinerea	Blight, grey mould	Carnation, geranium, hypericum, roses
Erysiphe spp.	Powdery mildew	Geranium
Fusarium oxysporum	Wilt	Gypsophylla
Pythium ultimum	Root & crown rot	Geranium
Erwinia chrysanthemi	Bacterial wilt	Chrysanthemum
Phytophthora parasitica	Crown rot	Gypsophylla

15 Economically Important Weeds and their Management

Taye Tessema,[1] Wondimagegnehu Mersie,[2] Amare Fufa,[3] Takele Negewo,[4] Firehun Yirefu[5] and Tsedeke Abate[6]*

[1]*Ethiopian Institute of Agricultural Research, Addis Ababa, Ethiopia;* [2]*Virginia State University, Petersburg, VA, USA;* [3]*Ethiopian Institute of Agricultural Research, Melkassa Research Center, Adama, Ethiopia;* [4]*Ethiopian Institute of Agricultural Research, Ambo, Ethiopia;* [5]*Kwale International Sugar Research, Nairobi, Kenya;* [6]*Homegrown Vision, Addis Ababa, Ethiopia*

Abstract

This chapter examines the outcomes of weed management research and development in Ethiopia and identifies future priority areas. So far, nearly 1100 weed species representing 552 genera and 126 plant families have been registered. Weeds in this country can cause a minimum of ~15% to a total crop failure. The research system has accomplished several milestones, including evaluation of new herbicides and development of herbicide recommendations, non-chemical methods (including weed biology and dynamics in crops and cropping systems), management of regular, parasitic and invasive species, and transfer of management technologies. There have also been successes in the biological control of the whitetop weed (*Parthenium hysterophorus*) and integrated management of the witchweed (*Striga hermonthica*) on sorghum, and management of herbicide-resistant annual and perennial grasses. Future emphasis will include risk analysis, invasive alien and parasitic weeds, biocontrol, herbicide resistance, herbicide-tolerant crops, herbicide residue and environmental safety.

Introduction

Weeds reduce crop yield and displace desirable pasture species, thereby adversely affecting livestock production and reducing the value of agricultural products. Some have detrimental health effects on humans as well as domestic animals. They are more abundant than all other pests in tropical African farms (Obuo *et al.*, 1997). The relatively high temperature and humidity facilitate the fast and extensive growth of weeds in tropical Africa (Akobundu, 1980). Weeds cause more damage to farms in tropical Africa than in temperate zones such as many regions of North America. Weeds compete with crops for moisture, sunlight and nutrients, thereby reducing crop yield. In addition, some weeds produce toxic substances and become allelopathic to crops, while others become parasitic and drive water and nutrients from the host crops. In Africa, the yield loss due to weeds ranges from 25% to total crop failure (Akobundu, 1987; van Rijn, 2000; Tessema and Ejeta, 2019). These losses in agricultural crops vary depending on weed species composition, weed infestation level, availability of improved weed control technology, the level of weed management practices by farmers, and the cost of control.

* Corresponding author: tsedeke@homegrownvision.org

 Untold Stories of African Agriculture: Lessons from Ethiopia (T. Abate ed.)
DOI: 10.1079/9781800626386.0015

In the Ethiopian farming systems, weeds have been identified as one of the main crop production constraints. Various reports (cf. Abate, 1986; Tadesse, 2008, 2009) suggest that, depending on the crop, losses due to weeds in this country range from 15% to almost total crop failure each year (Fig. 15.1). Here, as in many countries elsewhere in Africa, weeds are controlled primarily by manual weeding. Manual weeding includes hoeing, pulling weeds by hand and occasionally slashing them. According to Chikoye *et al.* (2007), farmers in Africa spend 50–70% of their time hand weeding their crop fields. This task of hand weeding mostly falls on women and children (Akobundu, 1987). In Ethiopia, additional labour is hired to control weeds in high-value crops such as *teff* (*Eragrostis tef*) during the peak growing period. Estimates of labour required for hand weeding of *teff* and wheat, for example, range from 40–100 workdays/ha and 20–75 workdays/ha, respectively (Franzel *et al.*, 1989).

Hand weeding has many limitations as an effective weed control method. These include a lack of available labour for hire during the peak planting season and a lack of money to pay for workers to weed crop fields (Johnson, 1995). Farmers first prepare seedbeds and finish planting before they start hand weeding. This causes delays in weeding. Such late weeding will cause major crop yield losses, especially if it is done after the critical weed competition period. In addition, farmers who grow different crops are faced with different planting dates that can delay the start of weeding. Thus, weeding must be carried out at the early crop growth stage before weeds cause permanent damage.

Though family labour and, occasionally, hired human power are used for weeding operations in small-scale agriculture in Ethiopia, the country also spends significant amounts of foreign currency to purchase herbicides to control weeds, primarily on commercial farms. Herbicide use in this country showed upward trends from the first available data in 1997 to 2020 (Fig. 15.2). For example, Ethiopia imported annually about >4700 MT of herbicides at an estimated cost of US$18.81 million between 2016 and 2020 (FAO, 2023).

In this chapter, we present highlights of progress made in weed management research and development over the years in Ethiopia.

Major Weeds

Weeds in Ethiopia are classified into three major categories: regular weeds, parasitic weeds and

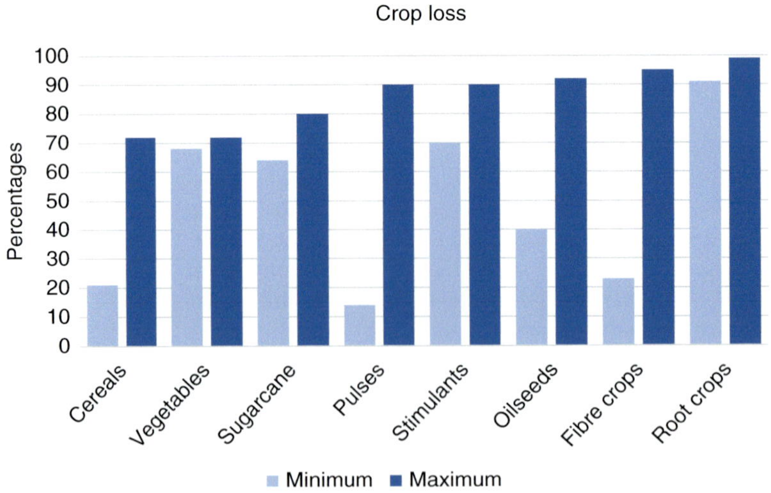

Fig. 15.1. Crop yield loss range in major crops due to weeds in Ethiopia: cereals (barley, rice, sorghum, wheat); vegetables (cabbage, carrot, chilli, garlic, onion, tomato); grain legumes (chickpea, common bean, faba bean, fieldpea, lentil, soybean); stimulants (coffee, tea); oilseeds (groundnut, linseed, *noug*, rapeseed, sesame); fibre crops (cotton, kenaf); root crops (sweet potato). Source: constructed by the authors from various sources.

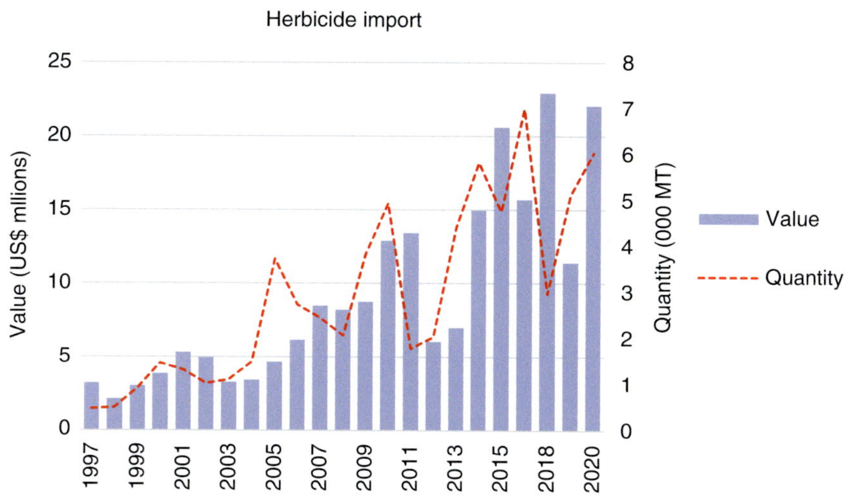

Fig. 15.2. Herbicide import quantity and value in Ethiopia. Source: calculated and constructed by the authors from FAO (2023).

invasive weeds. The Ethiopian agricultural research system has undertaken weed surveys, biological and ecological studies, and control methods since the late 1960s. Below we provide brief highlights for each category.

Regular weeds

So far, two critical reviews have been made on weeds and the other crop protection disciplines in Ethiopia. The first one was conducted in 1985 (Abate, 1986) and the second took place in 2006 (Tadesse, 2008, 2009). These two reviews have made significant contributions to the knowledge of weed management practices in the country. The major activities on regular weeds are summarized as follows:

- identification of weed problems on various major crops;
- development of national database of weeds;
- determining potential yield losses, critical periods for weeding and economic threshold of weed densities; and
- evaluation of effective chemicals.

Several surveys have been used to determine the distribution and impact of weeds on Ethiopia's agriculture.

Weeds cause most injury to crops during certain crop growth stages – a period termed 'the critical period of competition' – and control during this period is especially important. Critical periods have been established for several crops in Ethiopia (Table 15.1). Besides competitive ability and economic threshold, weed densities for the major grass and broadleaved weed species in wheat and barley were determined (Tessema and Tanner, 1997; Negewo *et al.*, 2006; Mola and Belachew, 2015). Numerous studies on the efficacy of different herbicides in different crops have been conducted in Ethiopia over the last 50 years.

The database provides information on the occurrence and diversity of major weeds in different crops and cropping systems as well as non-crop areas of the country. In Ethiopia, up to 1093 different weed species in 552 genera and 126 plant families have been collected from arable lands, range lands, forests and aquatic environments to date (APPRC, 2014).

Screening and efficacy trials of different selective and non-selective herbicides were conducted over the last 50 years in major crops such as *teff*, wheat, maize, sorghum, barley, faba bean (*Vicia faba*), fieldpea, common bean, coffee and cotton. Promising herbicides in terms of efficacy and productivity were identified for each crop type. Overall, there are about 154 recommended herbicides for use in different crops and cropping systems (MOANR, 2019).

Table 15.1. Critical period for weed competition and optimum weeding frequencies requirement for selected crops in Ethiopia. Sources: constructed by the authors from various sources.

Crop	Weed-free period required (WAE)	Length of competition tolerated (WAE)	Optimum weeding requirements (frequency)
Barley	3	6	1
Faba bean	5	8	2
Fieldpea	4	6	1
Common bean	5	8	2
Hot pepper	3	6	2
Lentil	4	8	2
Maize	3	8	2
Onion	3	6	2
Sorghum	3	8	2
Soybean	5	8	2
Teff	5	8	2
Tomato	3	6	2
Wheat	4	8	2

WAE = weeks after emergence.

As part of non-chemical and biological methods of weed management, investigations were carried out on the effects of frequency of tillage, planting density, rate and pattern, method of fertilizer application, and frequency of hand weeding on crop growth and yield, along with the reduction of weed infestation. Results of crop rotation and fertilizer effects on weed population in wheat- and barley-growing regions in Arsi indicated that such approaches proved to be satisfactory in reducing infestation of certain problematic weed species and improve crop yield (Taa and Tanner, 1998a). Also, in wheat- and maize-based systems, weed interference was minimized by changes in crop management practices such as crop rotation, tillage methods, crop residue management and soil fertility measures (Taa and Tanner, 1998b). Research on mechanical control methods using weeders was made at the Melkasa agricultural research centre by Agricultural Implements Research. An ox-drawn row-seeder has been developed to facilitate the placement of wheat and faba bean seed and basal fertilizer in rows, thereby facilitating manual and mechanical weed control (Taa *et al.*, 1997).

Parasitic weeds

Parasitic weeds can reduce host crop yield by up to 100%, depending on the levels of infestation, crop susceptibility and environmental conditions. The parasitic weeds affecting agricultural crops in Ethiopia are numerous and diverse. The major plant families of these parasitic weeds are Orobanchaceae, Convolvulaceae and Loranthaceae. Orobanchaceae has the most damaging species, such as the witchweed (*Striga* spp.) and broomrapes (*Orobanche* spp.). Dodders and mistletoes are members of the families Convolvulaceae and Loranthaceae, respectively.

Witchweed

The most important witchweeds in Ethiopia are *Striga hermonthica, S. asiatica and Alectra vogelii. S. asiatica* is a serious problem in the eastern and southern parts of the country. On the other hand, *S. hermonthica* is a serious problem in the western and northern parts of the country, while *Alectra vogelii is* lately reported in association with groundnut in lowland areas of Ethiopia.

Striga spp., commonly known as 'witchweed' because of their damage to the crop during their subterranean growth, i.e. before the plants emerge, are the major root hemiparasitic weeds affecting cereal crops. They are frequently referred to by their *genus* name as 'striga'. Sorghum, maize, finger millet, sugarcane, *teff* and some wild plants like *Rottboellia* spp. and wild sorghum are the major host plants in Ethiopia. They are particularly abundant in areas where soil fertility and moisture stress are the limiting factors. Although the exact distribution and coverage of these species are not yet known, it is estimated that over 50% of the area covered by sorghum, or about 1 million ha, is infested by *Striga* species in Ethiopia. Depending on infestation level, crop losses of 60–90% are common in sorghum-growing areas, and arable lands are often abandoned because of the prohibitive parasite populations (Tessema and Ejeta, 2019).

The most common striga species parasitic on sorghum and finger millet in Ethiopia are *S. hermonthica* and *S. asiatica*. They have pink flowers and red flowers, respectively. These two species represent the most devastating striga species in the world. The small striga seeds (270 million in 1 kg) are easily spread by wind, water or animals, and with infested tools and contaminated seed. A single plant may produce 20,000 seeds that are capable of remaining dormant and viable in the soil for over 20 years, which makes the parasite very difficult to control (Heide-Jorgensen, 2008).

Currently, striga is primarily controlled in Ethiopia by hand pulling. However, as reported by Reda *et al.* (2005), early weeding does not seem to influence grain yield. In fact, it leads to the re-sprouting of more shoots from underground buds, which further aggravates the problem. Thus, late weeding is recommended. It was also found that late weeding requires less than half the time needed for early hand pulling (Parker and Riches, 1993).

Not planting a susceptible cereal in a striga-infested field will prevent new seeding by striga and can help reduce the soil seed bank in the long run. Crops may vary in their effectiveness in reducing the striga seed population in the soil. Some so-called trap crops are known to produce striga germination stimulants but are not susceptible to them. These trap crops are particularly useful in causing suicidal germination and hastening the decline of the soil seed bank. Parker and Riches (1993) reported that the infestation level of striga on susceptible sorghum (variety 'IS 9302') following Sudan grass was five times lower, though this did not result in significant differences in yield.

Experiments conducted in Ethiopia showed that the effects of nitrogen varied with maize varieties (Reda *et al.*, 1997). Other experiments designed to develop integrated nutrient management showed that combining 41 kg/ha and 30 MT/ha resulted in a significant reduction in striga infestation and a considerable increase in sorghum yield (Esilaba *et al.*, 1998). However, the outcome of N fertilizer depended on weather patterns and the inherent fertility of a given location (Reda *et al.*, 2005). Intercropping is another practice that can help to maintain soil fertility, satisfy farmers' interest in getting different crops, and reduce damage due to weeds, insects and disease-causing agents. Results from the experiments conducted in Ethiopia showed that one row of legumes (cowpea and haricot bean) with every two rows of sorghum at Sirinka (Reda *et al.*, 1997) and the same planting arrangement of sorghum and cowpea at Adibakel (a dry highland in Tigray) resulted in higher crop yields and better striga control (Reda *et al.*, 2005).

Reda *et al.* (1997) reported that the most effective control method of *S. hermonthica* is through the integrated use of resistant varieties, row planting, fertilizer application and hand pulling of striga after flowering, or the use of 2,4-dichlorophenoxyacetic acid weed killer. At Adibakel, the use of sorghum row planting, fertilizers and hand pulling (during flowering) resulted in a 50% reduction in striga shoot counts and a 48% increased grain yield (Reda *et al.*, 1997).

There have been attempts to develop an integrated striga management (ISM) package that combines striga-resistant sorghum varieties with water conservation practice using tie-ridge and soil fertility amendments, i.e. organic and/or inorganic fertilizers (Tessema and Ejeta, 2019). Initially, the ISM package was demonstrated in striga-infested sorghum production areas from 2002 to 2005 in collaboration with Purdue University. The striga-resistant varieties developed are also drought-tolerant, early maturing and have good nutritional quality.

The ISM was widely demonstrated and popularized in Ethiopia from 2012–2016 by the project funded by the Bill & Melinda Gates Foundation entitled 'Integrated *Striga* Control in Sorghum in Ethiopia and Tanzania Project' by ensuring the production of quality seed of striga-resistant sorghum varieties. This project was developed by Purdue University in collaboration with the Ethiopian Institute of Agricultural Research (EIAR) and implemented with the respective bureaus of agriculture in Amhara, Oromia, SNNP and Tigray regional states. At the start, training was given to the selected farmers owning demonstration plots on the application of the ISM technology package before planting. Other agronomic practices, such as method of planting, spacing, thinning seedlings, weeding, and insect and disease control were made as per the recommended practices for sorghum production (Tessema and Ejeta, 2019).

In November 2013, a superimposed trial was made at two locations with high striga infestation in Babile district, eastern Harerge, to compare the performance of a striga-resistant variety ('*Gobiye*') with an improved non-striga-resistant sorghum variety ('*Teshale*') and a farmers' local variety. The result showed a significant difference in sorghum yield, stand count, and thousand seed weight among the varieties and in striga count. High relative yield advantages of 97.9% and 40.9% were obtained by using 'Gobiye' over the local and 'Teshale' varieties, respectively. Such a yield advantage of the ISM package was a highly commendable result for sorghum growers (Tessema and Ejeta, 2019). Furthermore, hundreds of ISM package demonstrations made on farmers' fields and farmer training centres during the project implementation period in the four project regions in Ethiopia depicted better performance of ISM practice as compared to farmers' practice. Field-day demonstrations were also made in project implementation regions at various levels, i.e. at the *kebele*, district, zonal and regional levels. Popularization of striga-resistant cultivars was also made to expand deployment of officially released striga-resistant and drought-tolerant sorghum cultivars through small packs of seeds and advisory practices. As a result, the ISM package generated a high demand for seeds of striga-resistant cultivars, especially during the drought seasons.

Nevertheless, none of the single current measures to control striga has proved to be effective. In addition, some of these control measures are not accessible or are too expensive for smallholder farmers. There is little or no information on the biodiversity and functional potential of indigenous microorganisms living on or in the roots of host crops and how these microorganisms impact plant growth and striga. Furthermore, no knowledge exists on the relationships between root architecture, root metabolome, root microbiome composition and functions, and resistance to striga. Microbes have different striga suppressive traits: (i) production of phytohormones that suppress or induce striga seed germination; (ii) being pathogenic towards the parasite; (iii) interference with host–parasite signalling (de-activation of strigolactone); (iv) enhancement of nutrient cycling and uptake of soil minerals (e.g. P and N); and (v) modulation of host physiology and inducing resistance (PROMISE, 2021).

To meet these challenges, a project entitled PROMISE (Promoting Root Microbes for Integrated Striga Eradication) was initiated as a discovery programme funded by the Bill & Melinda Gates Foundation. The goal of PROMISE was to suppress striga infestation by depleting the striga seed bank and enhance host crop productivity using microbes in conjunction with resistant/tolerant cultivars and agronomic practices in Africa. The project explored the potential of soil and plant microbiomes (i.e. the collective communities of microbes, their genomes and their interactions) as an effective, generic and environmentally sound approach to improving crop productivity.

For example, there is ongoing research on a new striga management approach based on soil and plant root-associated microorganisms having specific antagonistic effects via secondary metabolites and volatile organic compounds (Trognitz *et al.*, 2016; Masteling *et al.*, 2019). The amino acids L-leucine and L-tyrosine excreted by *Fusarium oxysporum* have been shown to be toxic to striga but innocuous to maize (Nzioki *et al.*, 2016).

Decomposition of methionine in soil will result in the production of sulfur volatiles such as dimethyl disulfide (DMDs) and dimethyl trisulfide (DMTs). Masteling *et al.* (2019) showed that these volatiles inhibit striga seed germination (DMDs) or even kill striga seed (DMTs). Greenhouse and field experiments in Ethiopia by the EIAR also showed a reduction in the number of striga plants emerging on a sorghum root when compared to the control (methionine untreated soil). The effect of adding methionine also varied with soil samples collected from different striga-infested sorghum-growing areas in Ethiopia.

Broomrapes

A characteristic of the broomrapes is that they have no green parts, with only the leafless flowering stem visible above ground. They are short-lived parasitic plants that grow on the roots of broadleaf host plants to obtain their growth requirements. Normal development of the parasite starts with seed germination, which comes in response to the reception of a chemical stimulus from host roots following a period of conditioning that is required for the dormancy–germination transition (Zhou *et al.*, 2004).

Most damage occurs before its emergence, and only 10–30% of the attached parasites emerge (Sauerborn, 1991). The broomrape can produce many tiny seeds ranging in size from 0.2–2.0 mm that can easily disperse over long distances by various means, and in most cases, remain viable in the soil for many years. It has been estimated that per season a given broomrape plant can bear up to half a million seeds, which can persist in the soil for up to 20 years (Gevezova *et al.*, 2012). Thus, a single broomrape plant is capable of significantly increasing the soil seed bank and/or rapidly infesting new fields via flood or irrigation water, wind, animals and humans.

In Ethiopia, *Orobanche crenata* is highly destructive on grain legumes like faba bean, fieldpea, chickpea and lentil. *O. cernua* and *Phelipanche ramosa*, on the other hand, are serious weeds of vegetables (tomato, tobacco and other solanaceous crops). They also parasitize numerous crop species in a wide range of plant families. The roots of *Orobanche* attach to the roots of the host crop and draw nutrients and water from them, reducing growth and yield. Severe problems with tomatoes have been reported, and the viability of some tomato juice factories has been threatened by the reduced yields. The average losses in tomato yield have been estimated at about 36%, while losses may reach up to 100% (Fessehaie, 1998).

Reference should be made to the volumes of the crop protection proceedings mentioned earlier and other selected publications, such as Sherif *et al.* (1990), Gebremariam (2005), Sahile *et al.* (2005), Parker (2009), Soto *et al.* (2010), Habimana *et al.* (2014), Abebe *et al.* (2015), and SARC (2015) for more details on broomrapes.

Invasive alien species

Invasive alien species (IAS) have been introduced to Ethiopia both accidentally and intentionally. Several alien species are reported to be spreading at a rapid rate. There is a long list of IAS occurring in Ethiopia. We have listed the top 20 in Table 15.2. The list includes some of the well-established species that are major threats to Ethiopian agriculture, such as the whitetop weed (*Parthenium hysterophorus*), mesquite (*Prosopis juliflora*) and the water hyacinth (*Eicchornia crassipes*). There are also emerging invasive plant species such as *Verbesina encelioides* (commonly known as gold weed, wild sunflower, or butter daisy) that are currently invading both crop and non-croplands in low and mid-altitude areas of the country (Fufa *et al.*, 2022). Below we present brief descriptions of two species as examples of IAS.

Whitetop weed/parthenium

The invasive whitetop weed, *Parthenium hysterophorus*, is suspected to have been introduced accidentally through aid grain shipments or from Somalia during the Ethio-Somali war (Tessema *et al.*, 2004). Its impact on natural habitats has not been assessed, but it clearly poses a major threat to rangelands and croplands (Tessema, 2002). It is known as '*kinche arem*' in Amharic and '*faramsissa*' (meaning 'sign-away your land') in the Oromo language. It entered Ethiopia in the late 1970s and quickly spread to all regions of the country (McConnachie *et al.*, 2010). Farmers consider it one of the most damaging weeds in both arable and grazing lands (Tana and Milberg, 2000). It reduces yields of food crops such as sorghum, ranging from 40% to 95%, depending on plant density (Tana and Milberg, 2000). This weed is found in crop fields such as *teff* until harvest, thereby competing with this important food crop for moisture and nutrients throughout the season. It displaces valuable pasture species (Nigatu *et al.*, 2010), thereby severely affecting livestock production. If livestock graze on this plant, it taints their meat and milk, thereby reducing their market value.

People also develop allergies such as dermatitis, fever and asthma when they come in contact with parts of the plant, and more seriously,

Table 15.2. The top 20 invasive alien species in Ethiopia. Source: constructed by the authors from various sources.

Names		Highly affected ecosystems[a]	Distribution
Scientific	Common (English, local)		
Argemone ochroleuca	Pale Mexican prickly pear	1, 2, 3, 4, 5, 6	Wide
Eichhornia crassipes	Water hyacinth	7	Wide
Lantana camara	Lantana, shrub verbena	1, 2, 3, 4, 5, 6, 8	Wide
Parthenium hysterophorus	Whitetop weed, Santa Maria	1, 2, 3, 4, 5, 6, 7, 8	Wide
Prosopis juliflora	Mesquite	1, 2, 3, 4, 5, 6, 7, 8	Wide
Senna didymobotrya	African senna, peanut butter cassia	1, 2, 3, 4, 5, 6, 8	Wide
Xanthium strumarium	Rough cocklebur, common cocklebur	1, 2, 3, 4, 5, 6	Wide
Acacia drepanolobium	Whistling thorn	1, 3, 4	Moderate
Calotropis procera	Milkweed, apple of sodom, Calotrope	1, 2, 3, 4, 5, 6, 7, 8	Moderate
Mimosa diplotricha	The giant sensitive plant	1, 2, 3, 4, 5, 6, 8	Moderate
Opuntia ficus-indica	Cactus pear, prickly pear	3, 4, 5, 6	Moderate
Opuntia stricta	Erect prickly pear	3, 4, 5, 6	Moderate
Senna occidentalis	Coffee senna, stinkweed, ant bush	2, 4, 5	Moderate
Xanthium spinosum	Spiny cocklebur	1, 2, 3, 4, 5, 6	Moderate
Acacia saligna	Orange wattle, golden wreath wattle	2, 3, 4, 5	Limited
Cryptostegia grandiflora	Rubber vine	2, 3, 4, 7, 8	Limited
Mimosa pigra	Black mimosa	3, 4, 7	Limited
Nicotiana glauca	Tree tobacco	1, 2, 3, 4, 5, 6	Limited
Parkinsonia aculeata	Jerusalem thorn, palo verde	2, 4, 5, 6	Limited
Verbesina encelioides	Gold weed, wild sunflower, butter daisy	1, 3, 4	Limited

[a]1 = cultivated land; 2 = roadside; 3 = grazing areas; 4 = non-cultivated land; 5 = rural villages; 6 = urban areas; 7 = waterways; 8 = forest areas.

the pollen causes respiratory problems such as asthma and bronchitis (Evans, 1997). These reactions could be attributed to the presence of secondary plant compounds (parthenin, chlorogenic acid, isocholorogenic acid, vanilic acid and caffeic acid), which are found in the plant with significant variation in their concentrations among the different plant parts, dependent on plant locality, moisture content and plant size (Wiesner *et al.*, 2007). The social cost of whitetop weed in Ethiopia was measured by Disability Adjusted Life Years and its equivalence in terms of monetary value was estimated at US\$2.5–4.4 million (Rupschus *et al.*, 2007). Thus, more resources have to be invested to tackle the weed, as the estimated loss is disproportionate to the cost of investment in its research and development activities.

Whitetop weed can dominate the landscape it invades due to its ability produce large numbers of seeds (up 28,000 per plant). They can germinate and grow anytime of the year if the soil moisture is adequate and temperature is ideal, flower and set seed within 4–6 weeks of germination, and can tolerate a variety of environmental conditions. Whitetop weed leaves and pollen can produce and release a toxic sesquiterpene lactone allelochemical known as parthenin to their surrounding environment (Kanchan and Chandra, 1980). This toxic allelochemical can inhibit the germination and growth of other plants (Mersie and Singh, 1987; Navie *et al.*, 1996; Evans, 1997), thus allowing the weed to form monocultures in previously disturbed and overgrazed fields.

There are several methods of whitetop weed control that can be used in crop and pasture fields, including hand weeding and chemical control. Hand weeding is not practical because the weed emerges throughout the season, requiring multiple visits to the field during the busy cropping season. Many people are also allergic to its leaves and pollen, thereby limiting the availability of labour for hand weeding. Herbicides require multiple applications to provide season-long effective control and are a cost-prohibitive solution for most smallholder African farmers.

Presently, the most sustainable, economical and effective method of managing whitetop weed is by using biological control agents. In its native range, natural enemies keep the weed under control. Several countries, including Australia (Dhileepan and McFadyen, 2012), India (McConnachie, 2015) and South Africa (Strathie *et al.*, 2011), have introduced natural enemies to control whitetop weed. One of these natural enemies used against the weed is *Zygogramma bicolorata* (Coleoptera: Chrysomelidae). The adult *Z. bicolorata* usually lays eggs on the underside of young parthenium leaves, either in a group or singly. The eggs hatch within 4–6 days, and both the larvae and the adults feed on the leaves of the weed. Fully developed larvae enter the soil and pupate in chambers before they emerge as adults.

The larval and pupal stages last 23 to 28 days, depending on the environmental temperature. Adults and larvae feeding on parthenium can cause complete defoliation of the weed. Australia and India tested and released *Z. bicolorata* in the field over 30 years ago (Dhileepan and Strathie, 2009). South Africa released the leaf-feeding beetle in 2013 (McConnachie, 2015) after conducting an extensive host range test. At high densities, the adults and larvae of *Z. bicolorata* can entirely defoliate stands of parthenium. Such severe defoliation will weaken the weed, thereby allowing the native vegetation to recover.

Another natural enemy used to control whitetop weed is the stem-boring weevil, *Listronotus setosipennis* (Coleoptera: Curculionidae). *L. setosipennis* is a nocturnal weevil that hides in leaflitter during the day. The adult *L. setosipennis* digs holes in newly opened parthenium flowers, lays its eggs and covers them with black frass. Thus, the oviposition spots on parthenium flowers appear as black marks. It can also lay eggs on the petioles of leaves on non-flowering parthenium plants. The eggs hatch within 3–5 days and the first instar larva bores into the flower base or leaf base and enters the stem. The larva feeds inside the stem for 18–30 days and exits at the bottom to enter the soil. Four to five larvae feeding inside a flowering parthenium stem, and as low as two at the rosette stage, can kill parthenium (Dhileepan, 2003). *L. setosipennis* has been field released in Australia (Dhileepan and McFadyen, 2012) and South Africa (Strathie *et al.*, 2016) to control whitetop weed.

Ethiopia has also introduced the leaf-feeding beetle *Z. bicolorata* and the stem-boring weevil *L. setosipennis* (Fig. 15.3) to manage parthenium (Mersie *et al.*, 2018). This was possible because of a project initiated in 2005 that was led by Virginia State University, USA, and funded by the United States Agency for International Development through the Integrated Pest Management Innovation Lab Program at Virginia Tech. Before

Fig. 15.3. Two natural enemies of parthenium. **(A)** *Zygogramma bicolorata* and **(B)** *Listronotus setosipennis*, released in Ethiopia, South Africa and Uganda. (Photos: from Ms. Lorraine Strathie and Dr Andrew J. McConnachie, with permission, ARC-PHP, South Africa.)

release, the host specificities of *Z. bicolorata* and *L. setosipennis* were assessed against 29 and 31 non-target plant species, respectively (Mersie *et al.*, 2018). The results showed that *Z. bicolorata* and *L. setosipennis* can only complete their life cycles on parthenium and would not attack other non-target plants.

Based on these host range test results, the Ethiopian government gave permission for the field releases of the two biocontrol agents. Both have started to establish and cause damage to the weed in Ethiopia. *L. setosipennis* is especially effective against the weed in the drier regions of Ethiopia (Alemayehu and Mersie, 2019).

Mesquite

Mesquite *(Prosopis juliflora)* is an evergreen tree native to South America, Central America and the Caribbean. It is a fast-growing, nitrogen-fixing tree that is tolerant to arid conditions and saline soils. Under the right conditions, *P. juliflora* can produce a variety of valuable goods and services, including construction materials, charcoal, soil conservation and the rehabilitation of degraded and saline soils (Mwangi and Swallow, 2005). *P. juliflora* has survived where other tree species have failed, and in many cases, become a major nuisance. The plant is a prolific producer of seeds with prolonged dormancy and can survive long periods of drought (secured within a protective seedpod). Seeds can be transported by floodwaters or ingested by animals and spread in their manure. Moreover, the mesquite tree is highly prolific, adaptable and resilient. It is invasive and difficult to eradicate once established, and inhibits the germination of seeds from other species, thereby enabling the tree to spread unopposed (Pasiecznik, 1999).

In Ethiopia, mesquite was intentionally introduced as an agroforestry species in the Awash River basin, but now threatens agricultural land and protected areas in the Awash National Park. It is aggressively invading pastoral areas in the Middle and Upper Awash Valley, eastern Harerge, vast areas in the Somali region, eastern Amhara and Tigray regions, and the southern Rift Valley, destroying natural pasture, displacing native trees, forming impenetrable thickets and reducing the grazing potential of rangelands (Shiferaw *et al.*, 2004;). Its invasion is estimated to be well above 1 million hectares and has taken over prime grazing and irrigable land in the Afar region alone, and the spread of the plant is advancing at the rate of about 50,000 hectares annually (PARDB, 2012). The species is known for its numerous harmful effects on the livelihood of the local pastoralists, specifically in the Afar and Somali Regions.

Further details on mesquite can be found in the volumes of the crop protection proceedings mentioned above, and CABI (2021).

Discussion and Conclusions

Just as in the other disciplines of crop protection – entomology and plant pathology – weed science research in Ethiopia has transitioned from weed surveys to establish the identity and economic importance, to screening and recommending effective herbicides, to sustainable management practices that rely on an integrated approach. The impact of the research can be gauged by the fact that there have been significant increases in saving potential losses in farmers' fields with improved technologies. The development and popularization of striga-resistant sorghum cultivars developed in the country is an important outcome of weed science research. The contributions made by weed science research in identifying and suggesting alternative methods of controlling herbicide-resistant annual grass weeds like *Snowdenia polystachya* in large-scale wheat and barley production areas are also worth mentioning (Fessahaie *et al.*, 2008). Realizing the public apprehension about the adverse effect of herbicides on the environment, research on the control of weeds with biocontrol agents and bioherbicides has been stepped up.

Going forward, the weed science research community in Ethiopia has identified several strategic areas as its focus in the coming years. First, to continuously upgrade the weed database to include non-crop and aquatic areas and weed shifts in species composition due to changes in cropping systems and weed control practices. The database shall provide reference material to weed species in each agroecology. This would mean preservation of the existing large national reference collection at the Ambo Plant Protection Research Center (APPRC, 2014) and further strengthening it with new collections.

Second, to prevent future accidental introductions, more weeds need to be subjected to rigorous weed risk analysis (WRA). The WRA process is a science-based quarantine risk analysis tool for determining the weed potential of proposed new imports. Increased exchange of grains and seeds following the globalization of agriculture is bound to result in further introductions. The sanitary and phytosanitary agreement of the World Trade Organization suggests that the countries not only update their quarantine laws, but also incorporate the elements of pest risk analysis for making regulatory decisions for both import and export. Therefore, there is an urgent need to analyse the risk factors associated with different exotic weeds to design safeguards and lower the risk of their entry. This entails generating data on weed biology and a list of quarantine weeds.

Third, IAS are the biggest threat to biodiversity, second only to human resettlement. Increasing trade and globalization, coupled with liberalization policies and huge future agricultural and industrial development plans in Ethiopia will further increase the risk of invasion by IAS. In addition, parasitic weeds are also posing problems in the productivity of some of the major crops and cropping systems and are economically important production constraints. Research interventions should continue to strengthen the existing integrated management of priority parasitic weeds. Furthermore, other research intervention is strengthening the exploration of the host range, host–parasite interaction pattern and host crops resistance to parasitic weeds. Increased efforts need to be in place to elaborate on control strategies that are efficient and easy to use. A single control method is not sufficient to control IAS and parasitic weed species. A combination of different methods in an integrated weed management approach seems to be the best solution to combat these species.

Fourth, priority research activities should target the nationwide release of the parthenium biocontrol agents mentioned earlier in this chapter. In addition, exploration of potential pathogens, insects and allelo-chemicals for use as biotic agents in the management of priority IAS require especial consideration. There is also an urgent need for the endorsement by policy makers of the national IAS strategy and action plan prepared by EIAR for meeting the challenges posed by these weeds.

Finally, other priorities include work on herbicide resistance in weeds, developing herbicide-tolerant cultivars and monitoring herbicide residue and environmental safety.

References

Abate T. (ed.) (1986) A Review of Crop Protection Research in Ethiopia: Proceedings of the First Ethiopian Crop Protection Conference, 4-7 Feb. 1985. IAR, Addis Ababa, Ethiopia.

Abebe, T., Nega, Y., Mehari, M., Mesele, A., Asefa Workineh, A., Beyene, H. (2015) Genotype by environmental interaction of some faba bean genotypes under diverse broomrape environments of Tigray, Ethiopia. *Journal of Plant Breeding and Crop Science* 7(3), 79–86.

Alemayehu, L. and Mersie, W. (2019) Survival and dispersal of the stem-boring weevil, *Listronotus setosipennis, and* the leaf-feeding beetle, *Zygogramma bicolorata,* one year after their release for the control of the invasive weed, *Parthenium hysterophorus,* at two locations in Ethiopia. *Journal of Biological Control* 33, 178–184.

Akobundu, I.O. (1980) Weed science research at the international institute of tropical agriculture and research needs in Africa. *Weed Science* 28, 439–445.

Akobundu, I.O. (1987) *Weed Science in the Tropics: Principle and Practices*. John Wiley & Sons, New York, 522 pp.

APPRC (Ambo Plant Protection Research Center) (2014) Annual Research Progress Report during 2013/14.

CABI (2021) *Prosopis juliflora* (Prosopis or Mesquite). BioNET-EAFRINET. Keys and Fact Sheet (accessed 23 September 2021).

Chikoye, D., Ellis-Jones, J., Riches, C. and Kanyomeka, L. (2007) Weed management in Africa: experiences, challenges and opportunities. In XVI International Plant Protection Congress (pp. 652–653), 15–18 October, Glasgow, Scotland.

Dhileepan, K. (2003) Current status of the stem-boring weevil *Listronotus setosipennis* (Coleoptera: Curculionidae) introduced against the weed *Parthenium hysterophorus* (Asteraceae) in Australia. *Biocontrol Science and Technology* 13, 3–12.

Dhileepan, K. and McFadyen, R.E. (2012) *Parthenium hysterophorus* L. – parthenium. In: Julien, M., McFadyen, R. and Cullen, J. (eds) *Biological Control of Weeds in Australia*. CSIRO Publishing, Melbourne, Australia.

Dhileepan, K. and Strathie, L. (2009) *Parthenium hysterophorus* L. (Asteraceae). In: Muniappan, R., Reddy, G.V.P. and Raman, A. (eds) *Biological Control of Tropical Weeds Using Arthropods*. Cambridge University Press, UK.

Esilaba, A.O., Mulatu, T., Reda, F., Ransom, J.K., Woldewahid, G. *et al.* (1998) A diagnostic survey on striga in the northern Ethiopian highlands. *Arem* 4, 13–27. EWSS, Addis Ababa,

Evans, H.C. (1997) *Parthenium hysterophorus:* A review of its weed status and the possibilities for biological control. *Biocontrol News and Information* 18, 89–98.

FAO (2023) Statistical databases and data-sets of the Food and Agriculture Organization of the United Nations. Available at: http://www.fao.org/faostat/en/#data (accessed 16 March 2023).

Fessehaie, R. (1998) Review of Broomrape (*Orobanche* spp.) in Ethiopia: NVRSRP Newsletter, Addis Ababa, pp. 20–22.

Fessahaie, R., Tessema, T., Nefo, K., Belle, A., Negewo, T. and Taa, A. (2008) Weed research in small cereals in Ethiopia. In: Abraham T. (ed.) *Increasing Crop Production through Improved Plant Protection: Vol. 1*. Plant Protection Society of Ethiopia, Addis Ababa, Ethiopia. pp. 431–474.

Franzel, S., Mekuria, M. and Yirga, C. (1989) Smallholder weed problem and weed control practices. In: Worku, B. and Sombat, C. (eds) Conservation Tillage-Alternative Systems of Teff (Eragrostis tef) Production for a Semi-Arid Central Rift Valley of Ethiopia. *Kamphaengsaen Academy Journal* 3(3), 1–10.

Fufa, A., Tessema, T., Bekeko, Z., Mesfin, T. (2022) Distribution and abundance of wild sunflower (*Verbesina encelioides*) and its impacts on plant biodiversity in the Central Rift Valley of Ethiopia. *Ethiopian Journal of Crop Science* 10(1), 89–106.

Gebremariam, E. (2005) Integrated control of branched broomrape (*Orobanche ramosa* L.) in tomato (*Lycopersicon esculentum* Mill.) in Central Rift Valley of Ethiopia. PhD Dissertation, Graduate School of Kasetsart University, Bangkok.

Gevezova, M., Dekalska, T., Stoyanov, K., Hristeva, T., Kostov, K., Batchvarova, R. and Denev, I. (2012) Recent advances in broomrapes research. *Journal of BioScience and Biotechnology* 91–105, 1314–6246.

Habimana, S.A., Nduwumuremyi, J.D. and Chinama, R. (2014) Management of *orobanche* in field crops – A review. *Journal of Soil Science and Plant Nutrition* 14(1), 43–62.

Heide-Jorgensen, H.S. (2008) *Parasitic Flowering Plants*. Brill, Leiden-Boston.

Kanchan, S.D. and Chandra, J. (1980) Allelopathic effects of *Parthenium hysterophorus* L. Part II. Leaching of inhibitors from aerial vegetative parts. *Plant and Soil* 55, 61–66.

Johnson, D.E. (1995) *Weed Management Strategies for Smallholder Rice Production*. Brighton Crop Protection Conference. Weeds, pp. 1171–1180.

Masteling, R., Lombard, L., De Boer, W., Raaijmakers, J.M. and Dini-Andreote, F. (2019) Harnessing the microbiome to control plant parasitic weeds. *Current Opinion in Microbiology* 49, 26–33.

McConnachie, A.J. (2015) Host range and risk assessment of *Zygogramma bicolorata*, a defoliating agent released in South Africa for the biological control of *Parthenium hysterophorus*. *Biocontrol Science and Technology* 25, 975–991.

McConnachie, A.J., Strathie, L.W., Mersie, W., Gebrehiwot, L., Zewdie, K. *et al.* (2010) Current and potential geographical distribution of the invasive plant *Parthenium hysterophorus* (Asteraceae) in eastern and southern Africa. *Weed Research* 51, 71–84.

Mersie, W. and Singh, M. (1987) Allelopathic effect of parthenium (*Parthenium hysterophorus* L.) extract and residue on some agronomic crops and weeds. *Journal of Chemical Ecology* 13, 1739–1747.

Mersie, W., Alemayehu, L., Strathie, L., McConnachie, A., Terefe, S., Negeri, N. and Zewdie, K. (2018) Host range evaluation of the leaf-feeding beetle, *Zygogramma bicolorata* and the stem-boring weevil, *Listronotus setosipennis* demonstrates their suitability for biological control of the invasive weed, *Parthenium hysterophorus* in Ethiopia. *Biocontrol Science and Technology* 29, 239–251.

MOANR (Ministry of Agriculture and Natural Resources) (2019) List of Registered Pesticides. Pesticide Registration and Control Office, Addis Ababa, Ethiopia.

Mola, T. and Belachew, K. (2015) Determination of critical period of weed-common bean (*Phaseolus vulgaris* L.) competition at Kaffa, Southwest Ethiopia. *Greener Journal of Agricultural Sciences* 5(3), 93–100.

Mwangi, E. and Swallow, B. (2005) Invasion of *Prosopis juliflora* and local livelihoods: Case study from the lake Baringo area of Kenya. ICRAF Working Paper – no. 3, World Agro-forestry Centre, Nairobi.

Navie, S.C., McFadyen, R.E., Panetta, F.D. and Adkins, S.W. (1996) The biology of Australian Weeds 27: *Parthenium hysterophorus* L. *Plant Protection Quarterly* 11, 76–87.

Negewo, T., Mekuria, M. and Hussien, T. (2006) Competition effects of major weed species at various densities on yield and yield components of barley (*Hordeum vulgare* L.). *Pest Management Journal of Ethiopia* 10, 53–59.

Nigatu, L., Hassen, A., Sharma, J. and Adkins, S. (2010) Impact of *Parthenium hysterophorus* on grazing land communities in north-eastern Ethiopia. *Weed Biology and Management* 10, 143–152.

Nzioki, H.S., Oyosi, F., Morris, C.E., Kaya, E., Pilgeram, A.L., Baker, C.S. and Sands, D.C. (2016) Striga biocontrol on a toothpick: a readily deployable and inexpensive method for smallholder farmers. *Frontiers in Plant Science* 7, 1121.

Obuo, J.E., Adipala, E. and Osiru, D.S.O. (1997) Weed suppression in a cowpea/sorghum intercrop. In: *16h East African Biennial Weed Science Conference Proceedings*, pp. 113–118.

PARDB (Pastoral Agriculture and Rural Development Bureau) (2012) Identified Agricultural Investment Areas, Afar Regional State. Special Report, Ministry of Agriculture and Rural Development (MOARD), MOARD, Addis Ababa, Ethiopia, pp. 3–53.

Parker, C. (2009) Observations on the current status of Orobanche and Striga problems worldwide. *Pest Management Science* 65(5), 453–459. DOI: 10.1002/ps. 1713.

Parker, C. and Riches, C.R. (1993) *Parasitic Weeds of the World: Biology and Control*. CAB International, Wallingford, UK. 332 p.

Pasiecznik, N. (1999) Prosopis - pest or providence, weed or wonder tree? *European Tropical Forest Research Network Newsletter* 28, 12–14.

PROMISE (Promoting Root Microbes for Integrated Striga Eradication) (2021) Progress update 2021. Available at: https://promise.nioo.knaw.nl/ (accessed 30 November 2022).

Reda, F., Weldewahid, G., Zemichael, B. and Bayou, W. (1997) An integrated approach to the control of striga in sorghum. *Pest Management Journal of Ethiopia* 1, 43–49.

Reda, F., Verkleij, J.A.C. and Ernst WHO (2005) Intercropping for the improvement of sorghum yield, soil fertility and striga control in the subsistence agriculture region of Tigray (Northern Ethiopia). *Journal of Agronomy and Crop Science* 191, 10–19.

Rupschus, C., Dieter, K., Carmen, B., Ulrich, C. and Tessema, T. (2007) *Economic Impacts of Invasive Weed Species in Developing Countries: The Case of Parthenium in Ethiopia*. Tropentag, Witzenhausen, Germany.

Sahile, G., Abebe, G. and Al-Tawaha, A.M. (2005) Effect of soil solarization on *Orobanche ramosa* L. and *Orobanche cernua* Loef control in tomato in the Central Rift Valley of Ethiopia. *World Journal of Agricultural Sciences* 1(2), 143–147.

SARC (Sirinka Agricultural Research Center) (2015) Progress Report for the Period 2014/15. Sirinka Agricultural Research Center, Sirinka, Ethiopia.

Sauerborn, J. (1991) Parasitic flowering plants in agricultural ecosystems of West Asia. *Flora et Vegetatio Mundi* 9, 83–91.

Sherif, A., Gelma, M. and Parker, C. (1990) Control of *Orobanche cernua* by soil fumigation at Nura Era Horticulture Enterprise. In: Parker, C. and Fessehaie, R. (eds) *Proceedings of the Sixth Annual Meeting of EWSC*. Addis Ababa, Ethiopia, pp. 67–74.

Shiferaw, H., Teketay, D., Nemomissa, S. and Assefa, F. (2004) Some biological characteristics that foster the invasion of *Prosopis juliflora* (Sw.) DC. at Middle Awash Rift Valley Area, north-eastern Ethiopia. *Journal of Arid Environments* 58, 135–154.

Soto, M.J., Fernandez-Aparicio, M., Castellanos-Morales, V., García-Garrido, J.M., Ocampo, J.A., Delgado, M.J. and Vierheilig, H. (2010) First indications for the involvement of strigolactones on nodule formation in alfalfa (*Medicago sativa*). *Soil Biology and Biochemistry* 42(2), 383–385.

Strathie, L.W., McConnachie, A.J. and Retief, E. (2011) Initiation of biological control of *Parthenium hysterophorus* L. (Asteraceae) in South Africa. *African Entomology* 19, 378–392.

Strathie, L.W., Den Breeyen, A., Sambo, S., Chidawanyika, F. and Gareeb, M. (2016) Recent developments in the management of the invasive plant *Parthenium hysterophorus* using natural enemies. *Proceedings of South African Sug Technol Association* 89, 307–313.

Taa, A. and Tanner, D.G. (1998a) Crop rotation and fertilizer effects on weed populations in the southeastern highlands of Ethiopia. *Arem* 4, 89–103.

Taa, A. and Tanner, D.G. (1998b) Stubble management, tillage and cropping sequence effect weed populations in the Southeastern highland of Ethiopia. *Arem* 4, 76–88.

Taa, A., Tanner, D.G., Temesgen, M. and Girma, K. (1997) Onfarm evaluation of an animal-drawn implement developed in Ethiopia for row placement of wheat seed and basal fertilizer. *African Crop Science Journal* 5(4), 359–369.

Tadesse, A. (ed.) (2008) Increasing crop production through improved plant protection – Volume I. *Plant Protection Society of Ethiopia (PPSE), 19–22 December 2006*. PPSE and EIAR, Addis Ababa, Ethiopia, 598 p.

Tadesse, A. (ed.) (2009) Increasing crop production through improved plant protection – Volume II. *Plant Protection Society of Ethiopia (PPSE)*. PPSE and EIAR, Addis Ababa, Ethiopia. 542 p.

Tana, T. and Milberg, P. (2000) Weed flora in arable fields of eastern Ethiopia with emphasis on the occurrence of *Parthenium hysterophorus*. *Weed Research* 40, 507–521.

Tessema, T. (2002) Investigation of pathogens for biological control of parthenium (*Parthenium hysterophorus* L.) in Ethiopia. PhD Thesis. Humboldt–Universitat zu Berlin, Landwirtschaftlich-Gartnerischen Fakultat, Berlin. 152 p.

Tessema, T. and Ejeta, G. (2019) *Sorghum Production in Transition through Striga Management*. ISBN: 789994466634. EIAR, Addis Ababa, Ethiopia. 110 p.

Tessema, T. and Tanner, D.G. (1997) Grass weeds competition and calculated economic threshold densities in bread wheat in Ethiopia. *African Crop Science Journal* 5, 371–384.

Tessema, T., Einhorn, G. and Metz, R. (2004) *Parthenium hysterophorus*, an invasive species in Ethiopia – Investigations on its occurrence and its pathogens. *Journal of Plant Diseases and Protection* 19, 271–278.

Trognitz, F., Hackl, E., Widhalm, S. and Sessitsch, A. (2016) The role of plant–microbiome interactions in weed establishment and control. *FEMS Microbiology Ecology* 92(10), fiw138. DOI: 10.1093/femsec/fiw138.

Van Rijn, P.J. (2000) *Weed Management in the Humid and Sub-humid Tropics*. Royal Tropical Institute (KIT), Amsterdam, 233 p.

Wiesner, M., Tessema, T., Hoffmann, A., Wilfried, P., Buttner, C., Mewis, I. and Ulrichs, C. (2007) *Impact of the Pan-tropical Weed* Parthenium hysterophorus *L. on Human Health in Ethiopia*. Witzenhausen, Germany.

Zhou, W.J., Yoneyama, K., Takeuchi, Y., Iso, S., Rungmekarat, S. *et al.* (2004) In vitro infection of host roots by differentiated calli of the parasitic plant *Orobanche*. *Journal of Experimental Botany* 55(398), 899–907.

16 The Emerging Seed Systems

Tsedeke Abate,[1]* Karta Kaske Kalsa,[2] Amsalu Ayana,[3] Maru Degefa[4] and Melaku Admasu[5]

[1]*Homegrown Vision, Addis Ababa, Ethiopia;* [2]*Ethiopian Institute of Agricultural Research, Addis Ababa Ethiopia;* [3]*Independent Consultant, Addis Ababa, Ethiopia;* [4]*Ethiopian Agricultural Business Corporation, Addis Ababa, Ethiopia;* [5]*Ethiopian Seed Association*

Abstract

The Ethiopian Seed Enterprise, a parastatal organization, was the first and sole formal seed source of the country since its establishment in 1979 until the 1990s. The number of private seed companies increased fast in the first half of 2010s. As of December 2021, there were a total of 54 companies comprising 4 public, 2 transnationals, 1 farmer's cooperative union and 47 small private seed companies that are licensed to produce and market seed. Of the overall average demand of about 729,000 MT of certified seed in 2016–2018, the formal seed system was able to supply about 60% (of which 57% was distributed) that included 64%, 24% and 18% supply, respectively, for cereals, grain legumes and oilseeds. Overall, public entities produced 98% of the total early generation seed and 80% of certified seed. Ethiopia has yet to recognize the role of community seed production to satisfy its growing seed demand.

Introduction

Seed plays a crucial role in improving crop productivity. Tarekegn and Mogiso (2020) suggest that opportunity losses in crop productivity due to a lack of use of improved seed could be as high as 35%. There are several proposed categories of seed systems in Ethiopia. Hirpa *et al.* (2010) and Sisay *et al.* (2017) recognize three categories – informal, formal and alternative (or intermediary). Louwaars *et al.* (2013) recognize five categories – farm saved, community based, public companies, commercial companies and closed-value chain. Welu (2015) and Hodson *et al.* (2020) use formal and informal to describe the seed systems in Ethiopia. Mulesa (2021) uses four categories: government-led formalization, private-led formalization, farmer-based localization and community-based integrative system development.

There is a general consensus that improved seed use in Ethiopia is low and the informal seed system is dominant (Welu, 2015; Minwagaw and Ejigu, 2021). Hodson *et al.* (2020) reported that the informal seed system accounted for 80% of the total wheat seed in 2016/17. O'Neill (2018) reported that only 4.7% of croplands were under improved seed in the 2007/08 *meher* season.

Several challenges to use of improved seed in Ethiopia have been cited. These include limited quantity and access, unaffordable price, untimely distribution, poor quality and market information and linkages (Ojiewo *et al.*, 2020; Tarakegn and Mogiso, 2020; Minwagaw and Ejigu, 2021;

* Corresponding author: tsedeke@homegrownvision.org

DOI: 10.1079/9781800626386.0016

Mohammednur and Megersa, 2021), as well as low level of engagement of private seed companies (Kifle and Atilaw, 2018).

As stated in earlier chapters, successive agricultural policy makers in Ethiopia have made sustained efforts to increase agricultural productivity and production over the past several decades. Initiatives to improve the seed systems in this country started with the establishment of the Ethiopian Seed Enterprise (now part of the Ethiopian Agricultural Business Corporation), a public organization, in 1979. The systems have undergone substantial changes since then. This chapter makes a critical review of Ethiopia's seed systems and provides suggestions for the future.

Approach

Crop production data were obtained from the Central Statistical Agency (CSA, 2007–2021) of the Government of Ethiopia. Information on seed was from various sources: public and private companies, the literature (e.g. Alemu, 2015; Atilaw *et al.*, 2017; Bishaw and Alemu, 2017; Bishaw *et al.*, 2018; Kalsa, 2019; Jaleta *et al.*, 2020), personal communications, observations and experiences.

We also calculated the seed replacement rate (SRR), i.e. the percentage of area sown out of the total area of crop planted in the season by using certified/quality seed other than the farm-saved seed. In other words, it is a measure of cropped area covered with quality seed.

We calculated the SRR for major cereals in Ethiopia using the following formula:

$$SRR = \left(\frac{C}{A * K}\right) * 100$$

where C is certified seed (C1), A is area under crop and K is the seeding rate per unit of area.

Results

Major sources of seed supply

The Ethiopian Seed Enterprise (ESE), now under the Ethiopian Agricultural Business Corporation (EABC) was the only formal seed provider between 1979 and 1990. The number of seed companies showed accelerated growth starting from the early 2000s. The system has undergone substantial changes over the years. There was a total of 54 seed companies licensed to produce seed in Ethiopia as of December 2021. Of these, 42 were registered under the Ethiopian Seed Association. These include 4 public, 2 transnational companies, 1 farmer's cooperative union and 35 small private seed companies that are owned by Ethiopian nationals (Table 16.1).

The majority of the domestic seed companies (both public and private) are engaged in cereals seed production (see Annex Fig. 16.1 and Annex Table 16.1). For example, 17 of 22 (77%) seed companies for which we have data were engaged in maize, while eight each were engaged in teff and wheat. Three companies each produced chickpea and onion. Two companies each produced common bean, soybean, *noug*, forages, pepper and mango. Only one company each was engaged in malt barley, black cumin, tomato, vegetables, fruits, papaya, sesame, potato, pigeonpea, alfalfa and sesbania.

Ethiopia uses the Organisation for Economic Co-operation and Development description of four generation-based seed classes: breeder seed, pre-basic seed, basic seed and certified seed. The first three seed classes are often referred to as early generation seed (EGS). Breeder seed is largely produced in small quantities (just enough to plant 0.5 to 1.0 ha, depending on the multiplication rate of the crop variety) and maintained by the breeder/breeding institute at research centres. The average amount of breeder seed produced between 2016 and 2019 in Ethiopia was about 35 MT. Of this, the Ethiopian Institute of Agricultural Research (EIAR) and Oromiya Agricultural Research Institute produced about 79.24% and 20.29%, respectively, with the Amhara Seed Enterprise covering the remaining 0.47%.

As shown in Table 16.2, there are eight public entities and three private companies that are engaged in EGS production in Ethiopia. The overall production showed declining trends for the 3 years between 2016 and 2019. For example, it was nearly 14,000 MT in 2016, compared to roughly 10,000 MT in 2017 and a little over 9000 in 2018.

The public organizations accounted for nearly 98% of the total EGS production, whereas the private companies accounted for just over 2% (Table 16.2). The Oromiya Seed Enterprise

Table 16.1. List of registered members of Ethiopian Seed Association (as of December 2021). Source: constructed by the lead author from Ethiopian Seed Association data.

Number	Organization	Type	Location	YOE
1	EABC	Parastatal	Nationwide	1979
2	DuPont-Pioneer (Corteva)	Multinational	Nationwide	1991
3	Anno Agro-Industry	Private	Oromiya	1993
4	Mohammed Awel Oumer Farm	Private	South	1994
5	Gadissa Gobena Farm	Private	Oromiya	1996
6	Abdela and Family PLC	Private	South	1997
7	Yimam Tesema Farm	Private	Amhara	1999
8	Hadiya Enterprise	Private	Oromiya	2000
9	Nono Agricultural Development	Private	Oromiya	2000
10	Meki-Batu	Farmers' Coop Union	Oromiya	2002
11	Taddese Genna Farm	Private	Oromia	2004
12	Zi-Andata	Private	South	2005
13	Bayih Mekonnen Agriculture	Private	Amhara	2006
14	Gasha Redda Lema Mechanized Farm	Private	Tigray	2006
15	Ayehu Zingini Farm PLC	Private	Amhara	2006
16	Ethio Agri-CEFT	Private	Amhara	2007
17	Avallo International PLC	Private	Amhara	2007
18	Oromiya Seed Enterprise	Parastatal	Oromiya	2008
19	Nile Seed Enterprise	Private	Amhara	2008
20	Amhara Seed Enterprise	Parastatal	Amhara	2009
21	Homa Seed PLC	Private	Oromiya	2009
22	Enaget International	Private	Amhara	2009
23	Amuari High Yield Varieties & Agric. Products PLC	Private	Oromiya	2012
24	Seid Hussen Pasture and Forage Seed	Private	Amhara	2013
25	Nunhems-Ethiopia PLC	Multinational	Nationwide	2015
26	Tsega Tegene Seed Grower	Private	Amhara	2016
27	E.A.R. PLC	Private	Addis Ababa	2019
28	Bumper Harvest Seed Farm	Regional	Addis Ababa	2020
29	Alemayehu Mekonnen Farm	Private	South	NA
30	Desta Berhe Farm	Private	Tigray	NA
31	Tetemke Tsegaye Hagos Farm	Private	Tigray	NA
32	Chemtex PLC	Private	Addis Ababa	NA
33	South Seed Enterprise	Parastatal	South	NA
34	Tsehay & Genet Farming Development PLC	Private	South	NA
35	ATM Trading PLC	Private	Tigray	NA
36	AFRI Seed G. PLC	Private	Amhara	NA
37	GAWT International Business PLC	Private	Addis Ababa	NA
38	Markos General Trading PLC	Private	Addis Ababa	NA
39	Green Life Trading PLC	Private	Addis Ababa	NA
40	Burka Agricultural PLC	Private	Oromiya	NA
41	Solomon Ayalew Girma Fruit and Vegetables Seed	Private	Addis Ababa	NA
42	Dembel Zer Abaze PLC	Private	Oromiya	NA

EABC = Ethiopian Agricultural Business Corporation; NA = not available; YOE = Year of establishment.

and EABC each produced roughly 40% of the total EGS. This was followed by EIAR, Southern Seed Enterprise, Oromiya Agricultural Research Institute, Amhara Regional Agricultural Research Institute and Southern Agricultural Research Institute. The private companies Amauri, Burka and Anno PLC (also referred to as Anno Agro-Industry) contributed roughly

1.2%, 1.0% and 0.2%, respectively, of the total EGS production.

The available data show that EGS production in Ethiopia is dominated by cereals (Annex Table 16.2). They constituted about 89% of the total EGS production between the 2006 and 2015 *meher* seasons. Grain legumes and oilseeds, on the other hand, formed approximately 9% and 3%, respectively. Furthermore, wheat, *teff* and maize take the largest share of EGS production in this country.

Certified seed production

Ethiopia has made substantial progress in the total amount of certified seed production in the 15 seasons between 2006/07 and 2020/21. Figure 16.1 depicts certified seed production for selected crops and groups of crops. The total production was roughly 18,000 MT in 2006/07 compared to about 117,000 MT in 2020/21.

Cereals accounted for about 88% of the total certified seed production, with pulses and oilseeds covering roughly 9% and 3%, respectively. Maize and wheat constituted 40% and 39%, respectively, of the total certified seed production. Seed production for these two cereals increased consistently over the above period. For example, the total amount of maize seed produced in 2006/07 was <7000 MT, compared to 52,000 MT in 2020/21; in a similar fashion, the total amount of certified seed for wheat production in 2006/07 was about 9000 MT in 2006/07 versus 47,000 in 2020/21 (Fig. 16.1).

Certified seed production for pulses and oilseeds fluctuated heavily; it surged in 2012–2014 and then in 2016. Production quantity of pulses was low for many of the remaining years. There were no records of oilseeds production for the 2007/08, 2008/09, 2015/16, 2017/18, 2019/20 and 2020/21 seasons.

Certified seed demand, supply and distribution

Certified seed demand, supply and distribution varied among crop groups and from crop to crop (Table 16.3). The Ethiopian seed system engaged mainly in cereals, grain legumes and oilseeds. The overall average certified seed demand in 2019 and 2020 for cereals, grain legumes, oilseeds, potato and cotton was more than 129,000 MT, whereas the supply was about 78,000 MT or roughly 60%. Of this, about 57% was distributed, mainly because of the mismatch between supply and demand of crops and varieties and poor promotion.

Table 16.2. Early generation seed production (MT) by different organizations in Ethiopia. Source: constructed by the authors from Ethiopian Institute of Agricultural Research database.

Organization	2016/17	2017/18	2018/19	Average	Share (%)
Public					
OSE	4,812	4,568	4,132	4,504	40.2
EABC	6,630	3,309	3,548	4,496	40.1
ASE	976	825	208	670	6.0
EIAR	420	543	408	457	4.1
SSE	521	334	390	415	3.7
OARI	209	367	123	233	2.1
ARARI	127	69	62	86	0.8
SARI	122	53	38	71	0.6
Sub-total	13,817	10,068	8,909	10,931	97.6
Private					
Amuari PLC	33	76	298	136	1.2
Burka PLC	66	150	124	113	1.0
Anno PLC	15	8	30	18	0.2
Sub-total	114	234	453	267	2.4
Grand total	13,932	10,301	9,361	11,198	100.0

ARARI = Amhara Regional Agricultural Research Institute; ASE = Amhara Seed Enterprise; EABC = Ethiopian Agricultural Business Corporation; EIAR = Ethiopian Institute of Agricultural Research; OARI = Oromiya Agricultural Research Institute; OSE = Oromiya Seed Enterprise; SARI = Southern Agricultural Research Institute; SSE = Southern Seed Enterprise.

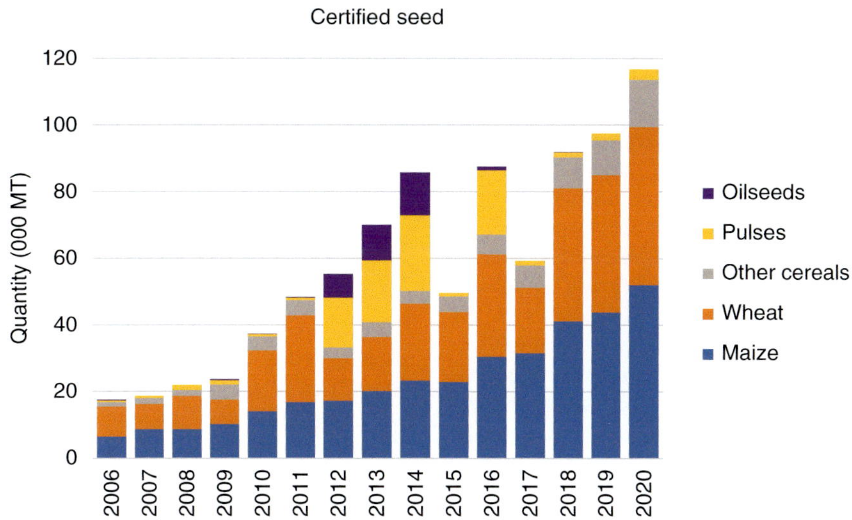

Fig. 16.1. Certified seed production for selected crops and groups of crops in Ethiopia. Source: constructed by the lead author from CSA data (www.csa.gov.et).

Table 16.3. Certified seed demand, supply and distribution in Ethiopia (2019 and 2020 average). Source: constructed by the authors from Ethiopian Institute of Agricultural Research database.

	Totals (MT)			Percentages	
Crop	Demand	Supply	Distribution	Supply	Distribution
Cereals					
Wheat	61,489	38,329	35,309	62.3	57.4
Maize	42,202	29,390	28,746	69.6	68.1
Barley	9,183	5,138	4,965	56.0	54.1
Teff	4,194	2,546	2,106	60.7	50.2
Sorghum	1,374	245	195	17.8	14.2
Rice	239	80	62	33.4	25.9
Millet	113	46	39	40.5	34.6
Sub-total/average	**118,794**	**75,774**	**71,421**	**63.8**	**60.1**
Grain legumes					
Common bean	5,050	1,794	1,536	35.5	30.4
Faba bean	2,273	224	198	9.9	8.7
Chickpea	732	70	58	9.6	8.0
Fieldpea	605	3	3	0.5	0.4
Lentil	155	5	5	3.5	3.5
Mungbean[a]	118	59	46	49.9	39.4
Sub-total/average	**8,932**	**2,156**	**1,846**	**24.1**	**20.7**
Oilseeds					
Sesame	270	52	41	19.1	15.2
Linseed	140	17	17	12.2	12.1
Groundnut	53	3	3	5.6	5.6
Soybean	83	29	29	35.6	35.6
Brassicas	3	0	0	0.0	0.0
Noug	<1	0	0	0.0	0.0
Sub-total/average	**550**	**101**	**90**	**18.4**	**16.5**
Others					
Potato	916	0	0	0.0	0.0
Cotton	103	9	9	8.9	8.9
Total/average	**129,242**	**78,011**	**73,343**	**60.4**	**56.7**

[a]Only 2020 data.

On average, about 63% of cereals, 24% of grain legumes and 18% of oilseeds demand was supplied. In a similar fashion, roughly 60%, 21% and 17% of the total supply of cereals, grain legumes and oilseeds, respectively, was distributed. Among cereals, maize, wheat, barley and *teff* received the highest amount of certified seed distribution, whereas sorghum was the lowest. Certified seed supply among the grain legumes was highest for common bean and mungbean, whereas lentil and fieldpea were the lowest (Table 16.3). Chickpea was seemingly low because the farmers mostly obtain their improved seed from specialized farmer seed growers or farmer's cooperatives in their communities rather than from seed companies. With about 36% of both the total supply and distribution, soybean was the highest among oilseeds, followed by sesame with about 19% supply and 15% distribution.

Seed replacement rate

The SRR varied very widely among the cereal crops. The maize area under certified seed increased progressively in the 15 years between 2006 and 2020. It jumped from 15% in 2006 to roughly 82% in 2020 (Table 16.4). This was followed by wheat, where the total area under certified seed stood at nearly 5% in 2006, compared to nearly 17% in 2020. In a similar fashion, the total area of *teff* under certified seed was <1% in 2006 vs nearly 7% in 2020. Barley area under certified seed was <1% in 2006 and nearly 5% in 2020. The area covered under certified seed for sorghum stayed below 1% and changed little over the years.

Variety turnover

The Ethiopian Agricultural Research System has released nearly 1200 crop and forage varieties through to 2017 (MOALR, 2017) but has not put in place the mechanism for removing obsolete varieties. Consequently, very old varieties are still being produced and marketed by several of the seed companies to the present day. For example, available data show that the average age of crop varieties in Ethiopia is about 26 years, which is higher than the average age of maize cultivars for Africa (Abate *et al.*, 2017). The average age of *teff*, maize, chickpea and oilseeds was 22, 20, 18, 15 and 15 years, respectively (Table 16.5).

The malt barley variety 'Holker' was released in 1979 (43 years old in 2022); another malt barley variety, 'Beka' was released in 1976. Both of these and many others for the mentioned crops above were reported as being under production

Table 16.4. Per cent area covered under certified seed for major cereals in Ethiopia. Source: constructed by the author from Central Statistical Authority sample surveys and crop management practices (www.csa.gov.et).

	Crop				
Season	Maize	Wheat	*Teff*	Barley	Sorghum
2006	15.35	4.89	0.64	0.51	NA
2007	19.60	4.26	1.04	0.62	0.07
2008	19.66	5.48	0.94	0.65	0.06
2009	23.05	3.48	2.49	1.01	2.39
2010	28.65	9.45	1.98	NA	0.18
2011	32.73	14.54	2.55	1.34	0.42
2012	34.20	6.34	1.68	1.13	0.09
2013	40.36	8.14	3.45	0.71	0.35
2014	44.22	11.09	2.85	0.71	0.35
2015	43.37	10.12	2.30	1.80	0.33
2016	57.17	14.51	3.35	1.79	0.89
2017	59.31	9.24	2.19	3.16	0.70
2018	69.41	18.31	3.95	4.46	0.22
2019	76.96	18.45	2.20	4.88	0.55
2020	82.34	16.66	6.62	4.51	0.60

NA = not available.

Table 16.5. Variety name, year of release (YOR) and age of various crops used for certified seed production in Ethiopia. Source: constructed by the lead author from Ethiopian Seed Association data.

Crop	Variety	YOR	Age (2021)
Maize	BH-140	1988	33
Maize	BH-660	1993	28
Maize	BH-540	1995	26
Maize	PHB-3253	1995	26
Maize	Melkassa-1	2001	20
Maize	BH-542Q	2002	19
Maize	BH-543	2005	16
Maize	PHB-30G19	2006	15
Maize	AMH-850	2008	13
Maize	BH-661	2011	10
Maize	AMH-760Q	2012	9
Average			*20*
Teff	Tsedey	1984	37
Teff	Quncho	2006	15
Teff	DZ-387	2007	14
Teff	DZ-138	–[a]	–
Average			*22*
Wheat	Kubsa	1995	26
Wheat	Galema	1995	26
Wheat	Kakaba	2010	11
Wheat	Mekelle-1	2012	9
Average			*18*
Chickpea	Arerti	1999	22
Chickpea	Shasho	1999	22
Chickpea	Habru	2004	17
Average			*20*
Potato	Jalane	2002	19
Potato	Gudane	2006	15
Potato	Belete	2009	12
Average			*15*
Noug	Shambu	2002	19
Noug	Ginchi	2010	11
Soybean	Ethio-Yugo	2007	14
Average			*15*
Barley (malt)	Holker	1979	42
Onion	Bombay Red	1980	41
Overall average			*26*

[a]Dash indicate data not available.

as of 2017 (MOALR, 2017). Similarly, the onion variety Adama Red (Bombay Red) was released in 1980 but still is the dominant onion variety under production (Table 16.5).

Seeding rate

Our study also demonstrated that the seeding rate smallholders employ for local seed of several of the major cereals is higher than the rate for improved cultivars (Table 16.6). For example, the seeding rate for local maize, *teff*, wheat and sorghum is higher by 20, 14, 8 and 5 kg/ha, respectively, than the rate for improved seed. Barley was the only exception, where the seeding rate for improved seed was higher by 21 kg/ha.

Discussion and Conclusions

The Ethiopian seed systems have undergone substantial changes over the years. The changes happened in terms of seed regulations, number

Table 16.6. Seeding rate of improved and local seed of major cereals in Ethiopia. Source: constructed by the author from Central Statistical Authority sample surveys and crop management practices (www.csa.gov.et).

	Maize		*Teff*		Wheat		Barley		Sorghum	
Year	Improved	Local	Improved	Local	Improved	Local	Improved	Local	Improved	Local
2006	24	45	35	51	188	185	241	164	–	26
2007	25	45	46	53	182	190	151	180	5	26
2008	25	44	42	53	178	189	157	174	7	26
2009	–[a]	40	43	56	193	140	197	152	24	25
2010	25	44	41	53	190	199	–	178	37	28
2011	27	43	42	52	217	197	208	181	20	26
2012	26	44	36	51	182	197	198	177	21	27
2013	25	43	33	48	181	198	169	181	19	27
2014	24	41	34	45	189	193	177	167	33	25
2015	26	44	33	47	195	198	227	181	21	25
2016	28	53	34	49	186	204	175	176	32	30
2017	27	53	35	49	176	179	244	161	23	25
2018	30	51	42	48	207	209	225	171	20	25
2019	33	67	15	54	133	224	151	178	–	32
Avg.	27	47	37	51	185	193	194	173	22	27

[a]Dashes indicate data not available, not representative or discarded.

of companies and the amount of seed demand, supply and distribution. Up until 1990, the ESE remained the sole provider of certified seed in Ethiopia. However, the number of seed companies increased fast from the early 2000s following the government seed policy to allow direct seed sales to smallholders by both public and private companies. Mekonnen *et al.* (2021) reported that the rolling out by government of direct seed sale – i.e. selling seed directly to the smallholders rather than through the government apparatus – has resulted in a 26% yield increase of maize and in a 5% increase in maize harvest sold in 2013; in addition, it led to improvements in seed availability for maize, wheat and *teff*, even though it had no effect on the yields of the latter two crops.

Benson *et al.* (2014) stated that 'While Ethiopian seed producers will have to significantly increase the scope and quality of their marketing efforts to develop profitable and reliable direct commercial seed supply relationships with farmers, we find that the suspicion on the part of government of competitive market forces in the formal seed subsector in Ethiopia is not justified'. Our current study does not support the above statement. The contribution of small private national companies is still very small even though some of them have been in business for close to 30 years.

The majority of private seed companies are more interested in cereals – particularly hybrid maize – than in open-pollinated crops because the former has better profit margin and the technology is more effectively promoted compared to other crops. For example, 18 of the 22 (82%) seed companies for which we have information (Annex Fig. 16.1), have maize occupying the largest proportion of their business portfolio. Only eight of the companies produce wheat and *teff*; three produce chickpea and onion; two produce common bean, soybean and *noug* and forages; and one each for the rest of the crops (including malt barley, black cumin, tomato, papaya, sesame, potato, pigeonpea, alfalfa and sesbania). Many of the seed-producer farmers are also partial to cereals.

Seed produced by farmers is often not counted even though it constitutes significant proportions of seed grown by smallholders. Farmer seed growers also do not receive adequate technical and material support. For example, smallholders who grow seed in the Ejere area of Lume woreda installed seed cleaning machines through support from Food and Agriculture Organization in 2008, but the machines have not been operational because the electric power authority did not provide transformers.

References

Abate, T., Fisher, M., Abdoulaye, T., Kassie, G.T., Lunduka, R., Marenya, P. and Asnake, W. (2017) Characteristics of maize cultivars in Africa: How modern are they and how many do smallholder farmers grow? *Agriculture & Food Security* 6, 30. DOI 10.1186/s40066-017-0108-6.

Alemu, K. (2015) Seed production and dissemination systems analyses: The case of Ethiopia. *Food Science and Quality Management* 35, 25–37.

Atilaw, A., Alemu, D., Bishaw, Z., Kifle, T. and Kaske, K. (2017) Early generation seed production and supply in Ethiopia: Status, challenges and opportunities. *Ethiopian Journal of Agricultural Sciences* 27(1), 99–119.

Benson, T., Spielman, D. and Kasa, L. (2014) Seed marketing program in Ethiopia in 2013: An operational evaluation to guide seed-sector reform. IFPRI discussion paper 01350 May 2014. IFPRI, Washington, DC, USA.

Bishaw, Z. and Alemu, D. (2017) Farmers' perceptions on improved bread wheat varieties and formal seed supply in Ethiopia. *International Journal of Plant Production* 11(1), 117–130.

Bishaw, Z., Molla, A., Ayana, A. and Atilaw, A. (2018) Strategies and innovative approaches for food legumes seed delivery in Ethiopia. *Ethiopian Journal of Crop Sciences* 6(1), 1–35.

CSA (2007) Agricultural Sample Survey 2006/2007 (1999 E.C.), Report on Farm Management Practices (Private Peasant Holdings, *Meher* Season) (September–December), Volume III, Statistical Bulletin. CSA, Addis Ababa, Ethiopia.

CSA (2008) Agricultural Sample Survey 2007/2008 (2000 E.C.), Report on Farm Management Practices (Private Peasant Holdings, *Meher* Season) (September–December), Volume III, Statistical Bulletin. CSA, Addis Ababa, Ethiopia.

CSA (2009) Agricultural Sample Survey 2008/2009 (2001 E.C.), Report on Farm Management Practices (Private Peasant Holdings, *Meher* Season) (September–December), Volume III, Statistical Bulletin. CSA, Addis Ababa, Ethiopia.

CSA (2010) Agricultural Sample Survey 2009/2010 (2002 E.C.), Report on Farm Management Practices (Private Peasant Holdings, *Meher* Season) (September–December), Volume III, Statistical Bulletin. CSA, Addis Ababa, Ethiopia.

CSA (2011) Agricultural Sample Survey 2010/2011 (2003 E.C.), Report on Farm Management Practices (Private Peasant Holdings, *Meher* Season) (September–December), Volume III, Statistical Bulletin. CSA, Addis Ababa, Ethiopia.

CSA (2012) Agricultural Sample Survey 2011/2012 (2004 E.C.), Report on Farm Management Practices (Private Peasant Holdings, *Meher* Season) (September–December), Volume III, Statistical Bulletin. CSA, Addis Ababa, Ethiopia.

CSA (2013) Agricultural Sample Survey 2012/2013 (2005 E.C.), Report on Farm Management Practices (Private Peasant Holdings, *Meher* Season) (September–December), Volume III, Statistical Bulletin. CSA, Addis Ababa, Ethiopia.

CSA (2014) Agricultural Sample Survey 2013/2014 (2006 E.C.), Report on Farm Management Practices (Private Peasant Holdings, *Meher* Season) (September–December), Volume III, Statistical Bulletin. CSA, Addis Ababa, Ethiopia.

CSA (2015) Agricultural Sample Survey 2014/2015 (2007 E.C.), Report on Farm Management Practices (Private Peasant Holdings, *Meher* Season) (September–December), Volume III, Statistical Bulletin. CSA, Addis Ababa, Ethiopia.

CSA (2016) Agricultural Sample Survey 2015/2016 (2008 E.C.), Report on Farm Management Practices (Private Peasant Holdings, *Meher* Season) (September–December), Volume III, Statistical Bulletin. CSA, Addis Ababa, Ethiopia.

CSA (2017) Agricultural Sample Survey 2016/17 (2009 E.C.), Report on Farm Management Practices (Private Peasant Holdings, *Meher* Season) (September–December), Volume III, Statistical Bulletin. CSA, Addis Ababa, Ethiopia.

CSA (2018) Agricultural Sample Survey 2017/18 (2010 E.C.), Report on Farm Management Practices (Private Peasant Holdings, *Meher* Season) (September–December), Volume III, Statistical Bulletin. CSA, Addis Ababa, Ethiopia.

CSA (2019) Agricultural Sample Survey 2018/19 (2011 E.C.), Report on Farm Management Practices (Private Peasant Holdings, *Meher* Season) (September–December), Volume III, Statistical Bulletin. CSA, Addis Ababa, Ethiopia.

CSA (2020) Agricultural Sample Survey 2019/20 (2012 E.C.), Report on Farm Management Practices (Private Peasant Holdings, *Meher* Season) (September–December), Volume III, Statistical Bulletin. CSA, Addis Ababa, Ethiopia

CSA (2021) Agricultural Sample Survey 2020/21 (2013 E.C.), Report on Farm Management Practices (Private Peasant Holdings, *Meher* Season) (September–December), Volume III, Statistical Bulletin. CSA, Addis Ababa, Ethiopia.

Hirpa, A., Meuwissen, M.P.M., Tesfaye, A., Lommen, W.J.M., Lansink, A.O., Tsegaye, A. and Struik, P.C. (2010) Analysis of seed potato systems in Ethiopia. *The American Journal of Potato Research* 87, 537–552. DOI: 10.1007/s12230-010-9164-1.

Hodson, D.P., Jaleta, M., Tesfaye, K., Yirga, C., Beyene, H. *et al.* (2020) Ethiopia's transforming wheat landscape: Tracking variety use through DNA fingerprinting. *Scientific Reports* 10:18532. DOI: 10.1038/s41598-020-75181-8.

Jaleta, M. *et al.* (2020) Misidentification by farmers of the crop varieties they grow: Lessons from DNA fingerprinting of wheat in Ethiopia. *PLoS ONE* 15(7), e0235484. DOI: 10.1371/journal.pone.02354 84.

Kalsa, K.K. (2019) Farmers' attitudes and practices towards variety and certified seed use, seed replacement and seed storage in wheat growing areas of Ethiopia. *African Journal of Science, Technology, Innovation and Development* 11, 107–120. DOI: 10.1080/20421338.2018.1550932.

Kifle, T. and Atilaw, A. (2018) Constraints to private seed sector development in Ethiopia: A case study. *African Journal of Agricultural Research* 13(48), 2748–2756. DOI: 10.5897/AJAR2018.13461.

Louwaars, N.P., de Boef, W.S. and Edeme, J. (2013) Integrated seed sector development in Africa: A basis for seed policy and law. *Journal of Crop Improvement* 27(2), 186–214. DOI: 10.1080/15427528.2012.751472.

Mekonnen, D.K., Abate, G.T., Yimam, S., Benfica, R., Spielman, D. and Place, F. (2021) The impact of Ethiopia's direct seed marketing approach on smallholders' access to seeds, productivity and commercialization. Selected Paper prepared for presentation at the 2021 Agricultural & Applied Economics Association Annual Meeting, Austin, TX, USA.

Minwagaw, G.W. and Ejigu, W.G. (2021) Determinants of seed distribution system: The case of Womberma district, northwest Ethiopia. *Advances in Agriculture*, 3656320, 9 p. DOI: 10.1155/2021/3656320.

Mohammednur, T. and Megersa, T. (2021) Seed system analysis of soybean, *Glycine max* L. Merrill, in southwestern Ethiopia. *Frontiers* 1(2), 19–25. DOI: 10.11648/j.frontiers.20210102.12.

MOALR (Ministry of Agriculture and Livestock Resource) (2017) Crop Variety Register, Issue No. 20. June 2017, Addis Ababa, Ethiopia.

Mulesa, T.H. (2021) Politics of seed in Ethiopia's agricultural transformation: Pathways to seed system development. *Frontiers in Sustainable Food Systems* 5:742001. DOI: 10.3389/fsufs.2021.742001.

Ojiewo, C.O., Omoigui, L.O., Pasupuleti, P. and Lenne, J.M. (2020) Grain legume seed systems for smallholder farmers: Perspectives on successful innovations. *Outlook on Agriculture* 49(4), 286–292.

O'Neill, M. (2018) Tilling the earth: Improved seed use in Ethiopia and Brazil. *Global Majority E-Journal* 9(1), 29–44.

Sisay, D.T., Verhees, F.J.H.M. and van Trijp, H.C.M. (2017) Seed producer cooperatives in the Ethiopian seed sector and their role in seed supply improvement: A review. *Journal of Crop Improvement* 31(3), 323–355. DOI: 10.1080/15427528.2017.1303800.

Tarekegn, K. and Mogiso, M. (2020) Assessment of improved crop seed utilization status in selected districts of southwestern Ethiopia. *Cogent Food & Agriculture* 6, 1816252. DOI: 10.1080/23311932.2020.1816252.

Welu, G.M. (2015) Challenges and opportunities of seed multiplication in eastern Tigray, Ethiopia. *Journal of Biology, Agriculture and Healthcare* 5(3), 42–52.

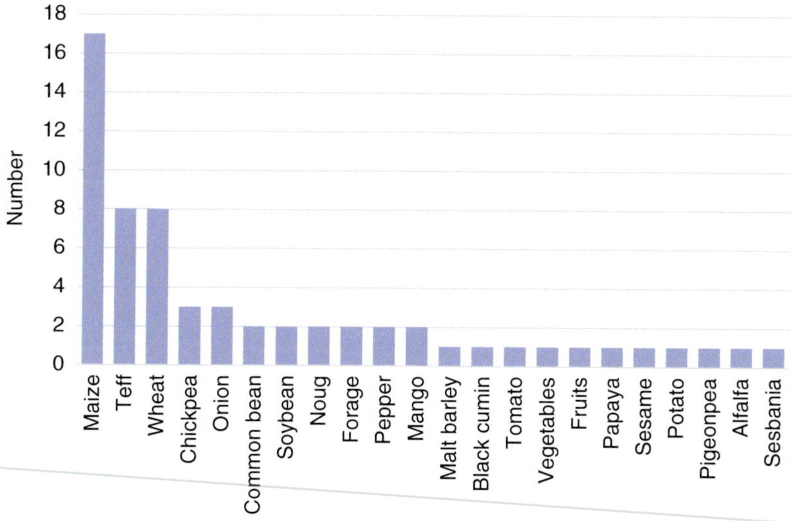

Annex Fig. 16.1. Number of seed companies producing certified seed of different crops in Ethiopia. Source: constructed by the lead author from Ethiopian Seed Association data.

Annex Table 16.1. Summary profiles of private seed companies registered with the Ethiopian Seed Association, as of August 2021. Source: constructed by the lead author from Ethiopian Seed Association data.

Number	Company name	YOE	Land (ha)	Crops	Varieties
1	Anno Agro-Industry PLC	1993	527	Hybrid maize, teff, common bean, soybean, *noug*, forage	BH-660, BH-661, Quncho, Tsedey
2	Mohammed Awel Agricultural Development	1994	1353	Maize, common bean, teff, chickpea,	BH-540, BH-660, DZ-1387, DZ-Cross-37
3	Gadissa Gobena Commercial Farm	1996	65	Hybrid maize, wheat, teff, chickpea, forage	BH-660, AMH-850, AMH-760Q
4	Yimam Tessema Seed Enterprise	1999	20	Maize, wheat	BH-660, BH-661, BH-540
5	Hadiya Seed Production and Agro-Industry	2000	452	Maize, soybean, *noug*	BH-660, BH-140, BH-540, BH-542, PHB-3253, 30G19, Ethio-Yugoslavia, Shambu, Ginchi
6	Tadesse Genna Farm	2004	214	Wheat, malt barley, black cumin	Qeqeba, Qubsa, Holker
7	Gasha Redda Lemma Mechanized Farm	2006	40	Teff, wheat, maize, onion, tomato, pepper, mango	Mekelle-1, Pica Flour, Melkassa-1
8	Safron Agricultural Farm PLC	2006	60	Wheat, teff, vegetables, fruits, maize	Mekelle-1, Pica Flour, Melkassa-1
9	Ayehu Zengeni Farm	2006	1631	Maize	BH-660, BH-540
10	Bayeh Mekonnen Integrated Agriculture Enterprise	2006	400	Maize, red pepper	BH-660, BH-540, BH-543, Mareko Fana
11	Ethio-AGRICEFT PLC	2007	4675	Maize, wheat	BH-540, BH-660, Pica Flour
12	Free Land General Trading PLC	2007	Contractual	Maize	BH-660, BH-661, Pica Flour
13	Loma and Azmer Seed Producer	2007	150	Maize	BH-660, BH-540
14	Avallo International Research and Development Center PLC	2007	800	Wheat, maize, teff	HAR-1685, HAR-604, BH-540, BH-660, DZ-387
15	Jema Agrotech PLC	2008	5	Maize, onion	BH-660, BH-543, Bombay Red
16	Gebru Hagos Integrated Irrigation Farm	2009	12	Wheat, chickpea, teff, fieldpea, maize, papaya, mango, sesame	Mekelle-1, Melkassa-1
17	Homa Seed PLC	2009	233	Maize, teff, chickpea	BH-140, BH-543, Quncho, Shasho, Arerti, Habro
18	Molla Bishaw Seed Producing Enterprise	2009	150	Maize, wheat	BH-660, BH-540
19	Yohannes Girmay Farm	2010	10	Onion	Bombay Red
20	Merkeb Union (cooperative)	2010	Contractual	Maize, wheat	BH-540, Pica Flour
21	Tesfaye Ararsa Potato	2012	12	Potato	Gudane, Jalane, Belete
22	Seid Hussien Forage Seed Producer	2013	20	Pigeonpea, alfalfa, sesbania	Not given

Annex Table 16.2. Early generation seed production in Ethiopia. Source: calculated and constructed by the lead author from various sources; 2006 to 2015 data (produced by Ethiopian Institute of Agricultural Research and Ethiopian Seed Enterprise only) from Atilaw *et al*. (2017); 2016 to 2018 data (OSE, ESE, ASE, SSE, EIAR, OARI, SARI, ARARI, Burka PLC and Amuari PLC) from EIAR.

	Amount (MT)				Percentages		
Season	Cereals	Pulses	Oilseeds	Total	Cereals	Pulses	Oilseeds
2006	2,161	343	87	2,592	83.4	13.2	3.4
2007	2,794	309	79	3,183	87.8	9.7	2.5
2008	3,183	419	63	3,664	86.9	11.4	1.7
2009	4,700	423	46	5,169	90.9	8.2	0.9
2010	5,202	468	158	5,828	89.2	8.0	2.7
2011	8,739	508	93	9,339	93.6	5.4	1.0
2012	4,344	376	77	4,797	90.6	7.8	1.6
2013	4,279	431	159	4,869	87.9	8.9	3.3
2014	3,751	432	290	4,472	83.9	9.7	6.5
2015	8,364	570	233	9,167	91.2	6.2	2.5
2016	13,932	–[a]	–	13,932	–	–	–
2017	10,301	–	–	10,301	–	–	–
2018	9,361	–	–	9,361	–	–	–

[a]Dashes show data not available.

ARARI = Amhara Regional Agricultural Research Institute; ASE = Amhara Seed Enterprise; EIAR = Ethiopian Institute of Agricultural Research; ESE = Ethiopian Seed Enterprise; OARI = Oromiya Agricultural Research Institute; OSE = Oromiya Seed Enterprise; SARI = Southern Agricultural Research Institute; SSE = Southern Seed Enterprise.

17 Ethiopia Has Yet to Meet the African Union Fertilizer Recommendation

Tsedeke Abate[1]* and Endeshaw Habte[2]
[1]*Homegrown Vision, Addis Ababa, Ethiopia;* [2]*Ethiopian Institute of Agricultural Research, Addis Ababa, Ethiopia*

Abstract

The use of mineral fertilizer to enhance crop productivity has become an essential component of farm management practices in Ethiopia following the imperial government's efforts to modernize agriculture in the early 1960s. Smallholder farmers in this country use both mineral and organic fertilizers. All of the national demand for mineral fertilizers is met through imports, which makes it highly vulnerable to price volatility similar to the situation after the COVID-19 pandemic and the war in Ukraine. The use of biofertilizers has not been accorded the due attention commensurate with its economic and environmental benefits. Overall, mineral and organic fertilizers covered about 62% and 12%, respectively, of the total cultivated area in 2017–2019. Mineral fertilizer use increased by 3.16% whereas organic fertilizer use declined by 2.24% annually between 2006 and 2019. Ethiopia's mineral fertilizer application rate of 35 kg/ha of arable land is below the African Union recommendation of 50 kg/ha.

Introduction

To provide food and nutrition security and economic wellbeing for its people, Ethiopia needs to double its current agricultural production by 2050. The current requirement for cereals of about 32 million MT is projected to increase by 25% by 2030. The minimum requirement for cereals by 2040 and 2050 is estimated at 40 MT and 57 million MT, respectively (Abate, 2021). The country needs to boost its agricultural productivity through the effective and efficient use of modern crop varieties and fertilizers, among others, to meet those needs.

Historically, Ethiopian farmers have used organic fertilizers (such as farmyard manure, compost, crop residue and household refuse) for agricultural production (Abate *et al.*, 2015). The use of mineral fertilizers started in the early 1960s following the imperial government's effort to modernize Ethiopian agriculture. This chapter aims to provide a critical review of fertilizer use in Ethiopian agriculture.

Approach

We used data from the Central Statistical Agency's (CSA; www.csa.gov.et) agricultural sample surveys for the 2006–2019 *meher* seasons and FAOSTAT data (www.fao.org/faostat/en/#data) for 1961–2019. Records of the former include both mineral and organic fertilizers; the parameters contain the number of smallholders, total area cultivated,

* Corresponding author: tsedeke@homegrownvision.org

 Untold Stories of African Agriculture: Lessons from Ethiopia (T. Abate ed.)
DOI: 10.1079/9781800626386.0017

total area receiving fertilizer, type of fertilizer and quantity of fertilizer products (not nutrients) used. The FAOSTAT provides data on the quantity of agricultural use of NPK nutrients. We used the land-use data to calculate application rates, i.e. divide the quantity of fertilizer nutrients used by arable land.

Organized country data from the CSA records are, however, available starting only from recent years (CSA, 2007–2020). FAOSTAT provides longer-term data but does not include organic fertilizer and does not specify crops. This study employed the latter's data to show long-term trends of both quantity and rate of application as well as comparisons with other countries. For comparisons among groups of crops and individual crops within each group, we used CSA's agricultural sample surveys for 14 years, from 2006–2019.

Results

Smallholder farmers in Ethiopia use both mineral and organic fertilizers to enhance crop productivity. The country relies totally on imports of mineral fertilizers from abroad. The FAO data for mineral fertilizer imports in Ethiopia are available from 2002 to 2020 (Fig. 17.1).

Import quantity and value showed substantial growth over this period. Import value peaked at US$620 million in 2012. The latest data show that Ethiopia imported nearly 1.6 million MT of mineral fertilizers (by product), valued at US$500 million in 2020. The annual average for 2016–2020 was >1.2 million MT valued at US$456 million. This means that import quantity and value increased at roughly 10% and 13%, respectively, annually in the 19 years between 2002 and 2020. Ethiopia is the third largest mineral fertilizer-importing country among the top ten in Africa after Morocco and South Africa, followed by Kenya, Zambia, Tanzania, Mali, Malawi, Zimbabwe and Ghana (Annex Fig. 17.1).

With more than 16 million ha arable land, Ethiopia ranks fourth (after Nigeria, Sudan and Niger) in Africa. It is followed by the Democratic Republic of the Congo, South Africa, Morocco, Algeria, Uganda and Mali. In the world scene, Ethiopia has the 20th largest area of arable land out of 166 countries and territories for which data were available in the 2016–2020 period. The leading countries include the USA, India, the Russian Federation, China and Brazil (FAO, 2022).

Figure 17.2 presents the long-term use and rate of mineral fertilizer nutrients in Ethiopia. Until recent times, this country imported diammonium phosphate (DAP) and urea as sources of nitrogen and phosphorus. Potassium fertilizers were not considered important in Ethiopian agriculture, as there is a perception that Ethiopian soils are not deficient in this element. With

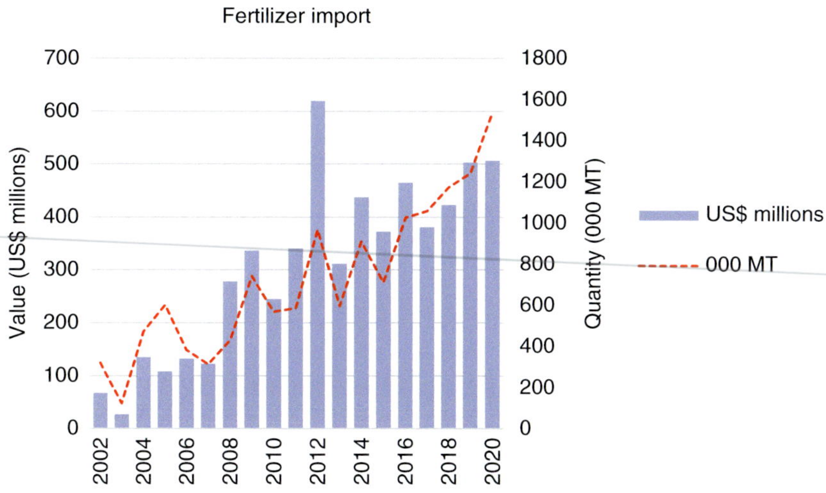

Fig. 17.1. Mineral fertilizer (by product) imports in Ethiopia. Source: constructed by the authors from FAO (2022).

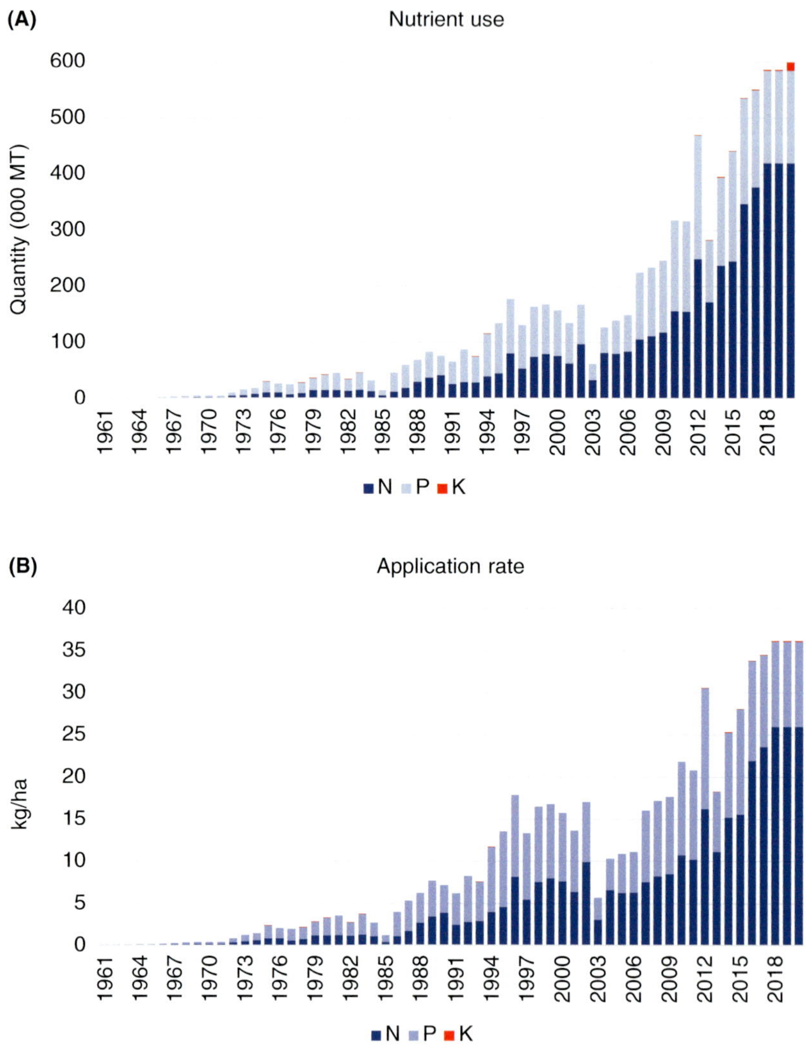

Fig. 17.2. (A) NPK nutrient consumption in Ethiopia. **(B)** Application rate of NPK nutrients per ha of arable land in Ethiopia, 1961–2020. Source: calculated and constructed by the authors from FAO (2022).

the exception of a few isolated cases (e.g. 1975, 1978–1980, 1982, 1983, 1993 and 1994), potassium was not regularly imported and applied. Since 2012, however, it has been included, albeit still in small quantities. In general, Ethiopia's mineral fertilizer use during the 2016–2020 period comprised roughly 70% N, 30% P and a negligible percentage of K (FAO, 2022).

There were also changes in the proportions of N and P over the years. N accounted for roughly 37%, 46% and 70%, respectively, of the total mineral fertilizer use in 1976–1980, 1996–2000 and 2016–2020 periods. By contrast, the share of P was about 63%, 54% and 30%, respectively, during the three periods. The proportion of K was 0.28%, 0.00% and 0.21%, respectively, in 1976–1980, 1996–2000 and 2016–2020, respectively.

Both the total consumption and application rate (kg/ha of arable land) of the total mineral fertilizer nutrients showed substantial increases over the years. Total consumption and application rate increased at an annual rate of 7.07% and 9.83%, respectively, between 2000 and

2020. N consumption and application rate increased by 11.56% and 7.17%, respectively, whereas that of P was 7.17% for consumption and 4.47% for application rate.

Table 17.1 depicts mineral fertilizer nutrient application rates in the top ten countries with the largest arable land area in Africa. The 2016–2020 average application rate for Ethiopia of 35 kg/ha of arable land ranked third among the top ten countries (after South Africa and Morocco). However, Ethiopia has yet to meet the African Union (AU) recommendation of the Abuja Declaration of 50 kg/ha.

Although the overall national average application rate per ha of arable land for Ethiopia is below the AU recommendation, its application rates for cereals (especially maize, *teff* and wheat) appeared to be close to that recommendation. In all of Africa, only Egypt, Zambia, Ghana, South Africa, Morocco, Algeria, Kenya and Tunisia have application rates exceeding the AU recommendation of 50 kg/ha (FAO, 2022). Botswana was nearly there with about 48 kg/ha application rate.

Annex Figure 17.2 presents the NPK application rates for the top ten countries in the world compared to Ethiopia. With its 35 kg/ha of mineral fertilizer nutrients, Ethiopia ranked 125th out of 166 countries and territories for which data were given in the 2016–2020 period (FAO, 2022). The leading countries were China, Brazil, Indonesia, India, France and Pakistan, with application rates ranging from 420–151 kg of mineral nutrients per ha of arable land.

According to the CSA, there were substantial differences among groups of crops and individual crops within the groups (Fig. 17.3) in terms of the area coverage by fertilizers. For example, the area covered by mineral fertilizers was 76% in vegetables versus 18% in oilseeds. In a similar fashion, organic fertilizer covered 58% of the total area in vegetables, compared to 7% in oilseeds. The highest area coverage of mineral fertilizers was in vegetables and cereals; roots and tubers (RTs) (not including *enset*) also receive relatively high area coverage of organic fertilizers, with potato accounting for the bulk of it. Vegetables and RTs receive the highest coverage of organic fertilizers while the reverse is the case in cereals and oilseeds that smallholders cultivated in the 2017–2019 main crop seasons (Fig. 17.3).

The average mineral fertilizer coverage for all crops increased from 44% in 2006 to 62% in 2019. By contrast, organic fertilizer coverage declined from 15% to 12% during the same period. This means that mineral fertilizers increased at an annual rate of 3.16% whereas organic fertilizers declined by 2.24%.

Within crop groups, mineral fertilizers covered about 86% of wheat area compared to 34% of emmer; 52% of avocado versus 32% mango; 42% rapeseed versus 13% *noug*; and 41% fieldpea versus 8% grasspea, according to the CSA data (not shown). By the same token, organic fertilizer coverage was 48% for yam versus 1% for beetroot; 19% for emmer versus 13%

Table 17.1. Application rate of NPK nutrients in the top ten countries with the largest arable land area in Africa (2016–2020 average). Please note that rankings for arable land and application rate do not necessarily match. Source: calculated and constructed by the authors from FAO (2022).

Country	Arable land (000 ha)	Application rate (kg/ha)			
		N	P	K	Total (NPK)
South Africa	12,000	32	21	11	64
Morocco	7,808	26	18	13	57
Ethiopia	16,078	25	11	0	35
Mali	6,411	18	7	4	29
Algeria	7,478	9	8	4	21
Nigeria	35,000	12	4	3	18
Sudan	20,812	7	1	0	8
Uganda	6,900	1	1	1	3
DRC	13,311	1	0	0	2
Niger	17,660	0	0	0	0

DRC = Democratic Republic of the Congo.

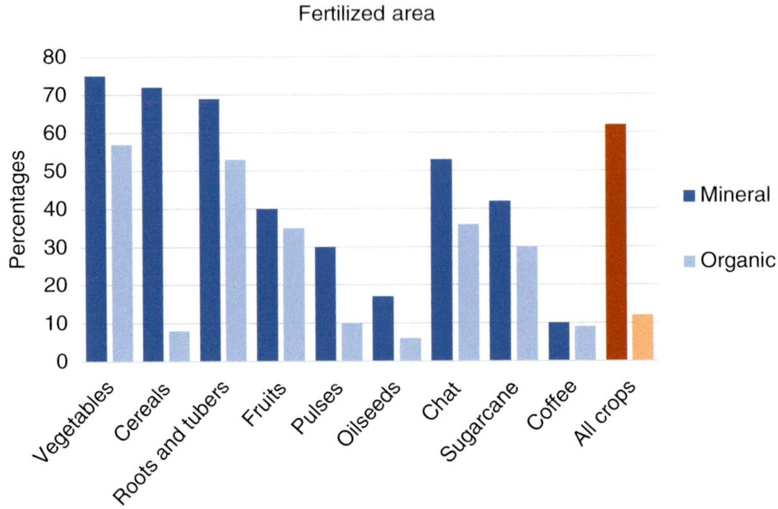

Fig. 17.3. Area covered under mineral and organic fertilizers of selected groups and individual crops in smallholder agriculture grown during the *meher* season in Ethiopia (2017–2019 weighted average). Source: calculated and constructed by the authors from CSA agricultural sample surveys (www.csa.gov.et).

for wheat; 17% for rapeseed versus 4% for safflower; and 14% for fieldpea versus 2% for soybean. In general, smallholder farmers use more organic fertilizer on the crops mostly produced for their own consumption. By contrast, they apply more mineral fertilizers on crops mostly produced for market. Further details are presented in Annex Table 17.1.

Below, we present further details of mineral and organic fertilizer use in smallholder agriculture in Ethiopia.

Mineral fertilizers

According to the CSA, mineral fertilizers in smallholder agriculture in Ethiopia comprised about 54% DAP, 38% DAP + urea mix and 8% urea (based on 2010–2012 avg.). Data on the use of NPS started in 2015 and at present, the composition included roughly 51 NPS + urea, 27% mixed + urea, 14% NPS, 3% each of mixed and urea, 2% DAP + urea and <1% DAP (2017–2019 average). This means that DAP is being phased out and replaced by NPS.

Mineral fertilizer consumption in Ethiopia has increased, especially following the introduction of improved crop varieties and hybrids over the last few decades. It showed fast growth starting from the second half of 2000s in particular (Fig. 17.2A). The annual mineral fertilizer (NPK nutrients) consumption in 2016–2020 was nearly 569,000 MT, making it the fourth largest consumer in Africa after Egypt, South Africa and Nigeria, followed by Morocco, Kenya, Zambia, Mali and Ghana. Put differently, the average consumption during 2016–2020 was ~569,000 MT compared to ~169,000 MT in 1996–2000 and 33,000 MT in 1976–1980. This means that mineral fertilizer consumption in Ethiopia increased at an annual rate of 9.83% between 2000 and 2020.

According to the CSA data, about 88% of the total mineral fertilizer products (not nutrients) consumption of nearly 1.4 million MT annually in 2017–2019 was applied on cereals; the share for vegetables, grain legumes, RTs and oilseeds was 5%, 4%, 2% and 1%, respectively. The share of cereals, grain legumes and RTs remained relatively constant for the 14-year period included in this study; vegetables share increased whereas the share for oilseeds tended to decline.

The CSA provides mineral fertilizer use information for 37 crops grown in the country. Maize, *teff* and wheat accounted for 29%, 27% and 22%, respectively, of the total mineral fertilizer consumption. These were followed by barley, dry pepper, finger millet, sorghum, faba bean, common bean and potato. The share of the remaining 27 crops each was less than 1%, and together they accounted for 4% of the total consumption.

The CSA data also show that there are substantial differences in the application rate of mineral fertilizers among the cereals (Annex Table 17.2). Wheat and maize have the highest application rate of about 200 kg/ha, followed by *teff* with >130 kg/ha. These showed continued growth over the 14 years included in this study. For example, the rates for wheat, maize and *teff* in 2006 was 80 kg/ha, 45 kg/ha and 48 kg/ha, respectively; by contrast, it was 200 kg/ha, 196 kg/ha and 134 kg/ha, respectively, in 2019. The rest of the cereals also showed increases, but the rates were much lower than the 'big three' mentioned here (Annex Table 17.2).

Organic fertilizer

As stated earlier, organic fertilizer use in smallholder agriculture in Ethiopia showed a declining trend – the opposite of mineral fertilizer. However, just as in the mineral fertilizer case, it was crop specific. It was highest in vegetables, RTs and, to some extent, fruit crops, whereas cereals and grain legumes are recipients of the least organic fertilizer coverage.

For example, the crop area showed a similar trend, with the crop area covered by mineral fertilizers increasing at an annual rate of 3.16% compared to an annual decline of 2.24% for organic fertilizers. Furthermore, smallholder farmers are more likely to apply mineral fertilizer in the crops destined for sale, while crops produced for their own consumption are more likely to receive more organic fertilizers (cf. Annex Fig. 17.3).

This study has also revealed another interesting observation: that is a comparison between coffee and *ch'at*. The coverage of mineral and organic fertilizers in the former was 21% and 19%, respectively, versus 54% and 37%, respectively, in the latter. Considering the economic importance of coffee and its preference for organic matter-rich soils, one would expect it to receive a better treatment than *ch'at*. Furthermore, our experience with indigenous practices suggests that sustainable soil fertility management in perennial crops such as *enset* and coffee can be achieved with optimal use of organic rather than mineral fertilizers. This calls for emphasis on increased production and appropriate use of biomass for soil fertility management, not only in perennial crops but also in annual crops, including cereals.

Biofertilizers

Research and development in the use of biofertilizers using *Rhizobium* inoculants in Ethiopia started in the mid-1990s, and so much knowledge and experience has been gained. There is abundant literature reporting the benefits of this technology (Tena *et al.*, 2016a,b; Samago *et al.*, 2018; Wolde-Meskel *et al.*, 2018; Belete *et al.*, 2019). Farmers are already using *Rhizobium* inoculants (with or without phosphorus supplements) to enhance the productivity of grain legumes in many parts of the country (Alemayehu, 2020; Mengesha, 2021). Small businesses are already producing and marketing *Rhizobium* inoculants in the country.

Emerging technologies in this field suggest that application of biofertilizer is not limited to grain legumes. For example, Asmamaw *et al.* (2019) demonstrated that application of dried and liquid cyanobacterial biofertilizer treatments significantly improved soil N, soil organic carbon, and available P, Fe and Zn. They recommended that this product be used as supplementary N source to inorganic fertilizer for maize and vegetables.

Discussion and Conclusions

The use of mineral fertilizer to improve crop productivity in Ethiopia goes back to the early 1960s, following the imperial government's efforts to modernize Ethiopian agriculture. It has become an essential component of farm management practices both for smallholder and commercial farming in this country over the last five decades or so.

Our study has clearly indicated that the overall mineral fertilizer use has shown continued growth, whereas coverage by organic fertilizer has been declining slowly. With mineral fertilizers, not only the total quantity growing but also the application rate has shown continued growth. However, the application rate for mineral fertilizers is still lower than the Abuja Declaration of 50 kg nutrients per ha of arable land. Only major cereals (particularly wheat and maize) may have achieved that threshold among all the crops mentioned in this study.

Several other studies (e.g. Endale, 2011; Ketema and Bauer, 2011; Yu and Nin-Pratt, 2014; Alemayehu and Jemberie, 2018; Abebe and Debebe, 2019; Engida *et al.*, 2019; Asfaw *et al.*, 2020; Iticha *et al.*, 2021) have shown that Ethiopian farmers use less than the optimum level of mineral fertilizers. High prices are among the major constraints limiting increased use of mineral fertilizers in this country.

Organic fertilizer use in the majority of crops has also been declining. Crop residues and farmyard manure are important sources of cash income in smallholder agriculture in Ethiopia. Cereal crop residue is used as animal feed, for construction and firewood; cow dung cake is also an important source of energy for cooking. Grain legumes residue is also totally removed from the field for animal feed. All this means that there is heavy biomass mining and little or no biomass that goes back to the soil.

The biggest gap for fertilizer use in Ethiopia is the of lack of effort by the policy makers to promote microbial fertilizers in spite of their huge benefits, especially for grain legumes. The technology is available and a good number of farmers are buying it from local small businesses, but it is high time for our policy makers to include biofertilizers in extension packages.

References

Abate, T. (2021) Critical issues in partnership for agricultural research and development in Ethiopia. Keynote presentation given at the 9th Annual Science Conference of Bahr Dar University, 21 May 2021.

Abate, T., Shiferaw, B., Menkir, A., Wegary, D., Kebede, Y. (2015) Factors that transformed maize productivity in Ethiopia. *Food Security* 7, 965–981. DOI: 10.1007/s12571-015-0488-z.

Abebe, G. and Debebe, S. (2019) Factors affecting use of organic fertilizer among smallholder farmers in Sekela district of Amhara region, northwestern Ethiopia. *Cogent Food & Agriculture* 5(1), 1669398. DOI: 10.1080/23311932.2019.1669398.

Alemayehu, N. (2020) The use of biofertilizer by smallholder farmers and its impact on productivity of pulse-cereal cropping system in Arsi Zone, Oromia Regional State, southeastern Ethiopia. *East African Journal of Sciences* 14(1), 1–12.

Alemayehu, M. and Jemberie, M. (2018) Optimum rates of NPS fertilizer application for economically profitable production of potato varieties at Koga Irrigation Scheme, Northwestern Ethiopia. *Cogent Food & Agriculture* 4, 1439663. DOI: 10.1080/23311932.2018.1439663.

Asfaw, A.G., Gedamu, S.A., Abushe, H.T., Mekonen, T.T., Muhamed, S.B. and Shibeshi, T.H. (2020) Fertilizer-nitrogen use optimization for tef in South Wollo Zone of Ethiopia. *African Journal of Agricultural Research* 16(6), 811–818. DOI: 10.5897/AJAR2020.14826.

Asmamaw, M., Wolde, G., Yohannes, M., Yigrem, S., Woldemeskel, E., Chala, A. and Davis, J.G. (2019) Comparison of cyanobacterial bio-fertilizer with urea on three crops and two soils of Ethiopia. *African Journal of Agricultural Research* 14(10), 588–596. DOI: 10.5897/AJAR2018.13707.

Belete, S., Bezabih, M., Abdulkadir, B., Tolera, T., Mekonnen, K. and Endalkachew Wolde-Meskel, E. (2019) Inoculation and phosphorus fertilizer improve food-feed traits of grain legumes in mixed crop-livestock systems of Ethiopia. *Agriculture, Ecosystems and Environment* 29, 58–64.

CSA (2007) Agricultural Sample Survey 2006/2007 (1999 E.C.), Report on Farm Management Practices (Private Peasant Holdings, *Meher* Season, September–December) Volume III, Statistical Bulletin. CSA, Addis Ababa, Ethiopia.

CSA (2008) Agricultural Sample Survey 2007/2008 (2000 E.C.), Report on Farm Management Practices (Private Peasant Holdings, *Meher* Season, September–December) Volume III, Statistical Bulletin. CSA, Addis Ababa, Ethiopia.

CSA (2009) Agricultural Sample Survey 2008/2009 (2001 E.C.), Report on Farm Management Practices (Private Peasant Holdings, *Meher* Season, September–December) Volume III, Statistical Bulletin. CSA, Addis Ababa, Ethiopia.

CSA (2010) Agricultural Sample Survey 2009/2010 (2002 E.C.), Report on Farm Management Practices (Private Peasant Holdings, *Meher* Season, September–December) Volume III, Statistical Bulletin. CSA, Addis Ababa, Ethiopia.

CSA (2011) Agricultural Sample Survey 2010/2011 (2003 E.C.), Report on Farm Management Practices (Private Peasant Holdings, *Meher* Season, September–December) Volume III, Statistical Bulletin. CSA, Addis Ababa, Ethiopia.

CSA (2012) Agricultural Sample Survey 2011/2012 (2004 E.C.), Report on Farm Management Practices (Private Peasant Holdings, *Meher* Season, September–December) Volume III, Statistical Bulletin. CSA, Addis Ababa, Ethiopia.

CSA (2013) Agricultural Sample Survey 2012/2013 (2005 E.C.), Report on Farm Management Practices (Private Peasant Holdings, *Meher* Season, September–December) Volume III, Statistical Bulletin. CSA, Addis Ababa, Ethiopia.

CSA (2014) Agricultural Sample Survey 2013/2014 (2006 E.C.), Report on Farm Management Practices (Private Peasant Holdings, *Meher* Season, September–December) Volume III, Statistical Bulletin. CSA, Addis Ababa, Ethiopia.

CSA (2015) Agricultural Sample Survey 2014/2015 (2007 E.C.), Report on Farm Management Practices (Private Peasant Holdings, *Meher* Season, September–December) Volume III, Statistical Bulletin. CSA, Addis Ababa, Ethiopia.

CSA (2016) Agricultural Sample Survey 2015/2016 (2008 E.C.), Report on Area and Production of Major Crops (Private Peasant Holdings, *Meher* Season, September–December) Volume III, Statistical Bulletin. CSA, Addis Ababa, Ethiopia.

CSA (2017) Agricultural Sample Survey 2016/17 (2009 E.C.), Report on Farm Management Practices (Private Peasant Holdings, *Meher* Season, September–December) Volume III, Statistical Bulletin. CSA, Addis Ababa, Ethiopia.

CSA (2018) Agricultural Sample Survey 2017/18 (2010 E.C.), Report on Farm Management Practices (Private Peasant Holdings, *Meher* Season, September–December) Volume III, Statistical Bulletin. CSA, Addis Ababa, Ethiopia.

CSA (2019) Agricultural Sample Survey 2018/19 (2011 E.C.), Report on Farm Management Practices (Private Peasant Holdings, *Meher* Season, September–December) Volume III, Statistical Bulletin. CSA, Addis Ababa, Ethiopia.

CSA (2020) Agricultural Sample Survey 2019/20 (2012 E.C.), Report on Farm Management Practices (Private Peasant Holdings, *Meher* Season, September–December) Volume III, Statistical Bulletin. CSA, Addis Ababa, Ethiopia

Endale, K. (2011) *Fertilizer Consumption and Agricultural Productivity in Ethiopia*. EDRI, Addis Ababa, Ethiopia.

Engida, L.E., Srivastava, A.M., Kuhn, A. and Gaiser, T. (2019) Household welfare implications of better fertilizer access and lower use inefficiency: Long-term scenarios for Ethiopia. *Sustainability* 11, 3952. DOI: 10.3390/su11143952.

FAO (Food and Agriculture Organization of the United Nations) (2022) Statistical databases and data-sets of the Food and Agriculture Organization of the United Nations. Available at: http://www.fao.org/faostat/en/#data (accessed 10 November 2022).

Iticha, M.D., Jaleta, M. and Mitiku, F. (2021) Determinants and profitability of inorganic fertilizer use in small-holder maize production in Ethiopia. *Cogent Food & Agriculture* 7(1), 1911046. DOI: 10.1080/23311932.2021.1911046.

Ketema, M. and Bauer, S. (2011) Determinants of manure and fertilizer applications in eastern highlands of Ethiopia. *Quarterly Journal of International Agriculture* 50(3), 237–252.

Mengesha, M. (2021) Review on the production practices and impacts of bio-fertilizers on legume production in Ethiopia. *Advances in Crop Science and Technology* 9, 462.

Samago, T.Y., Anniye, E.W. and Dakora, F.D. (2018) Grain yield of common bean (*Phaseolus vulgaris* L.) varieties is markedly increased by rhizobial inoculation and phosphorus application in Ethiopia. *Symbiosis* 75, 245–255. DOI: 10.1007/s13199-017-0529-9.

Tena, W., Wolde-Meskel, E. and Walley, F. (2016a) Response of chickpea (*Cicer arietinum* L.) to inoculation with native and exotic *Mesorhizobium* strains in southern Ethiopia. *African Journal of Biotechnology* 15(36), 1920–1929.

Tena, W., Wolde-Meskel, E. and Walley, F. (2016b) Symbiotic efficiency of native and exotic *Rhizobium* strains nodulating lentil (*Lens culinaris* Medik.) in soils of southern Ethiopia. *Agronomy* 6, 11; DOI: 10.3390/agronomy6010011.

Wolde-Meskel, E., van Heerwaarden, J., Abdulkadir, B., Kassa, S., Aliyi, A. *et al.* (2018) Additive yield response of chickpea (*Cicer arietinum* L.) to rhizobium inoculation and phosphorus fertilizer across smallholder farms in Ethiopia. *Agriculture, Ecosystems and Environment* 26, 144–152.

Yu, B. and Nin-Pratt, A. (2014) Fertilizer adoption in Ethiopia cereal production. *Journal of Development and Agricultural Economics* 6(7), 318–337. DOI: 10.5897/JDAE2013.0508.

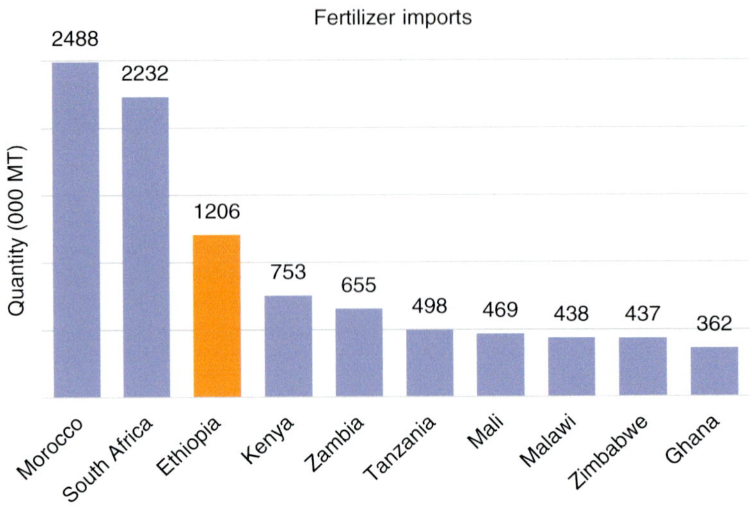

Annex Fig. 17.1. Top ten mineral fertilizer (by product) importing countries in Africa, 2016–2020 average. Source: constructed by the authors from FAO (2022).

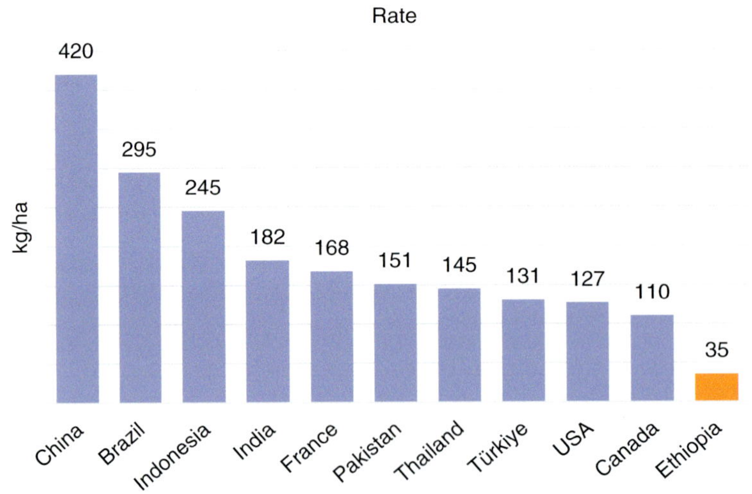

Annex Fig. 17.2. NPK nutrient application rate (kg/ha of arable land) in Ethiopia compared to the world's top ten countries with largest arable land, 2016–2020 average. Descending order of countries with largest arable land area: India, USA, China, Russian Federation, Brazil, Indonesia, Nigeria, Argentina and Ukraine; Ethiopia is 19th out of 223 countries and territories in the world. Source: calculated and constructed by the authors from FAO (2022).

Annex Table 17.1. Per cent share of mineral fertilizers (by product) among major crop groups grown under smallholder agriculture during the *meher* season in Ethiopia. Source: calculated and constructed by the authors from CSA (2007–2020).

Year	Cereals	Vegetables	Grain legumes	Roots and tubers	Oilseeds
2006	90.3	1.2	4.0	2.0	2.4
2007	90.2	1.7	3.6	1.3	3.1
2008	89.8	3.1	3.1	1.6	2.5
2009	81.4	1.9	10.0	2.4	4.3
2010	91.4	2.2	3.4	1.6	1.4
2011	88.9	0.3	7.3	3.1	0.3
2012	90.6	2.4	4.0	1.5	1.5
2013	91.5	1.9	3.6	1.4	1.5
2014	91.7	1.9	3.5	1.4	1.4
2015	90.0	3.2	3.5	1.8	1.6
2016	87.9	5.0	3.5	2.2	1.4
2017	88.2	3.3	4.1	2.8	1.6
2018	89.0	5.1	3.6	1.4	1.0
2019	87.6	6.9	3.3	2.2	0.8

Annex Table 17.2. Mineral fertilizer (by product) application rate (kg/ha) in major cereals grown under smallholder agriculture during the *meher* season in Ethiopia. Source: calculated and constructed by the authors from CSA (2007–2020).

	Crops								
Year	Wheat	Maize	*Teff*	Rice	Millet	Barley	Emmer	Sorghum	Weighted average
2006	80	45	48	–[a]	23	27	17	5	42
2007	85	54	52	2	30	32	17	3	45
2008	79	50	49	–	28	27	12	3	42
2009	57	42	43	5	31	27	36	6	36
2010	80	64	54	11	32	32	24	6	47
2011	50	–	54	–	34	33	–	–	40
2012	90	69	67	23	48	43	29	11	58
2013	101	83	73	–	–	44	34	12	65
2014	115	105	83	47	56	51	33	15	76
2015	127	92	83	56	54	59	27	14	75
2016	144	133	101	78	64	65	40	20	95
2017	139	160	107	159	65	68	72	16	101
2018	170	166	117	117	93	88	45	15	116
2019	200	196	134	114	97	80	34	14	131

[a]Dashes indicate data not available.

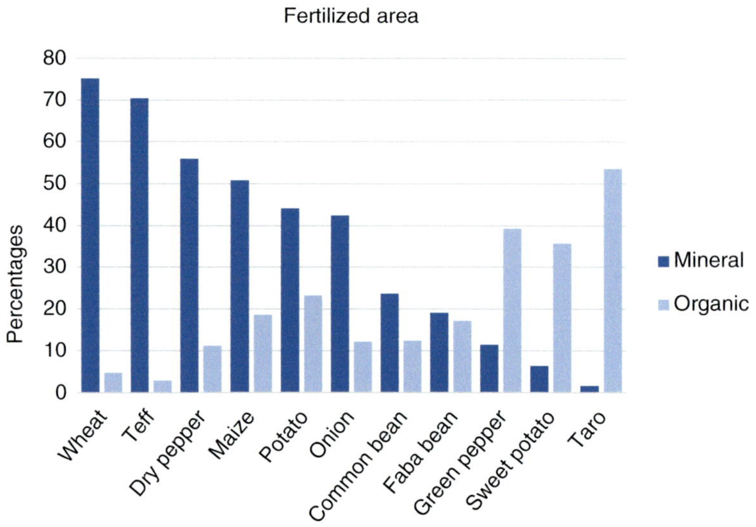

Annex Fig. 17.3. Area covered by mineral and organic fertilizers of selected crops in smallholder agriculture in Ethiopia, 2017–2019 *meher* season average. Source: calculated and constructed by the authors from CSA (www.csa.gov.et).

18 Agricultural Pesticide Use: Current Status and Critical Issues

Tsedeke Abate*
Homegrown Vision, Addis Ababa, Ethiopia

Abstract

This chapter reviews the status of pesticide usage in Ethiopia. As of August 2020, there were 184 active ingredients, represented by 586 trade names of synthetic agricultural pesticides, registered and in use. Of those, 22 are banned and another 21 are on the watch list in other countries. The largest number of active ingredients, 50, are registered for use against cut flower pests. Herbicides, insecticides and fungicides-bactericides constituted roughly 75%, 15% and 9%, respectively. The country imported close to 18,000 MT valued at US$108 million each year during the 2016–2020 period, which translates into 0.23 kg per ha of cropland, one of the lowest among its peers in Africa and elsewhere. Pesticides cover about 25% of the total cultivated area and roughly 11% of smallholders use them. Farmers selectively apply pesticides in those crops that they produce for sale, rather than those they produce for their own consumption.

Introduction

The term 'pesticides' includes all natural or man-made materials and products that kill pests (consisting of harmful insects and other arthropods, plant pathogens, weeds, and vertebrates such as rodents and migratory birds) that are in competition with humans for crop and livestock production or as disease vectors. According to the Food and Agriculture Organization of the United Nations (FAO), pesticides fall into four major categories: fungicides-bactericides, herbicides, insecticides and rodenticides. Acaricides (substances that kill mite pests) and nematicides are both counted under insecticides.

The use of modern agricultural pesticides in Ethiopia started with the control of desert locust plagues and other migratory pests such as the African armyworm (*Spodoptera exempta*) and the weaver bird (*Quelea quelea*) outbreaks, following the establishment of the Desert Locust Control Organization for Eastern Africa in 1962. The advent of commercial farms that produced mostly cotton in the Awash valley in the early 1970s saw increased use of pesticides in this country (Megenasa *et al.*, 1980). These farms were nationalized by the military/socialist government in the second half of the decade and administered by the now defunct Ministry of State Farms Development (MSFD). The MSFD and other parastatals such as the Grain Trade Enterprise and the Ethiopian Seed Enterprise were the major drivers of growth of pesticide use in Ethiopia (Gordon *et al.*, 1995; Abate, 1997; Abate and Fenta, 2003). The next wave of increases in pesticide use in Ethiopia followed the

* Corresponding author: tsedeke@homegrownvision.org

 Untold Stories of African Agriculture: Lessons from Ethiopia (T. Abate ed.)
DOI: 10.1079/9781800626386.0018

introduction of the cut flower industry in the early 2000s.

Recent reviews of pesticide use in Ethiopia deal with limited areas. For example, reports by Gesesew *et al.* (2016), Negatu *et al.* (2016), Mengistie *et al.* (2017), Loha *et al.* (2018) and Mequanint *et al.* (2019) focus on issues related to pesticide safety and handling. Mengistie (2016) reported low capacity for implementing its regulation. Agmas and Adugna (2020) pointed out that some pesticides might leave residue in the crops. Fikadu (2020) talks about the impact of pesticides on the honeybee.

This chapter aims to offer a more comprehensive and nuanced review of the subject and attempts to close the existing gap.

Approach

Records of pesticides registered in Ethiopia were obtained from the Plant Variety Release, Protection and Seed Quality Control Directorate of the Ethiopian Ministry of Agriculture (MOA). References were made to various sources in the literature regarding the status of pesticides registered in other parts of the world (mainly Europe and the USA). Our main sources of data for information on application of pesticides were the FAO (2022a, 2022b, 2023) and the agricultural sample surveys of the Central Statistical Agency (CSA) of the Government of Ethiopia (www.csa.gov.et). Information on banned and watch-list pesticides was obtained from UTZ (2015).

Results

We present the results under three topics: registration, consumption and spread among smallholders and share by crop.

Agricultural pesticides registered

The Plant Variety Release, Protection and Seed Quality Control Directorate of the MOA is responsible for pesticide registration in Ethiopia. Their records show that there were 184 active ingredients (AIs) registered and in use as of August 2020 (Annex Table 18.1). These comprised 67 herbicides, 57 fungicides-bactericides, 54 insecticides, 9 acaricides, 4 rodenticides, and 1 each of an avicide and a nematicide (note: some pesticides are used in more than one category). Insecticides accounted for the largest number of products or trade names (211), followed by herbicides (180), fungicides-bactericides (178), acaricides (10), avicides (1) and nematicides (1), for a total of 586 products. The largest number of AIs, 50 (or 27% of all pesticides), are registered for the control of pests on cut flowers.

A total of 22 pesticide AIs are banned in other countries but are still in use in Ethiopia (Table 18.1). The reasons for banning pesticides are: (i) international conventions; (ii) acute toxicity; and (iii) chronic toxicity (including carcinogenicity, mutagenicity, reproductive toxicity and endocrine disruption in humans). Some of the ingredients listed (e.g. benomyl, carbofuran and endosulfan) satisfy more than one criterion.

The herbicide atrazine and the insecticide carbaryl both qualify to be banned but are still in use because there are no suitable replacements for them.

In a similar fashion, a total of 21 pesticides are included on the watch list elsewhere but still in use in Ethiopia as of August 2020 (Table 18.2). The criteria for including pesticides under the watch list include acute toxicity, chronic toxicity (carcinogenicity and endocrine disruption) and environmental toxicity (toxicity to the honeybee). Just as in banned pesticides, some on the pesticides watch-list also satisfy more than one criterion. For example, lambda-cyhalothrin is on the watch list because of its acute toxicity, endocrine disruption and toxicity to bees. Similarly, epoxiconazole is implicated in carcinogenicity and endocrine disruption in humans. Neonicotinoids (also referred to neonics) such as acetamiprid, imidacloprid and thiamethoxam are in general highly toxic to bees.

Agricultural pesticide import

Ethiopia imports the bulk of its agricultural pesticides from abroad. The country imported about 20,000 MT of agricultural pesticides (by product), valued at US$129 million in 2020 (Fig. 18.1). This study has demonstrated that pesticide import in Ethiopia showed substantial growth over the years.

Ethiopia is the eighth largest importer (after Nigeria, Ghana, South Africa, Cote d'Ivoire, Morocco, Tanzania and Egypt; followed by Kenya and Angola, cf. Annex Fig. 18.1). The annual

Table 18.1. Pesticides banned in other countries but still in use in Ethiopia and indicators for banning. Source: Constructed by the author from www.utz.org/resource-library.org.

Active ingredient	International conventions[a]	Acute toxicity[b]	Chronic toxicity			
			Carcinogenic	Mutagenic	Reproductive toxicant	Endocrine disruptor
Acephate						
Alachlor	x					
Amitraz						
Atrazine[c]						x
Benomyl	x			x	x	
Brodifacoum		x				
Bromadiolone		x				
Carbaryl[c]						x
Carbofuran	x	x				
Carbosulfan		x				
Chlorfenapyr						
Dicofol						
Dimethenamid						
Endosulfan	x	x				
Fenthion						
Fluazifop-butyl					x	
Flucytrinate		x				
Methidathion		x				
Oxydemethon-methyl		x				
Paraquat dichloride		x				
Thiram	x					
Zinc phosphide	x					

[a]Refers to Stockholm convention on persistent organic pollutants, Rotterdam Convention (on highly hazardous toxins) and Montreal Protocol (on protecting the ozone depletion); [b]refers to the World Health Organization classification of 1A & 1B; [c]highly toxic to bees.

rate of growth for pesticide import quantity and value was 11.73% and 15.26%, respectively, between 2000 and 2020. Africa as a whole imported >1.42 million MT of pesticides (by product) for an estimated value of nearly US$5.58 billion each year during the 2016–2020 period (based on the 47 countries for which data were provided). The quantity and value for Africa increased at an annual rate of 9.13% and 9.21%, respectively, during the above period.

Agricultural pesticide use

The FAO data for agricultural pesticide use in Ethiopia started from 1993. Pesticides in this country comprised roughly 75% herbicides, 15% insecticides, 9% fungicides-bactericides and <1% others (including rodenticides), according to the FAO data for the 2016–2020 period. This is a complete reversal of the proportions of the various pesticide groups in the past. For example, Abate (1997) reported that insecticides, herbicides, fungicides, and rodenticides and others constituted 71%, 25%, 3% and 1%, respectively, of the total agricultural pesticides used in Ethiopia between 1983 and 1993. The overall share of herbicides, fungicides, insecticides and rodenticides for Africa as a whole was 52.29%, 25.09%, 20.20% and 0.42%, respectively (FAO, 2022b).

The current study shows that the annual average agricultural pesticide consumption in Ethiopia (2016–2020 average) was more than 4100 MT (in AIs). Pesticide (especially herbicide) use showed accelerated growth from the early 2000s but remained virtually constant after 2009 (Fig. 18.2).

Agricultural pesticide use per ha of cropland

Ethiopia has the third largest cropland area in Africa (after Nigeria and Sudan), followed by

Table 18.2. Pesticides in the UTZ watch list in other countries but in use in Ethiopia and indicators for their classification. Source: constructed by the author from www.utz.org/resource-library.org.

		Chronic toxicity		Environmental toxicity
Active ingredient	Acute toxicity	Probably likely carcinogenic (US EPA)	Endocrine disruptor chemicals (EU)	Highly toxic to bees
2,4-D			x	
Abamectin	x			
Acetamiprid[a]				x
Acetochlor			x	
Bromoxynil	x			
Carbendazim[b]				
Chlorpyrifos				x
Deltamethrin			x	
Dimethoate				x
Epoxiconazole[b]		x	x	
Fenitrothion			x	
Fipronil				x
Glufosinate-ammonium[b]				
Glyphosate		x		
Imidacloprid[a]				x
Lambda-cyhalothrin	x		x	x
Malathion				x
Mancozeb			x	
Metribuzin			x	
Thiacloprid[a]		x		
Thiamethoxam[a]				x

[a]Neonicotinoids; [b]qualify for banning but too challenging to replace.
2,4-D = 2,4-dichlorophenoxyacetic acid.

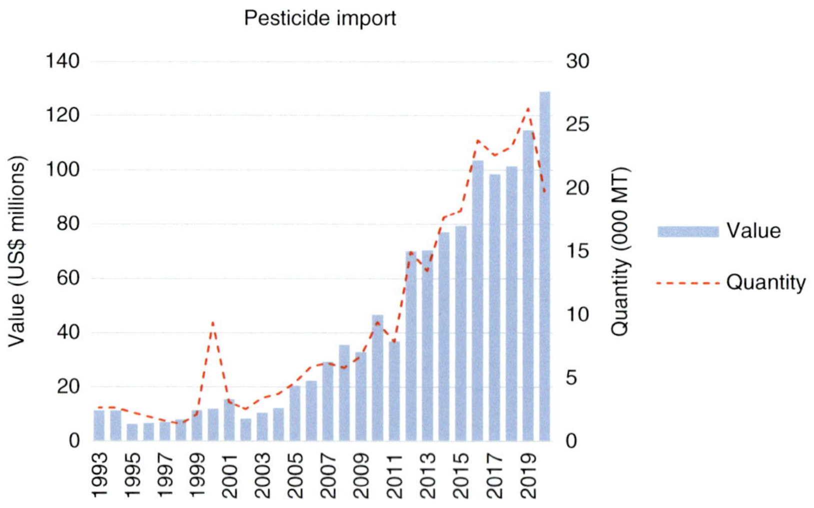

Fig. 18.1. Agricultural pesticide (by product) import into Ethiopia. Source: constructed by the author from FAO (2022b).

Niger, the Democratic Republic of the Congo (DRC), Tanzania, South Africa, Cote d'Ivoire, Morocco and Uganda among the top ten. Its pesticide use per ha of cropland is estimated at about 0.23 kg AI (2016–2020 average), as presented in Fig. 18.3. The pesticide use per ha of cropland showed somewhat similar trends as the total pesticide use. That is, there was an

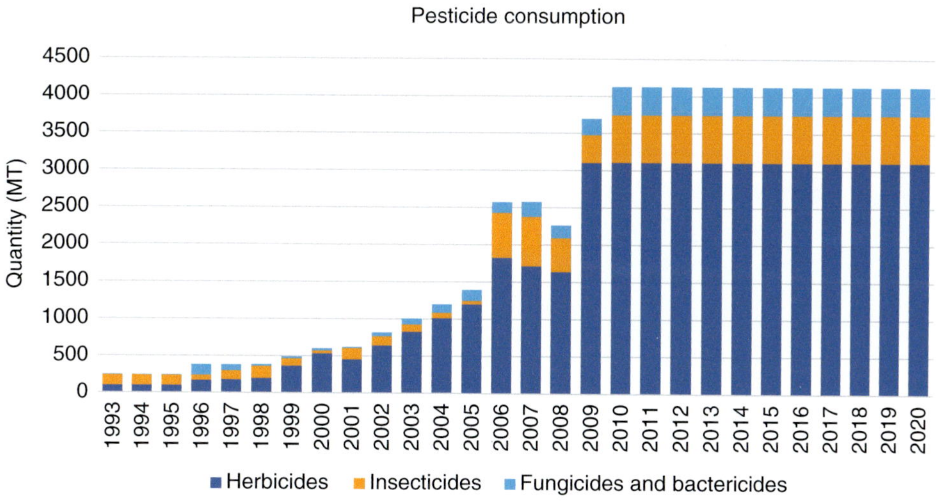

Fig. 18.2. Agricultural pesticide (active ingredient) consumption in Ethiopia. Source: constructed by the author from (FAO, 2022b).

upward trend through 2010 and then it levelled off, eventually taking a downward trend.

In comparison, the values for South Africa, Morocco, Nigeria, Uganda, Sudan, the DRC and Cote d'Ivoire were 3.14 kg/ha, 1.45 kg/ha, 1.03 kg/ha, 0.72 kg/ha, 0.18 kg/ha, 0.07 kg/ha and 0.01 kg/ha of crop land, respectively; the figures for Tanzania and Niger were negligible. The average for all of Africa was 1.05 kg/ha (Annex Fig. 18.2). This means that South Africa, Morocco, Nigeria and Uganda use 13.5-fold, 6.3-fold, 4.5-fold and 3.1-fold the amount used in Ethiopia (FAO, 2022b).

Figure 18.4 depicts pesticide use per ha of cropland in the top ten African countries and Africa between 1990 and 2020. As stated above, Ethiopia's pesticide use has slowly declined after 2010. Pesticide use in South Africa, Nigeria and Uganda showed upward trends; pesticide use for these three countries increased at an annual rate of 3.19%, 11.87% and 9.69%, respectively. The growth for Africa was 1.37%. The figures for Morocco, Africa, Cote d'Ivoire and the DRC have changed little or showed minor declines. The use for Niger and Tanzania was negligible.

On the world scene, India, the USA, China, the Russian Federation, Brazil, Indonesia, Nigeria, Argentina, Canada and the Ukraine are the top ten countries with the largest area of cropland. Ethiopia has the 19th largest area of cropland in the world (FAO, 2022b). Brazil and Indonesia have 8.98 kg and 5.32 kg pesticide AI per ha of cropland, respectively (Annex Fig. 18.2). This is followed by Argentina, the USA, Canada, China, Nigeria, the Ukraine, the Russian Federation and India. The world average was 3.68 kg/ha of cropland.

Pesticide use in smallholder agriculture

Smallholder farmers consider several factors when deciding to apply synthetic pesticides in their crops. Two of the major factors are the importance of the crop and the availability of pest control options. Accordingly, there were huge variations among groups of crops and individual crops within each group in the total area covered under synthetic pesticides and the number of farmers actually using them. For example, according to the CSA data, about 31.5% of cereals, 14.8% of grain legumes, 6.1% of spices, 5.9% each of oilseeds and vegetables, and 0.2% of fruit crops were covered under synthetic pesticides annually during 2015–2019.

Annex Fig. 18.3 depicts the area of various crops covered by synthetic insecticides during the above period. About 42% of wheat farmers, 41% of *teff* farmers, 32% of mungbean farmers, 29% of emmer farmers and 26% of millet farmers applied synthetic pesticides in their crop fields in the 2015–2019 period. By contrast, the number of smallholders applying pesticides on *gesho*, taro, avocado, banana and sugarcane was almost nil.

By the same token, about 56% of wheat, 48% of onion, 47% each of emmer and *teff*, and 36% of mungbean area was covered under synthetic pesticides, whereas banana, other perennials,

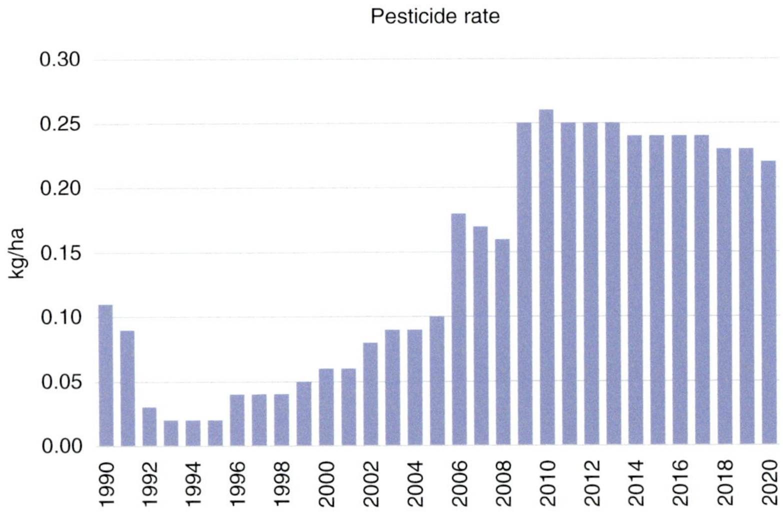

Fig. 18.3. Use of synthetic agricultural pesticides (active ingredient) per ha of cropland in Ethiopia. Source: constructed by the author from (FAO, 2022b).

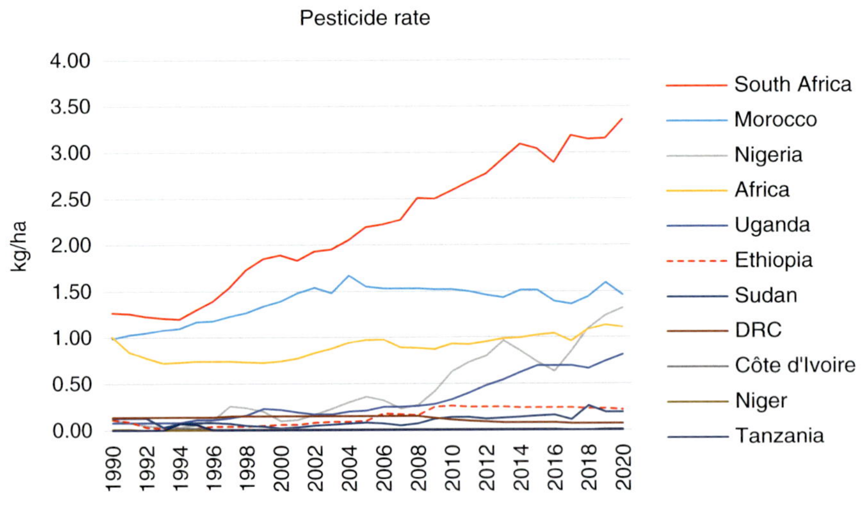

Fig. 18.4. Synthetic pesticide use (kg active ingredient per ha of cropland) in the top ten African countries with the largest cropland area and Africa. Source: constructed by the author from FAO (2023).

avocado, guava and *noug* received little or no pesticide treatments (Annex Fig. 18.4).

The type of pesticide applied also depends on the crop. For example, the major pesticides applied on *teff* are herbicides to control weeds and fungicides against rusts; in wheat it is mainly herbicides against weeds; in mungbean it is fungicides to control halo blight; in potato it is fungicides against late blight; and in onion it is insecticides for the control of onion thrips, or fungicides to control purple blotch.

Crops that receive high levels of pesticide application are those produced mainly for sale, rather than those for home consumption. In other words, smallholders in Ethiopia are willing to apply pesticides in those crops that they produce for commercial purposes, rather than those produced for domestic purposes. Overall, about 25% of the total crop area was covered under synthetic pesticides and 11% of smallholders used them annually during 2015–2019, according to the CSA data.

Discussion and Conclusions

This short chapter aimed to fill the existing gap on the status of agricultural pesticide use in smallholder agriculture in Ethiopia. We looked at the kinds of pesticides used, their growth pattern (especially in comparison with other countries) and the factors associated with agricultural pesticides use in this country.

The MOA has been in charge of pesticide registration in Ethiopia for close to four decades. Even though the information is available in its archives, it does not issue regular updates (similar to what is done for variety releases) to the public. This is perhaps why the country is still using about two dozen pesticides that are either banned or on the watch list in other countries. There is a need to build capacity for not only registration and record keeping, but also to do actual research on issues related to agricultural pesticides. Researchers need to clearly define what kind of research needs to be carried out on pesticide use in this country.

The overall pesticide consumption and application rate are much lower than many of Ethiopia's peers in Africa and elsewhere. This shows that there is no widespread contamination of the environment in smallholder agriculture. On the other, Ethiopian farmers have not been able to realize the full potential of their improved crop varieties through judicious use of pesticides and other inputs.

It is apparent that two factors dictate smallholder farmers' choice to use pesticides. These are: (i) value of the crop as source of cash; and (ii) the availability of options for controlling the pest. For example, all of the top ten crops receiving the highest-level pesticide application are produced more for the market rather than as main staples for home consumption. Mungbean was considered a minor crop and the CSA sample surveys did not include it until the 2014/15 *meher* season; today, it is an important export commodity for the country, worth US$35.8 million, and farmers are investing heavily in pesticides to control the pests associated with its production. Farmers in eastern Ethiopia invest more on the control of pests of *ch'at* than coffee because the former is more lucrative as it is harvested all year round. Finger millet is becoming more popular because of its *injera*-making quality, among other attributes (Anchala *et al.*, 2006), and farmers are willing to invest in its protection. On the other hand, smallholders' investment in maize protection is low despite its huge significance as a staple cash crop. This is because farmers use *shilshalo* (cultivation of maize [or other coarse grain] fields using ox-driven ploughs when the crop is knee-high) (Franzel and van Houten, 1992) and hand weeding to control weeds (instead of using herbicides), varietal resistance against diseases such as Maize Streak Virus and turcicum leaf blight (Tegegne *et al.*, 2008), and an integrated pest management approach for the control of major insects such as stalk borers (Gebre-Amlak, 1988; Bekele *et al.*, 2006; Abate, 2011) such that little or no pesticide application is necessary in smallholder farmers' fields.

There is a consensus that smallholder farmers in Ethiopia do not have adequate training and knowledge to properly handle pesticides (Gesesew *et al.*, 2016; Nigatu *et al.*, 2016; Mengistie *et al.*, 2017; Loha *et al.*, 2018; Mequanint *et al.*, 2019). The capacity to implement regulations is also low (Mengistie, 2016). This kind of situation is not unique to Ethiopia. Delays in implementing regulations also occur in economically advanced countries (Donley, 2019). According to the World Bank Group (IEG, 2016), Ethiopia disposed of only 35% of its inventoried, publicly held stockpiles of obsolete pesticides. Policy makers in Ethiopia can close the current gap in pesticide use and regulations by investing in capacity building for a more effective and efficient system.

References

Abate, T. (1997) Pesticides in Ethiopian agriculture: A researcher's view. *Pest Management Journal of Ethiopia* 1(1&2), 49–55.

Abate, T. (2011) Maize stalk borers of Ethiopia: Quantitative data on ecology and management. In: Worku, M. *et al.* (eds) *Meeting the Challenges of Global Climate Change and Food Security through Innovative Maize Research*. EIAR/CIMMYT, Addis Ababa, pp. 174–184.

Abate, T. and Fenta, T. (2003) Ethiopia Crop Production and Pesticide Evaluation Report (PERSUAP) and Safer Use Action Plan – Ethiopia Pest Management and Pesticide Evaluation Report. USAID/Ethiopia. USAID, Washington, DC, USA, 130 pp.

Agmas, A. and Adugna, M. (2020) Attitudes and practices of farmers with regard to pesticide use in Northwest Ethiopia. *Cogent Environmental Science* 6:1, 1791462, DOI: 10.1080/23311843.2020.1791462.

Anchala, C., Kidane, H. and Mulatu, T. (2006) The impact of improved finger millet technology promotion in the central Rift Valley of Ethiopia. In: Abate, T. (ed.) *Successes with Value Chain. Proceedings of Scaling up and Scaling out Agricultural Technologies in Ethiopia: An International Conference*, 9–11 May 2006, Addis Ababa. EIAR, Addis Ababa, Ethiopia, 275 pp.

Bekele, E., Azerefegne, F. and Abate, T. (eds) (2006) *Facilitating the Implementation and Adoption of Integrated Pest Management (IPM) in Ethiopia*. Drylands Coordination Group (DCG), Oslo.

Donley, N. (2019) The USA lags behind other agricultural nations in banning harmful pesticides. *Environmental Health* 18, 44. DOI: 10.1186/s12940-019-0488-0.

FAO (2022a) Statistical Databases and Data-Sets of the Food and Agriculture Organization of the United Nations. Available at: http://www.fao.org/faostat/en/#data (accessed 2 August 2022).

FAO (2022b) Statistical Databases and Data-Sets of the Food and Agriculture Organization of the United Nations. Available at: http://www.fao.org/faostat/en/#data (accessed 18 November 2022).

FAO (2023) Statistical Databases and Data-Sets of the Food and Agriculture Organization of the United Nations. Available at: http://www.fao.org/faostat/en/#data (accessed 21 September 2023).

Fikadu, Z. (2020) Pesticides use, practice and its effect on honeybee in Ethiopia: A review. *International Journal of Tropical Insect Science* 40, 473–481. DOI: 10.1007/s42690-020-00114-x.

Franzel, S. and van Houten, H. (1992) *Research with Farmers: Lessons from Ethiopia*. CAB International, Wallingford, UK.

Gebre-Amlak, A. (1988) Ecology and management of maize stalk borer, *Busseola fusca* (Fuller) (Lepidoptera: Noctuidae) in southern Ethiopia. PhD dissertation, Swedish University of Agricultural Sciences, Uppsala, Sweden.

Gesesew, H.A., Woldemichael, K., Massa, D. and Mwanri, L. (2016) Farmers knowledge, attitudes, practices and health problems associated with pesticide use in rural irrigation villages, Southwest Ethiopia. *PLoS ONE* 11(9), e0162527. DOI: 10.1371/ journal.pone.0162527.

Gordon, H., Chiri, A. and Abate, T. (1995) Environmental and economic review of crop protection and pesticide use in Ethiopia. Environmental and Natural Resources Policy and Training (EPAT) Project. Applied Research, Technical Assistance and Training. Winrock International Environment Alliance. Arlington, Virginia, USA. Deliver Order No. 33, Contract No. DHR-5555-Q-00-1085-00. 118 p.

IEG (Independent Evaluation Group of the World Bank Group) (2016) Project Performance Assessment Report. Ethiopia, Mali, Morocco, South Africa, Tanzania, Tunisia Stockpiles Program September 29, 2016. Report No. 108524. Available at: https://ieg.worldbankgroup.org/sites/default/files/Data/reports/ppar_africa_102416_2.pdf (accessed 14 May 2021).

Loha, K.M., Lamoree, M., Weiss, J.M. and de Boer, J. (2018) Import, disposal, and health impacts of pesticides in the East Africa Rift (EAR) zone: A review on management and policy analysis. *Crop Protection* 112, 322–331.

Megenasa, T., Gebremedhin, T. and Abate, T. (1980) A review of the state of agricultural pesticides use in Ethiopia. *Ethiopian Journal of Agricultural Sciences* 2, 29–38.

Mengistie, B.T. (2016) Policy-practice nexus: Pesticide registration, distribution and use in Ethiopia. *SM Journal of Environmental Toxicology* 2(1), 1006.

Mengistie, B.T., Mol, A.P.J. and Oosterveer, P. (2017) Pesticide use practices among smallholder vegetable farmers in Ethiopian Central Rift Valley. *Environment, Development and Sustainability* 19, 301–324. DOI: 10.1007/s10668-015-9728-9.

Mequanint, C., Getachew, B., Mindaye, Y., Amare, D.E., Guadu, T. and Dagne, H. (2019) Practice towards pesticide handling, storage and its associated factors among farmers working in irrigations in Gondar town, Ethiopia. *BMC Research Notes* 12, 709. DOI: 10.1186/s13104-019-4754-6.

Negatu, B., Kromhout, H., Mekonnen, Y. and Vermeulen, R. (2016) Use of chemical pesticides in Ethiopia: A cross-sectional comparative study on knowledge, attitude and practice of farmers and farm workers in three farming systems. *The Annals of Occupational Hygiene* 60(5), 551–566. DOI: 10.1093/annhyg/mew004.

Nigatu, A.W., Bråtveit, M. and Moen, B.E. (2016) Self-reported acute pesticide intoxications in Ethiopia. *BMC Public Health* 16, 575. DOI: 10.1186/s12889-016-3196-5.

Tegegne, G., Abebe, F., Hussien, T., Tilahun, T., Belete, E. *et al.* (2008) Review of maize, sorghum and millet pathology research. In: Tadesse, A. (ed.) *Increasing Crop Production through Improved Plant Protection Volume* I. Plant Protection Society of Ethiopia, Addis Ababa, pp. 245–302.

UTZ (2015) List of Banned Pesticides and Pesticides Watchlist. Version 1.0. Available at: www.utz.org/resource-library.org (accessed 15 May 2021).

Annex Table 18.1. List of pesticide active ingredients registered in Ethiopia (as of August 2020). Source: assembled by the author from Plant Variety Release, Protection and Seed Quality Control Directorate of the Ministry of Agriculture.

#	Active ingredient	#	Active ingredient	#	Active ingredient	#	Active ingredient
1	2,4-D[a]	47	Copper hydroxide	93	Fluopyram	139	Penconazol
2	Abamectin[a]	48	Copper oxychloride	94	Flurasulam	140	Pendimethalin
3	Acephate[b]	49	Copper thiodiazole	95	Fluroxypyr	141	Penoxasulam
4	Acetamiprid[a]	50	Cyazofamid	96	Flutriafol	142	Phenoxaprop-p-ethyl
5	Acetochlor[a]	51	Cyfluthrin	97	Folpet	143	Picoxystriobin
6	Acrinathrin	52	Cyhalofop-butyl	98	Fosetyl	144	Pinoxaden
7	Alachlor[b]	53	Cymoxanil	99	Glufosinate-ammonium[a]	145	Pirimiphos-methyl
8	Alazine	54	Cypermethrin	100	Glyphosate[a]	146	Polyoxin
9	Alphacypermethrin	55	Cyproconazole	101	Halauxifen-methyl	147	Potassium phosphate
10	Aluminium phosphide	56	Deltamethrin[a]	102	Hexazinone	148	Profenfos
11	Ametryn	57	Diafenithiuron	103	Imidacloprid[a]	149	Prometryn
12	Amicarbazone	58	Diazinon	104	Indoxacarb	150	Propamocarb hydrochloride
13	Amidosulfuron	59	Dicamba	105	Iodosulfuron-methyl-sodium	151	Propiconazole
14	Aminopyralid	60	Diclofop-methyl	106	Iprodione	152	Propineb
15	Amitraz[b]	61	Diclosulam	107	Iprovalcarp	153	Prothiconazole
16	Atrazine[b]	62	Dicofol[b]	108	Isoproturon	154	Pyraclostrobin
17	Azoxystrobin	63	Difenoconazole	109	Isoxaflutole	155	Pyrimethanil
18	Benomyl[b]	64	Dimethenamid-p[b]	110	kesoxim-methyl	156	Pyroxsulam
19	Bentazone	65	Dimethoate[a]	111	Lambda-cyhalothrin[a]	157	Quizalofop-p-ethyl
20	Beta cypermethrin	66	Dimethomorph	112	Lufenuron[a]	158	Rape seed oil
21	Bifenazate	67	Dithianon	113	Magnesium phosphide	159	Saflufenacil
22	Bifenthrin	68	Dodemorf acetate	114	Malathion[a]	160	S-metolachlor
23	Bispyribac	69	Emamectin benzoate	115	Mancozeb[a]	161	Spinetoram
24	Bixafen	70	Endosulfan[b]	116	Mandipropamid	162	Spinosad
25	Boscalid	71	Epoxiconazole[a]	117	MCPA	163	Spiromesifen
26	Brodifacoum[b]	72	Erbuthylazine	118	Mecoprop	164	Spiroxamine

Continued

27	Bromadiolone[b]	73	Ethoprophos/ethoprop	119	Mefenpyr diethl	165	Sulfoxaflor
28	Bromoxynil octanoate	74	Famoxadone	120	Mesosulfuron-methyl	166	Sulphur
29	Bupirimate	75	Femamidone	121	Mesotrione	167	Tebuconazole
30	Butachlor	76	Fenbutatin	122	Metalaxyl	168	Tebufenpyrad
31	Carbaryl[b]	77	Fenhexamid	123	Methidathion[b]	169	Tembotrion
32	Carbendazim[a]	78	Fenitrothion[a]	124	Methoxyfenozide	170	Terbutilazine
33	Carbofuran[b]	79	Fenoxaprop-p-ethyl	125	Metiram	171	Tetraconazole
34	Carbosulfan[b]	80	Fenpropimorph	126	Metribuzin[a]	172	Thiacloprid[a]
35	Carboxin	81	Fenthion[b]	127	Metryn	173	Thiamethoxam[a]
36	Chlorantraniliprole	82	Fipronil[a]	128	Metsulfuron methyl	174	Thiencarbazone-methyl
37	Chlorfenapyr[b]	83	Flocoumafen	129	Myclobutanil	175	Thiophanate-methyl
38	Chlorimuron-ethyl	84	Florasulam	130	Neem	176	Thiram[b]
39	Chlorothalonil[a]	85	Fluazifop-p-butyl[b]	131	Nicosulfuron	177	Topramezone
40	Chlorpyrifos[a]	86	Flubendiamide	132	Novaluron	178	Triadimenol
41	Chlorpyrifos-ethyl	87	Flucarbazone-sodium	133	Odosulfuron-methyl-sodium	179	Tribenuron methyl
42	Clodinafop propargyl	88	Flucythrinate[b]	134	Organosilicon	180	Trifloxystrobin
43	Clofentezine	89	Fludioxonil	135	Oxydemethon-methy[b]	181	White oil
44	Clomazone	90	Flufenoxuron	136	Oxyflorfen	182	XDF
45	Clopyralid	91	Flumetsulam	137	Oxymatrin	183	Zinc phosphide[b]
46	Cloquintocet-methyl	92	Fluopicolide	138	Paraquat[b]	184	Zoxamide

[a]On the watch list in other countries (cf. Table 18.2); [b]banned in other countries (cf. Table 18.1).

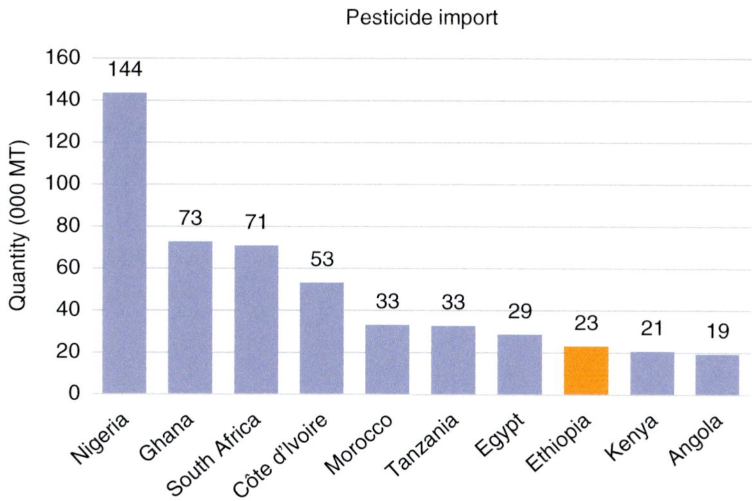

Annex Fig. 18.1. Top ten pesticide importing countries in Africa, 2016–2020 average. Source: constructed by the author from (FAO, 2022a).

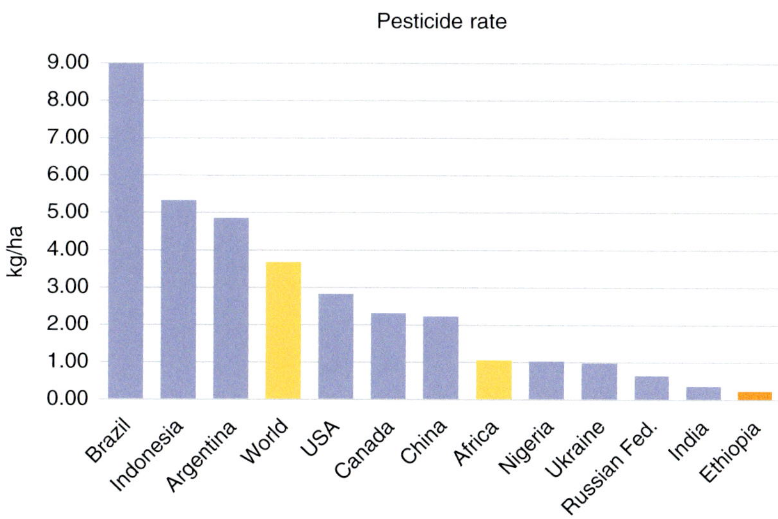

Annex Fig. 18.2. Pesticide use (kg active ingredient per ha of cropland) in Ethiopia, compared to the top ten countries with the largest cropland area, the world, and Africa, 2016–2020. Source: constructed by the author from FAO (2023).

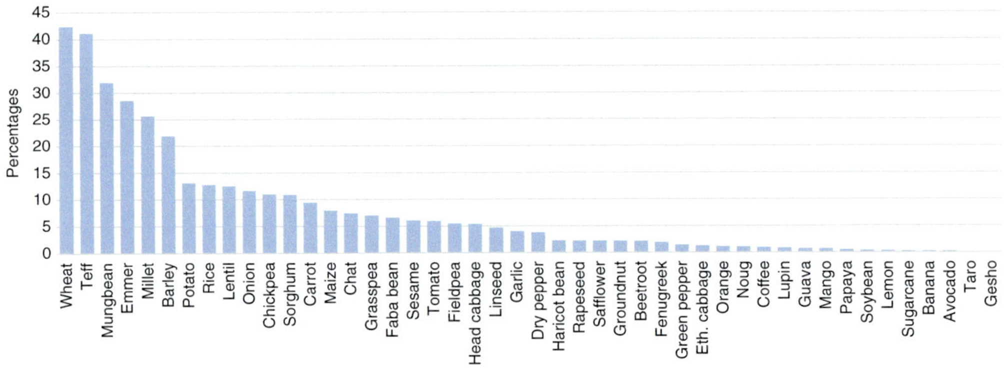

Annex Fig. 18.3. Proportion of smallholders using agricultural pesticides on crops grown in the *meher* season, 2014–2018. Source: constructed by the author from the Central Statistical Authority (www.csa.gov.et).

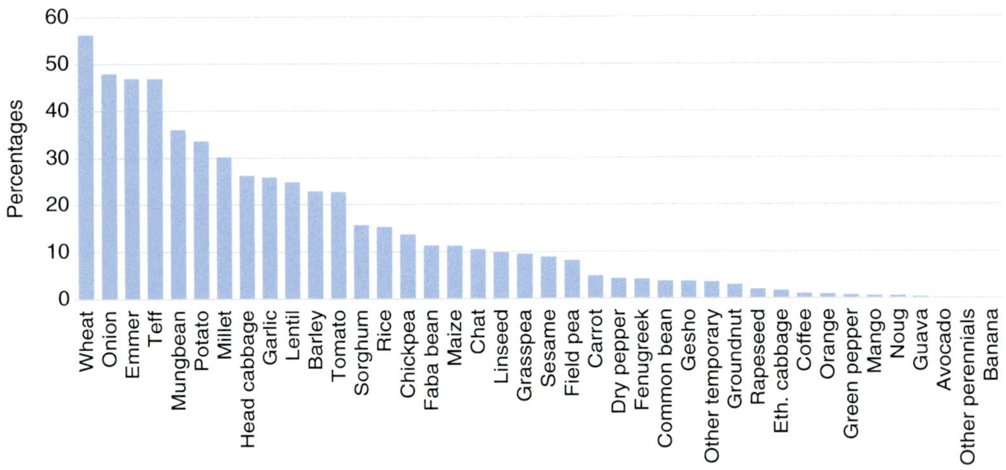

Annex Fig. 18.4. Crop area treated with agricultural pesticides in smallholder agriculture grown in the *meher* season in Ethiopia (2015–2019 average). Source: calculated and constructed by the author from the Central Statistical Authority agricultural sample surveys (www.csa.gov.et).

19 Status and Insights into the Future of Agricultural Mechanization

Friew Kelemu,[1] Tsedeke Abate[2]* and Mikael Fesseha Yohannes[2]
[1]*Formerly Ethiopian Institute of Agricultural Research, Melkassa Research Center, Adama, Ethiopia;* [2]*Homegrown Vision, Addis Ababa, Ethiopia*

Abstract

Successive governments in Ethiopia have tried to introduce agricultural mechanization to both smallholder and commercial agriculture. While the latter has enjoyed significant progress in terms of the introduction of tractors and related machinery, draft animal power and manual labour continue to be the dominant source of land preparation and other farm operations for smallholders. More than 90% of harvesting is done by manual labour, with the exception of wheat where 25% is done with combine harvester. There were approximately 8.65 tractors per 100 km^2 of arable land in 2020 versus 3.49 in 2010. However, the country still has among the lowest number of tractors compared to many of its peers in Africa. Often, but in our view incorrectly, tractors are taken as synonymous with farm mechanization. Developing the *enset*-processing device has remained a major priority over decades, but there is little to show for all the effort.

Introduction

The term agricultural mechanization is defined as the use of any kind of tools powered by humans, animals or engine/motor in the process of agricultural production. Hand-tool technology is the most basic level of agricultural mechanization, where a human being is the power source, using simple tools and implements such as hoes, machetes, sickles, wooden diggers and the like. The mechanized development of a country's agriculture is defined by different groups in different ways. Some define it as the available mechanical power as related to the total cultivated land, some understand it as the degree of farm operations using mechanical power to the total power used, using different means.

Perhaps one might wonder why we should bother about mechanization in agriculture in the first place. There are several positive reasons for agricultural mechanization, including saving time and labour, reducing crop loss, reducing drudgery and improving the quality of life for the farming community. For a start, mechanization as an increasing process of replacing human labour and hand tools – human work output is only about 25% of the energy consumed when relatively easy tasks such as pedalling, pushing or pulling is converted to actual human work output. Under more difficult work conditions, the efficiency of converting consumed energy to physical work may be as low as 5% or less. This means that, at the maximum continuous energy consumption rate of 0.30 kW and conversion

* Corresponding author: tsedeke@homegrownvision.org

DOI: 10.1079/9781800626386.0019

efficiency of 25%, the physical power output is approximately 0.075 kW sustained for an 8- to 10-hour workday. Naturally, higher rates can be maintained for shorter periods only (Inns, 1992). It should be noted that the problem is not necessarily with the tools used, but the inherent characteristics of smallholder farmers using hand-tool technology. Using hand-tool technology one cannot cultivate more than a hectare in a season. In Ethiopia, hand-tool technology is exhibited around the homestead in small-scale horticultural crops such as *enset* and other root crops.

The requirements for mechanization, however, depend on a number of factors, including farm size, slope of the land and the environment. Regarding the latter, for example, the amount, duration and distribution of rainfall are critical factors (Cheung *et al.*, 2008), considering the wide range of climatic variability Ethiopia has (Fig. 19.1). For example, according to a study conducted at Melkassa, a sub-moist zone with a short rainfall period, most farmers prepare their land and plant in June. They have around 11 days to complete the whole operation. This is especially the case with the small cereals, as the land has to be well pulverized and they cannot handle more than a hectare within this period or they risk doing substandard work, like planting late, which results in depressed yield as they will not be able to capture the available moisture for the whole growing period. In this zone, the land preparation and planting should be conducted in the shortest possible time. This is attainable if they are able to handle the work with a higher level of mechanization. If they opt for mechanical power technology for land preparation and settle with their traditional practice for the other operations, they will have three times the advantage. If they use mechanical power for both the operations, they will have a tenfold advantage over the traditional practice with that window of operation time.

Agriculture in Ethiopia started with the domestication of *enset* some 8000 years ago (Ehret, 1979). However, the country's implements are still dominated by the *maresha*, a single-tine oxen-drawn plough that has been used for the past 1500 years (McCann, 2018), along with hand-held tools such as hoes. Oxen are the major draft power source in Ethiopian agriculture, especially in field crops production. They are mainly used for land preparation, planting and threshing operations. Equines are mainly used in threshing and transportation, but horses are also used for draft power in some parts of the country. Draught animals produce more power than humans, yet their performance does not exceed 0.25 hectare per working day when working in a pair. In areas where the critical period for the last ploughing and planting operations is not more than 15 days, as in most sub-moist zones, one cannot cultivate more than three hectares per season.

Unlike the above animate power-based system, inanimate power-based mechanization enables the farmer to handle more hectares per season, depending on the horsepower of the machinery. It has more than 100-fold advantage over animate power sources. Mechanical power-based systems, especially tractors and combines, have been used by large state and commercial farmers since the 1960s. Despite these different approaches, the use of mechanical power in Ethiopian agriculture has been very low. It is below 0.5 hp/ha and the use of mechanical energy as related to the total energy expended (mechanical energy/hand tool + draught animal + mechanical) is less than 0.06 hp/ha.

Efforts to introduce agricultural mechanization in Ethiopia gained momentum in 1967 as part of the Chilalo Agricultural Development Unit with support from the Swedish International Development Agency. Successive governments in Ethiopia had different levels of priority on agricultural mechanization. The introduction of agricultural machinery and ancillary equipment during the emperor's reign in the 1960s and early 1970s helped to extend the agricultural production areas in the Awash basin, especially for the production of export commodities like cotton. The military-socialist government also introduced a large number of tractors from Eastern Europe in the late 1970s, exclusively for use in the state farms. It also established the Nazareth Tractor Assembly Plant (NTAP) in 1984. The Ethiopian Peoples' Revolutionary Democratic Front government de-emphasized the use of smallholder agricultural mechanization on the premise that Ethiopia is abundant in labour and short in capital (Ayele, 2022). Starting in 2013, however, agricultural mechanization has become a high priority for the country's agricultural development and there has been rapid uptake (Berhane *et al.*, 2017; Ayele, 2022).

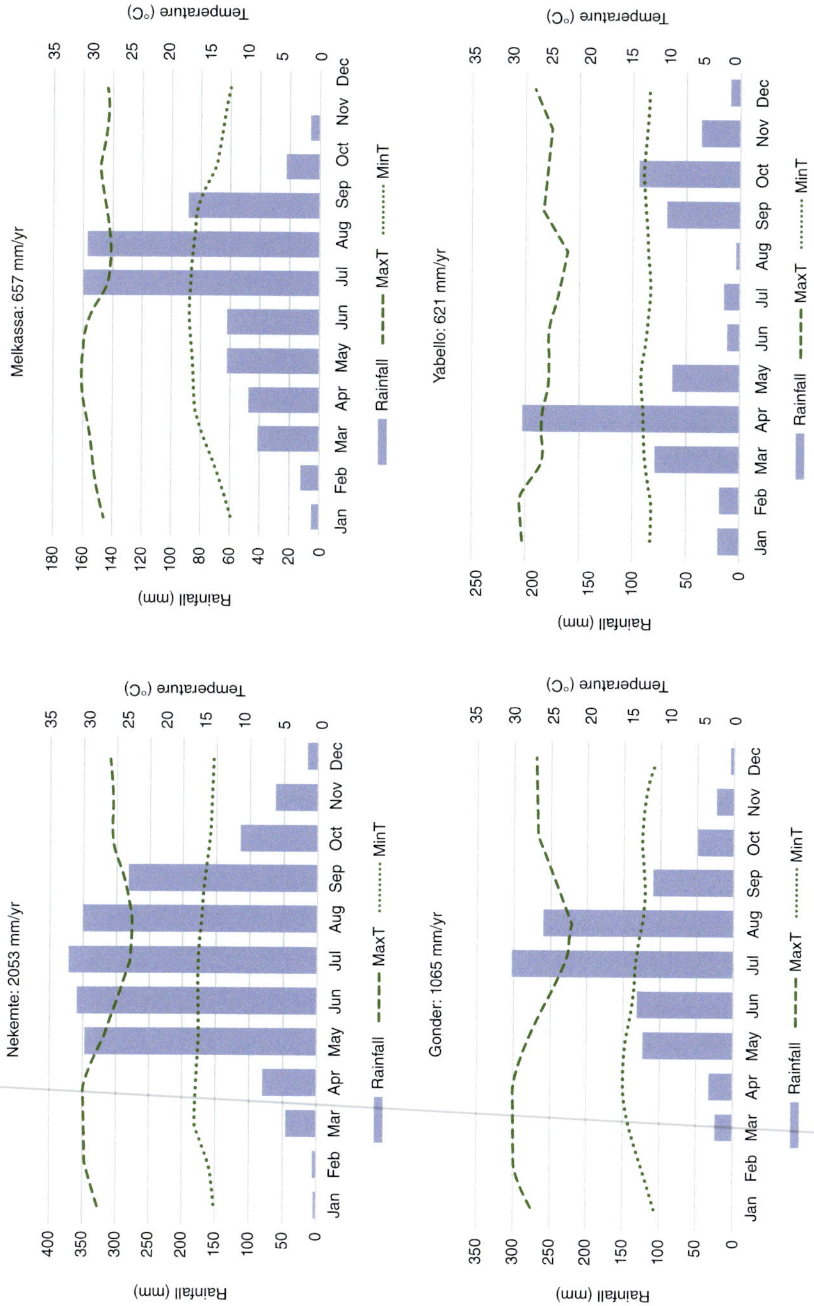

Fig. 19.1. Monthly average rainfall and maximum and minimum temperature for selected sites in Ethiopia, 2010–2016 averages. Source: constructed by the authors from the National Meteorological Agency.

Relevant scholarship on agricultural mechanization in Ethiopia can be found in Astatke *et al.* (2002), Kelemu (2015), Berhane *et al.* (2016, 2017), Klappoth (2018), Deribe and Getnet (2021), Deribe *et al.* (2021), Workneh *et al.* (2021) and Ayele (2022). This chapter provides a critical review of the status of agricultural mechanization and identifies the major challenges of and opportunities for mechanization, with a view to informing policy and priority setting as the country engages in accelerated transformation of its agricultural development.

Approach

There is no readymade datum on agricultural mechanization in Ethiopia. The Ethiopian Revenues and Customs Authority (ERCA) is perhaps the only organization in the country that keeps long-term records of agricultural machinery that are imported. However, the way those data are organized needs careful examination because there are items that are wrongly grouped together with tractors (e.g. pile extractors). The way the data are organized also makes it so tedious to extract. We used the ERCA data (2000–2020) to determine the number of tractors and related agricultural machinery. We also made references to both published and unpublished data from the Ministry of Agriculture (MOA) and the Ethiopian Institute of Agricultural Research (EIAR) on the number of tractors assembled in the country. To the best of our knowledge, the data on this aspect are incomplete. We also obtained expert opinion from senior MOA experts in the field. Kelemu (2015), Berhane *et al.* (2016, 2017), Klappoth (2018) and Ayele (2022) are some of the key references that contain relevant data.

Results

Klappoth (2018) suggests that Ethiopia followed three levels of agricultural mechanization: (i) improvement of small implements; (ii) adoption of two-wheeled tractors; and (iii) adoption of four-wheeled tractors. In our opinion, however, there are two major levels: (i) improvement of small farm implements; and (ii) adoption of tractors. The first approach builds mostly on private ownership in smallholder agriculture, whereas the other one relies more heavily on service provision. Further details on each are presented in the following sections, where we present and discuss the results of our findings under two broad topics of the current state of mechanization in Ethiopia and the research and development efforts to improve it.

The current state

Surveys of Feed the Future programme areas in Oromia, Amhara, SNNP and Tigray regional states showed that only 9.1% of smallholder farmers used mechanization for any farm operation: 5.5% for ploughing, 3.4% for harvesting and 2.1% for threshing (Berhane *et al.*, 2016). According to Ströh de Martínez *et al.* (2016), Ethiopia is the only country in Africa where draft animal power is the most dominant in farm operations.

Berhane *et al.* (2017) reported that less than 1% of smallholders' plots were ploughed using tractors in the 2016/17 *meher* (main crop) season. Table 19.1 shows the level of mechanization in Ethiopian agriculture. The various farm power sources used in the various farm operations for various crop groups/crops can be summarized as follows:

- Land preparation was done using draft animals for about 94% of oilseeds, 93% of cereals, 91% of grain legumes, 16% of fruits and vegetables and 7% of coffee.
- Manual harvesting was used for nearly 100% of coffee, fruits and vegetables, grain legumes and oilseeds, while it was 94% for cereals.
- About 3.3%, 1.1%, 0.2%, 0.1% and 0.1%, respectively, of oilseed, cereals, grain legumes, fruits/vegetables and coffee farmers employed machines for land preparation.
- About 55%, 55% and 26%, respectively, of cereal, grain legume and oilseed farmers used animal-drawn power for threshing.
- It has also been reported that about 25% of wheat is harvested using combine harvesters (Berhane *et al.*, 2017).
- The percentages of farmers using manual labour for threshing oilseeds, grain legumes

Table 19.1. Farm power sources (percentages) for different pre- and post-harvest operations of cereals, oilseeds and grain legumes in Ethiopia. Source: constructed by the authors from Berhane *et al.* (2017).

	Operation		
Power source	Land preparation	Harvesting	Threshing
Cereals			
Animal	93.0	0.0	55.4
Hoe only	5.9	NA	NA
Manual	0.0	98.4	42.4
Machine	1.1	1.6	1.2
Sheller	NA	NA	1.0
Oilseeds			
Animal	93.7	0.0	25.6
Hoe only	2.9	NA	NA
Manual	0.0	99.5	73.2
Machine	3.3	0.5	0.0
Sheller	NA	NA	1.3
Grain Legumes			
Animal	90.9	0.0	55.2
Hoe only	9.0	NA	NA
Manual	0.0	99.5	44.3
Machine	0.2	0.5	0.0
Sheller	NA	NA	0.4
Fruits/Vegetables			
Animal	15.6	0.0	NA
Hoe only	84.3	NA	NA
Machine	0.1	0.5	NA
Manual	0.0	99.5	NA
Coffee			
Animal	7.1	0.0	NA
Hoe only	92.8	NA	NA
Manual	0.0	99.6	NA
Machine	0.1	0.4	NA

NA = not applicable.

and cereals was 73%, 44% and 42%, respectively.

- Less than 2% of cereal farmers and less than 1% each of coffee, oilseeds, grain legumes and fruit/vegetable farmers employed machines for harvesting.
- About 1.3%, 1.0% and 0.4% of oilseeds, cereal and grain legumes farmers threshed their harvest with the help of shellers in spite of SG2000's continued attempts to promote the multiple crop sheller technology for cereals (Admassu *et al.*, 2014; Getachew *et al.*, 2022).

As stated earlier, draught animal power has been the dominant agricultural system in Ethiopia perhaps for more than 1500 years (McCann, 2018). The draught animal power-based technology is more related to cereal culture, whereas hoe-based technology is dominant in the southern and southwestern parts of the country, where root and tuber crops are more widely grown. Animal power (including oxen, horses, donkeys, mules and camels) is used not only for land preparation, but also used for threshing and, even more importantly, for transportation of farm products (Gebreab *et al.*, 2004).

Types and sources of agricultural machinery in Ethiopia

The use of tractors in Ethiopia started with the establishment of private commercial companies such as the Awash Valley Authority, the Tendaho Plantation Share Company that produced cotton, the Middle Awash Settlement Scheme (also producing cotton) and Setit Humera Enterprise that was mainly engaged in sesame growing during the imperial era. Up until the establishment of NTAP in 1984, Ethiopia relied solely on foreign imports for its agricultural machinery. NTAP changed its name to Adama Agricultural Machinery Industry (AAMI) in 1992. The ERCA is the main source of imported goods in the country. ERCA registers four types of tractors imported into the country: track-laying tractors, pedestrian-controlled tractors, standard tractors and road tractors, as presented in Table 19.2.

Ethiopia imported close to 41,000 different types of tractors between 2004 and 2019. These included more than 18,000 track-laying, 11,000 pedestrian-controlled, over 9000 standard and close to 2000 road tractors. Likewise, EIAR reported over 1800 locally assembled tractors by AAMI between 2005 and 2010 and a total of nearly 4000 tractors ranging in capacity from 18 HP to 375 HP between 2014 and 2019. This means that Ethiopia imported about 82% of the total tractors, whereas the remaining 12% were assembled locally.

However, according to MOA estimates, there were about 14,000 tractors in use in Ethiopia in 2020 (Fig. 19.2). Put differently, this is approximately 8.65 tractors per 100 km^2 (or 8.65 tractors for every 10,000 ha) of arable land. The estimated number of tractors in 2010 was roughly 5100 (or 3.49 tractors per 100 km^2 of arable land). This can be seen as some improvement.

It is important to note that these figures potentially include machinery imported for use by other sectors (such as construction) and/or double counted records. This is evident from the data presented in Fig. 19.2, which suggest that the different sizes and types of tractors in use (as of 2020) were only 14,000. Still, compared to previous decades, over the period 2010 and 2020, tractors in use increased by nearly threefold, from 5090 to 14,000. We argue that there have been multiple of drivers of mechanization over the decade or so. These include:

- increase in the number of large-scale commercial farms;
- improved policy environment for smallholder mechanization; and

Table 19.2. Estimated number of imported tractors and those assembled by Adama Agricultural Machinery Industry. Source: imported data from Ethiopian Revenues and Customs Authority and assembled data from Ethiopian Institute of Agricultural Research.

	Imported						
Year	Track-laying tractors	Pedestrian-controlled tractors	Standard tractors	Road tractors	Total imported	Assembled	Grand total
2004	103	7	133	32	275	NA	NA
2005	243	60	190	28	521	267	788
2006	181	65	374	22	642	300	942
2007	157	9	354	23	543	310	853
2008	152	263	238	44	697	315	1,012
2009	3,721	1,591	560	71	5,943	300	6,243
2010	360	149	535	156	1,200	318	1,518
2011	1,178	2,502	1,552	58	5,290	NA	NA
2012	2,062	7	1,774	20	3,863	NA	NA
2013	1,542	15	2,268	46	3,871	NA	NA
2014	1,179	13	1,088	898	3,178	849	4,027
2015	1,494	1,733	252	75	3,554	772	4,326
2016	247	165	18	12	442	517	959
2017	440	3,988	0	143	4,571	497	5,068
2018	591	13	0	32	636	472	1,108
2019	4,717	906	0	17	5,640	890	6,530
Grand total	18,367	11,486	9,336	1,677	40,866	NA	NA

NA = not available.

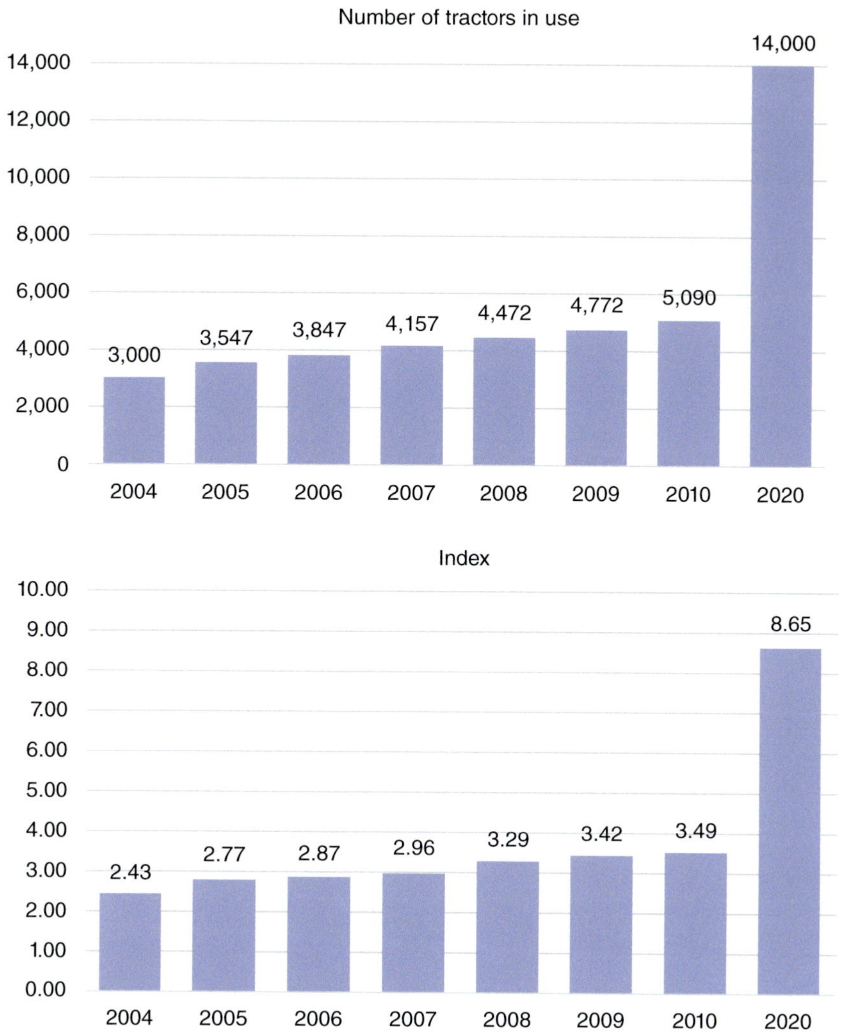

Fig. 19.2. The number of tractors in use (top) and number of tractors per 100 km^2 of arable land (bottom) in Ethiopia. Source: calculated and constructed by the authors from data from the Ministry of Agriculture and the Food and Agriculture Organization.

- a more suitable external environment, such as the increased availability of machinery that is more suitable to Ethiopian requirements, both in price and design specification.

Ethiopia started from a very low base of 0.23 tractors per 100 km^2 of arable land in 1961 and 2.58 tractors per 100 km^2 of arable land in 1972; it is still among the lowest of its peers. For example, the number of tractors per 100 km^2 for Morocco, Cote d'Ivoire, the African average, Kenya and Tanzania, respectively, is about 49, 32, 28, 25 and 25. Furthermore, the figures for the USA, India, Mexico, China and South Africa were about 271, 159, 102, 83 and 48, respectively (cf. Annex Table 19.1).

In addition to tractors, Ethiopia also imports various types of farm implements, accessories and spare parts from the outside world. These included disc harrows, ploughs, mowers, cultivators, threshers, balers, ridgers, soil chisels, combiners, choppers and mouldboards for the period from 2005 to 2015 (cf. Annex Fig. 19.1). According to the MOA, there were about 1300 combine harvesters in use between 2018 and 2020 (Tamiru Habte pers. com.)

Figure 19.3 presents the countries from where Ethiopia imports the various machinery. The agricultural machinery supplies came from a total of 58 countries between 2004 and 2019. China has been by far the largest supplier, accounting for 35% of the total; it was followed by Japan, Germany, the Netherlands, Brazil, India, Italy, the Czech Republic, Pakistan and Saudi Arabia, among others.

As shown above, the supply of mechanized tools, including tractors, has been increasing, but their effective use is hampered by a host of challenges. Two such challenges are the low level of repair and maintenance services and shortages of spare parts. The MOA and industry expertise note that breakage of large-scale farm machinery such tractors and combine harvesters are more frequent but, while quality maintenance helps increase the life span of the machinery, these services are barely organized and function poorly. There is also a dire shortage of spare parts that, again, limits the effective use of farm machinery. Finally, as many smallholder farmers are dispersed across the country, they are also less able to access repair and maintenance services and spare parts which are often located in central locations such as Addis Ababa.

Improvement of farm implements

Research on agricultural mechanization in Ethiopia started with a national programme at Melkassa, focusing on small-scale farm tools, animal-drawn farm implements and small engine harvest and postharvest agricultural machinery. Prior to Melkassa, there was a small farm implements research unit at Assela operating

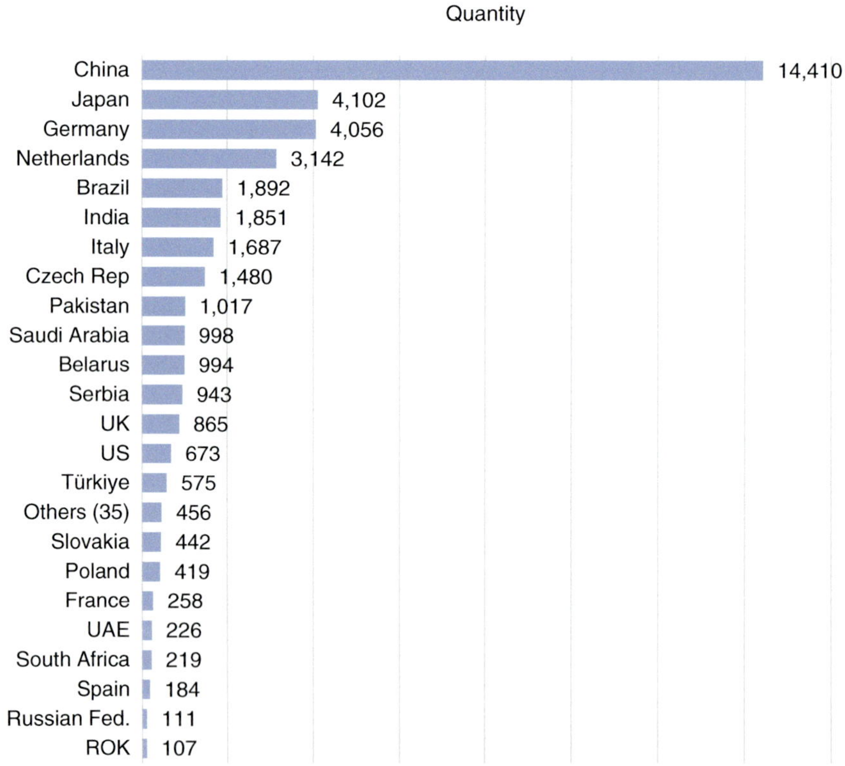

Fig. 19.3. Total number of various types of tractors Ethiopia imported from different countries between 2004 and 2019. ROK = Republic of Korea; Russian Fed. = Russian Federation; UAE = United Arab Emirates. Source: constructed by the authors from the Ethiopian Revenue and Customs Authority.

under the Arsi Agricultural Development Unit, which was subsequently transferred to NATP and then to Melkassa, when the Nazareth Research Center moved to the latter location. The centre started its work on two projects, *enset*-processing and water-lifting devices. At the same time, a hand-operated and animal-power driven chain and washer pumps were developed; their capacity was 180 L/min and 300 L/min from a depth of 4 m and 7 m, respectively. Furthermore, work on small-scale preharvest tools, animal-drawn implements and small engine-operated cereal threshers and shellers was initiated. The centre also had a strong testing and evaluation programme, where implements were tested using test procedures adopted from international groups but locally developed and recommended by a panel of local experts and approved as a guideline. Then there was an intent by the military-socialist government to establish a tractor-drawn farm implements factory at Nazareth, which required evaluation of the suitability of the implements before embarking on fully fledged production. The centre tested the implements, most of which were found unsatisfactory, and the idea was dropped by the government on the basis of these findings.

Despite these merits, there were shortcomings, starting with the transfer of the mechanization centre from Assela to Nazareth. At that time, though the office and workshop equipment were transferred, the highly skilled and experienced engineers did not join the group. Thus, the programme started with inexperienced new recruits that slowed down the progress of the programmes. The programme was strengthened with United National Development Project funding that capitalized on procurement of workshop equipment and training of staff and working more on the testing, evaluation and development of small-scale agricultural equipment. The expatriate staff spent the first phase on establishment, but when the trainees returned, they did not have much time to spend on the actual work. A robust second-phase programme could not be realized except for small extension programmes, and the expatriate staff left the centre without sharing the necessary expertise with the local staff. The testing and evaluation functions, despite their contribution to stop the introduction of substandard implements to the user community, was de-emphasized thereafter. Under all these setbacks, the programme has been struggling to meet its obligations, and it is hard to say it has been working to the expected level.

Besides the national research centres, the regional centres were the ones transferred from technology multiplication centres, with limited research culture. They have managed to do certain modifications on existing implements and demonstrate the implements to the user community in their vicinities. Inevitably, there was also duplication of efforts because of a lack of clearly defined mandates for the national and regional centres.

In addition to the research system, some projects (e.g. Sasakawa Africa Association) have small farm implements (such as the Multi-Crop Thresher) that have been demonstrated in major cereal-growing areas of the country. There is little or no private sector involvement in agricultural mechanization research and development.

Other stakeholders in agricultural mechanization in Ethiopia include federal and regional government agricultural engineering centres (Mekelle, Semera, Bahir Dar, Bako, Jimma, Assela, Melkassa, Fedis and Jigjiga), the MOA, higher-learning institutions, smallholder farmers, commercial farmers, non-governmental organizations, manufacturers and dealers.

Discussion and Conclusions

The use of modern machines in agriculture in Ethiopia started with the introduction of tractors for plantation crops like cotton and sesame in private commercial agriculture in the mid-1960s. The advent of the military-socialist government saw the end of large-scale private farming and tractor use was largely limited to the state farms it nationalized. The most recent past political administration did not prioritize mechanization on the premise that Ethiopia is labour abundant and short in capital. That policy changed in 2013 and agricultural mechanization is now back on the agenda. Expansion of mechanical power for agricultural production in Ethiopia is a necessary policy for enhancing agricultural productivity to feed the rapidly growing population and fostering economic development of the country. However, it needs to be tied to the bigger issue of the direction the country wants to follow with regards to elevating the landholding level from the current national average of less than 1 ha per household.

There are several positive attributes for the introduction of agricultural mechanization, including saving time and labour, reducing crop loss, reducing drudgery and improving the quality of life for the farming community. The requirements might depend on the type of the environment – i.e. the amount, distribution and duration of rainfall – considering the wide range of climatic variability within Ethiopia. Selection of the optimum mechanization level depending on land size and length of the growing period and the window of operation time for the most critical operation is also very important. Farmers should use the most efficient level of mechanization to maximize the small operational window available in most sub-moist zones in order to use the whole growing period.

The research system has produced several farm implements, including conservation tillage implements (Temesgen *et al.*, 2009), water-lifting devices, row weeders, donkey carts and tied ridgers. However, there is no information on the *enset*-processing device that was assigned high priority in the early days of the research system. The major breakthrough for agricultural mechanization research was that of the mouldboard plough for reclaiming water-logged areas in the 1980s (Astatke *et al.*, 2002).

The use of agricultural mechanization by smallholder farmers in Ethiopia varies from place to place and from crop to crop. The vast majority of mechanization here is on cereals – particularly combine harvesters for wheat, shellers for maize and threshers for *teff*. Smallholder farmers access these services through rental arrangements from private owners. We believe that increased farm machinery use in the foreseeable future in Ethiopia would rely more on service providers rather than smallholders owning the various tools. This would require the government putting in place an appropriate and fair regulatory system and incentive mechanisms.

References

Admassu, S., Bedaso, T., Wondwossen, T., Oumer, T., Aberash, T., Mesfin, K. and Fikadu, C. (2014) *Dissemination, Adoption and Impact of Multi-Crop Threshers in Ethiopia: Experience of Sasakawa Global 2000*. Research Report, No. 3. Sasakawa Africa Association, Addis Ababa.

Astatke, A., Saleem, M., Jabbar, M. and Erkossa, T. (2002) Development and testing of low-cost animal drawn minimum tillage implements: Experience on vertisols in Ethiopia. *Agricultural Mechanization in Asia, Africa and Latin America* 33(2), 9–14.

Ayele, S. (2022) The resurgence of agricultural mechanisation in Ethiopia: Rhetoric or real commitment? *The Journal of Peasant Studies* 49(1), 137–157. DOI: 10.1080/03066150.2020.1847091.

Berhane, G., Hirvonen, K. and Minten, B. (2016) Synopsis, agricultural mechanization in Ethiopia: Evidence from the 2015 feed the future survey. ESSP II Research Note 48. Washington, D.C. and Addis Ababa, Ethiopia: International Food Policy Research Institute (IFPRI) and Ethiopian Development Research Institute (EDRI). Available at: http://ebrary.ifpri.org/cdm/ref/collection/p15738coll2/id/130142 (accessed 18 March 2024).

Berhane, G., Dereje, M., Minten, B. and Tamru, S. (2017) The rapid – but from a low base – uptake of agricultural mechanization in Ethiopia: Patterns, implications and challenges. ESSP Working Paper 105. International Food Policy Research Institute, Washington DC, USA.

Cheung, W.H., Senay, G.B. and Singh, A. (2008) Trends and spatial distribution of annual and seasonal rainfall in Ethiopia. *International Journal of Climatology* 28(13), 1723–1734. DOI: 10.1002/joc.1623.

Deribe, Y. and Getnet, B. (2021) Agricultural mechanization in Ethiopia: Enabling policy, suppliers and inter-regional heterogeneity. *Preprint from Research Square* pp. 1–17. DOI: 10.21203/rs.3.rs-478878/v1.

Deribe, Y., Getnet, B., Kang, T.G. and Tesfaye, A. (2021) Benchmarking the status of agricultural mechanization in Ethiopia (October 15, 2021). Ethiopian Institute of Agricultural Research (EIAR). Research Report No. 133. DOI: 10.2139/ssrn.3968527.

Ehret, C. (1979) On the antiquity of agriculture in Ethiopia. *Journal of African History* 20, 161–177.

Gebreab, F., Gebre Wold, A., Kelemu, F., Ibro, A. and Yilma, K. (2004) Donkey utilisation and management in Ethiopia. In: Starkey, P. and Fielding, D. (eds) *Donkeys, People and Development. A Resource Book of the Animal Traction Network for Eastern and Southern Africa (ATNESA)*. ACP-EU Technical Centre for Agricultural and Rural Cooperation (CTA), Wageningen, The Netherlands. 244 pp.

Getachew, M., Fentahun, M. and, Fikadu, C. (2022) Economic analysis of threshing and shelling machine service provision to reduce post-harvest loss in Ethiopia. *African Journal of Food, Agriculture, Nutrition and Development* 22(6), 20701–20720. DOI: 10.18697/ajfand.111.22105.

Inns, F. (1992) Field power. In: *Tools for Agriculture,* 4th edn. Intermediate Technology Publication Ltd, London.

Kelemu, F. (2015) Agricultural mechanization in Ethiopia: Experience, status and prospects. *Ethiopian Journal of Agricultural Sciences* 25(1), 45–60.

Klappoth, B. (2018) Status and Impact of Different Levels of Agricultural Mechanisation on the Smallholder System in Ethiopia. Master's thesis. Institut für ökologischen Landbau, BOKU University, Vienna, Austria.

McCann, J.C. (2018) The history of agriculture in Ethiopia. Oxford Research Encyclopedia, African History. Online publication date: Feb 2017. DOI: 10.1093/acrefore/9780190277734.013.225.

Ströh de Martínez, C., Feddersen, M. and Speicher, A. (2016) *Food Security in Sub-Saharan Africa: A Fresh Look on Agricultural Mechanisation: How Adapted Financial Solutions Can Make a Difference*. Deutsche Institut für Entwicklungspolitik GmbH, Bonn, Germany.

Temesgen, M., Hoogmoed, W.B., Rockstrom, J. and Savenije, H.H.G. (2009) Conservation tillage implements and systems for smallholder farmers in semi-arid Ethiopia. *Soil and Tillage Research* 104(1), 185–191. DOI: 10.1016/j.still.2008.10.026.

Workneh, W.A., Ujiie, K. and Matsushita, S. (2021) Farmers' agricultural tractor preferences in Ethiopia: A choice experiment approach. *Discover Sustainability* 2, 11. DOI: 10.1007/s43621-021-00021-2.

Annex Table 19.1. Number of tractors per 100 km^2 of arable land in Ethiopia compared to selected countries. Source: calculated and constructed by the authors. Ethiopia data for 2020 is from MOA (Tamiru Habte, pers. com); the rest are from the World Bank Development indicators.

Country	Year 1961	Year 1972	Year Latest	Latest year for available data
Ethiopia	0.23	2.58	8.65	2020
Morocco	12.14	17.70	49.02	1999
Cote d'Ivoire	0.46	9.30	32.08	2001
Africa[a]	14.28	20.29	27.52	1983
Kenya	18.35	17.22	25.23	2002
Tanzania	31.83	22.57	24.66	2002
Nigeria	0.21	1.54	6.70	2007
Ghana	5.88	12.16	4.52	2005
USA	259.65	277.53	271.34	2007
India	1.99	10.61	158.51	2003
Mexico	28.71	53.33	101.55	2007
China	5.09	19.28	82.66	2000
South Africa	101.85	129.36	47.52	2004

[a]Excluding the North African region and South Africa

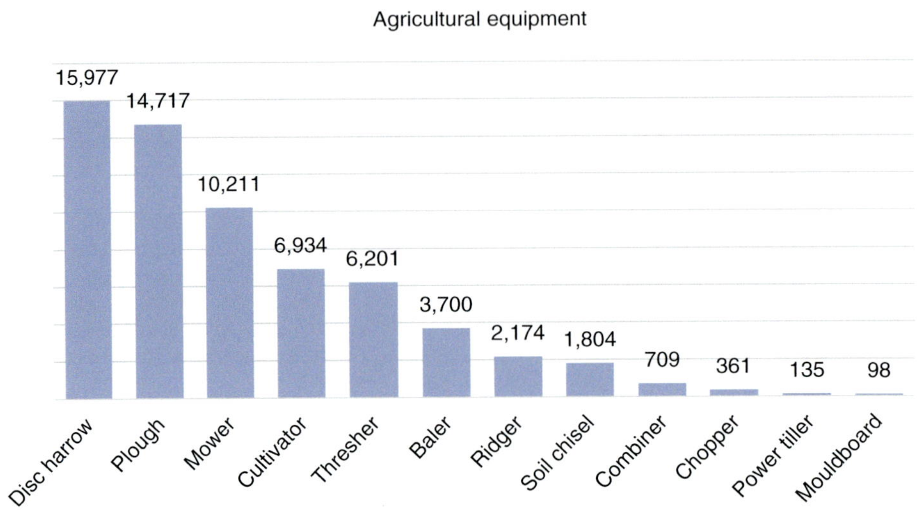

Annex Fig. 19.1. The number of agricultural equipment (not including tractors) imported into Ethiopia, 2005–2015. Source: constructed by the authors from the Ethiopian Revenue and Customs Authority.

20 A Call for Increased Investment in Climate Change

Rahwa Kidane*
Wageningen University and Research, Wageningen, the Netherlands

Abstract
This chapter assesses climate change adaptation policies, practices and implementation challenges in Ethiopia. The findings suggest that the risk of climate change is well recognized and significant progress has been made in formulating several climate-related policies, strategies and plans to reduce vulnerability and build resilience. Despite this progress, however, there is a deficit in government-led planned adaptation actions. Most of the adaptation projects and programmes are financed by international organizations and may not be sustainable. Smallholder farmers, on the other hand, are taking autonomous adaptation measures to combat the risk of climate change and improve their livelihoods, even though they are facing financial, sociocultural, infrastructural, land and labour constraints that impede their efforts. Overall, there is a need to simultaneously address the barriers to autonomous and government-led adaptation measures for Ethiopia to be ready to adapt to future impacts of climate change.

Introduction

In today's world, climate change is seen as one of the greatest global threats. Since the 1990s, the Intergovernmental Panel on Climate Change (IPCC) has reported evidence of climate change with a high level of accuracy. The latest IPCC report concluded that human-induced climate change has caused climate extremes such as heavy precipitation, heat waves, droughts and cyclones across the globe (IPCC, 2021a). For example, the report indicated that global surface temperature in 2011–2020 has increased by 1.09°C compared to 1850–1900 (IPCC, 2021a). On the other hand, the frequency and intensity of rainfall have increased globally. According to this report, Africa is warming faster than the rest of the world. Compared to the years 1850–1900, for example, the average temperatures and hot extremes have exceeded natural variability across all regions of Africa. Beyond the temperature changes, the frequency and intensity of heavy rainfall are expected to increase in most of the countries in Africa (IPCC, 2021b).

Among African countries, Ethiopia is highly exposed to the change in climate conditions (Conway and Schipper, 2011; López-Carr *et al.*, 2014; Gizaw and Gan, 2017). Since 1960, mean temperatures have increased by an average of 1°C, while an overall reduction in precipitation trends have been detected over the last three to four decades (World Bank Group, 2021). The

* Corresponding authors: rahwa.kidane@wur.nl; tsedeke@homegrownvision.org

DOI: 10.1079/9781800626386.0020

country has experienced serious drought events at various times, but the most notable ones occurred in 1983–1985, 1988, 2000, 2002–2003, 2006, 2011 and 2015 (Gummadi *et al.*, 2018). Flood events are also very common in Ethiopia, with the most frequent and intense events being recorded between 2001 and 2010 (Mamo *et al.*, 2019). The frequency and intensity of these extreme weather events are expected to increase further in the coming years (Liou and Mulualem, 2019; Shukla *et al.*, 2021; Tegegne *et al.*, 2021).

Climate-linked risks in Ethiopia have had a significant impact on agriculture, natural resources and human wellbeing (FDRE, 2019). Agriculture is the main livelihood source for almost 73% of the population and it generates about 90% of the country's export earnings (WFP and CSA, 2019). In 2019/20, the share of agriculture in Ethiopia's gross domestic product (GDP) was 32.7% (NBE, 2021). Yet, agriculture is one of the most sensitive sectors to the impacts of climate change and rural people are highly affected by its impacts due to their reliance on rainfed agriculture (Deressa *et al.*, 2009; Shukla *et al.*, 2021). Estimates indicate that by 2050, climate change will reduce Ethiopia's agricultural GDP by up to 31.1% and the production of crops such as *teff*, maize and sorghum will decline by 25.4%, 21.8% and 25.2%, respectively (Solomon *et al.*, 2021).

To reduce the impacts of climate-linked risks, there has been growing attention in recent years to adaptation policies and practices worldwide (Adger *et al.*, 2006; Eriksen *et al.*, 2011; O'Brien, 2016; Anderson *et al.*, 2021). The IPCC defines adaptation as 'the process of adjustment to actual or expected climate and its effects. In human systems, adaptation seeks to moderate harm or exploit opportunities' (IPCC, 2014, p. 5). This adaptation is called 'autonomous' when it is implemented by individuals (local communities) and 'planned' when it is initiated by external actors such as governments and development organizations (Smit *et al.*, 1999; Klein, 2003; Forsyth and Evans, 2013). These two types of adaptation measures are relevant (Spires *et al.*, 2014; Rahman *et al.*, 2021). However, it has been indicated that vulnerable communities often do not have the required resources and capacities to adapt on their own and autonomous adaptations will not be adequate to offset losses associated with variable climate conditions (Smit and Pilifosova, 2003; Clar *et al.*, 2013; Soubry, 2017). For this reason, it is now recognized that planned adaptation interventions are central to supporting local communities' adaptation efforts and reducing the impacts of climate change effectively (De Souza *et al.*, 2015; Tripathi and Mishra, 2017; Assan *et al.*, 2018).

To answer the question of how ready Ethiopia is to deal with future changes in climate, it is important to look at the various steps the country has taken so far in recognizing climate change as one of its development challenges and the extent to which climate-related policies, strategies and plans are translated into concrete actions at the local level. It is also equally important to assess how well smallholder farmers are adapting to climate change impacts and the barriers that impede their adaptation efforts. The objective of this review is, therefore, to provide an assessment of the existing institutional arrangements as well as the policies, strategies and plans formulated and implemented to address the issue of climate change in the country. The chapter will also provide a snapshot of agricultural adaptation measures taken by smallholders in Ethiopia and the key barriers they face in trying to adapt to climate change impacts.

The chapter includes five further sections covering climate policy. It begins by reviewing the institutional arrangements and climate-related policies, strategies and plans that are put in place to address the issue of climate change in Ethiopia, followed by an assessment of planned adaptation interventions. It will then present the agricultural adaptation measures taken by smallholder farmers. Section four provides the analysis of barriers to adaptation efforts experienced by smallholder farmers. The chapter concludes with an assessment of the adequacy of the existing autonomous and planned adaptation efforts in successfully reducing the risks of climate change.

Institutional Arrangements and Climate Policies of Ethiopia

This section provides an overview of the institutional arrangements that have been put in place to address the issue of climate change in Ethiopia. It will also present existing policies, strategies, plans and programmes that have been formulated at the national level to facilitate planned adaptation actions in the country.

Institutional analysis

Ethiopia ratified the United Nations Framework Convention on Climate Change (UNFCCC) in 1994 and the Kyoto Protocol in 2005 (EPA, 2012). The first institutional arrangement put in place in 1974 following the famine in the northern region of Ethiopia was the Relief and Rehabilitation Commission, later renamed to the Disaster Prevention and Preparedness Commission (DPPC) under the Ministry of Agriculture (Nachmany *et al.*, 2015). Although climate change is not directly stated, the DPPC was established with three main objectives: (i) preventing natural or human-made disasters (prevention); (ii) building capacity to reduce the impacts of disasters before their occurrence (preparedness); and (iii) ensuring timely emergency response to affected communities (Proclamation 10/1995). This effort exemplifies the country's first step in the commitment to take natural and human-made disasters seriously as early as the 1970s. However, in the early stages the climate-related efforts were not mainstreamed across the different sectors of the economy and there was no coordinated institutional arrangement (Eshetu *et al.*, 2014).

The National Meteorological Agency (NMA) under the Ministry of Water and Energy represented Ethiopia at the UNFCCC conference in Kyoto in 1997 and the Copenhagen climate change conference (COP15) in 2009. Following the 2009 Copenhagen negotiation, the Environmental Protection Authority (EPA) within the Ministry of Environment and Forest became the leading institution to oversee climate change issues (Bird and Zewdu, 2015). The EPA developed the Climate Resilient Green Economy (CRGE) strategy in 2011, which deals with climate change mitigation and adaptation efforts. Currently, the NMA is responsible for climate data recording, monitoring and projections, which is purely a technical task (Oates *et al.*, 2011). There is no clear justification for the change of NMA's leading institutional role on the issue of climate change in the country, but the NMA was criticized for not mainstreaming climate change matters into the country's overall development plans effectively (Eshetu *et al.*, 2014). Perhaps it was believed that the EPA has more capacity to deal with the wider political, socioeconomic and environmental issues related to climate change than the NMA (Oates *et al.*, 2011).

The EPA transformed into the Ministry of Environment and Forest in 2014, and in 2015 it evolved into the Ministry of Environment, Forest and Climate Change (MEFCC). The incorporation of a climate change component at a ministerial level shows the country's ambitious plan to take climate change seriously. In addition to the MEFCC, the government of Ethiopia has formed a second institution called the CRGE Facility under the Ministry of Finance and Economic Cooperation, which facilitates the administration of national and international climate funds to implement the CRGE strategy (Bird and Zewdu, 2015). The development of the CRGE Facility and the MEFCC institutions at the national level is promising. However, this is not taking place at a district and local level. Studies from least-developed countries show that there is limited capacity and knowledge on how to institutionalize the National Adaptation Plan of Action (NAPA) and other climate-related adaptation policies into regional and local development plans (Regmi *et al.*, 2016).

Climate policy context

The environmental policy

The Ethiopian government has designed various policies and strategies with the aim of reducing the vulnerability of the country to climate-linked risks. One of the prominent programmes is The Environmental Policy of Ethiopia which was issued in 1997 (EPCC, 2015). The Environmental Policy of Ethiopia (EPE) provides an overarching policy framework for numerous environmental problems in general, and climate change in particular. For example, Section 3.9 of the policy element particularly focuses on the challenges of atmospheric pollution and climate change. In this section, the policy aims to address the issue of climate change through: (i) promotion of climate monitoring programmes; (ii) active involvement in protecting the ozone layer; (iii) focus on enhancing the country's hydro, geothermal and solar energy potential as a means of reducing greenhouse gas emissions; and (iv) maximize the country's biomass by implementing agroforestry programmes and through reforestation and rehabilitation of degraded area (EPE, n.d.). It is clear that the EPE acknowledges the issue of climate change. However, apart from suggesting the promotion of a climate monitoring programme, it does not specify or prioritize alternative adaptation strategies needed to reduce the vulnerability of the country to climate change variability. This limitation might be because the policy was designed when the issues of climate change and the need for adaptation were not hot topics on the political agenda (Adem, 2011). In fact, the environmental policy in Ethiopia was designed not because of the Ethiopian government's own political will, but because of a push from donor agencies (Ruffeis *et al.*, 2010). The policy document dedicates a section where it sets out the mechanisms to ensure the impacts of climate change through monitoring, evaluation and reporting at the village, district, regional and federal level. However, there is no well-established reporting requirement that records the impact of the policy implementation in practice (Adem, 2011).

Ethiopia's growth and transformation plans (GTP I and II)

The government of Ethiopia has articulated two consecutive growth and transformation plans (GTPs), where the first was implemented during the period 2010/11 to 2014/15 and the second 5-year development plan was for the period 2015/16 to 2019/20. The first GTP document considers that climate change presents both a risk and opportunity for Ethiopia. Despite the low contribution of the country to global greenhouse gas emission, Ethiopia is hit hard by the impacts of climate change (MOFED, 2010). The GTP recognizes that specific economic sectors and agroecological zones are highly vulnerable to the effects of climate change, and as a result the country loses 2–6% of its total production. Because of this reason, the government gives attention to the need for adaptation during the GTP implementation period.

As one of the objectives of the environment and climate change programmes is to build a 'climate resistance green economy', the document highlights that appropriate adaptation and mitigation efforts are priority actions that need to be taken during the GTP implementation period, which will be realized through policies, strategies and action plans. As compared to the previous national climate policies, the GTP moves one step forward regarding environment and climate change issues, at least in theory, as evidenced in its clear objectives and targets outlined in the policy document. However, doubts were raised whether these ambitious objectives and targets will be realized in practice (Adem, 2011). Indeed, GTP II acknowledges that during the GTP I period 'limited implementation capacity, as well as inadequate adaptation and promotion of Green Technology Packages, have remained challenges in the course of implementation, which need to be taken and addressed in GTP II' (NPC, 2016, p. 62).

National Adaptation Plan of Action

Ethiopia actively engages in the global climate change process of the UNFCCC. As a member of the UNFCC, the country is required to tackle climate change issues by producing a NAPA and by mainstreaming climate change into sectoral development policy plans. As a result, Ethiopia has produced and submitted the first NAPA in 2007 (NMA, 2007). The rationale for developing the NAPA document is based on the fact that the country's adaptive capacity is low, hence urgent and immediate adaptation is required. In the preparation of the NAPA documents, 37 adaptation options were identified, of which 11 were prioritized using a multi-criteria analysis. Some of the prioritized strategies include: (i) promoting drought/crop insurance programmes; (ii) strengthening drought and flood early warning systems; and (iii) development of small-scale irrigation and water harvesting schemes (NMA, 2007, p. 11). The NAPA document was an initial step towards organizing adaptation activities across different government sectors, but it was not envisaged to be a long-term plan on its own (Oates *et al.*, 2011).

Ethiopia's Program of Adaptation to Climate Change

The NAPA was updated and replaced in 2010 by Ethiopia's Program of Adaptation to Climate

Change (EPACC), with the aim of adopting a more participatory approach in designing adaptation options at federal, regional and local levels (Eshetu *et al.*, 2014). The EPACC document recognizes that impacts of climate change will continue to threaten Ethiopia, hence the design and implementation of adaptation activities will play a key role to protect the country (EPA, 2010).

The programme identified about 20 problems associated with the effect of climate change – e.g. in the areas of health, agriculture, forestry, water and land management – along with 7 responses to tackle these problems. These priority measures to be taken for implementing adaptation to climate change are:

- identifying and mapping the areas prone to climate-linked risks;
- making information accessible for all;
- strengthening drought and flood early warning systems;
- mainstreaming adaptation into development activities;
- including adaptation to climate change into educational curricula;
- enhancing research capacity on adaptation to climate change; and
- seeking financial sources and new technologies from domestic and international donor agencies for the effective implementation of adaptation.

Unlike the NAPA document, the EPACC programme is more holistic in terms of identifying and recognizing the likely impact of climate change on various sectors such as human and animal health, agriculture and biodiversity, and land resources. Moreover, the preparation of EPACC documents involved various experts and, more importantly, farmers and pastoralists who bear the brunt of climate change.

Climate resilient green economy strategy

Ethiopia is among the few nations to have integrated its objective of developing a green economy and building resilience to climate change in a single policy framework entitled CRGE strategy in 2011 (Eshetu *et al.*, 2014). The main vision of the CRGE strategy is for Ethiopia to become a middle-income country by 2025, which entirely depends on carbon-neutral growth (FDRE, 2019). The CRGE has two components to realize this vision: the first one is the Green Economy Strategy which largely aims to address mitigation through reducing greenhouse gas emission and the second one is the Climate Resilience Strategy that focuses on the adaptation to climate change mainly in agriculture, forestry, and water and energy sectors.

The state of planned adaptation interventions in Ethiopia

To assess the state of planned adaptation interventions in Ethiopia, active and recently completed adaptation focused projects and programmes were reviewed through online sources. As indicated in Table 20.1, a total of 18 projects/programmes were identified between 2009 and 2021. These projects/programmes have different focus areas, including community-based adaptation, capacity building, research and communication. For example, the project named *Upscaling Community-Based Adaptation in Ethiopia* focuses on promoting new adaptation technologies to protect the livelihoods of local communities against the risk of climate change. On the other hand, some projects are mainly involved in capacity building efforts. The project titled *Building the National Capacity and Knowledge on Climate Change Resilient Actions in Ethiopia* aimed to increase the awareness and capacity of government officials (and local communities) to address climate change issues. Some projects also focus on advancing research. For example, the project named *Economics of Adaptation Studies* mainly works on drought analysis in the Afar and Somali regions to assist decision makers to be able to identify cost-efficient adaptation measures that can reduce climate-related losses and damages.

Almost all the reviewed adaptation projects/programmes are financed by multilateral organizations, including the United Nations Development Programme (UNDP), International Fund for Agricultural Development, The Adaptation Fund, The Green Climate Fund and the Global Environment Facility. In-country (domestic) sources of adaptation finances are very scarce. As seen in Table 20.1, the government of Ethiopia, through its ministerial offices and agencies, mainly plays the role of implementing the various adaptation projects and programmes

Table 20.1. List of adaptation projects and programmes in Ethiopia.

Project/programme name	Funding: Source	Funding: Amount (000)	Funding: Currency	Implementing agencies and partner organizations	Duration	Location(s)
Enabling pastoral communities to adapt to climate change and restoring rangeland environments	MDGAF	4,000	USD	FAO, UNDP, UNEP	2009–2013	Afar, Somali, SNNPR, and Oromia Regional states
Promoting Autonomous Adaptation at the Community Level	GEF	5,32	USD	EPA, UNDP, GEF	2011–2016	Amhara, Oromiya, Tigray, SNNPR
Africa Adaptation Programme	GOJ	2,468	USD	EPA, UNDP	2012–2012	National
Strategic Climate Institutions Programme	United Kingdom	32,68	GBP	DFID, EPA	2012–2016	National
Rural Resilience Enhancement Project	JICA	11,090	Japanese Yen	MOA, Oromia Somali Livestock	2012–2015	Rural Development Bureau
Building the national capacity and knowledge on climate change resilient actions in Ethiopia	EU	9,700	Euro	EPA, MOA, GIZ	2012–2016	National
PRIME	USAID	8,800	USD	Mercy Corps	2012–2019	Afar, Oromia and Somali regions of Ethiopia
Upscaling Community-Based Adaptation in Ethiopia	GEF	8,800	USD	National Government	N/A	Communality level (Rural)
	World Bank	27,000		UNDP		
	GIZ	2,000		GEF		
SCIP-EWS	GEF	20,750	USD	ENMA	2012–unknown	National
	UNDP	4,900	USD	GEF		
Strategic Climate Institutions Programme	DFID, Norway, Denmark	9,500	GBP	EPA	2012	National
Supporting CRGE Facility	Norway	13,000	USD	World Bank Group	2013–2015	National
REAALPA	USAID	6,000	USD	CRS, Cordaid, others	2014–2017	Oromia region
ASAP	IFAD	11,000	USD	GEO	2015–2021	National
Economics of Adaptation studies	KfW Development Bank (German government)	N/A	N/A	N/A	2019–2021	Afar and Somali Regions
CCAG	UNDP, GEF	5,965	USD	MEFCC	2017–2021	National

RIROD	GCF	45,000	USD	MOFEC, MEFCC	2017–2024	National
	GEO	5,000	USD			
CDSDRL	N/A	N/A	N/A	GIZ, MOA	2019–2022	Afar and Somali Regions
CCALEE	GEF	10,450	USD	ENMA, GEF, UNDP	2021–2027	National
	UNDP	5,838				

ASAP = Adaptation for Smallholder Agriculture Programme; CCAG = Climate Change Adaptation Growth; CCALEE= Climate Change Adaptation in the Lowland Ecosystems of Ethiopia; CDSDRL = Capacity Development to Strengthening Drought Resilience in the Lowlands of Ethiopia; CRGE = Climate Resilient Green Economy; CRS = Catholic Relief Services; DFID = Department for International Development (UK); ENMA = Ethiopia National Meteorological Agency; EPA = Environmental Protection Agency of Ethiopia; EU = European Union; FAO = Food and Agriculture Organization of the United Nations; GCF = Green Climate Fund; GEF = Global Environment Facility; GIZ = The Deutsche Gesellschaft für Internationale Zusammenarbeit; GOE = Government of Ethiopia; GOJ = Government of Japan; IFAD = International Fund for Agricultural Development; JICA = Japan International Cooperation Agency; KfW = The Kreditanstalt für Wiederaufbau; MDGAF = The Millennium Development Goals Achievement Fund; MEFCC = Ministry of Environ, Forest and Climate Change; MOA = Ministry of Agriculture; MOFEC = Ethiopia's Ministry of Finance and Economic Cooperation; N/A = not applicable; PRIME = Pastoralist Areas Resilience Improvement and Market Expansion; REAALPA = Resilience through Enhanced Adaptation, Action-learning and Partnership Activity; RIROD = Responding to the Increasing Risk of Drought: Building Gender-responsive Resilience of the Most Vulnerable Communities; SCIP-EWS = Strengthening Climate Information and Early Warning Systems in Ethiopia to Support Climate-Resilient Development; SNNPR = Southern Nations, Nationalities and Peoples Region; UNDP = United Nations Development Programme; UNEP = United Nations Environment Programme; USAID = United States Agency for International Development.

across the different regions of Ethiopia, including Amhara, Tigray, Oromiya, Afar and the Southern Nations Nationalities and Peoples Region (SNNPR).

The reach of the Ethiopian government to implement the various climate-related policies and strategies at the local level seemed to be limited. Complex and interlinked barriers constrain the planning and implementation of the government's policy-driven adaptation intervention in Ethiopia. The key barriers reported to be hindering adaptation policy implementation on the ground include resource (financial and human), institutional, technical and informational (Virtanen *et al.*, 2011; Eshetu *et al.*, 2014; Matewos, 2019; Paul and Weinthal, 2019). For example, the implementation of the CRGE strategy had limited success in Yabello (Oromia) and Mecha (Amhara) districts since the relevant sectoral offices are understaffed and receive limited financial resources from the national level (Eshetu *et al.*, 2014). On the other hand, in the Raya Azebo district of the Tigray region, institutional barriers that are mainly related to the challenge of coordinating relevant actors across the various line ministries, agencies and offices have hindered the translation of adaptation policies into concrete action (Kidane *et al.*, 2023). In Sidama, Borana and Guji zones, implementing actors' lack of technical knowledge on climate issues and their limited awareness of relevant climate-related policy documents were also found to be key barriers to implementing adaptation policies at the local level (Virtanen *et al.*, 2011; Matewos, 2019).

Agricultural adaptation efforts by smallholder farmers

To assess smallholder farmers' agricultural adaptation responses to climate change impacts, 22 household-level studies published between 2017 and 2021 were reviewed. Overall, smallholder farmers' adaptation responses can fall into 3 main categories and 12 sub-categories. The three main categories are: (i) crop and livestock management; (ii) land management; and (iii) irrigation and water management (Table 20.2). This section focuses on each adaptation measure practised by smallholder farmers.

As shown in Table 20.2, adjusting crop planting date is one of the key adaptation responses conducted by smallholder farmers in most regions of Ethiopia. This was reported by ten of the reviewed articles. This strategy is a low-cost and most practised adaptation response by smallholder farmers (Kidane *et al.*, 2019; Tessema *et al.*, 2019). The review suggests that farmers adjust planting dates to the onset of seasonal rain to protect sensitive crop development stages from drought-related water stress.

Crop diversification, which is reported by 12 of the reviewed studies, is also one of the most important adaptation measures practised by smallholder farmers in Ethiopia. As can be seen in Table 20.2, this strategy is commonly undertaken by smallholder farmers across the five regions of Ethiopia. Farmers diversify into two or more crops as a form of insurance against erratic rains and extreme weather events (risk-spreading measure). Despite its importance, not all Ethiopian farmers practice crop diversification, as the strategy requires a relatively large farm size (see Marie *et al.*, 2020; Eshetu *et al.*, 2021).

As part of their agricultural adaptation effort, farmers also plant drought resistant and early maturing crops (reported by n=10 and n=5 studies, respectively). For example, in SNNPR, farmers commonly grow *enset* (*Ensete ventricosum*) while in the Tigray region farmers plant cactus (*Opuntia ficus-indica*), as these two plants have high resilience or drought resistance and water-holding capacities (Nurebo, 2017; Kidane *et al.*, 2019). The reviewed articles further mentioned the use of improved seed varieties as one of smallholder farmers' adaptation actions in rural parts of Ethiopia (reported by n=10 studies). This strategy enables smallholder farmers to increase crop productivity while minimizing the impact associated with climatic factors (Cacho *et al.*, 2020).

Climate change poses a great risk to livestock production through its impact on the availability of forage and water resources (Rojas-Downing *et al.*, 2017). Livestock management is one of the adaptation measures that is commonly practised by pastoral and agropastoral communities of Ethiopia (indicated by n=13 studies). For example, the Afar pastoralists use livestock mobility as one of the key adaptation strategies during extended drought periods (Berhe *et al.*, 2020).

Table 20.2. Adaptation strategies implemented by smallholder farmers in Ethiopia

		Crop and livestock management							Land management			Irrigation and water management	
Author(s)	Region	AD	CD	DR	EM	FS	IS	LV	TP	SC	IN	IR	WH
Belay *et al.* (2017)	Oromia	✓	✓			✓		✓	✓	✓	✓		
Berhe *et al.* (2017)	Afar					✓		✓				✓	
Alemayehu and Bewket (2017a)	Amhara	✓	✓	✓	✓		✓	✓	✓	✓	✓	✓	
Yohannes *et al.* (2017)	Tigray			✓							✓		
Kidane *et al.* (2019)	SNNPR	✓	✓	✓					✓	✓	✓		✓
Asrat and Simane (2018)	Benishangul									✓			✓
Dika (2018)	Gambella	✓	✓	✓	✓			✓		✓			
Teklewold *et al.* (2019)	Nile Basin of Ethiopia		✓				✓		✓	✓	✓	✓	✓
Bedeke *et al.* (2019)	SNNPR			✓					✓	✓			✓
Kahsay *et al.* (2019)	Tigray	✓							✓	✓		✓	
Cholo *et al.* (2019)	SNNPR						✓				✓		
Tessema *et al.* (2019)	Amhara	✓	✓	✓	✓								
Marie *et al.* (2020)	Amhara	✓	✓	✓						✓			
Gebru *et al.* (2020)	Tigray	✓	✓	✓	✓		✓	✓	✓	✓	✓	✓	✓
Hirpha *et al.* (2020)	Oromia	✓					✓		✓	✓			✓
Dangia and Dara (2020)	Gambella	✓	✓	✓			✓		✓				
Darabant *et al.* (2020)	Amhara	✓	✓			✓	✓	✓		✓	✓	✓	✓
Habte *et al.* (2021)	SNNPR	✓	✓	✓	✓				✓	✓	✓		
Mekuyie and Mulu (2021)	Oromia						✓					✓	
Sertse *et al.* (2021)	Tigray		✓				✓		✓	✓			
Etana *et al.* (2021)	Oromia		✓				✓		✓	✓		✓	
Eshetu *et al.* (2021)	Oromia	✓	✓								✓		

AD = adjusting crop planting dates; CD = crop diversification; DR = drought-tolerant crops: EM = early maturing crops; FS = feed storage; IN = input application fertilizer; IR = modern irrigation system; IS = improved seed; LV = livestock management (destocking; restocking, use of cross-breeds; livestock mobility, etc.); SC = soil conservation; TP = tree planting; manure); WH = rainwater, drainage, groundwater

During a normal year, agropastoral communities are also known to be involved in the production and storage of livestock fodder as an emergency feed for drier conditions (Belay *et al.*, 2017).

Climate-related stressors such as flooding and extended moisture stress greatly contribute to land degradation (Sivakumar and Stefanski, 2007). In response, Ethiopian farmers are planting trees (reported by n=12 studies) and applying soil conservation techniques (e.g. building stone bunds) to prevent top soil from being washed away by heavy rainfall. Tree planting and soil conservation techniques were reported by n=12 and n=15 of the reviewed studies, respectively. Farmers also use various inputs such as manure, mulches and cover crops to improve soil quality, which was reported by ten studies.

The use of irrigation is vital in enhancing the resilience of farming systems as it minimizes farmers' reliance on rainfed agriculture. The reviewed studies indicated that smallholder farmers use irrigation to reduce the impact of climate variability and change (reported by n=10 studies). However, in most of these studies, the adoption of irrigation farming (particularly modern irrigation) is the least adopted adaptation strategy as it requires high investment costs. For this reason, some farmers use low-cost techniques such as rainwater harvesting and collecting water from rivers and hand-dug wells (reported by $n=7$ studies).

Barriers to the implementation of adaptation actions by smallholder farmers

To explore barriers that impede climate change adaptation actions by smallholder farmers, 11 empirical studies published between 2013 and 2021 were reviewed. Overall, barriers to smallholder farmers' adaptation action can be grouped into six categories: (i) financial; (ii) lack of information and knowledge; (iii) sociocultural; (iv) infrastructure; (v) institutional; and (vi) land/labour. The following section explores those barriers in detail.

A lack of finance is a major constraint impeding the implementation of key adaptation measures by smallholder farmers. As can be seen in Table 20.3, financial constraint is reported across all studies and regions of Ethiopia. Financial capitals (e.g. savings, remittance and credit access) enable farmers to acquire important technologies that are essential for climate change adaptation investments. However, particularly poor farmers' ability to adopt high-capital and high-return strategies (e.g. use of irrigation and improved seed) is constrained by financial resources

Table 20.3 Barriers experienced by smallholder farmers to adapt to climate change impacts.

		Barriers to climate change adaptation					
Author(s)	Region	Financial	Information/ knowledge	Sociocultural	Infrastructure	Institutional	Land/ labour
Gebrehiwot and van der Veen (2013)	Tigray	✓	✓				✓
Mersha and van Laerhoven (2016)	Tigray and Amhara	✓		✓		✓	
Alemayehu and Bewket (2017b)	Amhara	✓	✓				
Amare *et al.* (2018)	Nile Basin of Ethiopia	✓	✓			✓	✓
Kidane *et al.* (2019)	SNNPR	✓	✓		✓		✓
Debela *et al.* (2019)	Oromia	✓	✓			✓	✓
Etana *et al.* (2020)	Oromia	✓	✓			✓	
Marie *et al.* (2020)	Amhara	✓	✓				✓
Godebo *et al.* (2021)	Ethiopian Rift Valley	✓	✓				✓
Destaw and Fenta (2021)	Amhara	✓	✓				✓
Gebeyehu *et al.* (2021)	Lower Omo Valley of Ethiopia	✓	✓		✓	✓	✓

(Kidane *et al.*, 2019; Etana *et al.*, 2020). As a result, these groups of farmers are often forced to engage in riskier (e.g. temporary migration) or low-return adaptation strategies (e.g. adjusting planting dates, wage labour, etc).

In addition to financial constraints, ten of the reviewed studies indicated a lack of information and/or knowledge as a hindrance to smallholder farmers' adaptation efforts (Table 20.3). For instance, a lack of information about the appropriate adaptation measures was reported as a barrier by 21% of surveyed smallholder farmers in the Gondar Zuria District of the Amhara regional state (Marie *et al.*, 2020). Further, although climate information is a useful instrument that can be used to assist smallholder farmers to make important farming decisions (e.g. adjusting planting periods), 44% of the farmers in Oromia indicated lack of climate information as a significant constraint to adaptation (Debela *et al.*, 2019).

Beyond financial and information-related barriers, institutional factors present another challenge to smallholder farmers' adaptation efforts (reported by n=5 studies). It is widely recognized that formal and informal institutions can be key in facilitating adaptation actions among smallholder farmers through their influence on resource mobilization and skills development and capacity building potential (Makate, 2019). However, some of the reviewed studies indicate that the weakening of informal institutions coupled with inadequate support of formal institutions (e.g. development agents) hindered the adaptation potential of smallholder farmers (Debela *et al.*, 2019; Etana *et al.*, 2020). In some instances, institutions themselves can act as barriers rather than enablers to local adaptation efforts (see Milman and Arsano, 2014).

Pre-existing asset holdings are central to smallholder farmers' adaptive capability (Kidane *et al.*, 2019), and eight studies suggest that the shortage of basic livelihood assets impeded smallholder farmers' capacity to adapt to climate change impacts. Of those basic livelihood assets, land plays a significant role in smallholder farming systems. However, in the Degen district of Ethiopia, for example, over half (57%) of the respondents reported shortage of land as a major constraint to the adoption of crop diversification strategy (Amare *et al.*, 2018). Shortages of labour is also a key hindrance to smallholder adaptation, cited by 11% of the respondents in Oromia regional state (Debela *et al.*, 2019).

Two of the reviewed studies indicate how inadequate rural infrastructure constrains smallholder farmers' adaptation efforts (Kidane *et al.*, 2019; Gebeyehu *et al.*, 2021). Similar studies conducted in other parts of Africa show rural infrastructure as a key challenge hampering the successful implementation of adaptation strategies (Antwi-Agyei *et al.*, 2015; Chingombe and Musarandega, 2021). For example, poor rural roads and a lack of transport facilities could limit smallholder farmers from obtaining agricultural inputs that are vital for the local adaptation efforts.

Sociocultural barriers to adaptation did not receive much attention in the reviewed articles, except in one study (Mersha and van Laerhoven, 2016). Specifically, this study shows how gendered norms in Raya Azebo and Kobo districts of Ethiopia restricted female-headed households from ploughing and how that limited their ability to undertake farm-related climate change adaptation measures.

Discussion and Conclusions

The review suggests that smallholder farmers in Ethiopia are taking various agricultural adaptation measures that are broadly categorized into crop-livestock management, land management, and irrigation and water management. However, this does not necessarily mean that the implemented measures are effective in reducing climate-linked risks. There is evidence that climate change adaptation measures often do not work as intended – i.e. they may have little impact, no impact or they may even be maladaptive. Even if some of the farmers' adaptation measures might be effective, not all farmers can take the strategies due to the presence of various socioeconomic and cultural barriers. Without additional support, farmers cannot successfully adapt to climate change impacts on their own. International climate funds will not be sustainable and adequate to meet smallholder farmers' adaptation needs and priorities. For this reason, mobilizing domestic adaptation finance from public and private sources is crucial to support smallholder farmers' adaptation efforts. Obviously, the Ethiopian government has recognized the issue of climate change through the various policy documents and adaptation is set on the political agenda. Yet, there is limited progress in the translation of these policy documents into actual adaptation actions due to financial-, institutional-, technical- and information-related constraints. Overcoming these barriers, therefore, is key to effectively implement adaptation policies and help smallholder farmers build resilience against the risk of climate change.

References

Adem, A.B.W. (2011) *A Climate Change Country Assessment Report for Ethiopia*. Forum for Environment, Addis Ababa, Ethiopia. Available at: https://www.southernvoices.net/images/docs/6-Ethiopia_Country_Assessment_Report_ECSNCC.pdf (accessed 16 January 2021).

Adger, W.N., Paavola, J., Huq, S. and Mace, M.J. (eds) (2006) *Fairness in Adaptation to Climate Change*. MIT Press, Cambridge, MA, USA.

Alemayehu, A. and Bewket, W. (2017a) Smallholder farmers' coping and adaptation strategies to climate change and variability in the central highlands of Ethiopia. *Local Environment* 22(7), 825–839.

Alemayehu, A. and Bewket, W. (2017b) Determinants of smallholder farmers' choice of coping and adaptation strategies to climate change and variability in the central highlands of Ethiopia. *Environmental Development* 24, 77–85.

Amare, Z.Y., Ayoade, J.O., Adelekan, I.O. and Zeleke, M.T. (2018) Barriers to and determinants of the choice of crop management strategies to combat climate change in Degen District, Nile Basin of Ethiopia. *Agriculture & Food Security* 7(1), 1–11.

Anderson, R., Bayer, P.E. and Edwards, D. (2021) Climate change and the need for agricultural adaptation. *Current Opinion in Plant Biology* 56, 197–202.

Antwi-Agyei, P., Dougill, A.J. and Stringer, L.C. (2015) Barriers to climate change adaptation: Evidence from northeast Ghana in the context of a systematic literature review. *Climate and Development* 7(4), 297–309.

Asrat, P. and Simane, B. (2018) Farmers' perception of climate change and adaptation strategies in the Dabus watershed, North-West Ethiopia. *Ecological Processes* 7(1), 1–13.

Assan, E., Suvedi, M., Schmitt Olabisi, L. and Allen, A. (2018) Coping with and adapting to climate change: A gender perspective from smallholder farming in Ghana. *Environments* 5(8), 86.

Bedeke, S., Vanhove, W., Gezahegn, M., Natarajan, K. and Van Damme, P. (2019) Adoption of climate change adaptation strategies by maize-dependent smallholders in Ethiopia. *NJAS-Wageningen Journal of Life Sciences* 88, 96–104.

Belay, A., Recha, J.W., Woldeamanuel, T. and Morton, J.F. (2017) Smallholder farmers' adaptation to climate change and determinants of their adaptation decisions in the Central Rift Valley of Ethiopia. *Agriculture & Food Security* 6(1), 1–13.

Berhe, M., Hoag, D., Tesfay, G., Tadesse, T., Oniki, S., Kagatsume, M. and Keske, C.M. (2017) The effects of adaptation to climate change on income of households in rural Ethiopia. *Pastoralism* 7(1), 1–15.

Bird, N. and Zewdu, E. (2015) *Climate Change Spending in Ethiopia: Recommendations to Bridge the Funding Gap for Climate Financing*. Climate Science Centre of Addis Ababa University.

Cacho, O.J., Moss, J., Thornton, P.K., Herrero, M., Henderson, B., Bodirsky, B.L. and Lipper, L. (2020) The value of climate-resilient seeds for smallholder adaptation in sub-Saharan Africa. *Climatic Change* 162(3), 1213–1229.

Chingombe, W. and Musarandega, H. (2021) Understanding the logic of climate change adaptation: Unpacking barriers to climate change adaptation by smallholder farmers in Chimanimani District, Zimbabwe. *Sustainability* 13(7), 3773.

Cholo, T.C., Fleskens, L., Sietz, D. and Peerlings, J. (2019) Land fragmentation, climate change adaptation, and food security in the Gamo Highlands of Ethiopia. *Agricultural Economics* 50(1), 39–49.

Clar, C., Prutsch, A. and Steurer, R. (2013) Barriers and guidelines for public policies on climate change adaptation: A missed opportunity of scientific knowledge-brokerage. *Natural Resources Forum* 37(1), 1–18.

Conway, D. and Schipper, E.L.F. (2011) Adaptation to climate change in Africa: Challenges and opportunities identified from Ethiopia. *Global Environmental Change* 21(1), 227–237.

Dangia, M. and Dara, P.K. (2020) Analysis of climate variability and adaptive strategies of rural household: The case of Abobo District and Itang Special District, Gambella Regional State, Ethiopia. *Asian Journal of Agricultural Extension Economics & Sociology* 38(4), 82–93.

Darabant, A., Habermann, B., Sisay, K., Thurnher, C., Worku, Y., Damtew, S. and Abiyu, A. (2020) Farmers' perceptions and matching climate records jointly explain adaptation responses in four communities around Lake Tana, Ethiopia. *Climatic Change* 163(1), 481–497.

De Souza, K., Kituyi, E., Harvey, B., Leone, M., Murali, K.S. and Ford, J.D. (2015) Vulnerability to climate change in three hot spots in Africa and Asia: Key issues for policy-relevant adaptation and resilience-building research. *Regional Environmental Change* 15, 747–753.

Debela, N., McNeil, D., Bridle, K. and Mohammed, C. (2019) Adaptation to climate change in the pastoral and agropastoral systems of Borana, South Ethiopia: Options and barriers. *American Journal of Climate Change* 8(1), 40–60.

Deressa, T.T., Hassan, R.M. and Ringler, C. (2009) Assessing household vulnerability to climate change. IFPRI Discussion Paper 00935.

Destaw, F. and Fenta, M.M. (2021) Climate change adaptation strategies and their predictors amongst rural farmers in Ambassel district, Northern Ethiopia. *Jàmbá: Journal of Disaster Risk Studies* 13(1), 1–11.

Dika, G. (2018) Impacts of climate variability and households' adaptation strategies in Lare district of Gambella Region, South Western Ethiopia. *Journal of Earth Science & Climatic Change* 9(2), 7.

EPA (Environmental Protection Authority of Ethiopia) (2010) Ethiopia's Draft Programme of Adaptation to Climate Change. EPA, Addis Ababa, Ethiopia, pp. 1–29.

EPA (2012) National Report of Ethiopia, the United Nations Conference on Sustainable Development (Rio+20). National Report of Ethiopia. Addis Ababa, Ethiopia.

EPCC (Ethiopian Panel on Climate Change) (2015) First Assessment Report – An Assessment of Ethiopia's Policy and Institutional Frameworks for Addressing Climate Change. Addis Ababa, Ethiopia. Available at: https://www.unisdr.org/preventionweb/files/46791_anassessmentofpolicyandinstitutiona.pdf (accessed 20 April 2021).

EPE (Environmental Policy of Ethiopia) (n.d.) Environmental Policy of Ethiopia. Available at: http://thereddesk.org/sites/default/files/environment_policy_of_ethiopia_1.pdf (accessed 13 February 2021).

Eriksen, S., Aldunce, P., Bahinipati, C.S., Martins, R.D.A., Molefe, J.I., Nhemachena, C. and Ulsrud, K. (2011) When not every response to climate change is a good one: Identifying principles for sustainable adaptation. *Climate and Development* 3(1), 7–20.

Eshetu, Z., Simane, B., Tebeje, G.N., Amsalu, W., Berhanu, A. and Trujillo, N. (2014) Climate finance in Ethiopia, Overseas Development Institute, London and Climate Science Centre, Addis Ababa. Available at: https://www.odi.org/sites/odi.org.uk/files/odi-assets/publications-opinion-files/8995.pdf (accessed 20 April 2020).

Eshetu, G., Johansson, T., Garedew, W. and Yisahak, T. (2021) Determinants of smallholder farmers' adaptation options to climate change in a coffee-based farming system of Southwest Ethiopia. *Climate and Development* 13(4), 318–325.

Etana, D., Snelder, D.J., van Wesenbeeck, C.F. and de Cock Buning, T. (2020) Dynamics of smallholder farmers' livelihood adaptation decision-making in central Ethiopia. *Sustainability* 12(11), 4526.

Etana, D., Snelder, D.J., van Wesenbeeck, C.F. and de Cock Buning, T. (2021) The impact of adaptation to climate change and variability on the livelihood of smallholder farmers in central Ethiopia. *Sustainability* 13(12), 6790.

FDRE (Federal Democratic Republic of Ethiopia) (2019) Ethiopia's Climate Resilient Green Economy. National Adaptation Plan. Available at: https://www4.unfccc.int/sites/NAPC/Documents/Parties/Final%20Ethiopia-national-adaptation-plan%20%281%29.pdf (accessed 30 April 2020).

Forsyth, T.I.M. and Evans, N. (2013) What is autonomous adaption? Resource scarcity and smallholder agency in Thailand. *World Development* 43, 56–66.

Gebeyehu, A.K., Snelder, D., Sonneveld, B. and Abbink, J. (2021) How do agro-pastoralists cope with climate change? The case of the Nyangatom in the Lower Omo Valley of Ethiopia. *Journal of Arid Environments* 189, 104485.

Gebrehiwot, T. and van der Veen, A. (2013) Farm level adaptation to climate change: The case of farmers in the Ethiopian Highlands. *Environmental Management* 52(1), 29–44.

Gebru, G.W., Ichoku, H.E. and Phil-Eze, P.O. (2020) Determinants of smallholder farmers' adoption of adaptation strategies to climate change in Eastern Tigray National Regional State of Ethiopia. *Heliyon* 6(7), e04356.

Gizaw, M.S. and Gan, T.Y. (2017) Impact of climate change and El Niño episodes on droughts in sub-Saharan Africa. *Climate Dynamics* 49(1), 665–682.

Godebo, T.R., Jeuland, M.A., Paul, C.J., Belachew, D.L. and McCornick, P.G. (2021) Water quality threats, perceptions of climate change and behavioral responses among farmers in the Ethiopian Rift Valley. *Climate* 9(6), 92.

Gummadi, S., Rao, K.P.C., Seid, J., Legesse, G., Kadiyala, M.D.M. *et al.* (2018) Spatio-temporal variability and trends of precipitation and extreme rainfall events in Ethiopia in 1980–2010. *Theoretical and Applied Climatology* 134, 1315–1328.

Habte, A., Mamo, G., Worku, W., Ayalew, D. and Gayler, S. (2021) Spatial variability and temporal trends of climate change in southwest Ethiopia: Association with farmers' perception and their adaptation strategies. *Advances in Meteorology* 2021, 13.

Hirpha, H.H., Mpandeli, S. and Bantider, A. (2020) Determinants of adaptation strategies to climate change among the smallholder farmers in Adama District, Ethiopia. *International Journal of Climate Change Strategies and Management* 12(4), 463–476.

IPCC (Intergovernmental Panel on Climate Change) (2014) Climate Change 2014: Impacts, Adaptation, and Vulnerability (Working Group II Contribution to the Fifth Assessment Report of the Intergovernmental Panel on Climate Change), IPCC. Available at: https://www.ipcc.ch/site/assets/uploads/2018/02/ar5_wgII_spm_en.pdf (accessed 2 June 2021).

IPCC (2021a) Summary for policymakers. In: Climate Change 2021: The Physical Science Basis. Contribution of Working Group I to the Sixth Assessment Report of the Intergovernmental Panel on Climate Change [Masson, V., Zhai, P., Pirani, A., Connors, S.L., Péan, C. *et al.* (eds)]. Available at: https://www.ipcc.ch/report/sixth-assessment-report-working%20group-i/ (accessed 2 July 2024).

IPCC (2021b) Contribution of Working Group I to the Six Assessment Report of the Intergovernmental Panel on Climate change. Regional Fact sheet Africa. Available at: https://www.ipcc.ch/report/ar6/wg1/#Regional (accessed 5 July 2021).

Kahsay, H.T., Guta, D.D., Birhanu, B.S. and Gidey, T.G. (2019) Farmers' perceptions of climate change trends and adaptation strategies in semiarid highlands of Eastern Tigray, Northern Ethiopia. *Advances in Meteorology* 2019, 13.

Kidane, R., Prowse, M. and de Neergaard, A. (2019) Bespoke adaptation in rural Africa? An asset-based approach from Southern Ethiopia. *The European Journal of Development Research* 31(3), 413–432.

Kidane, R., Wanner, T. and Nursey-Bray, M. (2023) Overcoming barriers to climate change adaptation policy implementation: Insights from Ethiopia. *International Development Planning Review* 45(2), 121–147. DOI: 10.3828/idpr.2022.11.

Klein, R.J. (2003) Adaptation to climate variability and change: What is optimal and appropriate? In: Giupponi, C. and Shechter, M. (eds) *Climate Change in the Mediterranean: Socio-Economic Perspectives of Impacts, Vulnerability and Adaptation*. Edward Elgar Publishing, Cheltenham, UK.

Liou, Y.A. and Mulualem, G.M. (2019) Spatio–temporal assessment of drought in Ethiopia and the impact of recent intense droughts. *Remote Sensing* 11(15), 1828.

López-Carr, D., Pricope, N.G., Aukema, J.E., Jankowska, M.M., Funk, C., Husak, G. and Michaelsen, J. (2014) A spatial analysis of population dynamics and climate change in Africa: Potential vulnerability hot spots emerge where precipitation declines and demographic pressures coincide. *Population and Environment* 35, 323–339.

Makate, C. (2019) Local institutions and indigenous knowledge in adoption and scaling of climate-smart agricultural innovations among sub-Saharan smallholder farmers. *International Journal of Climate Change Strategies and Management* 12(2), 270–280.

Mamo, S., Berhanu, B. and Melesse, A.M. (2019) Historical flood events and hydrological extremes in Ethiopia. In: Melesse, A.M., Abtew, W. and Senay, G. (eds) *Extreme Hydrology and Climate Variability.* Elsevier, pp. 379–384.

Marie, M., Yirga, F., Haile, M. and Tquabo, F. (2020) Farmers' choices and factors affecting adoption of climate change adaptation strategies: Evidence from northwestern Ethiopia. *Heliyon* 6(4), e03867.

Matewos, T. (2019) Deconstructing institutional roles in climate change adaptation: The case of local public institutions in drought-prone districts of Sidama, Southern Ethiopia. *Environmental Science & Policy* 98, 47–53.

Mekuyie, M. and Mulu, D. (2021) Perception of impacts of climate variability on pastoralists and their adaptation/coping strategies in Fentale district of Oromia Region, Ethiopia. *Environmental Systems Research* 10(1), 1–10.

Mersha, A.A. and van Laerhoven, F. (2016) A gender approach to understanding the differentiated impact of barriers to adaptation: Responses to climate change in rural Ethiopia. *Regional Environmental Change* 16, 1701–1713.

Milman, A. and Arsano, Y. (2014) Climate adaptation and development: Contradictions for human security in Gambella, Ethiopia. *Global Environmental Change* 29, 349–359.

MOFED (Ministry of Finance and Economic Development) (2010) Growth and Transformation Plan 2010/11-2014/2015 (Vol. I). Addis Ababa, Ethiopia.

Nachmany, M., Fankhauser, S., Davidová, J., Kingsmill, N., Landesman, T., Roppongi, H. and Sundaresan, J. (2015) The 2015 Global Climate Legislation Study: A Review of Climate Change Legislation in 99 Countries, 2–13. Available at: https://www.lse.ac.uk/granthaminstitute/wp-content/uploads/2015/05/Global_climate_legislation_study_20151.pdf (accessed 2 July 2024).

NBE (National Bank of Ethiopia) (2021) Quarterly Bulletin. First Quarter 2020/21. Fiscal Year Series. Available at: https://nbebank.com/wp-content/uploads/2021/02/First-Quarter%20Report-2020-21.pdf (accessed 10 June 2021).

NMA (National Meteorological Agency, Ethiopia) (2007) Climate Change National Adaptation Programme of Action (NAPA) of Ethiopia. Addis Ababa, Ethiopia. Available at: https://unfccc.int/resource/docs/napa/eth01.pdf (accessed 14 March 2021).

NPC (National Planning Commission) (2016) Growth and Transformation Plan II (GTP II) (2015/16-2019/20). Federal Democratic Republic of Ethiopia. Addis Ababa, Ethiopia. Available at: https://europa.eu/capacity4dev/file/30510/download?token=efsF8UiP (accessed 12 February 2021).

Nurebo, C.T. (2017) Diversity and resilient varieties of enset for climate change adaptation: The case of different agro-ecological zones of Hadiya, southern Ethiopia. *Journal of Earth Science & Climatic Change* 8(5), 402.

Oates, B.N., Conway, D. and Calow, R. (2011) The "mainstreaming" approach to climate change adaptation: insights from Ethiopia's water sector. (Background Working Paper). Overseas Development Institute, London, UK.

Paul, C.J. and Weinthal, E. (2019) The development of Ethiopia's climate resilient green economy 2011–2014: Implications for rural adaptation. *Climate and Development* 11(3), 193–202.

Rahman, H.T., Albizua, A., Soubry, B. and Tourangeau, W. (2021) A framework for using autonomous adaptation as a leverage point in sustainable climate adaptation. *Climate Risk Management* 34, 100376.

Regmi, B.R., Star, C. and Leal Filho, W. (2016) Effectiveness of the local adaptation plan of action to support climate change adaptation in Nepal. *Mitigation and Adaptation Strategies for Global Change* 21(3), 461–478.

Rojas-Downing, M.M., Nejadhashemi, A.P., Harrigan, T. and Woznicki, S.A. (2017) Climate change and livestock: Impacts, adaptation, and mitigation. *Climate Risk Management* 16, 145–163.

Ruffeis, D., Loiskandl, W., Awulachew, S.B. and Boelee, E. (2010) Evaluation of the environmental policy and impact assessment process in Ethiopia. *Impact Assessment and Project Appraisal* 28(1), 29–40.

Sertse, S.F., Khan, N.A., Shah, A.A., Liu, Y. and Naqvi, S.A.A. (2021) Farm households' perceptions and adaptation strategies to climate change risks and their determinants: Evidence from Raya Azebo district, Ethiopia. *International Journal of Disaster Risk Reduction* 60, 102255.

Shukla, R., Gleixner, S., Yalew, A.W., Schauberger, B., Sietz, D. and Gornott, C. (2021) Dynamic vulnerability of smallholder agricultural systems in the face of climate change for Ethiopia. *Environmental Research Letters* 16(4), 044007.

Sivakumar, M.V. and Stefanski, R. (2007) Climate and land degradation—An overview. In: Sivakumar, M.V.K., Ndiang'ui, N. (eds) *Climate and Land Degradation. Environmental Science and Engineering* (Environmental science) (1st ed., pp. 105–135). Springer, Berlin/Heidelberg.

Smit, B. and Pilifosova, O. (2003) Adaptation to climate change in the context of sustainable development and equity. *Sustainable Development* 8(9), 9.

Smit, B., Burton, I., Klein, R.J. and Street, R. (1999) The science of adaptation: A framework for assessment. *Mitigation and Adaptation Strategies for Global Change* 4(3), 199–213.

Solomon, R., Simane, B. and Zaitchik, B.F. (2021) The impact of climate change on agriculture production in Ethiopia: Application of a dynamic computable general equilibrium model. *American Journal of Climate Change* 10(1), 32–50.

Soubry, B. (2017) How Can Governments Support Adaptation to Climate Change by Small-Scale Farmers? A Case Study from the Canadian Maritime Provinces. Environmental Change Institute, University of Oxford, Oxford, UK. Available at: https://papers.ssrn.com/sol3/papers.cfm?abstract_id=3035782 (accessed 24 March 2021).

Spires, M., Shackleton, S. and Cundill, G. (2014) Barriers to implementing planned community-based adaptation in developing countries: A systematic literature review. *Climate and Development* 6(3), 277–287.

Tegegne, G., Melesse, A.M. and Alamirew, T. (2021) Projected changes in extreme precipitation indices from CORDEX simulations over Ethiopia, East Africa. *Atmospheric Research* 247, 105156.

Teklewold, H., Mekonnen, A. and Kohlin, G. (2019) Climate change adaptation: A study of multiple climate-smart practices in the Nile Basin of Ethiopia. *Climate and Development* 11(2), 180–192.

Tessema, Y.A., Joerin, J. and Patt, A. (2019) Climate change as a motivating factor for farm-adjustments: Rethinking the link. *Climate Risk Management* 23, 136–145.

Tripathi, A. and Mishra, A. (2017) Knowledge and passive adaptation to climate change: An example from Indian farmers. *Climate Risk Management* 16, 195–207.

Virtanen, P., Palmujoki, E. and Gemechu, D.T. (2011) Global climate policies, local institutions and food security in a pastoral society in Ethiopia. *Consilience* 5(1), 96–118.

WFP and CSA (World Food Program and Central Statistical Agency) (2019) Comprehensive Vulnerability and Food Security Analysis. Available at: https://reliefweb.int/report/ethiopia/ethiopia-comprehensive-food-security-and-vulnerability-analysis-cfsva-2019 (accessed 13 March 2021).

World Bank Group (2021) Climate Risk Country Profile: Ethiopia. Available at: https://climateknowledgeportal.worldbank.org/sites/default/files/2021-05/15463A-WB_Ethiopia%20Country%20Profile-WEB.pdf (accessed 12 February 2021).

Yohannes, D.F., Ritsema, C.J., Solomon, H., Froebrich, J. and Van Dam, J.C. (2017) Irrigation water management: Farmers' practices, perceptions, and adaptations at Gumselassa irrigation scheme, North Ethiopia. *Agricultural Water Management* 191, 16–28.

21 Improved Water Use and Management for Enhanced Agricultural Productivity

Daneal Fekersillassie,[1] Mesfin Shenkut,[2] Engida Zemedagegnehu,[2] Minyahil Fikadu,[2] Tsedeke Abate[3]* and Messele Fisseha[3]
[1]Addis Ababa Institute of Technology, Addis Ababa, Ethiopia; [2]MS Consult, Addis Ababa, Ethiopia; [3]Homegrown Vision, Addis Ababa, Ethiopia

Abstract
There are 12 river basins in Ethiopia, with a total annual flow of 122 billion cubic metres (BCM) surface water and 2.65 BCM groundwater, suggesting that it has huge water potential for irrigation that could substantially increase crop productivity and production if properly planned and managed. The total suitable land of all the basins for irrigation is estimated at 37 million ha, of which highly, moderately and marginally suitable areas accounted for 8%, 58% and 34%, respectively. Recent studies estimate the total physical irrigation potential at 7.50 million ha. The Growth and Transformation Program II planned to bring more than 3 million ha (17%) of the total cropland under irrigation by 2019/20, but there is no indication how much of that has been achieved. The main challenge for the country is that only 41% of the total irrigation schemes are fully functional and the overall efficiency is ~45%.

Introduction

Agriculture is the lifeline for Ethiopia's development (Makombe *et al.*, 2011). It employs more than 80% of the country's population and makes significant contribution to gross domestic product. In addition to being the main source of food and nutrition security, it is the major source of export earnings, import substitution and raw material for the local industry. Even though this country has ample water and land resources that can be used for irrigation development, it is heavily dependent on rainfed agriculture that varies spatially as well as temporally.

Ethiopia has 12 river basins with an annual runoff volume of 122 billion cubic metres (BCM) of surface water and an estimated 2.65 BCM of groundwater potential (Awulachew *et al.*, 2007). Even if the potential and actual irrigated area is not accurately investigated, the Ministry of Water, Irrigation and Energy (MOWIE) very recently adopted a new approach to estimate irrigation area and espouse the initiative to pursue the water-based growth corridor methods for delineating long-term and strategic investments and interventions for irrigation development. A parametric evaluation approach was employed for land suitability analysis. The method evaluates the land suitability for surface and pressurized irrigation separately (considering surface irrigation for lands with slopes less than 8% and pressurized irrigation for lands with slopes between 8% and 15%).

* Corresponding author: tsedeke@homegrownvision.org

DOI: 10.1079/9781800626386.0021

Accordingly, a total area of 37 million ha was estimated to be suitable for cultivation under both surface and pressurized irrigation systems, of which 27 million ha would be for surface irrigation and 10 million ha would be for pressurized irrigation (MOWIE, 2019).

The government of Ethiopia has been mobilizing huge public investment towards the construction of irrigation infrastructure to avoid the risk of crop failure due to drought. However, the identification, selection and prioritization of irrigation projects has not been systematic. As a result, the impact of preceding investments could not bring about integrated and sustained development. Ethiopia has a water resources management policy and a number of proclamations and regulations to ensure the implementation of integrated water resources management at basin scale by acknowledging that water is an economic and a social good. The Ethiopian water resource policy (MOWIE, 2020) states that water should be recognized as a natural resource with an economic value and ensures that fees are paid for services rendered. Accordingly, water is recognized as a vulnerable and scarce natural resource, and it ensures that all pricing systems and mechanisms should be geared towards conservation, protection and efficient use of water as well as promoting equity of access.

This chapter reviews the status of irrigation in Ethiopia. We provide information that is desperately needed for setting the priorities in agricultural research and development of Ethiopia. It is also our hope that teaching staff and students in agricultural universities and development workers would find this chapter useful in their career.

Approach

We conducted a comprehensive study of irrigation use in Ethiopia, drawing secondary data from a variety of sources. We analysed and interpreted this data in detail, and we also conducted a literature review to synthesize the existing knowledge on this topic. The rest of the information in our study is based on our own personal experiences working in Ethiopia and with Ethiopian farmers.

Results

Potential for irrigation in Ethiopia

There were several studies that have tried to estimate and map the irrigation potential of the country's basins and sub-basins, but they were not very accurate or comprehensive. They are mainly focused on land suitability (soil and terrain) without paying much attention to the availability and sustainability of water sources. Water supply is a crucial factor for irrigation development, as it determines the quantity and quality of water that can be used for irrigation, as well as the technical, financial and environmental costs of harnessing and transporting it to the irrigable areas. Table 21.1 shows the different estimates of national irrigation potential from different reports (including groundwater and rainwater potential).

Many of the estimates are done as a desk review exercise using geographic information systems and other software facilities with little consideration of detail site-specific data collection. The large variations reported in recent years is due to the assumptions involved when considering adopting pressurized irrigation for sloping areas greater than 8%. If potential assessment has to meet the needs of irrigation development decision making, it has to be done on the basis of site-specific data inputs. This requires soil data (physical and chemical), terrain information, water availability in relation to location of land suitable for irrigation, etc.

According to a recent study (Dereje *et al.*, 2020), the total physical irrigation potential of Ethiopia is estimated to be 7.50 million ha. The available surface water in the major basin outlets is used as given in the Master Plan studies as this is the physical irrigation potential. The results show that Abbay Basin has the largest physical irrigation potential amounting to about 51% of the total. Abbay, Baro-Akobo and Omo-Gibe are the top three basins with the largest irrigation potential. These three basins together constitute about 78% of the total physical irrigation potential of the country.

Land suitability

The land suitability analysis shows that a total area of about 37 million ha is found to be suitable

Table 21.1. Estimates of irrigation potential in Ethiopia.

	Potential (million ha) using			
Source/Authority	Surface water	Groundwater	Rainwater	Total
MOWIE (1995/1997)	–	–	–	3.70
IWMI (2010)	3.70	1.10	0.50	5.30
MOA and ATA (2016)	4.30	4.70	4.40	13.40
PRI (2018)	–	–	–	41.00
Abeyou *et al.* (2018)	–	–	–	6.00
GIRDC (2018)	–	–	–	21.30
IAG (2019)	6.14	1.29	–	7.43

–A dash means that data is not available.
ATA = Agricultural Transformation Agency; GIRDC = Generation Integrated Rural Development Consultant; IAG = International Association of Geodesy; IWMI = International Water Management Institute; MOA = Ministry of Agriculture; MOWIE = Ministry of Water, Irrigation and Energy; PRI = Policy Research Institute.

for cultivation through surface- and pressurized-irrigated agriculture. This figure reflects only the land suitability regardless of water availability. Irrigation potential, on the other hand, is based on both land and water availability. Water is generally a limiting factor rather than land for irrigation in almost all basins of Ethiopia and, therefore, the irrigation potential is likely to be less than the above figure.

Land suitability evaluation for surface and pressurized irrigation follows different criteria. This is mainly because surface irrigation is generally practicable in flat and gentle sloping lands owing to the difficulty of water distribution in sloping lands. On the other hand, pressurized irrigation (sprinkler and drip) can be applied in sloping lands as water is conveyed in pipe systems where their design takes account of the land slopes.

The method evaluates the land suitability for surface and pressurized irrigation separately (considering surface irrigation for lands with slopes less than 8%, and pressurized irrigation for lands with slopes between 8% and 15%).

The total suitable land of all the basins for both surface and pressurized irrigation is estimated at about 37 million ha (Table 21.2), which is essentially the total agricultural land of the country, according to data from the Food and Agriculture Organization (FAO). These comprised 3.00 million ha, 21.34 million ha and 12.54 million ha, respectively, of highly suitable, moderately suitable and marginally suitable areas. Put differently, highly suitable, moderately suitable and marginally suitable areas accounted for 8%, 58% and 34%, respectively. The distribution of the suitable area among the river basins is variable.

Table 21.2. Land suitability for surface irrigation by river basins. Source: constructed by the authors from MOWIE (2019).

	Suitability (million ha)			
River basin	High	Moderate	Marginal	Total
Wabe-Shebelle	0.98	6.16	2.29	9.43
Genale-Dawa	0.52	4.64	2.38	7.54
Awash	0.48	2.61	1.25	4.34
Ogaden	0.28	2.32	0.96	3.56
Rift Valley	0.24	1.27	0.59	2.10
Baro-Akobo	0.20	0.88	0.55	1.62
Omo-Gibe	0.15	0.95	0.80	1.90
Abbay	0.08	1.70	3.08	4.86
Danakil	0.05	0.45	0.19	0.68
Tekeze	0.02	0.33	0.45	0.80
Aysha	0.00	0.01	0.04	0.05
Mereb-Gash	0.00	0.02	0.01	0.03
Total	3.00	21.34	12.58	36.92

Moreover, the basins also varied substantially in their suitability for surface and pressurized irrigation, as presented in Fig. 21.1. For example, Wabe-Shebelle, Genale-Dawa and Abbay accounted for a combined total of 66% of the total suitable area for irrigation in Ethiopia.

Even more importantly, Wabe-Shebelle, Genale-Dawa and Awash (in combination) comprised about 1.98 million ha of highly suitable areas for both surface and ground irrigation (cf. Table 21.2 above). This is 65% of the total area,

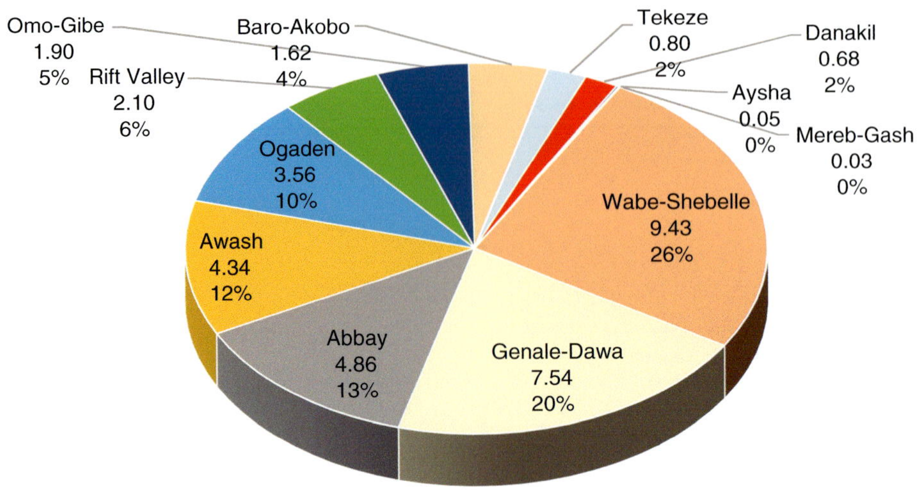

Fig. 21.1. Area (million ha) and percentages of total suitable agricultural areas for irrigation in different basins in Ethiopia. Source: calculated and constructed by the authors from MOWIE (2019).

consisting of 32%, 17%, and 16%, respectively, for the three river basins.

Physical irrigation potential

The physical irrigation potential reflects the area that could be potentially irrigated with the available land and water resources (both surface and groundwater) without considering the economics of water resources development. Suitable irrigable land and water are thus the two important resources for estimating the physical irrigation potential. In the MOWIE study of the physical irrigation potential, land suitability evaluation was made using the Suitability Index comprising seven land and soil parameters. On the other hand, surface water resources as obtained from the Master Plan studies (122 BCM) and ground water resources in each basin based on a recent study (Ayenew *et al.*, 2008) were considered. Irrigation water demand was estimated by selecting three major irrigated crops in each basin and data from the FAO (2022b). See Alemayehu (2006), Berhanu *et al.* (2014) and Yimere and Assefa (2022) for further details on groundwater and surface water potential of Ethiopia.

All the available surface water is not used for irrigation as there are multiple users within basins and environmental releases to main ecosystems. Globally, agriculture consumes about 70% of developed water resources, reaching up to 90% in some developing nations. For the Awash Basin in Ethiopia, irrigation consumes about 80% of the total abstracted water resources. Thus, it is assumed that 70% of the total surface water resources can be allocated for irrigation. The physical irrigation potential estimated for each river basin is shown in Fig. 21.2.

The physical irrigation potential, in general, is dictated by water availability as water is scarcer than land that is suitable for irrigation in Ethiopia. Accordingly, the Abbay Basin has the largest physical irrigation potential (about 46% of the total). The three basins: Abbay, Baro-Akobo and Omo-Gibe are the top three basins with largest irrigation potential. These three basins together constitute about 78% of the total irrigation potential of Ethiopia.

Irrigation status in Ethiopia

According to the FAO (2022a), Ethiopia has become a net importer of food between 2015 and 2020 and may continue to do so if the current agricultural production situation continues to prevail. The predominantly rainfed traditional agricultural system is highly vulnerable to climate variability, which in turn exacerbates the

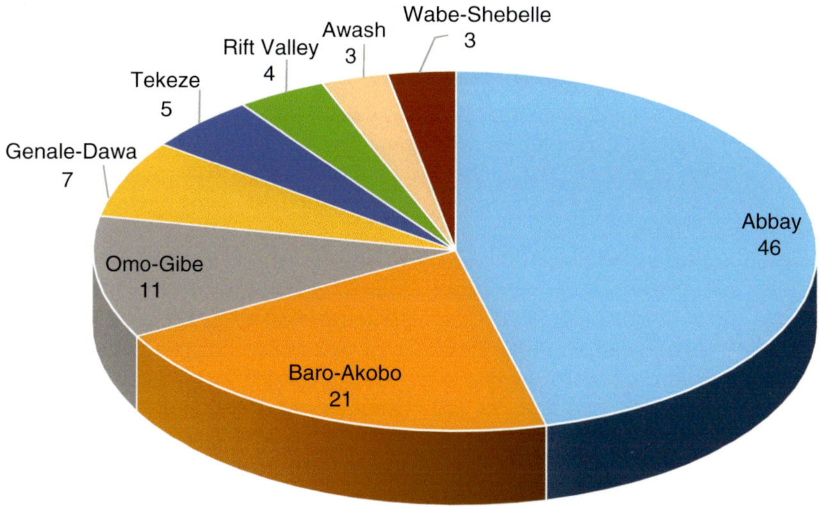

Fig. 21.2. Physical irrigation potential estimates (percentages) for river basins in Ethiopia. Source: calculated and constructed by the authors from MOWIE (2019).

conditions of food insecurity. On the other hand, the country has huge water resource potential that could be harnessed for irrigation to enhance agricultural productivity and production. Due to lack of water storage and large spatial and temporal variations in rainfall, there is not enough water for most farmers to produce more than one crop per year and hence there are frequent crop failures due to dry spells and droughts, which have resulted in an unending food shortage most of the time.

Irrigation is a significant input to agricultural practices in that it reduces climate risk and improves crop production. Reducing the risk of dry spells can also help to encourage the use of modern inputs such as fertilizers and improved seed, thereby further enhancing agricultural productivity. There is adequate evidence that irrigation has the potential to double or even triple cereal crop production in Ethiopia, compared to rainfed agriculture. Therefore, with increasing needs for the development and recognition of climate variability and its impact on highly vulnerable rainfed agriculture, the attention given to irrigation has increased from time to time. However, the pace of its development has been slow compared to increasing demand for food and raw materials for agricultural industries. To feed its increasing population and meet the raw material demand for the growing industries, Ethiopia must increase agricultural production by expanding irrigation, in addition to fully utilizing the production capacity created in the existing irrigation schemes and intensifying agricultural production under both irrigated and rainfed practices (Mekonen *et al.*, 2020).

Irrigation has been practiced in Ethiopia since ancient times for producing subsistence food crops. However, modern irrigation systems started in the late 1950s with the objective of producing industrial crops in the Awash valley. Private companies who operated farms for growing commercial crops such as cotton, sugarcane and horticultural crops started the first formal irrigation schemes around this time in the upper and lower Awash valley. In the 1960s, irrigated agriculture was expanded in all parts of the Awash valley and in the lower Rift Valley. The Awash valley saw the biggest expansion in view of the water regulation afforded by the construction of the Koka dam and reservoir that regulated flows with benefits of flood control, hydropower and assured irrigation water supply. In addition, the construction of the Addis-Assab road opened the Awash valley to ready markets in the hinterland as well as for export.

Since the mid-1980s, the Ethiopian government has responded to drought and famine through promoting and constructing an irrigation infrastructure aimed at increasing agricultural production. These were traditional, small-, medium- and large-scale irrigation schemes performing at different levels. Based on the MOWIE classification, irrigation projects in Ethiopia are categorized as large scale if the size of command area is greater than 3000 hectares; medium scale if it falls in the range of 200–3000 hectares; and small scale if it is covering less than 200 hectares.

Currently, the government is giving more emphasis to the irrigation sub-sector by way of enhancing the food security situation in the country. It continues to consider the development of the agriculture sector as a priority strategic focus area for economic growth. This was evidenced by the Agriculture Development Led Industrialization strategy that was implemented during different periods. In realization of the strategy, a series of development plans and programmes have been implemented: for example, the Sustainable Development and Poverty Reduction Program (from 2002/03–2004/05); Plan for Accelerated and Sustained Development to End Poverty (from 2005/06–2009/10); Growth and Transformation Program I (from 2010/11–2014/15) and Growth and Transformation Program II (from 2015/16–2019/20). All of these plans had ambitious targets that brought progress in the development of irrigation after long years of stagnation. Table 21.3 shows the irrigation development trends.

Most of the irrigation schemes in Ethiopia use water from perennial rivers through diversions using a weir or by direct pumping. The common discharge controlling structures are manually operated iron gates with a graduated pole to measure water level, which is then converted into discharge volume using standard tables. In most of the cases, trapezoidal earthen canals are used to convey water to the irrigation fields. Surface irrigation techniques, particularly furrow irrigation, are the most commonly used irrigation method, with the exception of Finchaa Sugar Estate where large-scale sugarcane cultivation has been practiced using sprinkler irrigation. Recently, there has been a gradual change in some regions from field ditches and siphons to a more labour- and water-saving technology called hydro flume for field irrigation water application.

The main crops cultivated in government-managed large-scale irrigation schemes include sugarcane, cotton, tobacco, and fruits and vegetables such as oranges, mandarin, guava and grape, while crops such as onion, potato, pepper, tomato and banana are commonly cropped in community-managed, small-scale irrigation schemes. Recently, some of the medium- and large-scale irrigation schemes have shown a significant shift in cropping patterns as a result of low market prices and crop damage due to pests. For example, the Middle Awash Agricultural Development Enterprise is now shifting from banana plantation to cotton production.

Government-managed irrigation schemes usually export their products such as sugar, cotton and green beans to the international market. However, there are also cases where some of these schemes are confined to the local market due to their relative location and distance to international markets. Community-managed schemes usually sell most of their products to local markets and use the remaining produce for household consumption. The most marketable crops in community-managed schemes are onion, tomato, cabbage and grain legumes like chickpea and lentil. At present, the total land equipped with irrigation is estimated at 3 million

Table 21.3. Irrigation development trends in Ethiopia. Source: constructed by the authors from MOA/MOWIE (2019).

	Area under irrigation (ha)		
Period	Small scale	Medium and large	Total
Base year (1991)	64,000	112,105	176,105
2010 (at the end of PASDEP)	853,000	134,221	987,221
GTP I (2010/11-2014/15)	2,353,000	280,000	2,633,000
GTP II (2015/16-2019/20)	2,528,000	490,000	3,018,000

GTP = Growth and Transformation Program; PASDEP = Plan for Accelerated and Sustained Development to End Poverty.

hectares (2.5 million ha under small scale and 0.5 million ha medium and large scale).

Existing and under study/construction of large- and medium-scale irrigation schemes

Most of the existing large- and medium-scale irrigation schemes in Ethiopia generally perform at much lower than their potential capacities. On average, the actually irrigated area for large-scale irrigation (LSI) schemes is only about 23% of their design potential, while the medium-scale irrigation (MSI) schemes operate at about 59%. There are also several LSI and MSI projects under study (feasibility, detail design) and construction, as summarized in Table 21.4.

At present, there are a total of 269 existing MSI and LSI schemes distributed across eight river basins of the country. Of those, 243 (or 90%) are MSI and 26 (about 10%) are LSI schemes. Furthermore, about 41%, 50% and 9%, respectively, of the existing total number of schemes are fully functional (FF), partially functional (PF) and non-functional (NF). In other words, 100, 135 and 24 of the existing total schemes are FF, PF and NF, respectively. Wabe-Shebelle, Rift Valley Lakes and Omo-Gibe basins each had 73, 72 and 55 of the total schemes, respectively. In other words, these three basins combined account for about 74% of the total schemes.

There are a total of 59 irrigation schemes that are currently under study or construction. Thirty-seven of those are LSI and 22 are MSI schemes. Wabe-Shebelle, Genale-Dawa and Abbay basins each have 13, 13 and 12 schemes, respectively. So, Ethiopia will have a total of 328 irrigation schemes by the time these are completed.

Table 21.5 gives a summary of the planned and implemented areas of existing schemes by river basin. The total planned potential of existing MSI and LSI schemes was close to 579,000 ha, while the actual irrigated areas is in the order of 259,000 ha, or roughly 45%, indicating the low performance of our irrigation schemes. Comparing the two categories, MSI executed about 60% of its planned irrigation area while LSI showed only 40% (Table 21.5).

There were also substantial differences among the basins. For example, Wabe-Shebelle, the Rift Valley Lakes and Tekeze implemented nearly 91%, 75% and 74%, respectively, of their planned irrigation area under MSI. In a similar fashion, the Rift Valley Lakes, Tekeze and Awash achieved about 86%, 60% and 49% of their planned area, respectively. Considering the performance of different basins for MSI and LSI combined, the Rift Valley Lakes, Tekeze and Wabe-Shebelle implemented about 77%, 60% and 52%, respectively, of their planned area whereas Baro-Akobo, Omo-Gibe and Abbay had achieved about 21%, 28% and 36%.

Figure 21.3 shows proportions of the actual irrigated area (MSI and LSI combined) for different river basins in Ethiopia. The largest proportion (31%) was reported for the Awash basin.

Table 21.4. Number of existing and under study/construction medium- (MSI) and large-scale (LSI) irrigation schemes and their status by river basin in Ethiopia. Source: constructed by the authors from MOWIE (2019).

	Fully functional		Partially functional		Non-functional		Under study/construction	
Basin	MSI	LSI	MSI	LSI	MSI	LSI	MSI	LSI
Wabe-Shebelle	61	0	6	1	5	0	11	2
Rift Valley	40	1	28	1	2	0	2	7
Omo-Gibe	0	0	44	6	5	0	0	1
Abbay	0	1	11	2	8	0	1	11
Awash	2	2	14	7	0	1	0	2
Genale-Dawa	1	0	8	0	0	2	6	7
Tekeze	1	0	2	1	1	0	1	7
Baro-Akobo	1	0	3	1	0	0	1	0
Total	106	4	116	19	21	3	22	37

LSI = large-scale irrigation; MSI = medium-scale irrigation.

Table 21.5. Summary of planned and implemented area (ha) of existing schemes by river basin. Source: calculated and constructed by the authors from MOWIE (2019).

	MSI			LSI			Total		
River basin	Planned	Implemented	Per cent	Planned	Implemented	Per cent	Planned	Implemented	Per cent
Rift Valley Lakes	43,013	32,081	74.6	10,920	9,420	86.3	53,933	41,501	76.9
Tekeze	1,313	970	73.9	50,000	30,000	60.0	51,313	30,970	60.4
Wabe-Shebelle	26,113	23,622	90.5	27,000	4,000	14.8	53,113	27,622	52.0
Awash	14,542	4,265	29.3	156,393	76,296	48.8	170,935	80,561	47.1
Genale-Dawa	2,857	1,297	45.4	0	0		2,857	1,297	45.4
Abbay	7,510	1,817	24.2	97,000	35,500	36.6	104,510	37,317	35.7
Omo-Gibe	31,580	12,352	39.1	97,983	24,494	25.0	129,563	36,846	28.4
Baro-Akobo	2,287	1,100	48.1	10,000	1,500	15.0	12,287	2,600	21.2
Total	129,215	77,504	60.0	449,296	181,210	40.3	578,511	258,714	44.7

LSI = large-scale irrigation; MSI = medium-scale irrigation.

This was followed by the Rift Valley Lakes (16%), Abbay and Omo-Gibe (14% each), Tekeze (12%) and Wabe-Shebelle (11%). Baro-Akobo and Genale-Dawa each had 1%.

The locations of existing and those under study/construction LSI and MSI schemes are shown in Fig. 21.4. Most of the schemes are concentrated in the middle part of the country. There are fewer irrigation schemes in the outlying areas in the west and east compared to areas along the Rift Valley Lakes and in the east-central parts, which follows the pattern of the major water basins.

Discussion and Conclusions

This section will discuss the major constraints associated with water-use efficiency improvement and the available promising opportunities. As mentioned in the preceding sections, the irrigation potential of the country is high. However, it has not been utilized to significantly improve the farmers' agricultural productivity and food security. Despite the fact that many rural farmers in various regions have been practicing traditional irrigation schemes for a long time, their crop yields could not exceed the traditional subsistence farming level (Mekonen *et al.*, 2020). This has been attributed to various reasons and challenges associated with the irrigation schemes.

Different constraints hinder the improvement of the water-use efficiency of irrigation schemes in Ethiopia. First, crop diseases appeared in different parts of the country affecting crop yields at different levels. From the point of view of plant disease occurrence, irrigation occasionally makes it possible to minimize disease outbreaks by growing crops out of season. For example, some farmers irrigate potato and wheat farms during the dry season to avoid destructive attacks of major diseases. More often, however, irrigation, by its effects on moisture and temperature conditions, will encourage the development of diseases to a greater or lesser extent. In most of the countries where irrigation is now practiced, diseases that used to be absent from crops grown without irrigation have become common ever since irrigation, especially in sprinkler form, has been applied. The irrigation water-use efficiency not only ensures the most efficient crop growth, but it is also essential for high-quality production of seed, food, textiles and other produce with increasing perception of the economic and environmental impacts. Adequate water supplied to roots of the plant is needed in order to warrant good yield or harvest of the crop. However, water applied in excess or deficient amounts may result in the development of many crop diseases (fungal and bacterial plant diseases) that compromise crop yield and crop quality and decrease the water-use efficiency.

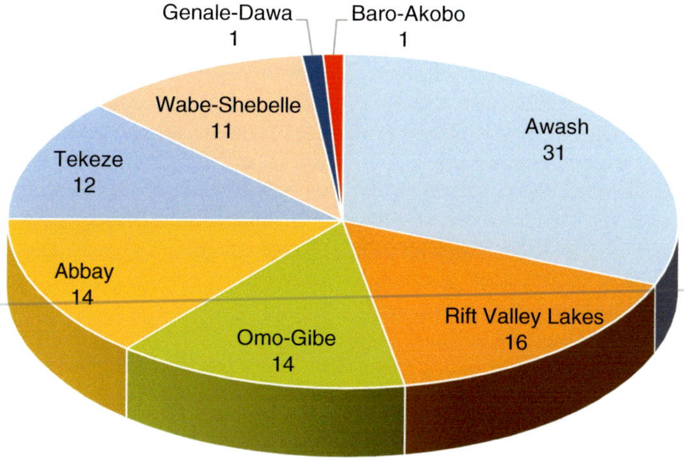

Fig. 21.3. Per cent share of area covered under medium- and large-scale irrigation in different river basins in Ethiopia. Source: calculated and constructed by the authors from MOWIE (2019).

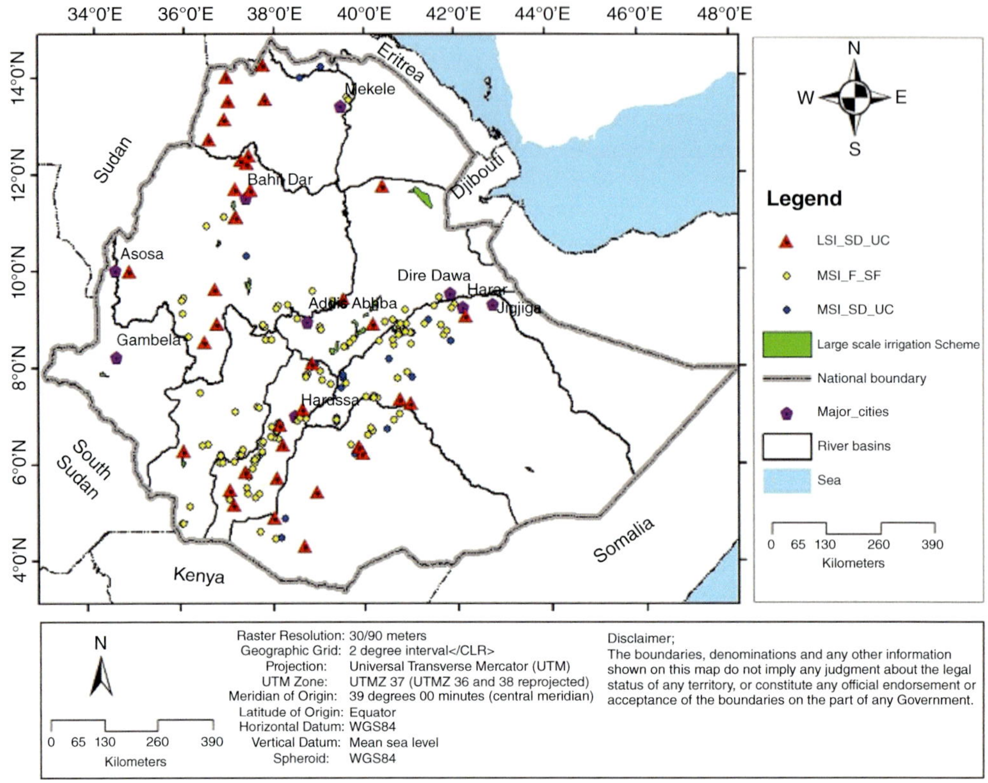

Fig. 21.4. Location map for existing large- (LSI) and medium-scale irrigation (MSI) schemes in Ethiopia and those under study/construction (SD-UC). Source: MOWIE (2019).

Second, since the economic background of most farmers is low income, there is a need to supply adequate budget from government and partners for the design and contractions of modern schemes and buying necessary appurtenances. The majority of farmer's activities in Ethiopia with regard to irrigation are constrained by a less than ideal environment, particularly in the case of smallholder farmers. The environment is not always conducive for them to interact with the private sector. Smallholder farmers might have challenges raising money from financial institutions to improve their farming business. They often lack access to the most suitable irrigation technology, as irrigation equipment suppliers have limited penetration in the countryside. This is where the Ethiopian government can have a critical role in making the surrounding environment more conducive for farmers.

Third, the marketing system of rural farmers also needs substantial attention. Developing markets and road infrastructure are required to improve farmers' access to markets. In Ethiopia, input supply in an organized form, through government or related sources, is often available only for a rainfed season. Farm inputs, especially fertilizer, are scarce and relatively expensive during the irrigation season and there is poor access to markets for both inputs and outputs (Awulachew *et al.*, 2005; MOWR, 2007).

Fourth, another consideration is the land tenure system in Ethiopia; it is generally believed that the existing land tenancy arrangements have acted as a major disincentive. Land is public property in Ethiopia. Fragmentation of land in smallholdings and the increasing trend of shrinking land holding influence the ability of land users to deal with their problems and invest in improved technologies. Due to the absence of clear landownership, farmers cultivate their land under temporary arrangements and may expect the land to be taken away from them during the next land redistribution. Under these conditions, farmers may perceive investment in long-term land improvement as inappropriate because they are unlikely to reap the benefit of their work in the future. Therefore, farmers prefer low or no investment in technologies, which may lower opportunities to improve agricultural productivity.

Smallholders may have been reluctant to develop irrigation infrastructure on the land that they are given as only 'use right'. The situation with private investors is not much different. According to the current land policy of Ethiopia, rural households are legally allowed to access agricultural land. Nonetheless, the difficulty of rural population in accessing farmlands involves controversial authenticity of this land tenure to solve problems of household farmland access (Tenaw *et al.*, 2009; Singirankabo and Ertsen, 2020; Beyene *et al.*, 2021).

Fifth, the capacity of farmers, development agents and extension workers to operate, maintain and manage irrigation structures needs to be improved. A clear organizational setup is required to improve the functionality of small-scale irrigation schemes and consequently increase water-use efficiency. The government of Ethiopia must build strong and relevant institutional arrangements that can easily cascade downwards to improve the coordination between institutions involved in irrigation development and to establish clear duties and responsibilities for various organs in the irrigation sector. Greater emphasis has to be given to constructing databases for irrigation schemes at the district level.

Finally, the government's commitment to irrigation development, availability and affordability for irrigation technology and the development of remote sensing, together with the proper use of various sensors and qualitative models, should be improved to improve water-use efficiency.

There are also policy challenges that need to be addressed. Adequate supply of irrigation water of suitable quality and quantity plays a major role in economic and social development. More than a million people, mostly the poor, still have no access to clean water in Ethiopia. Many parts of the country are constantly struggling with drought and floods; in addition, many areas face the problem of fast-growing populations, rapid urbanization, rising costs of providing water and pollution of fresh water. All of these problems have increased the urgency of formulating policies and associated strategies that will ultimately lead to effective measures to manage water as a social and economic resource with emphasis on its conservation.

Many countries, particularly those with acute shortages of irrigation water, have recognized the need to review their water policies and to develop comprehensive water resources management strategies. The new concepts in water resources management – all aimed at sustaining the resource – will take a long time to implement and may call for deep changes in national culture. Good policies will need to strike a balance between the different objectives of the challenge facing the global food system today. Good policies depend not only on the policy maker, but also on the input from many stakeholders. Given the scale and complexity of these challenges, policy makers may need to experiment with new techniques to develop policy solutions that are suitable to everyone.

The Ethiopian agricultural system is not significantly benefiting from water management technologies and irrigation, procedures that could improve productivity and reduce vulnerability to climate variability. Despite significant efforts by the government and stakeholders to improve agricultural water management and enhance irrigation, many constraints related to policy, institution, technology, capacity, infrastructure and the market still exist (MOWR, 2007).

Irrigation water-use services could also be provided through private investments in the development of irrigation and related water works construction (Regulation No. 162/2009). Investors in Ethiopia are allowed to engage in 'irrigation development...covering more than 50 hectares of land' (Article 2(3)). This legal framework is designed with 'the objectives of providing incentives to investors engaging in irrigation development to bring about an accelerated economic development of the country by increasing the role of private sector in developing irrigation through utilizing the huge available irrigable land and water resources of the country' (Article 4). In order to benefit from irrigation development incentives, it is a mandatory requirement to obtain an investment permit as required under investment proclamation and water-use permits in accordance with Proc. No. 197/2000 and Reg. No. 115/2005. Regulation No. 162/2009 stipulates incentives that can be granted to include 'exemptions from water-use charges and making available for further development major irrigation structures such as dams, main canals and access roads at government cost' (Article 6 (1) (3)). The period of exemptions from water-use charges for an investor who engaged in irrigation and related water works construction development lasts 'until the actual investment cost is recovered' (Article 7(1)). The period of exemption from water-use charges shall begin from the date of commencement of use of water resources as may be appropriate (Article 7(2)).

Recently, the irrigation and water-users' association proclamation No. 841/2014 was enacted to govern farmers' management of irrigation and drainage systems 'which are formed on irrigation infrastructures constructed by the federal government throughout Ethiopia' (Article 3). This proclamation envisages that irrigation water users that are not less than five may form an association (Articles 10(1)). An irrigation water-users association may be formed through two procedural mechanisms. First, those persons who possess land on the basis of landholding system and uses such land with water supplied from a traditional and drainage system may form a water-users association with a fulfilment of minimum requirements of membership through their own initiation. Second, the MOWIE may form irrigation water-users associations using irrigation water from a modern irrigation and drainage systems or traditional irrigation and drainage systems upgraded to modern irrigation and drainage systems (Articles 11 and 12).

The water-users association, among other things, shall have the objectives to manage, operate and maintain irrigation and drainage systems in its service area and provide equitably to its members for agricultural purpose; undertake construction and reconstruction works and issue internal rules for consumption of irrigation water and collect fees from its members for the services provided (Article 4).

Irrigation development in Ethiopia is facing multifaceted challenges at all levels of implementation and management that involves project identification, project study and design, project implementation (construction), and operation and maintenance rehabilitation (upgrading). There have been many challenges facing the development and management of irrigation projects at all phases of the irrigation project cycle. The complexity of the problem varies from project to project and sites.

- **Design related problems:** Irrigation system design and construction are carried out often on the basis of inadequate and unreliable data on water availability, soils, geology, climate information, crop yield responses, irrigation efficiency, irrigation management capacity, etc.
- **Constraints related to construction:** Irrigation project constructions have been facing the following major constraints:
    - most often the bidding and contract awards are based on incomplete design and geotechnical investigation, which had the consequences of design change (modification) and variation orders;
    - failures of consultants and contractors to meet their contractual obligations;
    - client's slow response to inquiries of consultant and contractors and inadequate monitoring and evaluation; and
    - significant time and cost overruns.
- **Underutilization of created potential:** There is a significant gap between the created potential and utilization. A recent study reveals that actual irrigated area is on average 72% of the created potential (Mekonen *et al.*, 2020).
- **Constraints related to operation, maintenance and management:** Weak irrigation management systems, lack of efficient and sustainable operation and maintenance, unreliable and inequitable water distribution, lack of proper seasonal and in-season water demand and supply planning have become a matter of increasing concern with regard to the performance of irrigation schemes in Ethiopia. Many of the schemes have fallen short of expectations. The major factor associated with this poor performance of the schemes is a lack of proper operation and maintenance. Considerable attention is often given to the planning, design and construction of the schemes while the operation aspect receives little attention.

- **Constraints related to technical and technological support:** First, it is important to acknowledge the irrigators' low technical know-how, inadequate technical support, and lack of flow measurement and monitoring. Then, the government has to give due attention to the agricultural sector to bring about a fundamental change to convert the farmlands and semi-arid regions into paradise and to increase productivity by implementing irrigation works extensively and in a coordinated manner, by investing intensively in highland and lowland areas, including pastoralist and semi-pastoralist areas.
- **Constraints related to sustainability:** The deterioration of physical structures such as watershed degradation in Ethiopia is still a menace despite considerable efforts to arrest soil erosion in many parts of the country. Excessive soil erosion of watershed causes sedimentation of downstream reservoirs and water conveyance structures. Borkena Dam in Amhara Regional State is an example where the reservoir was filled up with silt before the irrigation activities of the command area commenced. Since watershed conservation activities focus on a long-term perspective, they conflict directly with a farmer's or the investor's very short planning horizons. Action from different actors that brings together various inputs from a range of stakeholders is necessary because watershed management issues require varied technical and firm financial consideration. The collaboration has to be developed from the earliest stages of project identification instead of starting watershed management activities after the construction of the dam is completed (like the case of Tekeze Dam in Northern Ethiopia).

Other key constraints that have prevented successful development of commercial agriculture in Ethiopia are poor access to agriculture-supporting infrastructure (grid-connected electricity and all-weather feeder roads), lack of suitable finance, and insufficiently experienced agricultural entrepreneurs and senior managers.

In summary, Gebrehiwot (2015) identified irrigation development challenges as follows:

- inadequate awareness of irrigation water management as in irrigation scheduling techniques, water-saving irrigation technologies, water measurement techniques, operation and maintenance of irrigation facilities;
- inadequate knowledge on improved and diversified irrigation agronomic practices;
- shortage of basic technical knowledge on irrigation pumps, drip irrigation system, sprinkler irrigations, surface and spate irrigation methods;
- scheme-based approach rather than area/catchments-based approach for the development of small-scale irrigation schemes;
- inadequate baseline data and information on the development of water resources;
- lack of experience in design, construction and supervision of quality irrigation projects;
- low productivity of existing irrigation schemes;
- inadequate community involvement and consultation in scheme planning, construction and implementation of irrigation development; and
- poor economic background of users for irrigation infrastructure development, to access irrigation technologies and agricultural inputs, where the price increment is not affordable to farmers.

References

Abeyou, W.W., Jaehak, J., Dile, Y.T., Osorio, J., Schmitte, P., Gerik, T., Srinivasan, R. and Clark, N. (2018) Assessing potential land suitable for surface irrigation using groundwater in Ethiopia. *Applied Geography* 85, 1–13.

Alemayehu, T. (2006) *Groundwater Occurrence in Ethiopia*. Addis Ababa University Press, Addis Ababa, Ethiopia, pp. 1–105.

Awulachew, S.B., Yilma, A.D., Loulseged, M., Loiskandl, W.I., Ayana, M. and Alamirew, T. (2005) Water Resources and Irrigation Development in Ethiopia. IWMI Working Paper 123, Addis Ababa, Ethiopia.

Awulachew, S.B., Yilma, A.D., Loulseged, M., Loiskandl, W., Ayana, M. and Alamirew, T. (2007) Water Resources and Irrigation Development in Ethiopia. IWMI Working Paper 123. Colombo, Sri Lanka: International Water Management Institute, 78 pp.

Ayenew, T., Demlie, M. and Wohnlich, S. (2008) Hydrogeological framework and occurrence of groundwater in the Ethiopian aquifers. *Journal of African Earth Sciences* 52(3), 97–113.

Berhanu, B., Seleshi, Y., Melesse, A. (2014) Surface water and groundwater resources of Ethiopia: Potentials and challenges of water resources development. In: Melesse, A., Abtew, W., Setegn, S. (eds) *Nile River Basin*. Springer, Berlin, Germany. https://doi.org/10.1007/978-3-319-02720-3_6

Beyene, T., Berlie, A.B. and Ferede, M.B. (2021) Effects of the current land tenure on augmenting household farmland access in South East Ethiopia. *Humanities and Social Sciences Communications* 8, 35.

Dereje, H., Abiy, A.Z. and Setegn, S.G. (2020) Assessment of surface irrigation potential and land suitability analysis for irrigation in the Lake Tana basin, Ethiopia. *Hydrology* 7(4), 94.

FAO (Food and Agriculture Organization of the United Nations) (2022a) Statistical Databases and Data-Sets of the Food and Agriculture Organization of the United Nations. Available at: http://www.fao.org/faostat/en/#data (accessed 30 April 2022).

FAO (2022b) Ethiopia Country Profile. Available at: https://www.fao.org/countryprofiles/en/ (accessed 30 April 2022).

Gebrehiwot, H.G. (2015) Irrigation in Ethiopia – A Review. *Journal of Environment and Earth Science* 5(15), 141.

GIRDC (Generation Integrated Rural Development Consultant) (2018) Irrigation in Ethiopia: Challenges and opportunities. Government Irrigation Rehabilitation and Development Center, Ministry of Water, Irrigation and Energy, Addis Ababa, Ethiopia.

IAG (International Association of Geodesy) (2019) Beyond 100: The Next Century in Geodesy. *Proceedings of the IAG General Assembly*, Montreal, Canada.

IWMI (International Water Management Institute) (2010) *Banking on Groundwater in Times of Change*. Colombo, Sri Lanka: IWMI, 7p. (IWMI Water Policy Brief 032) DOI:10.5337/2010.240.

Makombe, G., Namara, R., Hagos, F., Awulachew, S., Ayana, M. and Bossio, D. (2011) A Comparative Analysis of the Technical Efficiency of Rain-fed and Smallholder Irrigation in Ethiopia. IWMI Working Papers.

Mekonen, Y., Woldeamanuel, A. and Mekonnen, A. (2020) Irrigation in Ethiopia: Status, challenges and opportunities. *Water Policy* 22(4), 625–644.

MOA and ATA (Ministry of Agriculture and Agricultural Transformation Agency) (2016) *The Agricultural Transformation Agency: A Strategy for Ethiopia's Agricultural Development*. Addis Ababa, Ethiopia: MOA and ATA.

MOWIE (Ministry of Water, Irrigation and Energy) (1995/1997) *National Irrigation Master Plan.* MOWIE, Addis Ababa, Ethiopia.

MOWIE (2019) *Water Sector Development Strategy 2019–2025*. MOWIE, Addis Ababa, Ethiopia.

MOWIE (2020) *National Water Resources Plan 2020–2030*. MOWIE, Addis Ababa, Ethiopia.

MOWR (Ministry of Water Resources) (2007) Managing Water Resources to Maximize Sustainable Growth. Country Water Resources Assistance Strategy, a World Bank Water Resources Assistance Strategy for Ethiopia. MOWR, Addis Ababa, Ethiopia, 91 p.

PRI (Policy Research Institute) (2018) Irrigation Sector Roadmap for Ethiopia: Enhancing Productivity and Sustainability to Accelerate Shared Prosperity. PRI, Addis Ababa, Ethiopia.

Singirankabo, U.A. and Ertsen, M.W. (2020) *Relations between Land Tenure Security and Agricultural Productivity: Exploring the Effect of Land Registration*. Water Management Department, Delft University of Technology, Delft, The Netherlands.

Tenaw, S., Islam, K.M.Z. and Parviainen, T. (2009) *Effects of Land Tenure and Property Rights on Agricultural Productivity in Ethiopia, Namibia and Bangladesh*. University of Helsinki, Department of Economics and Management, Helsinki, Finland.

Yimere, A. and Assefa, E. (2022) Current and future irrigation water requirement and potential in the Abbay River basin, Ethiopia. *Air, Soil and Water Research* 15, 1–15. DOI: 10.1177/11786221221097929.

22 Innovations in Agricultural Technology Transfer: Actors and Legacies

Tsedeke Abate[1]* and Endeshaw Habte[2]
[1]*Homegrown Vision, Addis Ababa, Ethiopia;* [2]*Ethiopian Institute of Agricultural Research, Addis Ababa, Ethiopia*

Abstract

This chapter offers a critical assessment of the status of agricultural extension in Ethiopia. The number of smallholders and crop area receiving extension packages has at least doubled between 2006 and 2019. This country has one of the densest development agents to farmer ratio of 1:230 in the world. The overall average proportion of smallholders and crop area under extension increased from 7% to 26% and from 13% to 36%, respectively, between 2006 and 2019. The biggest beneficiaries of the extension service were farmers engaged in cereal production where the proportion of farmers receiving extension packages increased from 19% in the base year to 54%. The relative success with the extension system in Ethiopia could be attributed to its homegrown approach to agricultural research and development.

Introduction

Efforts to modernize agricultural technology transfer in Ethiopia go back to the establishment of the Imperial Ethiopian College of Agriculture and Mechanical Arts (IECAMA) at Haramaya in 1952 (OSU, 1969). IECAMA was initially mandated to carry out teaching, research and extension – based on the Land Grant University system of the United States – in partnership with the Oklahoma State University. The agricultural extension system in this country has undergone a series of restructuring (Berhanu, 2009) since that time, under different political systems following different models and approaches.

Kassa (2003) made the first critical review of agricultural extension system in Ethiopia where he enumerated the constraints associated with the system, including the lack of inclusiveness, inadequate training and skills of staff, and disregard for traditional knowledge of farmers (among others). Similar reviews are also available from Abate (2007) and Kassa (2008).

Mogues *et al.* (2009) provide analysis of agricultural extension delivery, with an emphasis on gender. Their study showed that agricultural extension services were generally accessible, but there are gender differences with women receiving less attention than men. For example, their study showed that 27% of men received visits from development agents (DAs) compared to 20% for women (there is no indication whether there was significant difference). They reported that the disparity varied from region to region – i.e. some regions were better than others – and suggested that targeting women groups may be an alternative approach to reach women with extension services.

* Corresponding author: tsedeke@homegrownvision.org

DOI: 10.1079/9781800626386.0022

Davis *et al.* (2010) assessed the strengths and constraints of the public extension system and provided suggestions on what they termed 'best fit' solutions and their scale-up opportunities. Some of the constraints they pointed out included frequent restructuring, a top-down and supply-driven approach and too much emphasis on cereals – with cash crops, natural resources and livestock given low priority.

Berhanu and Poulton (2014) point out that investment in agricultural extension in Ethiopia is the largest and fastest growing on the continent but is driven by the 'twin imperatives' of economic growth and political control. Alemu *et al.* (2015) provide an overview of technical vocational education and training (TVET) and agriculture TVET (ATVET) in Ethiopia.

The recent government extension strategy document (MOANR, 2017) states that, with approximately 21 DAs per 10,000 smallholders, Ethiopia has one of the densest agricultural extension systems in the world. However, the system suffers from '... poor implementation and insufficient strategic interventions'. It also lacks effective linkage with research.

Berhane *et al.* (2018) provide a comprehensive and coherent review of the Ethiopian agricultural extension system. Leta (2018) reviewed the Ethiopian extension system as part of his PhD dissertation. His findings reinforced many of the challenges pointed out by the preceding authors above. He has also pointed out that the participatory extension system has been adopted by the Ministry of Agriculture since 2010.

This review presents an assessment of the status of agricultural extension in Ethiopia and identifies the work to be done in the near future.

Approach

The main source of data for agricultural extension packages delivered to smallholder farmers in Ethiopia is the agricultural sample survey of the Central Statistical Agency (CSA; www.csa.gov.et) of the government of Ethiopia. The data covering the period between 2006/07 and 2019/20 was analysed to determine the number and proportions of smallholders receiving extension packages and the total cultivated area covered. The rest of the information is from the literature and from our personal experiences.

Results

The results of this study are presented in two parts. The first part deals with the various interventions that have been undertaken over the years to improve agricultural technology transfer. The second deals with the actual technology delivery aspect.

Interventions

As indicated in the introduction, the agricultural extension system in Ethiopia has undergone a series of restructuring over the past six decades. The country has seen three regime changes including the imperial Ethiopian government, the military-socialist government and the Ethiopian People's Revolutionary Democratic Front. Each of the regimes followed different approaches and extension models. Table 22.1 summarizes examples of the major agricultural extension interventions in Ethiopia.

The initial agricultural extension efforts started with the establishment of IEACAMA that followed the Land Grant University system through Oklahoma State University partnership. The mandate of agricultural extension was later moved to the Ministry of Agriculture.

The first major intervention during this period was the introduction of 'comprehensive package projects'. The reasoning being that results obtained from the project zones could be replicated and scaled in other areas across the country. The Chilalo Agricultural Development Unit (CADU), later renamed as Arsi Agricultural Development Unit (ARDU) was one such high-profile project supported by the Swedish International Development Agency (SIDA) that was carried out between 1967 and 1986.

The main target of CADU/ARDU was to improve crop productivity using improved varieties, inputs (mainly mineral fertilizers) and crop management. Overall, this intervention can be declared as successful in terms of improving crop productivity and creating awareness about modern agricultural technologies to farmers. Its initiative has left behind a sustainable legacy in the areas where it was implemented. The major criticism about this initiative was that it benefited wealthier farmers rather than the

Table 22.1. Examples of major interventions and innovations in agricultural research and extension in Ethiopia.

Intervention/innovation	Location	Main target	Year/period	Source
Establishment of IECAMA	Haramaya	Education-Research-Extension	1952	OSU (1969)
Comprehensive package projects (CADU/ARDU), WADU, ADDP, TAHADU, SADP, HADP)	Chilalo/Arsi, Wolaita Sodo, Ad'a, Tigray, Hawassa, Humera	Increasing crop productivity	1967–1986	Kassa (2003)
Minimum package projects (MPP-I, MPP-II)	Countrywide	Increasing crop productivity	1971–1974; 1981–1985	Kassa (2003)
Establishment of Extension Programs Implementation Department (EPID)	MOA HQ	Administering MPPs	1971–1979	Kassa (2003)
Research Extension Liaison Committees (RELCs)	Countrywide	Coordinating research and extension stakeholders	1986–1991	Kassa (2003)
Peasant Agriculture Development Extension Program (PADEP)	Countrywide	Increasing crop productivity	1989–1991	Kassa (2003)
Establishment of ATVETs	Countrywide	Training personnel	1991–present	Mulugetta and Mekonnen (2016)
Sasakawa Africa Association (SAA)	Countrywide	Improving crop productivity	1993–present	https://www.ata.gov.et
AMAREW Project	Amhara Regional State	Increasing rural household income, thereby improving food security	2002–2007	Gebrekidan *et al.* (2007)
Sida-Amhara Development Program (SARDP)	Amhara Regional State	Improving food security and reducing poverty	1997–2010	SARDP (2010)
Participatory demonstration and training extension system (PADETES)	Countrywide	Improved productivity, capacity, linkages, grain storage	1993–1999	Kassa (2003)
Ejj-Le-Ejj (hand-in-hand) – EIAR initiative	Countrywide	Scaling technologies via working with stakeholders (value chain)	2004	Abate (2006)
Agricultural Transformation Agency (ATA)	Countrywide	Promoting agricultural sector transformation	2010–2022	https://www.ata.gov.et
Participatory extension systems (PES)	Countrywide	Increasing crop productivity	2010	Leta (2018)

ADDP = Ada District Development Project; AMAREW = Amhara Micro-enterprise development, Agricultural Research, Extension and Watershed management; ATVET = agriculture technical vocational education and training; CADU/ARDU = Chilalo Agricultural Development Unit/ Arsi Agricultural Development Unit; EIAR = Ethiopian Institute of Agricultural Research; IECAMA = Imperial Ethiopian College of Agriculture and Mechanical Arts; HADP = Humera Agricultural Development Project; MOA HQ = Ministry of Agriculture headquarters; SADP = Southern Agricultural Development Project; TAHADU = Tach Adiabo and Hedekti Agricultural Development Unit; WADU = Wolaita Agricultural Development Unit.

smallholders it was intended to help, but that had to do more with the land holding system and the government development policy and priorities than with the programme itself. It was also said that the project was expensive in terms of both finance and human resources (Kassa, 2003).

The rest of the comprehensive package projects (Wolaita Agricultural Development Unit, Ada District Development Project, Tach Adiabo and Hedekti Agricultural Development Unit, Southern Agricultural Development Project and Humera Agricultural Development Project) did not have much success to report, even though the Wolaita Agricultural Development Unit was perceived to have fared relatively better than the remaining initiatives. The minimum package projects also did not fare well in terms of benefitting the less well-off farmers.

The participatory demonstration and training extension system encouraged active participation by farmers. This and many of the subsequent interventions put much focus on cereals without paying much attention to livestock, natural resources management and cash crops in general. Kelemework and Kassa (2006) observed that more than 80% of cereals-related package activities were on maize, wheat and *teff*. According to these authors, the research–extension–farmer linkage failed due to several factors, including frequent restructuring, poor farmer representation, staff turnover, limited budget, lack of commitment by staff and rivalry among partner institutions.

The Sasakawa Africa Association has been involved in Ethiopian agriculture extension since 1993. Its main initial focus was improving crop productivity, but it included postharvest technologies since 2010. It implemented more than 11 different projects funded by different international development communities between 2010 and 2018. The initiative focuses on ten crops: maize, quality protein maize, wheat, barley, *teff*, sorghum, millet, potato, faba bean and sesame.

The introduction of the '*Ejj-Le-Ejj*' (hand-in-hand) project tried to break the perennial problem of lack of linkage among stakeholders. Its main target was to scale-up newly developed agricultural technologies using a value chain approach (Abate, 2006). The Ethiopian Institute of Agricultural Research championed this approach starting in 2004 and brought together all stakeholders to achieve a common goal. The systems approach followed under this initiative has led to accelerated adoption of new and high-yielding or low-risk varieties (Abate *et al.*, 2011), and eventually resulted in improved export earnings from grain legumes such as the common bean, chickpea and lentil. Its legacy was extended to other crops (Abate *et al.*, 2015) and it continues to be an important factor for improved agricultural transformation in Ethiopia.

Another major milestone in the attempt to improve agricultural productivity was the introduction of the Agricultural Transformation Agency (ATA; https://www.ata.gov.et). The primary aim of the ATA was to promote agricultural sector transformation by supporting existing structures of government, private sector and other non-governmental partners to address systemic bottlenecks in delivering on a priority national agenda for achieving growth and food security. The ATA was challenged from two fronts. First, there was the problem of governance. Various experts insisted that the leadership should be someone with rich experience in and knowledge of agricultural research and development in Ethiopia, but this was not to be as the appointment was more political than technical. Second, there appeared to be a lack of clarity of purpose. Many Ethiopian experts (including the senior author of this chapter) believe that the agency's main function should be to build capacity and bargaining power for the national system rather than doing experimentation on its own. This was a huge, missed opportunity and we hope that policy makers have learned some lessons from it.

Delivery

Farmers reached

The total number of smallholders receiving extension packages grew at a rapid rate during the 14 years of the study period. For example, the number of smallholders benefitting was estimated to be ~9.8 million in 2019 compared to 2.2 million in 2006 (Fig. 22.1A). Cereals accounted for by far the largest number of smallholders receiving extension packages. These were followed by grain legumes, roots and tubers, vegetables, oilseeds and fruit crops. All showed varying levels of rapid growth over the study period.

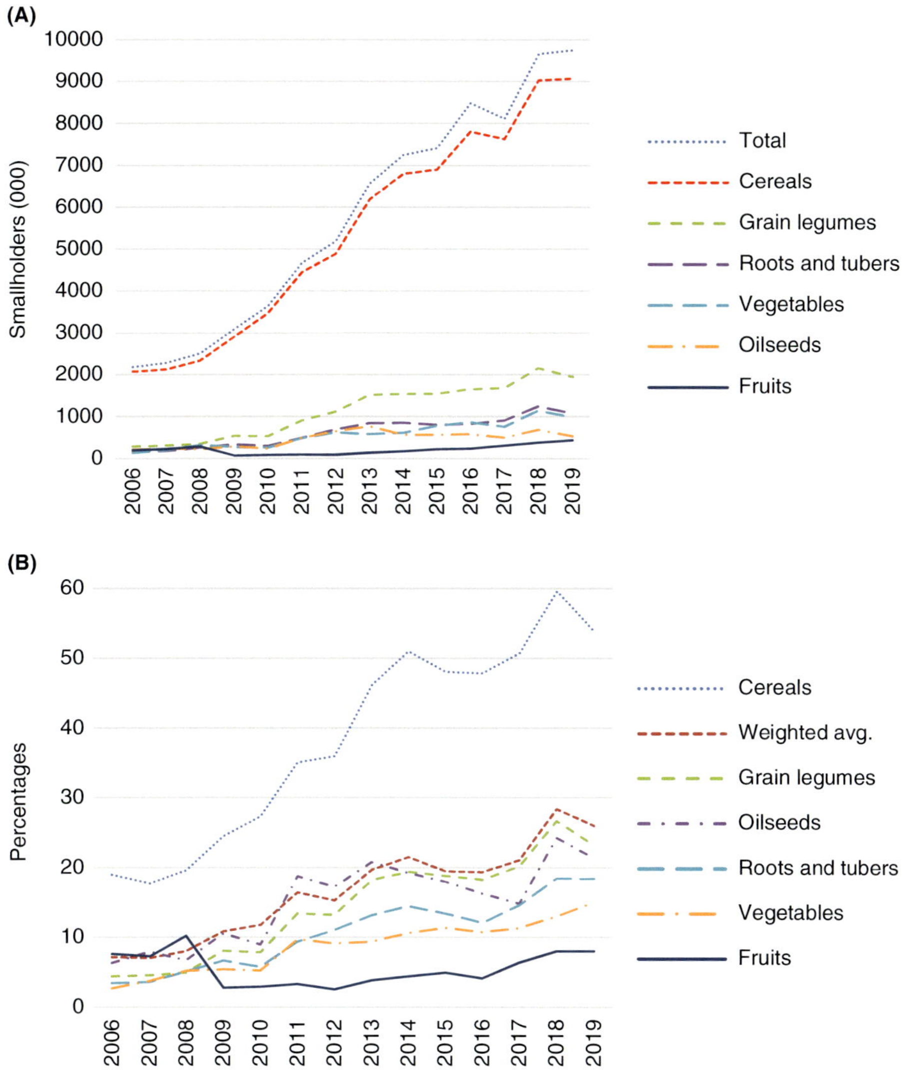

Fig. 22.1. (A) Number of smallholders receiving extension packages for different crop groups grown in the *meher* season in Ethiopia. **(B)** Proportions of smallholders receiving extension packages for different crop groups grown during the *meher* season in Ethiopia. Source: constructed by the authors from CSA (2007–2020).

The proportion of smallholders receiving extension packages also showed rapid growth (Fig. 22.1B). The numbers varied from crop group to crop group and within each group.

The overall average proportion of farmers receiving extension packages in 2019 was 26%, compared to 7% in 2006. An average of 54% of smallholders growing cereals received extension packages, the highest among all crop groups. These were followed by grain legumes (23%), oilseeds (21%), roots and tubers (does not include *enset*) (18%), vegetables (15%) and fruit crops (8%). All crop groups showed rapid growth except fruit crops, which remained virtually constant over the study period (Fig. 22.1B).

Perennials (including *ch'at*, coffee and *gesho*) also showed varying levels of growth over the study period. The proportion of *ch'at (Catha edulis)* growers increased from 4% in 2006 to 18% in 2019, whereas that for *gesho* (*Rhamnus prinoides*) (common [English] name is buckthorn) and coffee increased from 2% to 16% and 2% to 9%, respectively (not shown here).

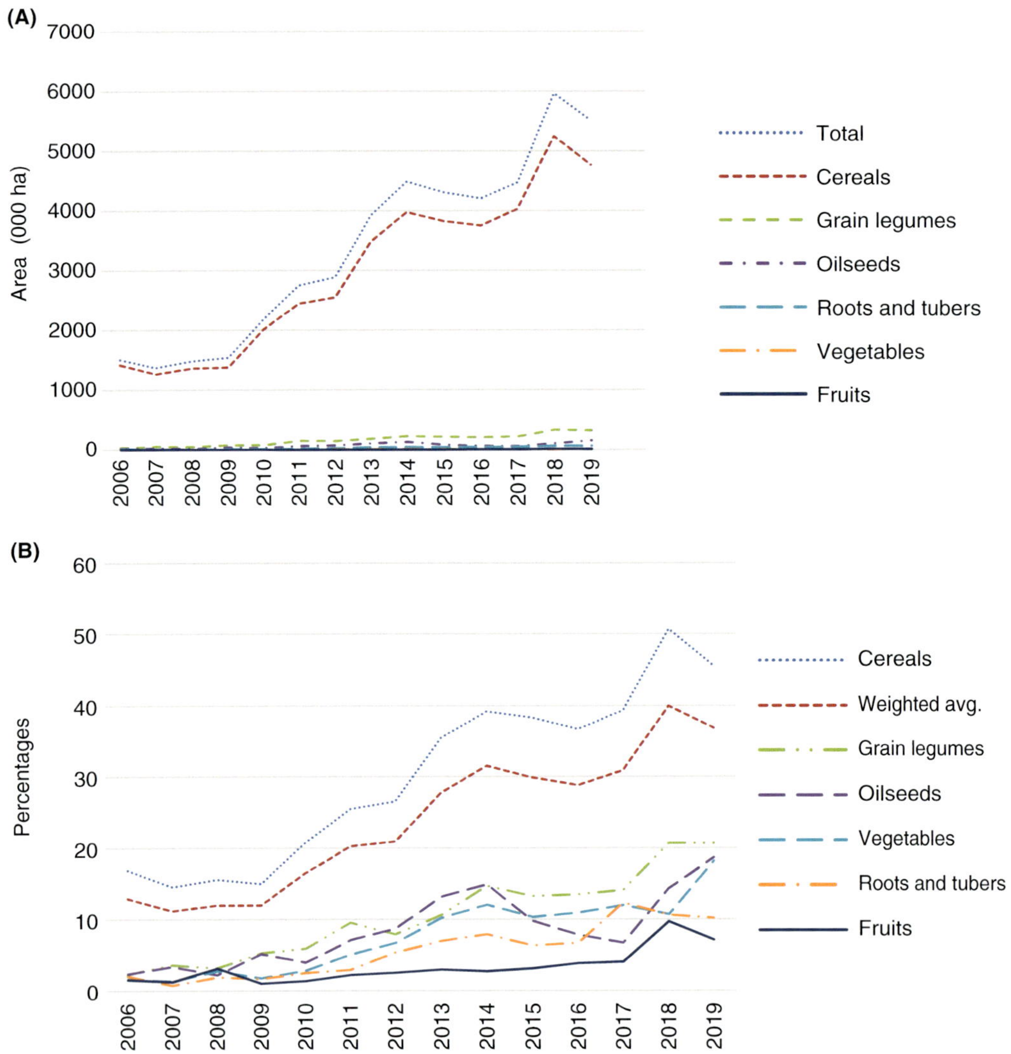

Fig. 22.2. (A) Smallholder area for different crop groups covered by extension packages during the *meher* season in Ethiopia. **(B)** Proportions of area of different smallholder crop groups covered under extension packages during the *meher* season. Source: constructed by the authors from CSA (2007–2020).

Overall, 25% of total smallholder farmers in Ethiopia received extension packages. However, this varied considerably from crop to crop. For example, 55%, 50%, 47%, 45% and 42%, respectively, of wheat, maize, rice, *teff* and millet farmers were provided with extension packages in the 2017–19 period, whereas only 3% of smallholders growing yam, 4% each of avocado and sugarcane, and 5% each of taro and papaya received extension packages during the same period (Annex Fig. 22.1).

Area covered

Figure 22.2A presents the cultivated area covered under extension package in Ethiopia. About 5.5 million ha of the total cultivated area was under extension packages in 2019. Nearly 87% of this was cereals, with grain legumes, oilseeds, and roots and tubers covering about 6%, 3% and 1%, respectively. Vegetables, fruits and others accounted for less than 1% each.

The area coverage by extension package showed rapid growth during the study period. Overall, it increased from about 1.6 million ha in the base year to 5.5 million ha in 2019, a growth rate of 12.88% per year in the 14 years between 2006 and 2019. All crop groups and individual crops witnessed varying levels of rapid growth.

Figure 22.2B shows the proportions of cultivated crop areas covered by extension packages between 2006 and 2019. Overall, 36% of the total cultivated area was under extension package in 2019, compared to 17% in 2006 – a 2.8-fold increase. In a similar fashion, cereals area coverage was 46% in 2019 versus 17% in the base year. The area coverage by extension packages for grain legumes, oilseeds, roots and tubers, vegetables and fruits increased from 2% to 18%, 2% to 13%, 2% to 11%, 2% to 14% and 2% to 7%, respectively (Fig. 22.2B).

In general, the proportion of smallholder crop area receiving extension packages varied considerably from crop group to crop group and within individual crops in each crop group. We illustrate this with the cereals example, shown in Annex Table 22.1. Maize, wheat, millet and *teff* had the highest proportion of area covered under extension package.

The coverage for the above four crops in 2006 was 22%, 25%, 23% and 18%, respectively, while it reached 63%, 57%, 48% and 43%, respectively, in 2019. The coverage for rice was also high but it accounts only for small proportions of the total cereals area in Ethiopia. With the exception of emmer wheat, the area under extension package more than doubled between the base and final years.

Annex Table 22.2 depicts extension coverage for miscellaneous crops during the 14-year period included in this study. Just as in cereals, the proportion of area coverage by extension package for dry pepper, garlic, fenugreek, *gesho* and *ch'at* also more than doubled in the 14 years. The coverage for coffee also doubled, but it was still extremely low. Data for sugarcane were missing for several years.

In summary, the CSA data for 2007–2019 show that 49 crops received varying levels of extension packages (Annex Fig. 22.1B). The overall average was 36%. There were huge differences among the different crops. These ranged from 64%, 59%, 50%, 49% and 49%, respectively, for maize, wheat, millet, potato and rice, to 3% for avocado and 4% each for yam, taro, coffee and *noug* (Annex Fig 22.1B).

Discussion and Conclusions

This chapter presents the analysis of nationally available data from the CSA on the number of smallholders reached and the total area covered under extension packages for all crops in the country during 14 *meher* seasons, 2006/07 to 2019/20. Our interpretation of the outcome on extension efforts by the public sector are based on these analyses and our own personal experiences.

The agricultural extension system in Ethiopia has undergone frequent restructuring and followed different approaches and models, depending on the nature of the political system of the day: from the imperial to quasi-socialist/military, to essentially dictatorial regimes of the last five decades or so. This involved several projects, innovations and actors at different periods. The common denominator for all the approaches and models tried so far is the fact that they relied more on copying models from abroad (e.g. training and visit from India) rather than creating a homegrown approach that would have taken indigenous knowledge into consideration (Abate *et al.*, 2006; Krishnan and Patnam, 2012).

All systems invariably favoured cereals over other traditional staples such as *enset* and oilseeds such as Ethiopian mustard and safflower, and livestock. At present, there are no decent statistics available for *enset* (a crop that feeds close to a third of Ethiopia's population and has great potential to expand further both within the country and beyond given the right kind of priority) let alone extension packages.

Even though the government's priority is cereals, it appeared that the priorities of the smallholders are not necessarily always aligned with government policy. For example, smallholders invest more in *ch'at* than coffee; proportions of area covered by extension packages for potato are not very far from those for major cereals; the level of farmers' investment on rapeseed, tomato, onion, dry pepper and garlic is comparable to that of *teff*.

According to the recent government extension strategy document (MOANR, 2017), the system suffers from '... poor implementation and insufficient strategic interventions' and lacks effective linkage with research, even though Ethiopia has achieved significant increases in its extension staff (NEPAD, 2013; Berhane *et al.*, 2018) and has one of the densest (a DA to farmer ratio of 1:230) agricultural extension systems in the world.

Ethiopia missed a big opportunity to transform its technology delivery system and hence its agricultural productivity when it introduced the very high-profile ATA in 2011. The ATA failed to deliver what it was intended to deliver primarily because its implementation followed a top-down approach from the very beginning. Its governance was not merit- and knowledge-based and it ended up doing field experimentation, whereas its main focus should have been enabling the national programme staff and building bargaining power for them. We hope policy makers have learned good lessons from this when planning future strategies and setting priorities.

References

Abate, H. (2007) *Review of Extension Systems Applied in Ethiopia with Special Emphasis to the Participatory Demonstration and Training Extension System*. FAO, Rome.

Abate, T. (ed.) (2006) *Successes with Value Chain: Proceedings of Scaling up and Scaling out Agricultural Technologies in Ethiopia: An International Conference, 9–11 May 2006*. EIAR, Addis Ababa, 275 pp.

Abate, T., Assefa, S., Hagmann, J., Ahmed, S., Kumsa, T. and Stroud, A. (2006) Agricultural innovation systems and partnership in practice: Value chain successes in Ethiopia. Presented at Innovation Africa Symposium, 20–23 November 2006. Kampala, Uganda.

Abate, T., Shiferaw, B., Gebeyehu, S., Amsalu, B., Negash, K. *et al.* (2011) A systems and partnership approach to agricultural research for development: Lessons from Ethiopia. *Outlook on Agriculture* 40(3), 213–220.

Abate, T., Shiferaw, B., Menkir, A., Wegary, D., Kebede, Y. *et al.* (2015) Factors that transformed maize productivity in Ethiopia. *Food Security* 7, 965–981. DOI 10.1007/s12571-015-0488-z.

Alemu, D., Cook, S. and Gubo, Q. (2015) Chinese agricultural expertise support in Ethiopia: Approaches, motives and perspectives. *Future Agricultures*. China and Brazil in Africa Agriculture (CBAA) Project work stream. Working Paper 114. Future Agricultures Consortium. Brighton, UK.

Berhane, G., Ragasa, C., Abate, G.T. and Assefa, T.W. (2018) The state of agricultural extension services in Ethiopia and their contribution to agricultural productivity. Strategy Support Program Working Paper 118. IFPRI/EDRI Addis Ababa, Ethiopia.

Berhanu, A. (2009) The Ethiopian extension and the farmer: A view from the farm. In: Ege, S., Aspen, H., Teferra, B. and Bekele, S. (eds.) *Proceedings of the 16th International Conference of Ethiopian Studies,* Trondheim, Norway.

Berhanu, K. and Poulton, C. (2014) The political economy of agricultural extension policy in Ethiopia: Economic growth and political control. *Development Policy Review* 32(S2), s197–s213.

CSA (2007) Agricultural Sample Survey 2006/2007 (1999 E.C.), Report on Farm Management Practices (Private Peasant Holdings, *Meher* Season) (September–December), Volume III, Statistical Bulletin. CSA, Addis Ababa, Ethiopia.

CSA (2008) Agricultural Sample Survey 2007/2008 (2000 E.C.), Report on Farm Management Practices (Private Peasant Holdings, *Meher* Season) (September–December), Volume III, Statistical Bulletin. CSA, Addis Ababa, Ethiopia.

CSA (2009) Agricultural Sample Survey 2008/2009 (2001 E.C.), Report on Farm Management Practices (Private Peasant Holdings, *Meher* Season) (September–December), Volume III, Statistical Bulletin. CSA, Addis Ababa, Ethiopia.

CSA (2010) Agricultural Sample Survey 2009/2010 (2002 E.C.), Report on Farm Management Practices (Private Peasant Holdings, *Meher* Season) (September–December), Volume III, Statistical Bulletin. CSA, Addis Ababa, Ethiopia.

CSA (2011) Agricultural Sample Survey 2010/2011 (2003 E.C.), Report on Farm Management Practices (Private Peasant Holdings, *Meher* Season) (September–December), Volume III, Statistical Bulletin. CSA, Addis Ababa, Ethiopia.

CSA (2012) Agricultural Sample Survey 2011/2012 (2004 E.C.), Report on Farm Management Practices (Private Peasant Holdings, *Meher* Season) (September–December), Volume III, Statistical Bulletin. CSA, Addis Ababa, Ethiopia.

CSA (2013) Agricultural Sample Survey 2012/2013 (2005 E.C.), Report on Farm Management Practices (Private Peasant Holdings, *Meher* Season) (September–December), Volume III, Statistical Bulletin. CSA, Addis Ababa, Ethiopia.

CSA (2014) Agricultural Sample Survey 2013/2014 (2006 E.C.), Report on Farm Management Practices (Private Peasant Holdings, *Meher* Season) (September–December), Volume III, Statistical Bulletin. CSA, Addis Ababa, Ethiopia.

CSA (2015) Agricultural Sample Survey 2014/2015 (2007 E.C.), Report on Farm Management Practices (Private Peasant Holdings, *Meher* Season) (September–December), Volume III, Statistical Bulletin. CSA, Addis Ababa, Ethiopia.

CSA (2016) Agricultural Sample Survey 2015/2016 (2008 E.C.), Report on Farm Management Practices (Private Peasant Holdings, *Meher* Season) (September–December), Volume III, Statistical Bulletin. CSA, Addis Ababa, Ethiopia.

CSA (2017) Agricultural Sample Survey 2016/17 (2009 E.C.), Report on Farm Management Practices (Private Peasant Holdings, *Meher* Season) (September–December), Volume III, Statistical Bulletin. CSA, Addis Ababa, Ethiopia.

CSA (2018) Agricultural Sample Survey 2017/18 (2010 E.C.), Report on Farm Management Practices (Private Peasant Holdings, *Meher* Season) (September–December), Volume III, Statistical Bulletin. CSA, Addis Ababa, Ethiopia.

CSA (2019) Agricultural Sample Survey 2018/19 (2011 E.C.), Report on Farm Management Practices (Private Peasant Holdings, *Meher* Season) (September–December), Volume III, Statistical Bulletin. CSA, Addis Ababa, Ethiopia.

CSA (2020) Agricultural Sample Survey 2019/20 (2012 E.C.), Report on Farm Management Practices (Private Peasant Holdings, *Meher* Season) (September–December), Volume III, Statistical Bulletin. CSA, Addis Ababa, Ethiopia.

Davis, K., Swanson, B., Amudavi, D., Mekonnen, D.A., Flohrs, A. *et al.* (2010) In-Depth Assessment of the Public Agricultural Extension System of Ethiopia and Recommendations for Improvement. IFPRI Discussion Paper 01041 December 2010.

Gebrekidan, B., Alemayehu, N., Ashine, Y., Bayafers, G. and Kassahun, S. (2007) AMAREW Final Report. Amhara Microenterprise Development, Agricultural Research, Extension and Watershed Management Project in Ethiopia. July 2000–December 2007, Bahir Dar, Ethiopia.

Kassa, B. (2003) Agricultural extension in Ethiopia: the case of participatory demonstration and training extension system. *Journal of Social Development in Africa* 18(1), 49–83.

Kassa, H. (2008) Agricultural extension in Ethiopia: Historical evolution, relevant policies and challenges. In: Assefa, T. (ed.) *Digest of Ethiopia's National Policies, Strategies and Programs*. Eclipse Printers, Addis Ababa.

Kelemework, F. and Kassa, H. (2006) Assessment of the current extension system of Ethiopia: A closer look at planning and implementation. Ethiopian Economic Association/ Ethiopian Economic Policy Research Institute (EEA/EEPRI). Issue Paper Series No. 2/2006.

Krishnan, P. and Patnam, M. (2012) Neighbours and extension agents in Ethiopia: Who matters more for technology diffusion? Working Paper, International Development Center.

Leta, G.D. (2018) The Ethiopian Agricultural Extension System and Its Role as a Development Actor: Cases from Southwestern Ethiopia. PhD dissertation. Friedrich-Wilhelms-Universität, Bonn.

MOANR (2017) *Agricultural Extension Strategy of Ethiopia*. Ministry of Agriculture and Natural Resources. Addis Ababa, Ethiopia.

Mogues, T., Cohen, M.J., Regina Birner, R., Lemma, M., Randriamamonjy, J., Tadesse, F. and Paulos, Z. (2009) Agricultural Extension in Ethiopia through a Gender and Governance Lens. Development Strategy and Governance Division, International Food Policy Research Institute – Ethiopia Strategy Support Program 2, Ethiopia. Discussion Paper No. ESSP2 007.

Mulugetta, M. and Mekonnen, T. (2016) Implementation of technical and vocational training strategy in agricultural sector in Ethiopia: Practices, challenges and the way forward. *Ethiopian Journal of the Social Sciences and Humanities* 12(2), 57–80.

NEPAD (2013) Review of agricultural technical vocational education and training (ATVET) in Africa – Best practices from Benin, Ethiopia, Namibia and Sierra Leone. Comprehensive Africa Agriculture Development Programme, Addis Ababa, Ethiopia.

OSU (1969) Oklahoma State University in Ethiopia, Terminal Report. January 1969, Stillwater, Oklahoma, USA.

SARDP (Sida-Amhara Rural Development Programme) (2010) *Building Ethiopia's Future: The Sida-Amhara Rural Development Programme.* Sida-Amhara Rural Development Programme. Bahir Dar, Ethiopia.

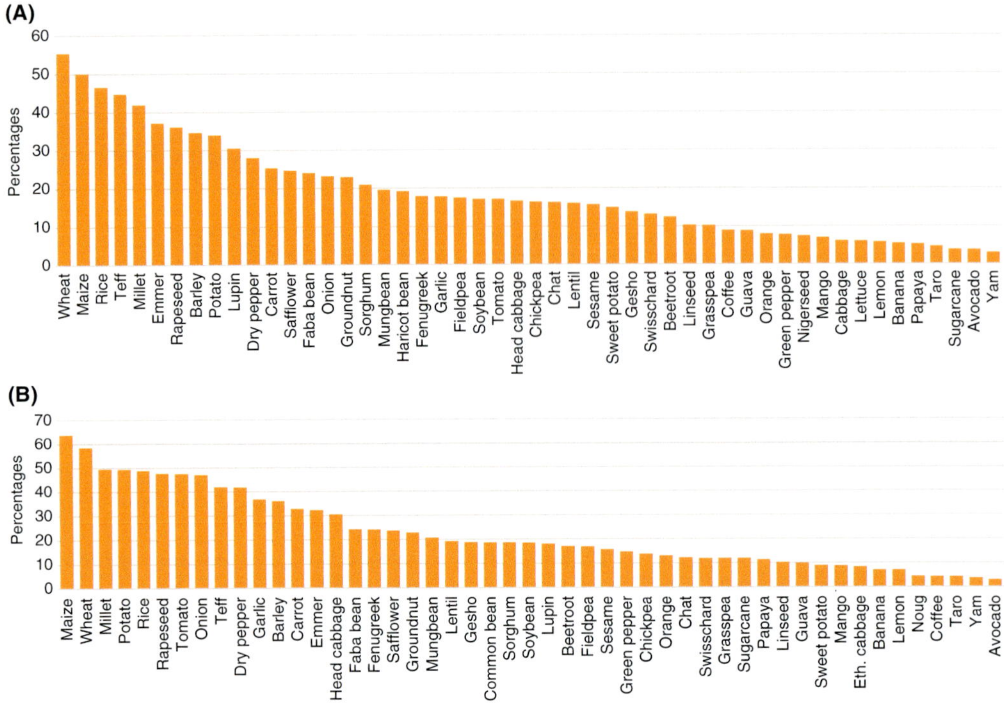

Annex Fig. 22.1. **(A)** Proportions of smallholder farmers receiving extension packages for different crops grown during the *meher* season in Ethiopia, 2007–2020. **(B)** Proportions of area covered under extension packages in different crops grown during the *meher* season in Ethiopia, 2007–2020. Source: constructed by the authors from CSA (2007–2020).

Annex Table 22.1. Per cent smallholder area covered under extension package of different cereal staples grown during the *meher* season in Ethiopia. Source: constructed by the authors from CSA (2007–2020).

Year	Maize	Wheat	Millet	*Teff*	Barley	Sorghum	Emmer	Rice
2006	22	25	23	18	12	2	16	—
2007	21	22	23	14	11	1	13	2
2008	23	21	29	16	9	3	10	2
2009	16	18	26	19	11	5	6	—
2010	31	27	32	20	16	6	20	—
2011	36	35	37	26	19	7	22	—
2012	42	31	36	24	19	10	25	14
2013	52	43	44	36	22	13	38	36
2014	57	49	47	41	25	15	32	—
2015	54	49	46	38	26	15	36	27
2016	58	49	37	34	25	13	28	—
2017	62	51	40	36	26	16	48	30
2018	67	68	61	47	45	20	30	52
2019	63	57	48	43	38	20	19	64

Dashes indicate data not available.

Annex Table 22.2. Per cent smallholder area covered under extension package of miscellaneous crops grown during the *meher* season in Ethiopia. Source: constructed by the authors from CSA (2007–2020).

Year	Dry pepper	Garlic	Fenugreek	*Gesho*	Chat	Coffee	Sugarcane
2006	18	—	4	4	4	2	—
2007	12	5	2	4	2	2	2
2008	22	7	1	6	4	2	—
2009	17	10	2	7	2	2	4
2010	17	10	6	8	7	2	—
2011	25	16	7	10	13	3	4
2012	32	16	7	10	14	3	4
2013	29	22	—	17	15	3	—
2014	40	18	11	17	14	3	5
2015	37	31	25	19	16	3	8
2016	39	23	14	16	11	2	5
2017	39	34	—	14	4	3	—
2018	49	40	24	20	15	4	—
2019	37	37	24	22	19	6	12

Dashes indicate data not available.

23 Agri-Curriculum: The Foundational Role of Indigenous Knowledge in Early Childhood Education and Beyond

Hawani Negussie[1] and Tsedeke Abate[2*]
[1]*Early Childhood Education Ethiopia, Addis Ababa, Ethiopia;* [2]*Homegrown Vision, Addis Ababa, Ethiopia*

Abstract
The natural landscape within Ethiopia is undervalued and increasingly redacted by the education system. The current curriculum and instructional approaches are neither derived from Ethiopia's history, culture, or social infrastructure nor does it convey indigenous, homegrown or resemble the child entering the classrooms during the early years or in later grades. This chapter, in part, examines the critical role of including indigenous knowledge in the curriculum and instruction, focusing on integrating Ethiopia's farming systems in early childhood education and fostering a sustainability mindset beginning in pre-primary education. College-level agricultural education in Ethiopia has undergone substantial changes since its establishment at Haramaya in the early 1950s. Currently, at least 28 universities in Ethiopia offer up to PhD degrees in various fields of agricultural sciences. Inadequate funding and lack of ability to attract the best quality students because of limited job opportunities for agriculture graduates are some of the significant constraints.

Introduction

This chapter discusses education at two different stages. The first stage deals with early childhood education, while the second introduces experiences in Ethiopian agricultural college education.

African countries abide by implicit and explicit cultural expectations that govern most indigenous communities' education, care and childrearing practices. These natural experiences that translate into mastery of skills are gained through what is commonly referred to as indigenous knowledge or informally constructed education (Mason and Sinha, 1992). Informal education is the inexplicable knowledge passed down from generation to generation, generally embedded in daily rituals. Prochner and Kabiru (2008) described it as a phenomenon in that 'children learned and were taught as they participated in the daily living activities in the home, through ceremonies, direct instructions, observation and apprenticeship' (p. 124). The community becomes the classroom, and interactions with the environment serve as a unique, culturally saturated pedagogical experience. Vygotsky (as cited in Greenfield *et al.*, 2000) shares that 'development is constructed through social interaction, cultural practices, and the internalization or cognitive appropriation of symbolic tools.' Research supports the inclusion of indigenous knowledge in the construction, implementation and application stage of early childhood care and education in Africa as the most critical element for ensuring success in education programmes.

* Corresponding author: tsedeke@homegrownvision.org

 Untold Stories of African Agriculture: Lessons from Ethiopia (T. Abate ed.)
DOI: 10.1079/9781800626386.0023

In her article, Fleer (2003) notes that this is a far contrast from the Western, early education value system, which puts the child 'in segregated spaces, and are given toys and materials that do not relate to real-world activities'; but, in the context of indigenous communities, children are 'situated within the family and community as central beings' (p. 66) enabling them to interact with members and real situations daily. The need to include the community and culturally specific socialization and integration efforts of indigenous practices in education was born from many factors. Primarily, the recent spotlight and acknowledgement of indigenous communities, the impact of globalization and continued education system failures by marginalized and underserved children worldwide have amplified the need to start from what is near, known and passed down from generations. With this contextual knowledge, this chapter explores the generational benefits of embedding indigenous knowledge and traditional practices in Ethiopia's early learning environment.

Efforts to align agricultural education in Ethiopia with that of the Western world started with establishing the Ambo Agricultural Technical School, followed by the Jimma Agricultural Technical School. Brief accounts of progress made so far and prospects for the future are presented here.

Currently, 28 universities offer BSc, MSc and PhD degrees in various agriculture-related fields, including agriculture/fisheries, forestry/natural resources and veterinary medicine.

Theoretical framework

Ethiopia's childrearing practices ascribe to Vygotsky's (Mahn, 1999) idea of community as having an implicit and explicit role in how children gather early codes that later become knowledge. Vygotsky's work highlights two states of children's development: nature is primarily associated with maturation, and nurture refers to interactions between children and their community, which 'allows mastery not only of the items of cultural experiences but the habits and forms of cultural behaviour, the cultural methods of reasoning' (as cited in Mason and Sinha, 1992). Associated with this notion of cultural development is Vygotsky's theoretical perspective on the development of language, thought and the use of symbols as it relates to children's cognitive development and the role of the environment. Learning during the early years, defined as the period from birth to 5 years of age, takes on the form of spontaneous imitation from parents or caregivers included in the child's daily world. Daily life in Ethiopia imitates the theory of Vygotsky, in which the community sets the educational foundation for later learning.

Frequently, children under 3 years are carried on their mothers' backs and engaged in looking at open market exchanges, where they pick up social cues first by observing and, later, by participating fully when development permits. During these formative years, 'children acquire the social skills that can be learned only through experiences' (Gestwicki, 2010) with other members and peers in their surroundings. These social experiences, guided by unpronounced cultural exchanges, help shape children's ideas of who they are and where they belong in their growing world. In defining the influence of social generation, it is exposed to the past.

In the context of early childhood care and education (ECCE), 'community consists, in essence, of such connections between expressed thought and lived experience: a dynamic cyclical relationship between the stories people tell about themselves and the ways they relate to one another and their environments' (Fettes, 1999). Vygotsky believed members found in children's immediate world, such as elders, teachers, parents, priests and neighbours, assist children by engaging in what he calls the zone of proximal development, where children are involved on two levels of the cognitive process. On the lower end, children can function independently; on the higher end, children can achieve more complex tasks with help from those in the community (as cited in Doolittle, 1997). Figure 23.1 depicts the learning progression.

Literature review

Indigenous knowledge begins with using local materials, including hiring teachers in the community, practicing oral storytelling, singing traditional songs and enabling informal education to

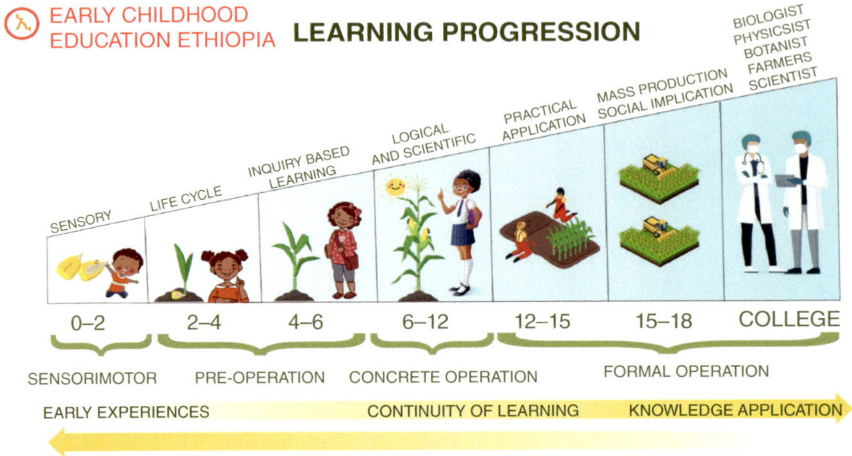

Fig. 23.1. The 'learning progression' as visualized by the Early Childhood Education Ethiopia. Source: Hawani Negussie.

be reflective in the education paradigm. According to Neisser (1993), 'young children are actively engaged with the people, objects and physical properties of their environments. They are curious and willing to explore their world from a very early age'. In the past, the prominence placed on indigenous knowledge in Africa concentrated on extracting alternative medicine for individuals and organizations seeking non-Western-based cures for illness and other ailments. However, increasingly, 'the value of indigenous knowledge has been recognized in other fields, including early childhood development, as development agencies and institutions of higher learning come to appreciate its contribution' (Pence and Schafer, 2006).

The importance of including indigenous knowledge and cultural practices in the education process of young children has been extensively documented (Schafer *et al.*, 2004; Banda, 2006; Nsamenang, 2008; Pence and Soudee, 2009) as having not only a positive effect, but also as a qualitative measurement for the advancement of ECCE in the right direction. However, current practices, including the pressure of meeting unilateral international goals, continue to undermine the importance of indigenous education, mainly viewed as informal in ECCE programmes. Scholars May and Aikman (2003), in their article on indigenous education strategies, note that:

> Achieving acceptance and recognition for indigenous alternatives to state-run formal schooling is a slow and challenging path-such initiatives have to overcome national policies aimed at assimilation and homogenization, as well as trends towards standardization which smother innovation and diversity for the sake of accountability and supposed 'equality'.

When the interest to invest in children's education took the world stage, the vision to meet this need stemmed from a Western educational framework. In discussing this approach, Nsamenang (2008) states:

> When Euro-American early childhood development programmes are applied as the gold standards by which to measure forms of Africa's early childhood development, they deny equity to and recognition of Africa's ways of provisioning for its young, and they deprive the continent of a niche in global early childhood development knowledge waves. (p. 136).

Much of the interventions proposed by international and national agencies are constructed to change the ideas and perceptions of current situations children and families face in Ethiopia. International non-governmental organizations' single focus on increasing access presents a far contrast from the tenets of ECCE that specify children's environment, use of native language and a culturally enriched curriculum as the primary disposition to promote the progress of

ECCE programmes in Ethiopia and beyond. Since the primary focus continues to be on just getting children in the classroom, the significant representation of who goes to school under 5 years of age in Africa is gathered from formal schools in 'urban areas and easier to reach populations' (Soudee, 2009, p. 16), disqualifying the informal education institutions and those who attend it. Pence and Nsamenang (2008) contend that informal education institutions that exist throughout Africa 'deserve investigative discovery and enhancement rather than neglect and erasure' (p. 21). In the case of Ethiopia, approximately 80% of the population lives in rural areas, where indigenous farming is a way of life. Children are more likely to have access to informal education and indigenous knowledge mediums of instruction dissimilar to those programmes highlighted, funded and supported by international agencies, broadly disconnected from such demographics.

The intervention efforts also become cultural adjustment procedures: baptizing countries from the previous ways (indigenous education) towards the new acceptable socially dominant Western education paradigm. Yosso's (2005) cultural wealth theory challenges these perceptions by asserting that communities, children and families from disadvantaged communities possess strength and inherent solutions and are cultural contributors to the education system they participate in daily. However, these assets, even in the international infrastructure, are often discounted, muted and ignored as contenders for solutions. In their study, Fleer *et al.* (2004) state that Western development, enacting a thought curriculum, positioned children from Western cultures as the dominant and privileged group. Their and their families' expectations aligned more closely with our early childhood educational settings. As such, the 'taken for granted' practices of early childhood education have become the accepted community practice, the predominant way things are done (Fleer *et al.*, 2004).

Several studies have challenged the perception that indigenous education systems are inadequate or that intergeneration knowledge passed down by traditional and cultural practices has a lesser impact on the community and classroom environment (Negussie and Slater, 2019). These studies remind us of the need to integrate and reteach our pupils about the past to navigate their future better. More specifically, the decisive role of integrating traditional farming systems in curriculum and instruction starting at early childhood education.

Research

Indigenous knowledge in education and instruction

In recent years, many countries have asserted the right to educate their youngest population and the right to be educated using the cultural capita found in indigenous knowledge (Smith and Smith, 2009). In a study of First Nations or Native American groups residing in British Columbia, Canada, explorative research examined the First Nations Partnership Program (FNPP) between the local university and the area tribal communities. The basis of the study was to understand the role of the First Nations Indigenous Knowledge in the early childhood development sector. The research concentrated on gathering perspectives of what culture and indigenous knowledge means to the First Nation and how such practices were included in the education paradigm. The execution of this model took a collective approach where the local university, community members and instructors took active ownership of this programme: 'There is an assumption that culturally valued and useful knowledge about childhood and childcare is embedded within the community and that this knowledge needs to be afforded a central place in the development of training curricula' (Ball and Simpkins, 2004, p. 482).

Using the concepts found in the generative curriculum model, which operates by using the community as the principal educational foundation, the FNPP gives power, obligation and support to the elders, including prominent community members. The generative curriculum model included:

- draw upon community and individual strengths;
- ensure a broad ecological perspective (awareness of the child in the context of family and community);
- support community initiatives in a community-based setting;

- promote respect 'all ways' (multicultural inputs);
- provide education and career laddering for students, such that credit for this coursework will be fully applicable to future study and practice;
- engage in the co-construction of a bicultural curriculum in which elders and other community resource people figure prominently; and
- provide the basis for a broader child, youth, family and community-serving training and services, while the immediate focus is early childhood care and development (www.fnpp.org).

During the study, researchers asked how indigenous knowledge was perceived and carried out in the First Nation community. The tendency to look at culture from this diluted and tourist perspective intensely compromises the value, prominence and distinction found in many earlier established countries worldwide. Ball and Simpkins (2004) reiterate the degradation of such a view, stating:

> Western tendency to look for concrete, identifiable practices as signifiers of 'culture' may overlook the more durable and deeply significant underlying attitudes, meaning, memories and values being passed on by First Nations childcare practitioners to the children and parents whom they serve (p. 484).

Participants shared the value of having elders as their guide throughout the training cycle and beyond. Elders helped them reconnect to colonization, bridging the knowledge gap and affirming the young childcare workers in training that their history was preserved. They enacted songs, imparted knowledge of food preparation and secret recipes to consume when pregnant. The information conveyed during the training period among the community members and childcare workers exhumed the ways of the past, that generational practice that only survives by keeping culture and traditions. In the FNPP study, elders resurrected rituals and customs indigenous to the First Nation communities, such as discipline, respect and sharing. They shared with teachers the meanings behind the words of daily songs heard in the classroom, and they pressed upon the importance of healthy eating by consuming indigenous meals like salmon, berries and dried meat.

An emerging concept discussed in the study was the idea of 'cultural programming', or an aspect of the curriculum that embodies the traditions of the First Nation community. Specific activities, such as hunting and preparing animals, songs, dances, language programmes, storytelling and elder involvement, were integrated (Ball and Simpkins, 2004). An early notation from the study was how teachers connected the training they received in the classroom environment. Many teachers gravitated toward a specific knowledge base, such as music or history, to expand their programmes. At the same time, for others, it was essential to teach children who their relatives were, their aunts and uncles, and how they are uniquely connected with each family member to add to the pot of the cultural knowledge to which they all belonged. In the First Nation belief system, knowledge is a constant wave, and it is accepted that what is 'true or necessary may give way to new, more fitting knowledge' (Ball and Simpkins, 2004, p. 486).

Another study examined the familial makeup of preschool-age children from Australia's aboriginal population and the importance of integrating indigenous knowledge and cultural practices in ECCE programmes. Selected families from the community were given video recorders to share the cultures and traditions that were most important to them. Results showed that the 'child is a composite of an extended family, community and set cultural beliefs' (Fleer *et al.*, 2004, p. 59), but previous misconceptions of one's culture and the belief of indigenous children as being deficient had built up to an educational paradigm that did not include the community, cultural spine and values in everyday curriculum and instruction.

This particular research, through the implementation of professional development, reported making teachers into observers of children's activity and restructuring the classroom curriculum to allow teachers to facilitate learning to occur inside children's zone of proximal development and include forms of nonverbal communication as part of this population's custom of learning in daily living (Fleer, 2003). The outcome of this study indicates that children, especially children from culturally enriched environments as those found in developing countries, need to be welcome in a classroom where they see part and parcel of themselves in the curriculum, teaching strategies and language used to transmit formal knowledge.

Ethiopia's Cultural and Traditional Knowledge

The highlands of Ethiopia boast a remarkable landscape and topography that supports an astonishing diversity of approximately 7000 plant species (Giday *et al.*, 2003). Within this vibrant ecosystem, Ethiopians have cultivated a long-standing tradition of harnessing the healing properties of plants and natural resources found in their environment to address ailments and combat various illnesses. Traditional medicine is significant in Ethiopian culture, with an estimated 80% of the population relying on and utilizing these age-old remedies (Abebe and Ayehu, 1993). This profound reliance on traditional medicine highlights the deep-rooted connection between the Ethiopian people and their natural surroundings, signifying the importance of including indigenous knowledge in healthcare practices. The vast array of plant species and the widespread utilization of traditional medicine signify the invaluable role of Ethiopia's natural resources in supporting the wellbeing and health of its population.

Traditional medicine

Traditional medicine plays a significant role in the healthcare practices of indigenous communities, preserving centuries-old knowledge of plant-based remedies and healing techniques. In an ethnobotanical study conducted to understand the indigenous traditions of the Zay people, researchers examined the community elders' knowledge of plants for medicinal use. The study's findings indicated that the Zay people commonly administered medicines orally, followed by topical and nasal applications (Giday *et al.*, 2003). These traditional restorative practices passed down through generations serve as a cornerstone for preserving indigenous knowledge and cultural heritage. However, the sustainability of these practices faces challenges such as deforestation and cultural shifts influenced by the growing prominence of Western-based treatment approaches. As the Zay people and others like them continue to grapple with environmental changes and the encroachment of modern influences, their invaluable relationship with their environment and traditional healing practices is gradually eroding.

In Ethiopian culture, the relationship between health and spirituality is deeply intertwined, with religious beliefs and practices significantly shaping people's perceptions of illness and wellbeing. Many Ethiopians believe that their health conditions and outcomes are profoundly influenced by prayer, dedicating a substantial amount of time to supplication and making religious commitments. The connection between God and religion is pervasive in Ethiopian society, extending to various aspects of daily life. This is evident in the customary morning greetings, where individuals express their wish for the wellbeing of others with the phrase '*tena yistlign*', which translates to 'may God give you health'. Such religious formalities highlight the importance of spirituality and the acknowledgement of a higher power in the Ethiopian approach to health and wellbeing, reflecting the deep-rooted connection between culture, faith and the pursuit of good health.

The relationship between health and God is entwined in Ethiopian culture and religious norms. Many Ethiopians believe their conditions and outlook on illness are prayer bound, spending much of their time pleading and promising, which involves religious sacraments. God and religion in Ethiopia are associated with a more noticeable and processional aspect of daily life. These formalities are pervasive in the morning greetings, which begin with a wish to the on-comer, '*tena yistlign*'.

Cultural practices

Greetings between two people begin with a bow or men taking off their hats in salute and respect for another man in their presence. These cultural practices are the silhouettes of everyday life, and children grow up looking at these exchanges, first, as newborns hearing and listening to the sounds and sights and, later, as able participants, imitating others and using words to make sense of all that is deposited into their young life. The cultural milieu envelops children, guiding their development and understanding of the world. They unknowingly mirror the actions of their community elders, whether it is paying respects to the church or mosque or seeking healing from local healers. These surroundings become the ecological foundation of a child's world, fostering curiosity and nurturing a sense of identity and belonging.

Religion is an integral part of daily life in Ethiopia, permeating the rituals and practices of its people. It is common to witness passersby halting amidst bustling traffic to pay reverence and offer prayers at churches and mosques. The ceremonies intertwined between religion and culture are seen in the 'distinction between clean and unclean food; the method employed for slaughtering animals; the circumcision of males after the eighth day' (Lipsky and Blanchard, 1962, p. 108). Boys are baptized on the 40th day and girls on the 80th day, until which time most children are not given names.

Fasting holds universal significance in Ethiopian culture and religious observance. Families, including children, adhere to fasting traditions, particularly during Lent. This period involves abstaining from meat and, for some, refraining from eating until noon each day (Lipsky and Blanchard, 1962; Buxton, 1970). When a death occurs in the family, a mourning of significant proportion pours from the home and the community at large. Ethiopia's collective society mourns with those closest to the deceased, marked by individuals donning all-black attire, women shaving their hair and men growing out their beards. These daily occurrences shape the child's environment, allowing them to derive meaning and integrate them into their cognitive understanding.

Just as Western cultural nuances influence children in their respective countries, Ethiopia's traditional systems serve as moral guidelines, providing ethical direction for informal education and nurturing the foundation of indigenous knowledge. Children grow up amidst these rich cultural practices and religious values, observing and deriving meaning from their surroundings, shaping their worldview and personal development.

The Role of Agriculture and Farming Systems in Early Childhood Education

Traditional agricultural farming systems and practices are a persistent sociocultural facet of daily life for the Ethiopian populous. Agroforestry practices allow smallholder farmers to live dynamically within their social and ecological communities (Abate, 2016). As a result, the daily operational practices of agrarian life are significantly correlated with all domains of early childhood education and socialization for children growing up in these environments. Faced with climate change, social and ecological adaptability is central to preserving and maintaining Ethiopian agrarian communities. This gives young children growing up in indigenous pastoral communities an expansive cultural space to support holistic development, one that is informed by traditions, culture, collective practices and generationally passed down wisdom.

A recent study by Wondimu and Kebede (2022) explored the multifaceted use of *enset* within the Gurage population in south-central Ethiopia. This study delved into the sociocultural and economic dimensions of *enset* utilization, shedding light on its significance in the daily lives of the locals. The study findings revealed that *enset* is deeply 'ingrained in almost every aspect of Gurage life; the Gurage's most arable land is set aside for *enset* cultivation' (Wondimu and Kebede, 2022, p. 6). Additionally, the cultivation and successful production of *enset* emerged as a symbol of wealth and prosperity among farmers, representing their ability to provide for their families adequately. This recognition highlights the economic significance of *enset* within the Gurage society.

Furthermore, the society's labour division and communal working infrastructure are divided between men and women in a traditionally constructed labour party called '*jege*' (Wondimu and Kebede, 2022). Moreover, the *enset*-harvesting process transmits indigenous practices and knowledge to the younger generation, 'allowing Chaha girls to learn and practice the traditional ways of the *enset*-harvesting process or work culture' (Wondimu and Kebede, 2022, p. 9). This aspect emphasizes the intergenerational transmission of skills and cultural identity through *enset* cultivation.

The studies highlighted above underline the critical role of indigenous knowledge in early childhood education, specifically the integration of agricultural systems and farming wisdom in early learning curriculum and instruction. By incorporating foundational knowledge of local crops and farming techniques such as *enset* in early childhood education, it is possible to conserve and propagate traditional knowledge and cultural creativity, paving the way for a viable and enduring future for generations.

Children raised in such environments have a rich cultural space supporting their holistic development. They actively engage with their social environment and, through agricultural practices, participate in cultural, traditional and religious customs. These conventional systems serve as the bedrock for early conceptual knowledge and lay the foundation for lifelong learning, offering significant benefits for early childhood education while ensuring a sustainable future. However, in the face of local and international educational reforms, the universal policy of increasing access without questioning 'how' and shortsighted outreach efforts, incorporating these valuable segments of the informally assembled world, may be perceived as irrelevant or even primitive from the perspective of Western education systems. Nonetheless, embracing the culture and social identities in less industrialized countries worldwide is imperative to achieve a sustainable transformation in early education in Ethiopia and beyond.

Agricultural college education

Ethiopia's first tertiary education in agriculture started with establishing the then Imperial Ethiopian College of Agriculture and Mechanical Arts (IECAMA) at Alemaya (now Haramaya University) in the early 1950s. Initially, the two agricultural technical schools – Ambo and Jimma – were the sources of students entering the agricultural college. However, the opportunities were further extended to all students who passed the 12th-grade matriculation examination from high schools all over Ethiopia in later years.

The College of Agriculture at Alemaya (now Haramaya University) was based on the United States Land Grant University, which was engaged in teaching, research and extension. The original US government grant was contracted to and implemented by Oklahoma State University from 1952 to 1968 (OSU, 1969).

Ethiopia's agricultural college education system has undergone substantial changes over the seven decades since the establishment of IECAMA. First, the research and extension mandates were transferred to the Ethiopian Institute of Agricultural Research and Ministry of Agriculture in the mid-1960s and the mandate and focus of the college was on teaching, even though the staff still did some research.

Second, the number of universities offering agricultural degrees has increased substantially, especially over the past decade. The work of Kassa (2015) shows that 28 universities in Ethiopia offered different levels of agriculture-related degrees by that time (Fig. 23.2). Agriculture/fisheries, forestry/natural resources and veterinary medicine accounted for 66%, 24% and 10%, respectively, of the total programmes carried out at agricultural universities and colleges in Ethiopia.

Agricultural education programmes

The type and number of agricultural education programmes varied according to the universities. Agricultural education in Ethiopia is divided into three broad categories: agriculture/fisheries (including crops, livestock and fisheries), forestry/natural resources and veterinary medicine. Seven of the 28 universities offer degrees in all three areas. Hawassa and Haramaya have by far the largest number of programmes. The top ten – Hawassa, Haramaya, Jimma, Mekelle, Wollega, Bahir Dar, Addis Ababa, Gondar, Ambo and Wolaita-Sodo – account for 71% of the total programmes. Hawassa, Haramaya, Jimma and Addis Ababa offer up to PhD-level degrees.

Challenges for agricultural education

There are several challenges facing agricultural college education in Ethiopia and elsewhere in Africa. The work of Eicher (2006) suggests that the main challenge for African agriculture is

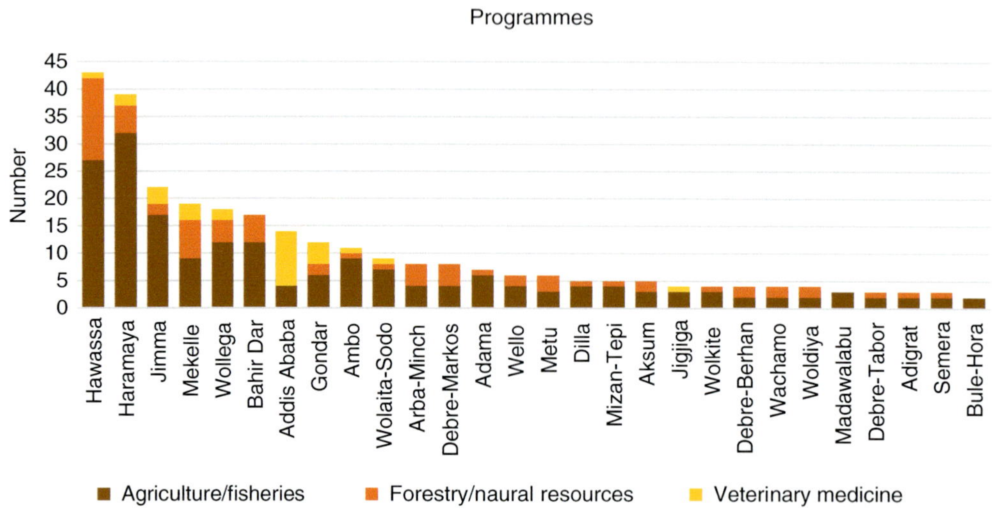

Fig. 23.2. Number of agricultural programs in different universities in Ethiopia. Source: constructed by the authors from Kassa (2015).

inadequate spending on agriculture. He argued that Asian governments invested three times more than African governments in the 1960s and 1970s; according to this author's experience with agricultural development in the USA, Brazil and Japan, it would take 40–60 years to build a financially sustainable system of agricultural research, education and extension. On the contrary, there is a tendency in Africa to look for quick fixes and shortcuts, with the expectation that the international development community would come up with solutions.

Hulluka (2013), Kassa (2015) and Shibru *et al.* (2016) suggest more specific challenges for college agricultural education in Ethiopia, including:

- poor capacity to enrol large numbers of students;
- decline in the quality of education;
- decline in relevance;
- high attrition rate of highly qualified staff;
- poor practical skills;
- lack of job opportunities for graduates from agricultural colleges; and
- failure to attract students.

Hulluka (2013) states, 'public sector educational institutions are not always autonomous enough to make bold and rational decisions required to effect improvements on the existing system'. Shibru *et al.* (2016) emphasize the need to align higher education in agriculture with employment opportunities. One constraint that has not been given enough attention in agricultural college education in Ethiopia is the lack of coordination and collaboration among various institutions. The agricultural universities and colleges mentioned here are publicly funded institutions but have little collaboration among themselves and with the research system.

Discussion and Conclusions

In early childhood education, the environment is considered the third teacher for its influential and contributing role to the learning paradigm. Unfortunately, Ethiopia's rich historical, cultural and traditional past has yet to be formally translated and utilized, as indigenous knowledge is missing from the curriculum of early education programmes. The concept of culture, customs and traditions as leading resources in the education of young children is still a general idea in the Ethiopian educational milieu.

Culturally relevant, developmentally appropriate and environmentally inclusive models of approaching early childhood development are fundamental to educational intervention's success, retention and impact. Researchers and educators must critically embed themselves in the community's cultural context. The tenets of early childhood education reiterate that children's early education experience is moulded by the social experiences extended from the child's community. This knowledge expressed in daily customs and traditions within the Ethiopian culture can cultivate cognitive, linguistic, physical and social-emotional abilities in the early education environment.

One organization leading the way in full embrace of such practice is Early Childhood Education Ethiopia. Through its whole-child approach, leveraging local knowledge and leading education reforms through community-led practices, it is poised to change the landscape of providing asset-based instruction while fully embedding children's background knowledge in curriculum and instruction. Ethiopia can construct an education infrastructure that serves the population of children in Ethiopia and could be an exemplar for the rest of Africa.

References

Abate, T. (2016) Contribution of indigenous knowledge to climate change and adaptation response in Southern Ethiopia. *Journal of Earth Science and Climatic Change* 7(11), 377. DOI: 10.4172/2157-7617.1000377.

Abebe, D. and Ayehu, A. (1993) *Medicinal Plants and Enigmatic Health Practices of Northern Ethiopia*. BSPE, Addis Ababa, Ethiopia, 511 p.

Ball, J. and Simpkins, M. (2004) The community within the child: Integration of indigenous knowledge into First Nations childcare process and practice. *American Indian Quarterly* 28(3/4), 480–498.

Banda, D. (2006) Education for all (EFA) and 'African indigenous knowledge systems (AIKS)': The case of the Chewa people of Zambia (Doctoral dissertation). Available at: http://etheses.nottingham.ac.uk/525/ (accessed on 11 June 2023).

Buxton, D.R. (1970) *The Abyssinians*. Praeger, New York.

Doolittle, P.E. (1997) Vygotsky's zone of proximal development as a theoretical foundation for cooperative learning. *Journal on Excellence in College Teaching* 8(1), 83–103.

Eicher, C. (2006) The Evolution of Agricultural Education and Training: Global Insights of Relevance for Africa. World Bank (AFTHD) Commissioned Report.

Fettes, M. (1999) Indigenous education and the ecology of community. *Indigenous Community-Based Education* 11(3), 20–40.

Fleer, M. (2003) Early childhood education as an evolving 'community of practice' or as lived 'social reproduction': Researching the 'taken-for-granted'. *Contemporary Issues in Early Childhood* 4(1), 64–79.

Fleer, M., Anning, A. and Cullen, J. (2004) A framework for conceptualizing early childhood education. In: Anning, A., Cullen, J. and Fleer, M. (eds) *Early Childhood Education: Society and Culture*. Sage Publishers Ltd, London, UK, 175–189.

Gestwicki, C. (2010) *Developmentally Appropriate Practice: Curriculum and Development in Early Education*. Wadsworth, Belmont, CA.

Giday, M., Asfaw, Z., Elmqvist, T. and Woldu, Z. (2003) An ethnobotanical study of medicinal plants used by the Zay people in Ethiopia. *Journal of Ethnopharmacology* 85(1), 43–52.

Greenfield, P.M., Maynard, A.E. and Childs, C.P. (2000) History, culture, learning, and development. *Cross-Cultural Research* 34(4), 351–374. DOI: 10.1177/106939710003400404.

Hulluka, M. (2013) Agricultural higher education in Ethiopia: Challenges and perspectives. *JAD* 3(2), 69–78.

Kassa, B. (2015) Potentials for agricultural university graduates to go into commercial farming in Ethiopia: Challenges and opportunities. *International Journal of Science and Technology* 4(1), 185–200. DOI: 10.4314/stech.v4i1.16.

Kassaye, K.D., Amberbir, A., Getachew, B. and Mussema, Y. (2006) A historical overview of traditional medicine practices and policy in Ethiopia. *Ethiopian Journal of Health Development* 20(2), 127–134.

Lipsky, G.A. and Blanchard, W. (1962) *Ethiopia: Its People, Its Society, Its Culture*. Hraf Press, New Haven, CT.

Mahn, H. (1999) Vygotsky's methodological contribution to sociocultural theory. *Remedial and Special Education* 20(6), 341–350.

Mason, J.M. and Sinha, S. (1992) Emerging literacy in the early childhood years: Applying a Vygotskian model of learning and development. *Center for the Study of Reading Technical Report*; no. 561.

May, S. and Aikman, S. (2003) Indigenous education: Addressing current issues and developments. *Comparative Education* 39(2), 139–145.

Negussie, H. and Slater, C.L. (2019) Indigenous knowledge and early childhood care and education in Ethiopia. *Journal of Educational Leadership, Policy and Practice* 33(2), 4–16.

Neisser, U. (1993) *The Self-perceived*. Cambridge University Press, Cambridge, England, pp. 3–21.

Nsamenang, A.B. (2008) Culture and human development. *International Journal of Psychology* 43(2), 73–77.

OSU (Oklahoma State University) in Ethiopia (1969) Terminal Report, 1952–1969. Stillwater, Oklahoma, USA. 84 p.

Pence, A. and Nsamenang, B. (2008) A case for early childhood development in Sub-Saharan Africa (Working Papers in Early Childhood Development No. 51). Retrieved from Bernard van Leer Foundation. Available at: http://www.bernard vanleer.org/a case for _early _childhood_development_in_sub-Saharan_Africa (accessed 11 June 2023).

Pence, A. and Schafer, J. (2006) Indigenous knowledge and early childhood development in Africa: The early childhood development virtual university. *Journal for Education in International Development* 2, 3.

Prochner, L. and Kabiru, M. (2008) ECD in Africa: A historical perspective. *Africa's Future, Africa's Challenge*, 117–133.

Schafer, J., Ezirim, M., Gamurorwa, A., Ntsonyane, P., Phiri, M., Sagnia, J. and Bairu, W.W. (2004) Exploring and promoting the value of indigenous knowledge in early childhood development in Africa. *International Journal of Educational Policy, Research, and Practice: Reconceptualizing Childhood Studies* 5(3), 61–80.

Shibru, A.K., Osiru, M. and Aklilu, H.M. (2016) Higher agricultural education in Ethiopia: Current status and future prospects. *African Journal of Rural Development* 1(2), 151–158.

Soudee, A.R. (2009) Incorporating indigenous knowledge and practice into ECCE: A comparison of programs in the Gambia, Senegal and Mali. *Current Issues in Comparative Education* 11, 15–23.

Wondimu, H. and Kebede, A. (2022) Cultural, economic, and symbolic value of the enset crop in Sebat-Bet Guraghe, Southwest Ethiopia: a qualitative study. *Humanities and Social Sciences Communications* 9, 164. DOI: 10.1057/s41599-022-01179-4.

Yosso, T.J. (2005) Whose culture has capital? A critical race theory discussion of community cultural wealth. *Race Ethnicity and Education* 8(1), 69–91. DOI: 10.1080/1361332052000341006.

24 The Changing Food Balance Sheet: What are the Policy Implications?

Tsedeke Abate,[1]* Workneh Ayalew,[2] Mintewab Bezabih,[3] Mulat Demeke,[4†] Girma Tesfahun Kassie,[5] Amare Tegbaru[6] and Konjit Hailu[7]

[1]Homegrown Vision, Addis Ababa, Ethiopia; [2]International Centre for Insect Physiology and Ecology, Addis Ababa, Ethiopia; [3]Environment and Climate Research Center, Policy Studies Institute, Addis Ababa, Ethiopia; [4]Formerly Food and Agriculture Organization of the United Nations, Nairobi, Kenya; [5]ICARDA-Morocco, Rabat, Morocco; [6]Formerly International Institute of Tropical Agriculture, Ibadan, Nigeria; [7]Department of Business Administration and Commerce, Addis Ababa University, Addis Ababa, Ethiopia

Abstract

Ethiopia's dietary energy intake increased from ~1700 kcal/person/day in the 1990s to ~2500 kcal/person/day in 2015–2019, exceeding the World Health Organization (WHO) minimum requirement of 2100 kcal/person/day. However, Ethiopian diets are low in fat, protein and mineral nutrients, compared to many of their peers in Africa. Complex carbohydrates, protein and fat constituted 78%, 14% and 4%, respectively, of the total dietary energy versus the WHO recommendation of 55–70% carbohydrates and 15–30% fat. Maize has displaced *teff* (*Eragrostis tef*) and sorghum as the staple food crops while *enset* (*Ensete ventricosum*) and other root crops have shown the fastest growth rate; their production and usage has expanded as part of farmers' innovation and increased demand by people who cannot afford high-priced cereals like *teff*. Ethiopia needs to diversify its heavy reliance on limited cereals and upgrade its investment in objectively identified native staples and harness their full potential to achieve sustainable food and nutrition security.

Introduction

Ensuring food security and economic wellbeing for a rapidly growing population has been a key priority for policy makers of successive governments in Ethiopia, especially following the great drought and famine that ensued in 1972–1975, resulting in the deaths of hundreds of thousands of smallholder farmers (Kumar, 1987; Cochrane, 2013) and causing influx of people from rural areas to urban centres. Another drought and mass starvation occurred a decade later and the military-socialist government at the time assigned the highest priority to the expansion and improvement of the productivity of cereals, particularly maize, because of its high potential for yield per unit area of land, compared to the hitherto widely grown traditional cereals such as *teff* (*Eragrostis tef*), sorghum, wheat, barley and millet.

As is the case in many developing countries, such policies were preoccupied with meeting the energy requirements without much consideration

* Corresponding author: tsedeke@homegrownvision.org

† Recently deceased.

DOI: 10.1079/9781800626386.0024

to the nutritional quality and balance of the diets (Remans *et al.*, 2011; Vasileska and Rechkoska, 2012; Pingali, 2015; Cochrane, 2018) or their diversity and sustainability. The government effort to increase food production and improve calorie intake started in the early 1980s with the launch of the National Development Campaign (ONCCP, 1987) but comprehensive reviews on its outcomes and impacts are not available. Nevertheless, several authors (Amare *et al.*, 2012; Headey, 2014; Kennedy *et al.*, 2015; Sibhatu *et al.*, 2015; FTF, 2018; Gebru *et al.*, 2018; Sheehy *et al.*, 2019) suggest that Ethiopian diets are low in fat, protein and micronutrients. According to Gebru *et al.* (2018), diet diversity in Ethiopia is low and nearly 82% of adult women have insufficient intake of vitamin A and 50% have zinc deficiency. According to Sheehy *et al.* (2019), Ethiopian diets were composed of nearly 76% carbohydrates, 13% fat and 11% protein in 2011, even though the World Health Organization (WHO) recommends 55–70% energy from complex carbohydrates and 15–30% energy from fats, for a healthy diet.

Abebe *et al.* (2006, 2007) reported micronutrient deficiencies in the traditional diets consumed in southern Ethiopia. Amare *et al.* (2012) reported that nearly 13% of the 356 individuals who participated in their experiment in Gondar, northwest Ethiopia, were undernourished and their daily protein intake was inadequate according to the WHO recommendation. They further noted that the energy intake in women was 1929 kcal/person/day versus the recommended level of 2031 kcal/person/day, whereas men consumed 3001 kcal/person/day versus the recommended amount of 2510 kcal/person/day. Studies from other countries suggest that there is a large gap between intakes and recommended requirements across countries, and reproductive women in low-income settings are at particularly high risk of low micronutrient intake when their diets are dominated by staple foods (Arimond *et al.*, 2010) that are only rich in complex carbohydrates.

The Ethiopian Ministry of Agriculture is currently undertaking a policy revision, with particular emphasis on food security. A comprehensive national nutrition survey is lacking to this date. Most of the presently available literature and data on nutrition in Ethiopia has dealt with either limited areas of the country and/or limited number of crops (Abebe *et al.*, 2006, 2007; Atlabachew and Chandravanshi, 2008; Arimond *et al.*, 2010; Koisha, 2013; Woldemariame, 2015; Melaku and Duguma, 2016). There are limited reports (Demeke, 2012; Worku *et al.* 2017) and anecdotes regarding changes in the level of energy sources and the food balance sheet. Sheehy *et al.* (2019) analysed the Food and Agriculture Organization of the United Nations (FAO) food balance sheets for Ethiopia from 1961 to 2011 to determine what changes have taken place in the country's dietary energy and nutrient supply. However, empirical data linking the changes and their policy implications are lacking. This chapter aims to provide such data to inform the policy decision that is needed for priority setting and determining the future direction of agricultural research and development in Ethiopia.

Approach

We used the latest available data under the domains of Food Supply – Crops Primary Equivalent and Livestock and Fish Primary Equivalent from FAOSTAT (http://www.fao.org/faostat/en/#home). The latest available data as of 8 September 2022 cover the period from 1961 to 2019. We make graphical presentations of the long-term data covering all years or use 5-year averages of three periods, namely 1975–1979, 1995–1999 and 2015–2019, wherever there was a need to compare the data for Ethiopia with several peer countries in Africa (several countries), Asia (China and India) and Latin America (Brazil and Mexico).

We analysed the daily per capita intake of energy, protein and fat, and the per capita consumption of the various staples or staple groups that constitute the daily diets of Ethiopians. We paid particular attention to measuring changes in the sources of the various parameters mentioned here.

Results

Nutritional status

Ethiopia's nutritional status has undergone substantial changes over the years. Figure 24.1 summarizes the overall long-term trends of dietary energy, dietary protein and dietary fat intake in

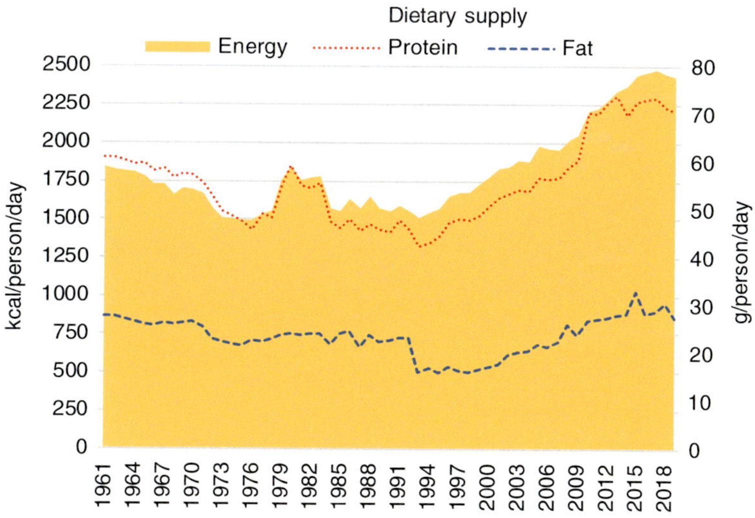

Fig. 24.1. Dietary energy, dietary fat and dietary protein supply in Ethiopia, 1961–2019. Source: constructed by the lead author from FAO (2022).

this country for the period from 1961 to 2019. The average dietary energy intake was roughly 1600 kcal/person/day in 1975–1979; it increased to 1700 kcal/person/day in 1995–1999 and reached close to 2500 kcal/person/day in 2015–2019, an increase of nearly 800 kcal/person/day.

Similarly, protein intake was estimated at 49 g/person/day in 1975–1979, 47 g/person/day in 1995–1999 and 72 g/person/day in 2015–2019, whereas the values for fat were approximately 23 g/person/day, 16 g/person/day and 30 g/person/day in 1975–1979, 1995–1999 and 2015–2019, respectively.

As of 2015–2019, the FAO (2022) database lists a total of 92 crop and livestock products as sources of nutrition for Ethiopia. Our study has shown that Ethiopians obtain 96% of dietary energy, 91% of dietary protein and 71% of dietary fat from crops and crop products while livestock products account for the balance, i.e. 4% of dietary energy, 9% of dietary protein and 29% of dietary fat (Table 24.1).

Cereals (mainly maize, wheat, *teff*, sorghum and barley) covered ~67% of dietary energy, ~61% of dietary protein and ~14% of dietary fat in 2015–2019. Other major sources of dietary energy included grain legumes, roots and tubers, and oilseeds and their products (oil). Cereals and grain legumes constituted nearly 61% and 21%, respectively, of the total protein intake. The bulk of dietary fat supply in Ethiopia comes from oil and oilseeds, cereals and dairy products, which together accounted for ~73% of the total (Table 24.1).

Please note that the FAO data reported here do not include *enset* and are therefore underestimates for all parameters measured. The FAOSTAT (accessed on 30 July 2020) indicated that root crops in total contributed 509 kcal/person/day, nearly 20 g/person/day protein and about 1 g/person/day fat during the 2009–2013 period. The contribution of *enset* was reported as 205 kcal/person/day, 2.19 g/person/day protein and 0.27 g/person/day fat.

Comparing Ethiopia's nourishment status with other countries

With approximately 2500 kcal/person/day, Ethiopia's overall dietary energy intake has fulfilled the WHO minimum requirement of 2100 kcal/person/day (ONCCP, 1987). However, this figure compares unfavourably with those in the majority of African countries. For example, Ethiopia ranked 29th out of 50 countries for which data were available for 2015–2019. In a similar fashion, it was 46th and 47th in fat and total per capita food supply, respectively, during the same period. The country fared better with its protein ranking of 17th out of the 50 countries in Africa for which data were available. Please refer to Annex Table 24.1 for more details regarding

Table 24.1. Crop and livestock sources and their relative importance as dietary energy, dietary protein and dietary fat supply in Ethiopia (2015–2019 average). Source: constructed by the lead author from FAO (2022).

	Energy[a]		Protein[a]		Fat[a]	
Item	kcal/cap/d	Per cent	g/cap/d	Per cent	g/cap/d	Per cent
Cereals	1632	66.56	43.83	60.71	4.12	13.92
Grain legumes	225	9.16	15.32	21.22	0.67	2.27
Roots	209	8.54	2.42	3.35	0.15	0.51
Oil/oilseed	104	4.23	0.98	1.36	14.27	48.16
Sugar	80	3.26	0.00	0.00	0.00	0.00
Vegetables	29	1.17	1.24	1.72	1.06	3.58
Spices	23	0.94	0.77	1.06	0.02	0.07
Beverages	21	0.87	0.20	0.28	0.00	0.00
Fruits	17	0.69	0.27	0.37	0.69	2.32
Coffee	3	0.14	0.47	0.65	0.00	0.00
Tea	0	0.00	0.02	0.03	0.00	0.00
Crop total	2343	95.56	65.52	90.76	20.99	70.83
Dairy	55	2.25	2.93	4.06	3.13	10.56
Bovine meat	23	0.93	1.51	2.09	1.82	6.15
Other meat	13	0.52	1.07	1.48	0.88	2.96
Mutton & goat meat	9	0.38	0.62	0.86	0.73	2.46
Honey	4	0.16	0.00	0.01	0.00	0.00
Chicken	2	0.09	0.24	0.34	1.07	3.62
Eggs	2	0.07	0.12	0.16	0.11	0.36
Fish	1	0.04	0.16	0.22	0.03	0.11
Pork	0	0.00	0.01	0.01	0.87	2.94
Miscellaneous	0	0.00	0.02	0.02	0.00	0.00
Livestock total	109	4.44	6.67	9.24	8.64	29.17
Grand total	2452	100.00	72.19	100.00	29.64	100.00

[a]kcal/cap/d and g/cap/d represent kcal/person/day and g/person/day, respectively.

comparisons of dietary energy intake for Ethiopia with selected African, Asian and Latin American countries.

In general, like many of its peers in Africa, Ethiopian diets are dominated by complex carbohydrates compared to many other countries outside the continent (Fig. 24.2). For example, complex carbohydrates, protein and fat in Ethiopia constituted roughly 78%, 14% and 4%, respectively, of the total dietary energy intake. This compares unfavourably with the WHO recommendation of 55–70% carbohydrates and 15–30% fat for a healthy diet.

Ethiopian diets are particularly low in fruits, vegetables, dairy products, red meat, poultry and fish. For example, a close examination of Fig. 24.2 demonstrates that the per cent composition of:

- fruits in Ethiopia is 0.69% versus 14.00% and 13.72%, for Uganda and Ghana, respectively;
- vegetables in Ethiopia is 1.17% versus 7.70% and 4.18% for Brazil and Algeria, respectively;
- dairy products in Ethiopia is 2.25% versus 9.47% and 7.99% for India and Brazil, respectively;
- red meat in Ethiopia is 1.31% versus 12.74% and 7.98% for China and Brazil, respectively;
- poultry in Ethiopia is 0.16% versus 7.39% and 6.33% for Brazil and Mexico, respectively; and
- fish in Ethiopia is 0.04% versus 1.79% and 1.55% for Ghana and Angola, respectively.

Similarly, the per capita consumption was as follows:

- vegetables 14 kg/person/yr in Ethiopia versus 189 kg/person/yr in Algeria, 109 kg/person/yr in Mali, 374 kg/person/yr in China;
- fruit 9 kg/person/yr in Ethiopia versus 191 kg/person/yr in Ghana, 120 kg/person/yr in Uganda, 116 kg/person/yr in Mexico;
- milk 27 kg/person/yr in Ethiopia versus 78 kg/person/yr in Kenya, 47 kg/person/yr in South Africa, 143 kg/person/yr in Brazil;
- red meat 5 kg/person/yr in Ethiopia versus 26 kg/person/yr in South Africa, 18 kg/person/yr in Mali, 52 kg/person/yr in Brazil;

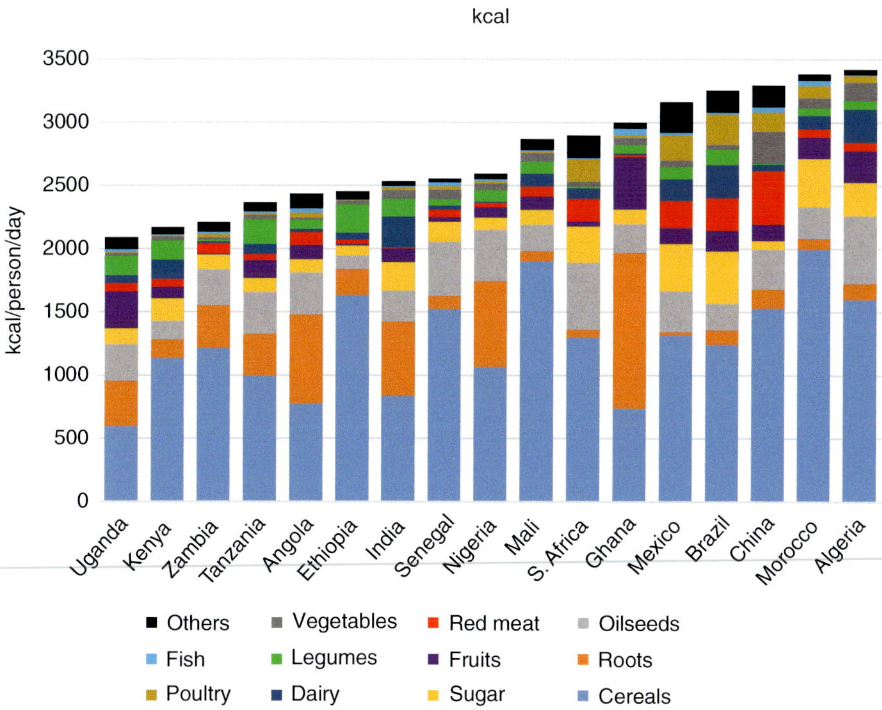

Fig. 24.2. Composition of dietary energy sources in Ethiopia compared to selected countries in Africa, Asia and South America. Source: constructed by the lead author from FAO (2022). Figures are 2015-2019 averages.

- chicken <1 kg/person/yr in Ethiopia versus 37 kg/person/yr in South Africa, 20 kg/person/yr in Morocco, 47 kg/person/yr in Brazil;
- eggs <1 kg/person/yr in Ethiopia versus 6 kg/person/yr in South Africa, 8 kg/person/yr in Morocco, 20 kg/person/yr in China; and
- fish <1 kg/person/yr in Ethiopia versus 25 kg/person/yr in Ghana, 20 kg/person/yr in Angola, 23 kg/person/yr in China.

Please refer to Annex Table 24.2 for more details.

Even though the per capita consumption of vegetables and fruits (sources of vitamins and minerals) showed slight growth, their composition as a percentage of the total dietary energy showed a declining trend. For example, they accounted for 2.51% in 1975–1979, compared to 1.94% and 1.64%, respectively, in 1995–1999 and 2015–2019.

Shifts in the food balance sheet

Substantial changes have occurred in the food balance sheet of Ethiopia over the last several decades. This can be demonstrated by changes in the food balance sheet with two examples: first, changes in the intake of major staples, and second changes in the intake of oilseeds and their products.

Shifts in major staples

Cereals, roots and tubers, grain legumes and oilseeds are the major sources of dietary energy in Ethiopia (Table 24.2). Roots and tubers in general showed the fastest growth among the major crop and livestock product groups. The FAO (2020) data showed that the dietary energy supply for *enset* was 56 kcal/person/day, 169 kcal/person/day and 205 kcal/person/day, respectively, in 1969–1993, 1989–1993 and 2009–2013. This means that there was a 265% increase from the first to the last period. Meanwhile, the overall contribution of roots and tubers was estimated at 94 kcal/person/day, 197 kcal/person/day and 254 kcal/person/day, respectively, at that time. Put differently, root crops accounted for roughly 20%, 8% and 2% of dietary energy, dietary protein and dietary fat, respectively, in 2015–2019.

Table 24.2. Contributions of crop and livestock items to dietary energy supply during different periods in Ethiopia. Source: constructed by the lead author from FAO (2022).

	Energy (kcal/person/day)			Percentages		
Item	1975–1979	1995–1999	2015–2019	1975–1979	1995–1999	2015–2019
Cereals	1010	1134	1632	64.42	68.29	66.56
Grain legumes	131	105	225	8.34	6.33	9.16
Roots	147	233	209	9.36	14.01	8.54
Oil	21	26	89	1.31	1.58	3.62
Sugar	32	34	80	2.05	2.03	3.26
Vegetables	30	23	29	1.93	1.36	1.17
Spices	0	0	23	0.00	0.01	0.94
Beverages	15	6	21	0.97	0.37	0.87
Fruits	12	12	17	0.78	0.75	0.69
Oilseed	41	3	15	2.64	0.17	0.60
Coffee/tea	4	2	3	0.26	0.10	0.14
Crop totals	1443	1578	2343	92.07	95.00	95.56
Dairy	40	31	55	2.55	1.84	2.25
Meat	71	45	45	4.52	2.70	1.83
Honey	4	4	4	0.27	0.24	0.16
Poultry	9	3	4	0.60	0.20	0.16
Fish	0	0	1	0.00	0.00	0.04
Livestock total	124	83	109	7.93	4.98	4.44
Grand total	1568	1661	2452	100.00	100.00	100.00

However, the latest data (covering through 2019) suggest much lower figures of 9%, 14% and 9%, respectively, in 1975–1979, 1995–1999 and 2015–2019. This underestimation stems from the fact that the most recent data on roots and tubers do not include *enset*, the biggest crop of the group.

While the overall contribution of crops to the dietary energy intake remained around 94%, there were declines in livestock products from ~7.93% in 1975–1979 to 4.44% in 2015–2019 (Table 24.2). There were also substantial declines in meat and poultry products.

Shifts in oilseeds and their products

Another major shift in the food balance sheet of Ethiopia occurred in the consumption patterns (and therefore their contributions to the dietary energy and dietary fat) of oilseeds and oil. Over the years, the per capita consumption of oilseeds declined constantly, whereas that of manufactured oil increased (Fig. 24.3).

The per capita consumption of oilseeds declined from 3.32 kg/person/year in 1975–1979 to 1.44 kg/person/year in 2015–2019. By contrast, the per capita consumption of manufactured oil increased from 0.85 kg/person/year in 1975–1979 to 3.65 kg/person/year in 2015–2019. This means that the Ethiopian public is consuming less whole food today than they did decades ago. Furthermore, the traditional oilseeds such as *noug*, linseed, Ethiopian mustard and sesame dominated the oilseed market in the earlier years, whereas palm oil commands the largest share of the cooking oil market at present (cf. Chapter 6).

Shifts in per capita consumption patterns of selected items

In addition to adequate dietary energy, it is important that the diets also contain adequate amounts of protein, fat, vitamin and mineral sources, for a balanced and healthy diet. Here we present the status of vegetables and fruits (vitamin and mineral sources); sugar (energy); and red meat (consisting of bovine meat [47%], other meats [31%], and mutton and goat meat [18%]), raw animal fats [4%] and pork [<1%]), chicken and egg, as important sources of protein.

Vegetables

Ethiopia's per capita vegetable consumption started from a low base. With 13.89 kg/person/yr, 13.63 kg/person/yr and 15.69 kg/person/yr in 1975–1979, 1995–1999 and 2015–2019, respectively, it showed slight upward trends but remained still one of the lowest among its peers in Africa and elsewhere. Ethiopia ranked 47th

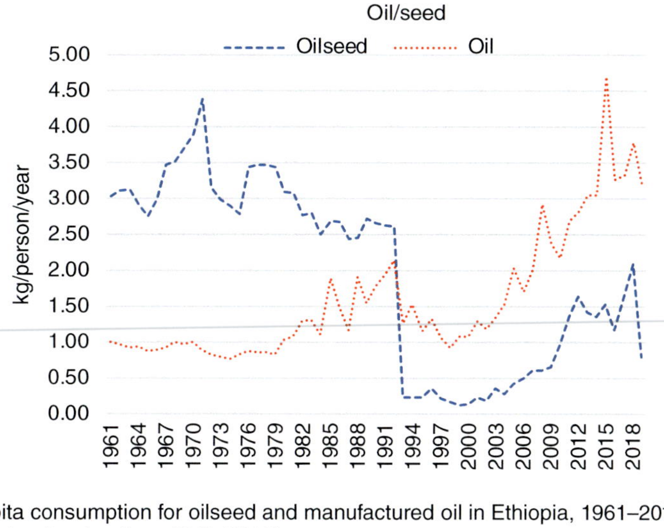

Fig. 24.3. Per capita consumption for oilseed and manufactured oil in Ethiopia, 1961–2019. Source: constructed by the lead author from FAO (2022).

out of 50 African countries in 2015–2019. The per capita for the top five countries, namely, Tunisia, Libya, Algeria, Egypt and Cameroon was roughly 263.28 kg/person/yr, 196.54 kg/person/yr, 189.43 kg/person/yr, 160.66 kg/person/yr and 117.76 kg/person/yr, respectively. The overall African average was 58.13 kg/person/yr. Ethiopia was ahead of only the Comoros, Chad and the Democratic Republic of the Congo (DRC). China, India, Mexico and Brazil registered 373.89 kg/person/yr, 88.77 kg/person/yr, 68.19 kg/person/yr and 51.53 kg/person/yr, respectively.

Fruits

Just as in vegetables, the per capita fruit consumption in Ethiopia started from a very low base. It averaged 5.37 kg/person/yr, 4.74 kg/person/yr and 9.20 kg/person/yr in 1975–1979, 1995–1999 and 2015–2019, respectively. Ethiopia ranked 45th out of the 50 African countries for which data were available in 2015–2019. Sao Tome and Principe, Ghana, Malawi, Rwanda and Uganda were the top five countries with 215.88 kg/person/yr, 191.11 kg/person/yr, 167.73 kg/person/yr, 163.01 kg/person/yr and 135.84 kg/person/yr, respectively. The African average was 64.10 kg/person/year. Ethiopia was ahead of only Togo, Zambia, Chad, Gambia and Burkina Faso. The 2015–19 per capita consumption for Mexico, Brazil, China and India was 115.99 kg/person/yr, 107.93 kg/person/yr, 100.29 kg/person/yr and 68.37 kg/person/yr, respectively (FAO, 2022).

Sugar

Ethiopia's per capita sugar consumption was 3.30 kg/person/yr, 3.44 kg/person/yr and 8.16 kg/person/yr in 1975–1979, 1995–1999 and 2015–2019, respectively. In other words, it more than doubled in the two decades between 1995–1999 and 2015–2019 (FAO, 2022). However, this is still among the lowest in Africa and the world. This country ranked 43rd out of 50 countries in Africa for which data were available in the 2015–19 period. Eswatini, Seychelles, Mauritania, South Africa and Libya, with 58.91 kg/person/yr, 45.62 kg/person/yr, 44.47 kg/person/yr, 40.91 kg/person/yr and 39.75 kg/person/yr, respectively, were the top consuming countries in Africa. Ethiopia was ahead of only Sierra Leone, Central African Republic (CAR), Madagascar, Guinea-Bissau, Niger and the DRC. The figures for Brazil, Mexico, India and China were 58.35 kg/person/yr, 37.88 kg/person/yr, 33.15 kg/person/yr and 6.72 kg/person/yr, respectively.

Milk

According to the FAO, Ethiopia has made slow progress in its per capita consumption of milk. The average per capita consumption was 20.30 kg/person/yr, 16.35 kg/person/yr and 28.64 kg/person/yr in 1975–1979, 1995–1999 and 2015–2019, respectively. This country ranked 17th out of 50 in Africa and 113th out of 179 countries and territories in the world in 2015–2019 (FAO, 2022).

The leading countries in Africa are Botswana, Sudan, Tunisia, Mauritania and Algeria, with averages of 118.72 kg/person/yr, 95.49 kg/person/yr, 92.6 kg/person/yr, 80.69 kg/person/yr and 77.76 kg/person/yr, respectively. The African average is 28.40 kg/person/yr. In a similar fashion, the top five leading countries in the world were:

- Montenegro – 353.92 kg/person/yr;
- Finland – 348.44 kg/person/yr;
- Albania – 313.05 kg/person/yr;
- Netherlands – 300.19 kg/person/yr; and
- Switzerland – 286.97 kg/person/yr.

The world average was estimated at 83.00 kg/person/yr. The share for Brazil, Mexico, India and China was 143.05 kg/person/yr, 91.91 kg/person/yr, 71.98 kg/person/yr and 23.00 kg/person/yr, respectively.

Red meat

Ethiopia's per capita consumption of red meat (consisting of bovine meat, mutton and goat meat, and pork) has shown declining trends over the years. It was 8.95 kg/person/yr, 5.43 kg/person/yr and 5.40 kg/person/yr in 1975–1979, 1995–1999 and 2015–2019, respectively (FAO, 2022). Ethiopia ranked 45th out of 50 countries in Africa in its red meat per capita consumption. The leading consumers were CAR, South Africa, Chad, Mali and the Seychelles, with averages of 40.95 kg/person/yr, 30.55 kg/person/yr, 27.44 kg/person/yr, 22.64 kg/person/yr and 22.22 kg/person/yr, respectively. Ethiopia was ahead of only Ghana, Rwanda, Burundi, Libya and Cote d'Ivoire in Africa. The African average was 11.30 kg/person/yr. The share for Brazil, China, Mexico and India was 51.57 kg/person/yr, 47.23 kg/person/yr, 33.20 kg/person/yr and 1.84 kg/person/yr, respectively. Ethiopia ranked 163rd out of 179 countries in the world. Hong Kong, Mongolia, Spain, Macao and Australia, with 78.85 kg/person/yr, 74.89 kg/person/yr, 68.78 kg/person/yr, 67.62 kg/person/yr and 67.29 kg/person/yr, respectively, are the leading per capita consumers in the world. The world average was estimated at 26.57 kg/person/yr.

Chicken

Ethiopia's per capita consumption of chicken showed declining trends over the years. For example, it was 1.63 kg/person/yr, 0.60 kg/person/yr and 0.72 kg/person/yr in 1975–1979, 1995–1999 and 2015–2019, respectively. Ethiopia ranked 48th out of 50 countries in Africa. The leading countries were Mauritius, Seychelles, South Africa, Gabon and Libya, with averages of 39.53 kg/person/yr, 37.06 kg/person/yr, 37.04 kg/person/yr, 33.02 kg/person/yr and 31.49 kg/person/yr, respectively. Ethiopia was ahead of only Burundi and Chad, among African countries. The African average was 9.19 kg/person/yr. The average per capita for Brazil, Mexico, China and India was 46.94 kg/person/yr, 33.85 kg/person/yr, 13.76 kg/person/yr and 2.89, respectively.

Egg

According to the FAO (2022), Ethiopia's per capita egg consumption was 1.64 kg/person/yr, 0.40 kg/person/yr and 0.40 kg/person/yr in 1975–1979, 1995–1999 and 2015–2019, respectively. This country was ahead of only Cameroon, Niger, Chad, Burundi, Angola and the DRC. With 9.86 kg/person/yr, 7.87 kg/person/yr, 7.79 kg/person/yr, 7.59 kg/person/yr and 7.38 kg/person/yr, Libya, Morocco, Tunisia, Algeria and Seychelles, respectively, were the leading egg consumers in Africa. The African average was 2.24 kg/person/yr. The averages for China, Mexico, Brazil and India were 19.99 kg/person/yr, 19.81 kg/person/yr, 10.81 kg/person/yr and 3.04 kg/person/yr, respectively.

Fish

The per capita consumption for fish in Ethiopia has shown upward trends over the years, but it started from very low base. For example, it was 0.09 kg/person/yr, 0.18 kg/person/yr and 0.51 kg/person/yr in 1975–1979, 1995–1999 and 2015–2019, respectively. According to the FAO (2022), Ethiopia has the lowest per capita consumption of fish in all of Africa. The leading countries in Africa were the Seychelles, Gabon, Sao Tome and Principe, Congo and Gambia, with averages of 286.11 kg/person/yr, 158.22 kg/person/yr, 146.42 kg/person/yr, 138.66 kg/person/yr and 133.72 kg/person/yr, respectively.

Ethiopia ranked 177th out of 179 countries and territories in the world in 2015–2019 (FAO, 2022). It was only ahead of Tajikistan and Afghanistan. With 105.38 kg/person/yr, 92.29 kg/person/yr, 75.82 kg/person/yr, 67.65 kg/person/yr and 61.69 kg/person/yr, the island states and territories of Maldives, Iceland, Kiribati, Hong Kong and Macao, respectively, were the leading fish consumers in the world. The share for China, Mexico, Brazil and India was 22.86 kg/person/yr, 10.93 kg/person/yr, 8.25 kg/person/yr and 6.21 kg/person/yr, respectively.

Discussion and Conclusions

Policy makers in Ethiopia initiated massive efforts to provide the rapidly growing population with adequate food and economic wellbeing, especially following the great drought and famine that ensued in 1972–1975, a repeat of which occurred a decade later, resulting in mass migration to urban centres. Like many of its peers on the continent, and elsewhere in the developing world, this country concentrated on fulfilling the calorie requirement of its diet, while the nutrition aspects have not received such a high priority. Furthermore, its priorities to achieve food security centred round major cereals while other traditional crops have not received commensurate attention even where they have demonstrated advantages in nutrition, productivity, resilience, sustainability and environmental protection (Abate, 2021). For example, a recent study (Powell *et al.*, 2015; Tamrat *et al.*, 2020) shows that *enset* is superior to all major staples in Ethiopia in its Ca, Fe, Mg, K and Zn content and is superior to all roots and tubers and similar to many major cereals in P, Cu and Mn contentment, but it has not received the due priority it deserved. This country also has a large number of traditional crops (Mekbib and Deressa, 2016; Melaku and Duguma, 2016) and wild fruits with the potential to contribute to food security (Molla *et al.*, 2011) and nutritional diversity as has been reported for other countries (Remans *et al.*, 2011; Jones *et al.*, 2014; Powell *et al.*, 2015; Tirado-von der Pahlen, 2017).

Even though the government focus has been on cereals, root crops showed the fastest growth in their contribution to the country's food security. *Enset* benefited from farmers' own innovations rather than government interventions in spite of its huge importance to the food and nutrition security of the country (Nurfeta *et al.*, 2008; Mohammed *et al.*, 2013, Tamrat *et al.*, 2020). Until about 50 years ago, the crop was little known in Oromo agriculture, but it has become an increasingly important component of food security for the region. With its higher yield per unit area of land than any other staple in Ethiopia and several other advantages mentioned above, it is considered the crop of the future for Ethiopia and perhaps for other African countries. It is our hope that the government would allocate resources for its research and development commensurate with its economic importance to the country.

Another major change occurred in Ethiopia in the number of traditional oilseeds vis-à-vis manufactured oil consumed. Traditional oilseeds such as *noug*, sesame and groundnut comprised 67% of the total oilseed products consumed during 1975–1979, but their proportion shrunk to 38% during 1995–1999 and to a mere 12% 20 years later. Such shifts have occurred in other countries as well, but what is of concern is the dominance of palm oil with its health issues. It does not seem to be logical for Ethiopia to export its unprocessed oilseeds that have superior quality oil (mainly sesame and *noug*) and import palm oil.

The unintended consequence of the decline of the local oilseed industry that led to the major shift in the contribution of oilseeds and manufactured oil to dietary energy and fat was the sharp drop in the supply of oilseed cake to the feed industry. By continuing to export raw oilseeds (sesame, *noug*, rapeseed/mustard, groundnut and soybean), the local economy loses cumulative opportunities for developing the local feed industry and improving supply of formulated feeds. Now, and into the foreseeable future, Ethiopia's rapidly evolving economy needs a robust domestic food oil industry that builds on processing locally produced oilseeds that will also play multiple strategic roles – supplies valuable raw materials for the struggling livestock feed industry, supports rural agroindustries, promotes food self-sufficiency and generates gainful employment.

Perhaps with a possible exception of dairy, the contribution of all livestock products to nutrition in Ethiopia has shown declining trends through the years. It is ironic for a country with the largest head of cattle in Africa and 10th largest in the world, and fastest growing economy (double digit for many years), to rank 113th in dairy products and 163rd in red meat per capita consumption in 2015–2019. Furthermore, given the large amount of knowledge on poultry science in Ethiopia accumulated over the years (Forsido, 1986; Yami, 1995; Dessie, 2003; Asgedom, 2007; Hassen *et al.*, 2008a,b; Pagani and Wossene, 2008; Dana *et al.*, 2010; Mohammed *et al.*, 2013; Ibrahim *et al.*, 2019a,b), one wonders why it has not been possible to improve its productivity in this country, whereas many countries in Africa, Asia and Latin America have made significant improvement over the last four decades.

Several causes can be mentioned regarding the lack of achieving improved nutrition, especially animal products, in Ethiopia. First, as indicated by Shapiro *et al.* (2015), productivity of the livestock sector has shown little change over the years. For example, the draft revised agriculture and rural development policy (MOA, 2020) acknowledges that while the human population continued to grow at about 3% per annum, per capita production of milk could grow only at 1.2%, which must have limited the per capita milk consumption to only 28.40 kg, compared to 118.72 kg for Botswana, 95.49 kg for Sudan, 92.61 kg for Tunisia, 80.69 kg for Mauritania and 77.76 kg for Algeria (FAO, 2022).

Second, there is a persistent rise in retail prices of animal products (milk, meat and egg), which coincides with the relatively high population growth, rapid urbanization and expansion of the middle-income category of the population. Currently, the weak operations of private commercial livestock farms have been unable to sufficiently respond to the strong growth in demand for high quality products. A recent study in Ethiopia showed that between 2007 and 2016, the real prices of grains and roots crops declined while real prices of vitamin A-rich dark green leafy vegetables, fruits and animal source food increased markedly. The price of non-health food items such as oils and fats as well as sugar and honey also decreased (Bachewe and Minten, 2019).

Third, gender inequalities in food and nutrition access have long been salient features of rural households, and the concern for this under a changing climate has long been the subject of empirical work (FAO, 2011; Beuchelt and Badstue, 2013). Many women in developing countries, especially those in female-headed households, have less access to information, financial services and other resources needed to improve food and nutrition quality. Empirical evidence has shown that female-headed households in Malawi had increased dietary diversity in the presence of high levels of crop and livestock diversity (Jones *et al.*, 2014). Similar trends were observed between dietary diversity and vegetable production diversity in Tanzania and Kenya (Herforth, 2010). Hence, it is important to understand how women's access to resources and rights differentially affects the access to nutrition in male- and female-headed households. Fourth, nutrition interventions are traditionally delivered through the health sector. However, in recent years, a number of collaborative research and development work between other sectors, including agriculture, have emerged to address the underlying determinants of malnutrition and its direct and underlying determinants (Bhutta, 2013). However, there is still lack of evidence on how, and how much, agriculture-, health- and nutrition-sensitive projects that target women and promote the production and consumption of nutrient-rich foods can also have a positive impact on gender relations and contribute to empowerment in these communities (Neha *et al.*, 2018).

Fifth, several studies have indicated that undernourishment can be overcome by educating people on the importance of nutrition for a healthy and productive life, and by pointing out the need to diversify eating habits to more nutritious food products (such as *kocho* from *enset* [Tamrat *et al.*, 2020]) that are relatively easy to access. This means that increased efforts are needed to create public awareness on nutrition. Schools need to include nutrition in their curricula, as has been done in countries like Kenya. Agricultural and health extension services must be reviewed to lay more emphasis on a balanced and healthy diet.

Finally, agriculture has become more knowledge-intensive, and access to and use of this knowledge is often mediated by gender, ethnicity, household wealth, health and nutrition status, among other factors. Agricultural research and extension service providers need to embrace such knowledge to impact change in food security and encouraging a balanced and healthy diet.

References

Abate, T. (2021) Africa's compelling case for a different mode of partnership in agricultural research and development. Online presentation at Swedish University of Agricultural Sciences, Alnarp, Sweden, 9 February 2021. Available at: https://www.slu.se/en/ew-calendar/2021/2/africas-compelling-case-for-a-different-mode-of-partnership-in-agricultural-research-and-development/ (accessed 17 March 2024).

Abebe, Y., Hinds, M.J., Stoecker, B.J. and Gates, G.E. (2006) Nutritive value and sensory acceptability of corn- and kocho-based foods supplemented with legumes for infant feeding in Southern Ethiopia. *African Journal of Food, Agriculture, Nutrition and Development* 6(1), 1–19. DOI: 10.4314/ajfand.v6i1.19172.

Abebe, Y., Hambidge, K.M., Stoecker, B.J., Bailey, K. and Gibson, R.S. (2007) Phytate, zinc, iron and calcium content of selected raw and prepared foods consumed in rural Sidama, southern Ethiopia, and implications for bioavailability. *Journal of Food Composition and Analysis* 20(3–4), 161–168. DOI: 10.1016/j.jfca.2006.09.003.

Amare, B., Moges, B., Moges, F., Fantahun, B., Admassu, M., Mulu, A. and Kassu, A. (2012) Nutritional status and dietary intake of urban residents in Gondar, northwest Ethiopia. *BMC Public Health* 12, 752.

Arimond, M., Wiesmann, D., Becquey, E., Carriquiry, A., Daniels, M.C. *et al.* (2010) Simple food group diversity indicators predict micronutrient adequacy of women's diets in 5 diverse, resource-poor settings. *Journal of Nutrition* 140, 2059S–2069S. DOI: 10.3945/jn.110.123414.

Asgedom, A.H. (2007) Village poultry in Ethiopia: socio-technical analysis and learning with farmers. Wageningen University, Wageningen, the Netherlands. 178 p. (Ph.D. thesis).

Atlabachew, M. and Chandravanshi, B.S. (2008) Levels of major, minor and trace elements in commercially available enset (*Ensete ventricosum* (Welw.), Cheesman) food products (kocho and bulla) in Ethiopia. *Journal of Food Composition and Analysis* 21, 545–552. DOI: 10.1016/j.jfca.2008.05.001.

Beuchelt, T.D. and Badstue, L. (2013) Gender, nutrition-and climate-smart food production: Opportunities and trade-offs. *Food Security* 5(5), 709–721.

Bachewe, F.N. and Minten, B. (2019) The rising costs of nutritious foods: The case of Ethiopia. Strategy Support Program, Working Paper 134. IFPRI Ethiopia and Policy Research Institute, Addis Ababa, Ethiopia.

Bhutta, Z.A. (2013) Early nutrition and adult outcomes: Pieces of the puzzle. *The Lancet* 382, 486–487. DOI: 10.1016/ S0140-6736(13)60716-3.

Cochrane, L. (2013) Strengthening Food Security in Rural Ethiopia. PhD Dissertation, the University of British Columbia, Vancouver, Canada.

Cochrane, L. (2018) Food security in Ethiopia: Review of research, 2005–2016. *Ethiopian Journal of Applied Science & Technology* 9(1), 1–10.

Dana, N., Dessie, T., van der Waaij, L.H. and van Arendonk, J.A.M. (2010) Morphological features of indigenous chicken populations of Ethiopia. *Animal Genetic Resources* 46, 11–23.

Demeke, M. (2012) Analysis of incentives and disincentives for maize in Ethiopia. Technical notes series, MAFAP, FAO, Rome.

Dessie, T. (2003) Phenotypic and genetic characterization of local chicken ecotypes in Ethiopia. Humboldt University of Berlin, Berlin, 209 p. (Ph.D. thesis).

FAO (2011) Women in Agriculture. Closing the gender gap for development. In: *The State of Food and Agriculture.* FAO, Rome.

FAO (2020) FAOSTAT. Available at: http://www.fao.org/faostat/en/#home (accessed 30 July 2020).

FAO (2022) FAOSTAT. Available at: http://www.fao.org/faostat/en/#home (accessed 8 September 2022).

Forsido, T. (1986) Studies on the meat production potential of some local strains of chickens in Ethiopia. J. L. University of Giessen, Germany, 186 p. (PhD thesis).

FTF (Feed the Future) (2018) Global Food Security Strategy (GFSS) Ethiopia Country Plan. USAID.

Gebru, M., Remans, R., Brouwer, I., Baye, K., Melesse, M.B. *et al.* (2018) Food Systems for Healthier Diets in Ethiopia: Toward a Research Agenda. IFPRI Discussion Paper 01720 April 2018.

Hassen, H., Neser, F.W.C. and van Marle-Koster, E. (2008a) Village based indigenous chicken production system in north-west Ethiopia. *Tropical Animal Health and Production* 39, 189–197. DOI: 10.1007/ s11250-007-9004-6.

Hassen, H., Neser, F.W.C., van Marle-Koster, E. and de Kock, A. (2008b) Phenotypic variation of indigenous chicken populations in northwest Ethiopia. *Tropical Animal Health and Production* 39, 507–513. DOI: 10.1007/s11250-007-9032-2.

Headey, D. (2014) An analysis of trends and determinants of child undernourishment in Ethiopia, 2000–2011. ESSP 2014; Working Paper 70. Ethiopian Development Research Institute, Addis Ababa Ethiopia.

Herforth, A. (2010) *Promotion of Traditional African Vegetables in Kenya and Tanzania. A Case Study of an Intervention Representing Emerging Imperatives in Global Nutrition*. Cornell University, New York.

Ibrahim, D., Goshu, G., Esatu, W. and Cahaner, A. (2019a) Dual-purpose production of genetically different chicken crossbreeds in Ethiopia. 1. Parent stocks' feed intake, body weight, and reproductive performance. *Poultry Science* 98(8), 3119–3129. DOI: 10.3382/ps/pez136.

Ibrahim, D., Goshu, G., Esatu, W. and Cahaner, A. (2019b) Dual-purpose production of genetically different chicken crossbreeds in Ethiopia. 2. Egg and meat production of the final crossbreed females and males. *Poultry Science* 98(9), 3405–3417. DOI: 10.3382/ps/pez137.

Jones, A.D., Shrinivas, A. and Bezner-Kerr, R. (2014) Farm production diversity is associated with greater household dietary diversity in Malawi: Findings from nationally representative data. *Food Policy* 46, 1–12. DOI: 10.1016/j.foodpol.2014.02.001.

Kennedy, E., Tessema, M., Hailu, T. *et al.* (2015) Multi-sector nutrition program governance and implementation in Ethiopia: Opportunities and challenges. *Food and Nutrition Bulletin* 36(4), 534–548.

Koisha, B.S. (2013) The assessment of root crops contribution to household food security: The case of Soddo Zuria Woreda, Wolaita Zone, SNNPR, Ethiopia. MSc thesis, Addis Ababa University, 92 p.

Kumar, G. (1987) Ethiopian famines 1973–1985: A case study. Wider Working Papers; World Institute for Development Economic Research of the United Nations University.

Mekbib, Y. and Deressa, T. (2016) Exploration and collection of root and tuber crops in East Wollega and Ilu Ababor zones: Rescuing declining genetic resources. *Indian Journal of Traditional Knowledge* 15(1), 86–92.

Melaku, Y. and Duguma, T. (2016) Proximate composition, phytochemical screening, and antioxidant activities of the tubers of *Plectranthus edulis*. *International Research Journal of Biological Sciences* 5(12), 13–17.

MOA (Ministry of Agriculture) (2020) Draft Revised Agriculture and Rural Development Policy of the Federal Republic of Ethiopia. MOA (unpublished).

Mohammed, B., Gabel, M. and Karlsson, L.M. (2013) Nutritive values of the drought tolerant food and fodder crop enset. *African Journal of Agricultural Research* 8(20), 2326–2333. DOI: 10.5897/AJAR12.1296.

Molla, E.L., Asfaw, Z., Kelbessa, E. and Van Damme, P. (2011) Wild edible plants in Ethiopia: A review on their potential to combat food insecurity. *Afrika Focus* 24(2), 71–121.

Neha, K., Nguyen, P.H., Harris, J., Harvey, D., Rawat, R. and Ruel, M.T. (2018) What it takes: Evidence from a nutrition- and gender-sensitive agriculture intervention in rural Zambia. *Journal of Development Effectiveness* 10(3), 341–372. DOI: 10.1080/19439342.2018.1478874.

Nurfeta, A., Tolera, A., Eik, L.O. and Sundstøl, F. (2008) Yield and mineral content of ten enset (*Ensete ventricosum*) varieties. *Tropical Animal Health and Production* 40, 299–309.

ONCCP (Office of the National Committee for Central Planning) (1987) The food and nutrition strategy for Ethiopia. Executive summary (unpublished final report). Government of Ethiopia, Addis Ababa.

Pagani, P. and Wossene, A. (2008) Review of the new features of the Ethiopian poultry sector: Biosecurity implications. Consultancy Report, FAO, Rome.

Pingali, P. (2015) Agricultural policy and nutrition outcomes – getting beyond the preoccupation with staple grains. *Food Security* 7, 583–591.

Powell, B., Thilsted, S.H., Ickowitz, A., Termote, C., Sunderland, T. and Herforth, A. (2015) Improving diets with wild and cultivated biodiversity from across the landscape. *Food Security* 7, 535–554. DOI 10.1007/s12571-015-0466-5.

Remans, R., Flynn, D.F.B., DeClerck, F., Diru, W., Fanzo, J. *et al*. (2011) Assessing nutritional diversity of cropping systems in African villages. *PLoS ONE* 6(6), e21235. DOI: 10.1371/journal.pone.0021235.

Shapiro, B.I., Gebru, G., Desta, S., Negassa, A., Nigussie, K., Aboset, G. and Mechal, H. (2015) Ethiopia livestock master plan. ILRI Project Report. International Livestock Research Institute (ILRI), Nairobi, Kenya.

Sheehy, T., Carey, E., Sharma, S. and Biadgilign, S. (2019) Trends in energy and nutrient supply in Ethiopia: A perspective from FAO food balance sheets. *Nutrition Journal* 18, 46. DOI: 10.1186/s12937-019-0471-1.

Sibhatu, K.T., Krishna, V.V. and Qaim, M. (2015) Production diversity and dietary diversity in smallholder farm households. *PNAS* 112(34), 10657–10662.

Tamrat, S., Borrell, J.S., Biswas, M.K., Gashu, D., Wondimu, T. *et al*. (2020) Micronutrient composition and microbial community analysis across diverse landraces of the Ethiopian orphan crop enset. *Food Research International* 137(November 2020), 109636.

Tirado-von der Pahlen, C. (2017) *Sustainable Diets for Healthy People and a Healthy Planet*. United Nations System Standing Committee on Nutrition (UNSCN), Rome.

Vasileska, A. and Rechkoska, G. (2012) Global and regional food consumption patterns and trends. *Procedia – Social and Behavioural Sciences* 44, 363–369.

Woldemariame, M.H. (2015) Comparison of the nutritional and microbial composition of kocho from wild and cultivated enset from Bonga, Ethiopia. MSc thesis, Addis Ababa University.

Worku, I.H., Dereje, M., Minten, B. and Hirvonnen, K. (2017) Diet transformation in Africa: The case of Ethiopia. *Agricultural Economics* 48(supplement), 73–86.

Yami, A. (1995) Poultry production in Ethiopia. *World's Poultry Science Journal* 51, 197–201. DOI: 10.1079/WPS19950014.

Annex Table 24.1. Dietary energy, dietary protein, dietary fat, and per capita food supply in Ethiopia compared with selected countries. Source: constructed by the lead author from FAO (2022).

Country	Energy[a] (kcal/cap/d)	Protein[a] (g/cap/d)	Fat[a] (g/cap/d)	Food supply[a] (kg/cap/yr)
Uganda	2091	47	46	532
Kenya	2169	62	45	488
Zambia	2210	60	52	410
Tanzania	2365	61	54	575
Angola	2435	53	59	628
Ethiopia	2452	72	30	368
India	2533	64	58	521
Senegal	2555	63	67	471
Nigeria	2596	60	59	626
Mali	2869	80	56	611
S. Africa	2900	80	88	532
Ghana	3001	61	41	862
Mexico	3165	91	101	698
Brazil	3255	94	127	775
China	3296	104	102	977
Morocco	3384	100	69	668
Algeria	3420	90	95	760

[a]kcal/cap/d, g/cap/d and kg/cap/yr represent kcal/person/day, g/person/day and kg/person/year, respectively. Figures are 2015-2019 averages.

Annex Table 24.2. Per capita consumption (kg/cap/yr) of selected dietary sources in Ethiopia compared with selected countries in Africa, Asia and Latin America (2015–2019 average). Source: constructed by the lead author from FAO (2022).

Country	Vegetables	Fruit	Milk	Red meat	Chicken	Egg	Fish
Ethiopia	13.85	8.66	28.64	5.40	0.72	0.40	0.51
Zambia	25.05	7.95	6.86	14.07	3.62	3.20	11.87
Nigeria	73.19	56.55	1.25	5.47	1.31	3.04	9.41
Angola	26.52	76.28	6.92	13.06	10.36	0.16	20.19
Senegal	96.64	18.43	11.67	10.02	6.05	1.91	18.35
Tanzania	52.96	75.66	27.95	8.55	1.68	1.23	6.97
S. Africa	40.05	23.39	47.23	25.61	37.04	6.38	6.35
Kenya	47.19	55.92	77.65	12.09	1.61	1.45	3.29
Uganda	33.58	120.60	36.26	8.86	1.61	0.79	11.35
India	88.77	68.32	68.77	1.85	2.89	3.04	6.21
Mali	109.34	69.45	44.63	18.00	2.68	0.56	8.63
Ghana	29.93	191.11	1.98	3.70	7.01	1.15	24.73
Morocco	105.07	97.63	24.81	12.12	20.36	7.87	19.65
Mexico	68.19	115.99	91.51	33.20	33.85	19.81	10.93
Algeria	189.43	106.62	77.76	12.71	7.09	7.59	3.88
Brazil	51.53	107.93	142.51	51.56	46.94	10.81	8.25
China	373.89	100.29	22.79	47.23	13.76	19.99	22.86

25 Agricultural Commodity Trade: What are the Major Issues?

Tsedeke Abate,[1]* Mikael Fesseha Yohannes[1] and Dessie Ambaw[2]
[1]Homegrown Vision, Addis Ababa, Ethiopia; [2]University of South Australia, Adelaide, Australia

Abstract

Ethiopia depends heavily on agricultural commodity export as a major source of foreign currency earning and imports large amounts of agricultural items to compensate for the shortfall in food and nutrition security and raw material supply for the local food industry. It reached several milestones over recent decades. There was rapid growth of export earnings between mid-2000 and mid-2010 along with commodity diversification, which meant reduced reliance on coffee alone and increased contribution of oilseeds, grain legumes and meat. The country transitioned from a net importer to net exporter of tea and exported more meat than live animals and more processed than raw hides and skins. Despite those successes, however, 92% of the total export is not value added and there has been a continued decline of earnings between 2016 and 2020, making the country a net food importer, even with its huge potential for increased agricultural productivity and production.

Introduction

Ethiopia depends heavily on agriculture for food and nutrition security, raw material for the local industry, export earnings, import substitution and overall socioeconomic wellbeing. The country is well known for its crop and livestock diversity. According to the Food and Agriculture Organization of the United Nations (FAO, 2022), more than 70 species of crops were cultivated annually during the 2016–2020 period. The Central Statistical Agency (CSA) collects annual data on 52 cultivated crops (www.csa.gov.et), but there are still many crops neither FAO nor CSA collects data on. Both the crops and livestock in this country are produced for local consumption as well as export earnings.

Policy makers in the early 1970s were aware of the danger associated with reliance on a limited number of export commodities, coffee in this case, and made significant efforts to diversify via the introduction of new or expansion of existing alternatives such as oilseeds, grain legumes (e.g. common bean) and horticultural crops. Successive governments have also made attempts to improve value addition of export commodities and raw material supply for the local industry import substitution. The number of Ethiopia's export and import destinations has also grown over the years, but there has not been a clear policy direction on this aspect. Nor has there been a critical evaluation of the outcomes of those efforts.

Despite its huge importance, there have not been well-researched, comprehensive coherent

* Corresponding author: tsedeke@homegrownvision.org

DOI: 10.1079/9781800626386.0025

studies and reports on this subject of agricultural trade to inform policy decision and priority setting for the future of agricultural commodity trade in Ethiopia. In this current study, we conduct a critical assessment of the sector and provide evidence-based suggestions and lessons learnt.

Approach

We obtained long-term data (1961 to 2020) for export and import quantity and value of agricultural commodities from the FAO (http://www.fao.org/faostat/en/#data) accessed at different times. We also accessed the Ethiopian Revenue and Customs Authority (ERCA; www.erca.gov.et) database for 2000 to 2020 data, mainly for establishing export destinations and import sources. To calculate the annual rate of growth (ROG) for each commodity or group of commodities, we employed the method followed for Chapter 3 in this book (also see Abate *et al.*, 2015).

Results

Performance of the export market

The FAO (2022) data show that Ethiopia exported varying amounts of 195 items, derived from crops and livestock, with an estimated value of >US$1.45 billion each year during 2016–2020. These can be grouped into 24 major categories, as presented in Annex Table 25.1. Of those (in descending order), coffee, oilseeds, grain legumes, meat and vegetables constituted nearly 96% of the total earnings. Overall, approximately 92% of the export earnings were from plant products, while livestock products accounted for the remaining 8%. Meat was the largest livestock export from Ethiopia.

There have been shifts in the proportions of export commodities over the last several decades. Figure 25.1 summarizes the proportions of earnings from selected major commodities export during different periods: 1976–1980, 1996–2000 and 2016–2020. For example, coffee accounted for about 77%, 82% and 45%, and oilseeds constituted 2%, 6% and 27%, respectively, in the first, second and third periods. In a similar fashion, earnings from grain legumes rose from <5% and <3% in 1976–1980 and 1996–2000, respectively, to 12% in 2016–2020 (cf. Annex Table 25.1 for more details).

About 99% of earnings from coffee are obtained from green (unprocessed, not value added) coffee; oilseeds include 85% sesame; grain legumes, 63% common bean and 22% chickpea; vegetables, 40% fresh and 28% dehydrated vegetables; and meats, 80% goat meat, 14% cattle (beef and veal) and 5% mutton (see Annex Table 25.1 for further details).

Of the 195 individual export items listed, complete data for 21 years (2000–2020) were available for 33 items to calculate their annual ROGs. Of those, 29 showed positive growth, while four showed declines in growth. Overall, agricultural export earnings increased at an annual rate of nearly 11% during the abovementioned period. The ROGs varied from commodity group to commodity group from commodity to commodity. Fresh vegetables, faba bean, natural honey, alcoholic beverages and pastry increased at an annual rate of about 45%, 37%, 35%, 34% and 29%, respectively, between 2000 and 2020. Goat meat, beer of barley, chickpea, common bean, mutton, tomato, live sheep, potato, papaya, banana, herbs, sesame, black pepper, other spices, wine, dry onion, cardamoms and millet also showed double-digit growth rates that ranged from 13–26%. The growth rate for green bean, coffee, orange, other oilseeds, cereal flour and tea ranged from 3–9%. Growth rates were rapid, especially between 2000 and 2015. On the other hand, ginger, mustard, frozen vegetables and non-alcoholic beverages had negative growth rates, ranging from −19% to −1%.

There were major changes in the importance of the export commodities over the years (Fig. 25.2). This was true particularly for coffee, oilseeds, grain legumes, vegetables and meat. As stated earlier, coffee dominated the Ethiopian export market through the early 2000s. The trends are now showing substantial change. Oilseeds, grain legumes, meat and vegetables are gaining in prominence. Spices have also emerged in recent times.

Although vegetables showed fast growth between 2008 and 2015, they showed declining trends thereafter. We tried to get explanations for this from the Ethiopian Horticulture Producers and Exporters Association, but our efforts did not yield results.

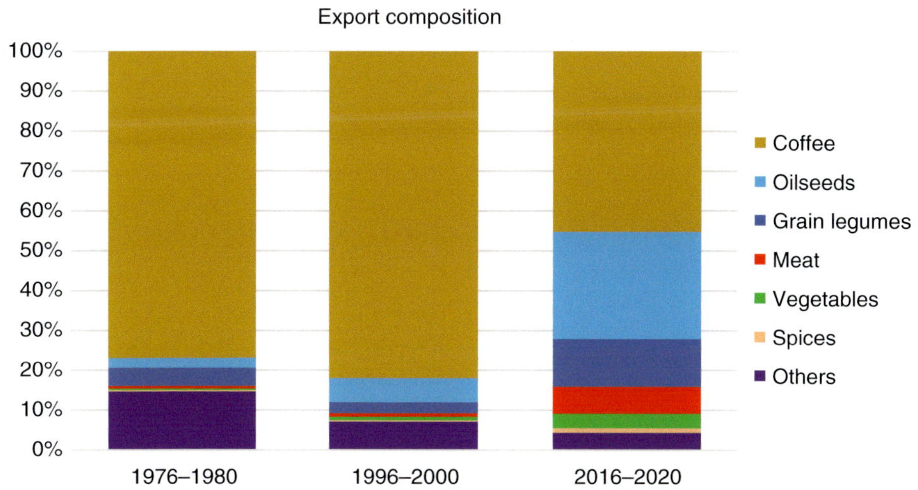

Fig. 25.1. Proportions of Ethiopia's major agricultural export commodities during different periods. Source: constructed by the lead author from FAO (2022).

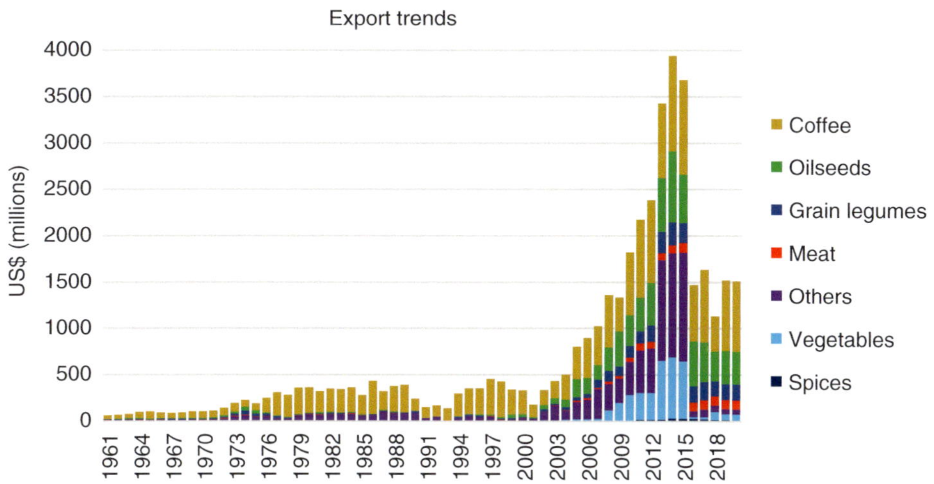

Fig. 25.2. Export trends for value of major agricultural export commodities in Ethiopia, 1961–2020. Source: calculated and constructed by the lead author from FAO (2022).

The total number of exported commodities rose from 76 in 1976–1980 to 81 in 1996–2000 to 195 in 2016–2020, a nearly 2.57-fold increase from 1976–1980 to 2016–2020 and a 2.41-fold increase from 1996–2000 to 2016–2020. These facts indicate that the country has moved away from total reliance on coffee to more diversification of its agricultural commodities.

Approximately 89% of all agricultural products in Ethiopia are exported raw, while only 11% are processed or semi-processed. This has remained similar throughout the years. Looking at specific examples, green coffee accounted for 98.62% of the total quantity exported in 2016–2020, whereas coffee substitutes, extracts, roasted coffee and husks covered 0.81%, 0.55%, 0.01% and 0.01%, respectively (FAO, 2022). In addition, all the oilseeds and other major commodities are exported raw.

Trends in the import market

The FAO (2022) data show that Ethiopia imported a total of 275 agricultural products in 2016–2020. These are grouped into 24 major categories, as presented in Annex Table 25.2. The total import value was estimated at US$2.09 billion per annum, in 2016–2020, compared to US$193.45 million in 1996–2000 and US$66.14 million in 1976–1980. Ethiopia's agricultural commodity imports include about 95% food items and 5% non-food items (see Annex Fig. 25.1 for further details).

Cereals, cooking oil, sugar and prepared food together constituted >87% of the total expenditure. Wheat, rice and sorghum accounted for about 50%, 37% and 10%, respectively, of the total cereals value. Palm oil accounted for about 88% of the total cooking oil value, with sunflower, soybean and vegetable oils accounting for 9%, 2% and 1%, respectively.

Similar to export products, import items showed fast growth in terms of both number and quantity, starting from the mid-2000s. According to the FAO (2022), the number of items imported to Ethiopia was 120 in 1976–1980 compared to 184 in 1996–2000 and 275 in 2016–2020 (as mentioned above). The overall import expenditure increased at an annual rate of nearly 13% per annum between 2000 and 2020.

Just like in export, there were shifts in the composition of commodities and groups of commodities (Fig. 25.3). For example, wheat accounted for 48% and 49% of the total expenditure in 1976–1980 and 1996–2000, respectively, while its proportion in 2016–2020 was less than 24%. By contrast, cooking oil, rice, sugar, prepared food and sorghum showed successive increases from 1976–1980 to 1996–2000 to 2016–2020. Further details are available in Annex Table 25.2.

We calculated the ROGs for those crops with uninterrupted data points for the period between 2000 and 2020. Expenditures on sorghum, rice, dry peas, sugar, vegetables and fruits increased at annual rates of 35%, 31%, 27%, 25%, 21% and 20%, respectively. In a similar fashion, the ROGs for cooking oil, industrial products, prepared food, beverages and manufactured goods were 16%, 14%, 14%, 13% and 10%, respectively, whereas dairy products, wheat, maize and miscellaneous products (cf. Annex Table 25.2) increased at 9%, 7%, 7% and 1%, respectively.

The majority of import commodities showed substantial growth starting from late 2000s, but started to slow down after 2017. Sorghum, rice, dry peas, sugar, vegetables and fruits, in particular, showed fast growth.

Figure 25.4 presents the long-term trends of imports of major agricultural commodities in Ethiopia. Wheat dominated the import market until the mid-2010s but started to decline after that. By contrast, cooking oil, rice and prepared food continued to grow throughout the above period.

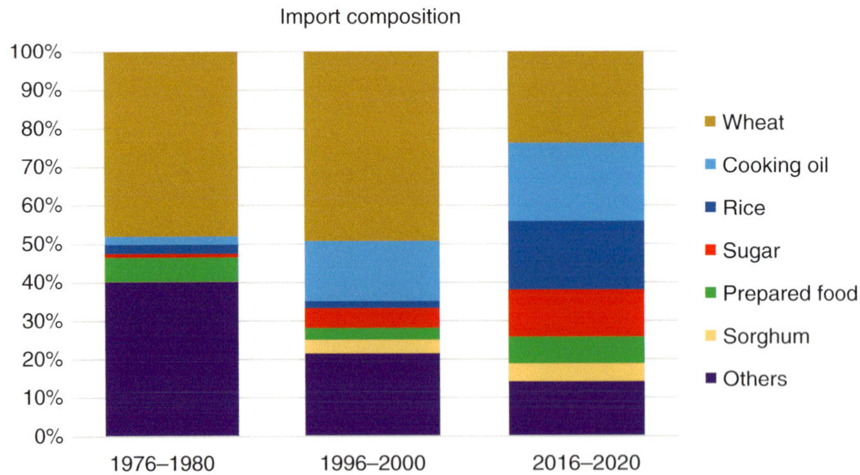

Fig. 25.3. Proportions of Ethiopia's major agricultural import commodities during different periods. Source: calculated and constructed by the lead author from FAO (2022).

Trading partners

Export partners

Ethiopia trades with a large number of countries across the globe. The 2018 and 2019 data from ERCA show that this country's export market included a total of 133 countries and territories in five continents. Ethiopia's average export earnings during the above period was estimated at US$1.51 billion (FAO estimate [2016–2020 average] was US$1.45 billion). Of those, Asia, Europe and the Middle East, combined, accounted for about 78% of the total (Fig. 25.5). Africa's share was estimated at about 11%, whereas North America was 9%. Oceania, the Russian Federation, Central America/Caribbean and South America covered the balance.

The top 20 destinations for Ethiopia's agricultural commodity export are: Saudi Arabia, the

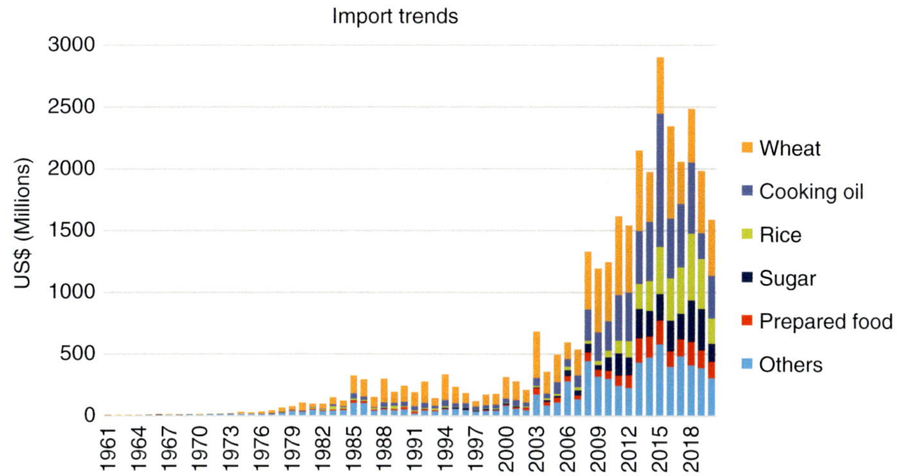

Fig. 25.4. Import trends of major agricultural import commodity values in Ethiopia, 1961–2020. Source: calculated and constructed by the lead author from FAO (2022).

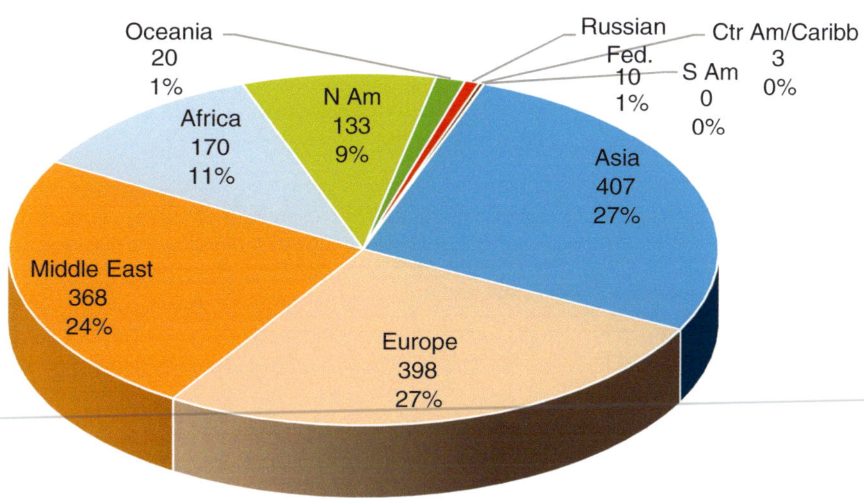

Fig. 25.5. Ethiopia's agricultural export commodity to different world regions. Source: constructed by the authors from ERCA (data for 2018 and 2019 average); values are US$ 000. Caribb = Caribbean; Ctr Am = Central America; N Am = North America; S Am = South America.

USA, Germany, the Netherlands, Japan, China, Israel, the United Arab Emirates (UAE), Somalia, India, Belgium, Viet Nam, Republic of Korea, Yemen, Italy, Djibouti, Sudan, UK, Türkiye and Indonesia. Together, these countries accounted for nearly 86% of Ethiopia's total agricultural export market. None of them covered more than 10% of the total export earnings.

There were substantial differences in the amount of earnings from commodities exported to different countries (Table 25.1). We discuss here eight crops for which data were available from ERCA. Coffee is the most widely traded agricultural commodity in Ethiopia, with a total of 74 destinations during the 2018–2019 period. Germany, the USA, Saudi Arabia, Japan and Belgium were the top five importers of Ethiopian coffee, with a combined share of about 63%. By contrast, there were only six countries that imported meat from Ethiopia: the UAE, Saudi Arabia, Bahrain, Oman, Kuwait and Qatar. The UAE and Saudi Arabia accounted for 56% and 40%, respectively, of the total meat export earnings.

India, the UAE, Türkiye, Iran and Pakistan together had a 92% share of chickpea from Ethiopia, out of 19 export destinations; India's share was 43%. There were 30 countries that imported common bean from Ethiopia. Of those, Viet Nam, Indonesia, Egypt, India and Singapore accounted for approximately 91% of the total. The first two countries had 37% and 36%, respectively, of the total export earnings from common bean.

There were 21 countries that imported sesame and other oilseeds from Ethiopia in 2018 and 2019. The top five (Israel, China, Viet Nam, the UAE and the USA) accounted for 86% of the total earnings out of 21 countries importing oilseeds from this country. Out of a total of 13 countries that imported soybean from Ethiopia, India's share was about 84%. Sudan, Egypt, Viet Nam, Saudi Arabia and Türkiye were the top five countries that imported nearly 93% of faba bean. The emerging spices market is well spread out among 28 countries. The USA, Sudan, Israel, Eritrea and India accounted for nearly 59% of the total spices export earnings for Ethiopia.

Import partners

According to ERCA, imports of agricultural items to Ethiopia originated from 124 countries and territories, representing five continents in the 2018–2021 period. Countries in Asia, the Middle East, Europe, Africa, North America, Oceania and South America accounted for 54%, 21%, 12%, 8%, 3%, 1% and 1%, respectively, of the total import bills for this country.

The top 20 of the countries where the imports originated are: India, Malaysia, Türkiye, Indonesia, China, the UAE, South Africa, the USA, Egypt, Djibouti, Ireland, France, the Netherlands, Pakistan, the UK, Belgium, Italy, Germany, the Ukraine and Saudi Arabia. These top 20 countries, together, accounted for 93% of the total import value in the 2018–2021 period. The top five of them covered nearly 68% of the total import bill.

ERCA provides import data only for a limited number of commodities. Just like in the export partners discussed above, there were substantial differences in the amount of import expenditure from different countries and commodities (Table 25.2).

Table 25.1. Value of selected agricultural export commodities and number of trading partners in Ethiopia (2018–2019 average). Source: constructed by the authors from the Ethiopian Revenue and Customs Authority.

Commodity	Value (US$ 000)	Partners (no.)	Share of top five (%)	Top five partners
Coffee	689,489	74	63	DE, US, KS, JP, BM
Meat	38,903	6	100	UA, KS, BH, OM, KT
Chickpea	21,974	19	92	IN, UA, TK, IR, PK
Common bean	20,245	30	91	VT, IO, EG, IN, SG
Sesame and others	17,597	21	86	IS, CH, VT, UA, US
Soybean	13,814	13	96	IN, KY, TK, SP, UA
Faba bean	12,364	16	93	SD, EG, VT, KS, TK
Spices	4,376	28	59	US, SD, IS, ER, IN

BH = Bahrain; BM = Belgium; CH = China; DE = Germany; EG = Egypt; ER = Eritrea; IN = India; IO = Indonesia; IR = Iran; IS = Israel; JP = Japan; KS = Saudi Arabia; KT = Kuwait; KY = Kenya; OM = Oman; PK = Pakistan; SD = Sudan; SG = Singapore; SP = Spain; TK = Türkiye; UA = United Arab Emirates; US = United States of America; VT = Viet Nam

Table 25.2. Value of selected major agricultural import commodities and the number of trading partners in Ethiopia (2018–2021 average). Source: constructed by the authors from the Ethiopian Revenue and Customs Authority.

Commodity	Value (US$ 000)	Partners (number)	Share of top five (%)	Top five partners
Cooking oil	486,588	42	95	MA, TK, IO, DJ, UA
Meslin flour	31,203	20	91	EG, UR, TK, FR, UA
Malt[a]	26,749	11	84	FR, NT, BM, PL, DE
Maize	25,942	24	94	ZA, AR, CH, UA, IN
Dry peas	14,703	12	96	US, UR, TK, UA, IN
Sugar	8,476	33	77	IN, CH, PK, TK, US
Prepared food	7,920	27	93	US, IT, ZA, BM, DE
Rice husk	2,956	18	95	IN, CH, US, KS, RK

[a]Malt data are 2018–2019 average.
AR = Argentina; BM = Belgium; CH = China; DE = Germany; DJ = Djibouti; EG = Egypt; FR = France; IN = India; IO = Indonesia; IT = Italy; MA = Malaysia; NT = Netherlands; PK = Pakistan; TK = Türkiye; UA = United Arab Emirates; UR = Ukraine; US = United States of America; ZA = South Africa.

Here we discuss eight commodities (for which ERCA data are available) as examples: cooking oil, meslin flour, malt, maize, dry peas, sugar, prepared food and rice husk. Ethiopia paid close to US$487 million annually between 2018 and 2021 to purchase cooking oil; the expenditure reached a record high of over US$1.13 billion in 2020. Malaysia, Türkiye, Indonesia and Djibouti accounted for 95% of the annual import value.

The annual expenditure on meslin flour was estimated at US$31 million, of which Egypt alone accounted for 51% of the total out of 20 countries that supplied the commodity. Other major suppliers included the Ukraine, Türkiye, France and the UAE. Ethiopia was reported to spend close to US$27 million on malt (2018–2019 average), with France, the Netherlands, Belgium, Poland and Germany being the major suppliers covering about 84% of the total.

Ethiopia spent nearly US$26 million annually on maize in the 2018–2021 period, where South Africa, Argentina, the UAE and India covered 94% of the total supply. The maize import was in the form of grain (88%), starch (7%) and flour (5%). Dry peas import cost the country close to US$15 million (the FAO estimates for 2016–2020 were close to US$50 million) annually during the above period. About 64% of the total supply came from the USA. Other major suppliers included the Ukraine, Türkiye, the UAE and India, which together covered 96% of the total expenditure.

According to ERCA, Ethiopia's annual expenditure on sugar was about US$8.5 million, way below the FAO estimate of US$255.5 million (cf. Annex Table 25.2). The major suppliers (out of the total of 33 countries) were India, China, Pakistan, Türkiye and the USA. The expenditure on prepared food was estimated at nearly US$8 million (versus the FAO estimate of US$146.4 million in 2016–2020). The USA accounted for about 41% of the total expenditure; other suppliers included Italy, South Africa, Belgium and Germany. India, China, Pakistan, Türkiye and the Republic of Korea are said to be the major suppliers of rice to Ethiopia.

Discussion and Conclusions

Past studies on agricultural trade in Ethiopia (e.g. Minten *et al.*, 2014, 2016a,b; Reardon *et al.*, 2015) focused on agricultural marketing, value chain and food transformation. This current study emphasizes agricultural trade per se targeting the trends in agricultural export and import commodities over the years. We understand that this is not an exhaustive coverage of the topic and expect that a more comprehensive sequel will follow in the immediate future.

After showing fast growth from the late 2000s, Ethiopia's export earnings showed declining trends from 2015 onwards, but there are signs of recovery in 2019 and 2020. The major challenge for the country with its export market is that approximately 89% of all agricultural products are exported without value addition.

Ethiopia is the world's fifth largest producer of coffee (after Brazil, Viet Nam, Colombia and Indonesia), but 98.58% of it is sold as green coffee; roasted coffee accounted for a mere 0.55%. By contrast, Italy, France, the Netherlands and Germany produce zero coffee, but roasted coffee accounted for approximately 93%, 88%, 76% and 41%, respectively, of their total export in 2016–2020 (FAO, 2022); the USA produces less than 1% of its total export, of which 69% is roasted.

Ethiopia had good examples of value addition with its sheep skins and goat meat exports. For example, the country exported more than 4300 MT of the commodity at an estimated value of US$20.3 million in 1976–1980; similarly, it exported about 1000 MT for an estimated value of nearly US$16.9 million in 1996–2000; there was a negligible amount of these products exported in 2016–2020 because it was processed or semi-processed and either used for the local leather industry or exported. By contrast, the country earned nearly US$78.4 million from the export of close to 14,000 MT of goat meat in 2016–2020; the earning in 1996–2000 was more than US$2.9 million from 1400 MT, whereas there was no export in the 1970s at all.

This country has also made substantial progress with diversification of its export commodities. For example, the number of export commodities grew to 195 in 2016–2020, compared to 81 and 76 in 1996–2000 and 1976–1980, respectively. Coffee accounted for 77% and 82% of Ethiopia's total export earnings in 1976–1980 and 1996–2000, respectively, whereas its share in 2016–2020 was roughly 45%. By contrast, the contribution of oilseeds grew from about 2% in 1976–1980 and 6% in 1996–2000, to approximately 27% in 2016–2020. Vegetables and meat have also showed substantial growth in recent years.

Ethiopia has also managed to reduce the level of wheat import in recent years. For example, the country's wheat import expenditure was nearly US$748 million in 2016 compared to a little over US$453 in 2020. Put differently, wheat import values accounted for about 48% and 49%, respectively, of the total during the 1976–1980 and 1996–2000 periods. By contrast, it was nearly 24% during 2016–2020 (cf. Fig. 25.3).

There is one good example of import substitution in Ethiopia. The country imported about 1000 MT of tea in 1976–1980, compared to 522 MT and 47 MT, respectively, in 1996–2000 and 2016–2020. Export earnings from tea were estimated at about US$2.2 million annually in 2016–2020, compared to US$617,000 20 years earlier. This was made possible by the introduction of tea research and development initiatives in the late 1970s.

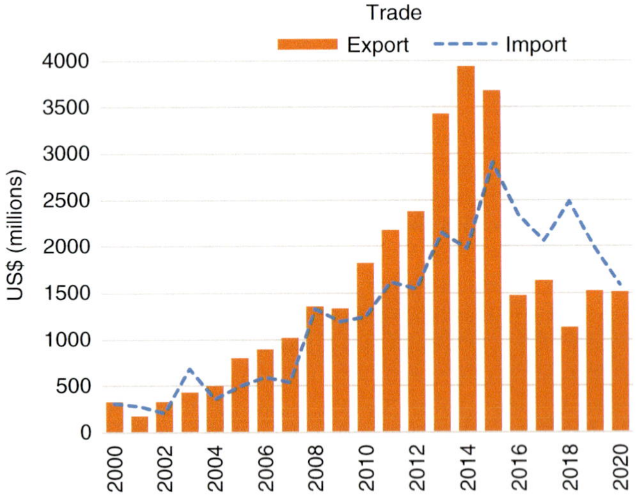

Fig. 25.6. Export and import values for Ethiopia, 2000–2020. Source: constructed by the lead author from FAO (2022).

This study has revealed that, in spite of many successes in agricultural export earnings, Ethiopia's import values are growing faster than its export earnings in recent years. The country has become a net importer of agricultural commodities from 2016 to 2020 (Fig. 25.6). Policy makers have to introduce measures to reverse this situation as a matter of high priority.

References

Abate, T., Shiferaw, B., Menkir, A., Wegary, D., Kebede, Y. *et al.* (2015) Factors that transformed maize productivity in Ethiopia. *Food Security* 7, 965–981. DOI: 10.1007/s12571-015-0488-z.

FAO (2022) Statistical databases and data-sets of the Food and Agriculture Organization of the United Nations. Available at: http://www.fao.org/faostat/en/#data (accessed 14 August 2022).

Minten, B., Tamru, S. and Stifel, D. (2014) Structural transformation in cereal markets in Ethiopia. *Journal of Development Studies* 50(5), 611–629.

Minten, B., Assefa, T., Engida, E. and Tamru, S. (2016a) Food processing, transformation and job creation: The case of Ethiopia. ESSP Working Paper 96. Addis Ababa, Ethiopia.

Minten, B., Tamru, S., Engida, E. and Kuma, T. (2016b) Transforming staple food value chain in Africa: The case of teff in Ethiopia. *Journal of Development Studies* 52(5), 627–645.

Reardon, T., Tschirley, D., Minten, B., Haggblade, S., Liverpool-Tasie, S. *et al.* (2015) Transformation of African agrifood systems in the new era of rapid urbanization and the emergence of the middle class. In: *Proceedings of the ReSAKSS Annual Conference: Beyond a Middle-Income Africa. Trends and Outlook Report.* 1–3 September 2015. ReSAKSS, Addis Ababa, Ethiopia.

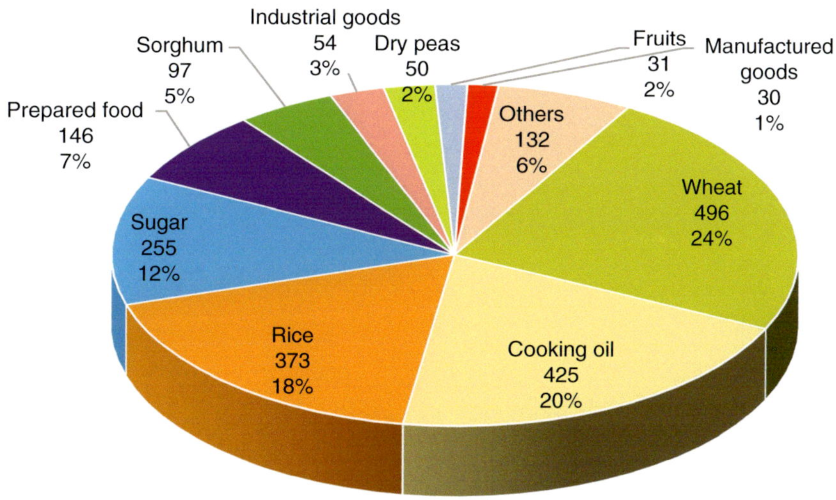

Annex Fig. 25.1. Value (US$ millions) of Ethiopia's major agricultural commodity imports, 2016–2020. Source: constructed by the author from FAO (2022).

Annex Table 25.1. Value and composition of major agricultural export products in Ethiopia during different periods. Source: calculated and constructed by the lead author from FAO (2022).

	Value (US$ 000)			Composition (%)		
Commodity	2016–2020	1996–2000	1976–1980	2016–2020	1996–2000	1976–1980
Coffee	658,034	313,282	240,544	45.29	82.01	76.89
Oilseeds	389,518	23,128	7,722	26.81	6.05	2.47
Grain legumes	175,002	10,384	14,529	12.04	2.72	4.64
Meat	97,469	3,363	1,887	6.71	0.88	0.60
Vegetables	53,290	3,394	1,040	3.67	0.89	0.33
Spices	16,336	913	1,116	1.12	0.24	0.36
Crude materials	12,619	0	6,944	0.87	0.00	2.22
Food	12,145	240	111	0.84	0.06	0.04
Live animals	9,544	711	3,556	0.66	0.19	1.14
Fruit	7,268	510	1,348	0.50	0.13	0.43
Industrial goods	3,610	475	904	0.25	0.12	0.29
Insect products	3,120	1,245	1,975	0.21	0.33	0.63
Roots and tubers	2,993	580	259	0.21	0.15	0.08
Cereals	2,864	415	215	0.20	0.11	0.07
Beverages	2,342	58	4	0.16	0.02	0.00
Tea	2,206	617	0	0.15	0.16	0.00
Fibre	2,137	3,585	3,330	0.15	0.94	1.06
Feed	1,612	81	3,891	0.11	0.02	1.24
Cooking oil	631	1	0	0.04	0.00	0.00
Hides and skins	149	16,880	20,323	0.01	4.42	6.50
Dairy	63	10	37	0.00	0.00	0.01
Sugar	58	2,132	3,016	0.00	0.56	0.96
Poultry	57	0	74	0.00	0.00	0.02
Manufactured goods	5	0	0	0.00	0.00	0.00
Total	1,453,069	382,004	312,824	100.00	100.00	100.00

Annex Table 25.2. Value and composition of major agricultural commodity groups import in Ethiopia during different periods. Source: constructed by the lead author from FAO (2022).

	Value (US$ 000)			Composition (%)		
Items	2016–2020	1996–2000	1976–1980	2016–2020	1996–2000	1976–1980
Cereals	997,858	118,634	35,161	47.76	61.33	53.16
Cooking oil	424,634	30,326	1,407	20.32	15.68	2.13
Sugar	255,480	9,847	679	12.23	5.09	1.03
Prepared food	146,352	5,830	4,164	7.00	3.01	6.30
Industrial goods	54,133	6,196	3,407	2.59	3.20	5.15
Grain legumes	53,335	2,445	6	2.55	1.26	0.01
Fruits	31,299	1,645	696	1.50	0.85	1.05
Manufactured goods	29,929	5,121	4,682	1.43	2.65	7.08
Beverages	23,333	3,587	1,636	1.12	1.85	2.47
Vegetables	18,037	173	298	0.86	0.09	0.45
Dairy	12,987	2,829	5,106	0.62	1.46	7.72
Crude materials	8,567	0	525	0.41	0.00	0.79
Live animals	7,435	1,904	495	0.36	0.98	0.75
Spices	5,877	301	307	0.28	0.16	0.46
Fibre crops	5,651	2,357	3,523	0.27	1.22	5.33
Roots and tubers	4,083	1,195	1,078	0.20	0.62	1.63
Meat	3,884	69	449	0.19	0.04	0.68
Poultry	1,906	89	16	0.09	0.05	0.02
Oilseeds	1,868	446	26	0.09	0.23	0.04
Hides and skins	1,230	0	0	0.06	0.00	0.00
Coffee	985	15	12	0.05	0.01	0.02
Tea	238	429	2,439	0.01	0.22	3.69
Feeds	149	1	3	0.01	0.00	0.00
Insect products	93	8	18	0.00	0.00	0.03
Total	2,089,343	193,447	66,137	100.00	100.00	100.00

Index

Note: Page numbers in **bold** type refer to **figures** and page numbers in *italic* type refer to *tables*.